Mechanical Behaviour of Metal–Organic Framework Materials

Inorganic Materials Series

Series editors:
Duncan W. Bruce, *University of York, UK*
Dermot O'Hare, *University of Oxford, UK*
Richard I. Walton, *University of Warwick, UK*

Titles in the Series:

1: Pre-combustion Carbon Dioxide Capture Materials
2: Post-combustion Carbon Dioxide Capture Materials
3: Solar Energy Capture Materials
4: Bioinspired Inorganic Materials: Structure and Function
5: Energy Storage and Conversion Materials
6: Two-dimensional Inorganic Nanomaterials for Conductive Polymer Nanocomposites
7: The Chemistry of Inorganic Biomaterials
8: Computer Simulation of Porous Materials: Current Approaches and Future Opportunities
9: Inorganic Thermoelectric Materials: From Fundamental Concepts to Materials Design
10: Biomedical Applications of Inorganic Materials
11: Photocatalysis Using 2D Nanomaterials
12: Mechanical Behaviour of Metal–Organic Framework Materials

How to obtain future titles on publication:
A standing order plan is available for this series. A standing order will bring delivery of each new volume immediately on publication.

For further information please contact:
Book Sales Department, Royal Society of Chemistry, Thomas Graham House, Science Park, Milton Road, Cambridge, CB4 0WF, UK
Telephone: +44 (0)1223 420066, Fax: +44 (0)1223 420247
Email: booksales@rsc.org
Visit our website at www.rsc.org/books

Mechanical Behaviour of Metal–Organic Framework Materials

Edited by

Jin-Chong Tan
University of Oxford, UK
jin-chong.tan@eng.ox.ac.uk

ROYAL SOCIETY OF CHEMISTRY

Inorganic Materials Series No. 12

Print ISBN: 978-1-83916-408-8
PDF ISBN: 978-1-83916-659-4
EPUB ISBN: 978-1-83916-660-0
Print ISSN: 2472-3819
Electronic ISSN: 2472-3827

A catalogue record for this book is available from the British Library

The Royal Society of Chemistry is a charity, registered in England and Wales, Number 207890, and a company incorporated in England by Royal Charter (Registered No. RC000524), registered office: Burlington House, Piccadilly, London W1J 0BA, UK, Telephone: +44 (0) 20 7437 8656.

For further information see our website at www.rsc.org

Printed in the United Kingdom by CPI Group (UK) Ltd, Croydon, CR0 4YY, UK

Preface

This book tackles the very fast-evolving subject of 'MOF Mechanics' and systematically exposes the reader to the core principles and cutting-edge science of the mechanical behaviour of metal–organic framework (MOF) materials. The body of work pertaining to MOF mechanics has reached a critical mass, which justifies the need for a first book entirely devoted to the topic. Our goals are manifold: to consolidate the knowledge uncovered thus far, to illuminate general trends, to predict future directions, and to stimulate new research on mechanical phenomena in framework materials.

MOFs are crystalline or amorphous nanoporous hybrid compounds, with highly tuneable chemical and physical properties relevant to many potential technological applications (*e.g.*, sensors, sorption, catalysis, photonics, dielectrics). In this book, the key experimental and theoretical techniques for studying MOF mechanical properties are treated in detail. The mechanical properties considered cover elastic and plastic deformations, dynamical response, structural flexibility, high-pressure phenomena, time-dependent behaviour, cracking and fracture. Materials selection charts and extensive tables are presented throughout the book. Formulated by curating the latest data, they not only give the reader the *big picture* but help to identify trends and unveil gaps in the field. Representative examples are chosen to illustrate the latest advances made and highlight challenges faced in several emergent topics: framework anisotropy and terahertz dynamics, anomalous mechanical properties, structural defects, guest-mediated structural behaviour, and mechanisms of mechanical energy absorption. In essence, our emphasis is on establishing fundamental structure–property relationships, as the ability to link the chemical structures of MOFs to their resultant mechanical characteristics will allow the engineering of bespoke properties and functions.

Inorganic Materials Series No. 12
Mechanical Behaviour of Metal–Organic Framework Materials
Edited by Jin-Chong Tan

Published by the Royal Society of Chemistry, www.rsc.org

The contents of the book have been tailored to be accessible to advanced undergraduate students, postgraduates, and current researchers, as well as new researchers entering the field. To this end, each chapter has been carefully structured to blend in an appropriate level of introductory materials and basic concepts, which are progressively built upon in complexity and developed through the book. A consistent set of mathematical notations is adopted (or redefined where appropriate). Extensive references are provided at the end of each chapter to point the interested readers to the original sources for further reading. Readers of this book will be well positioned to embark on an exciting journey, to explore and discover the fascinating mechanics of an *almost* infinite array of framework materials and hybrid structures.

Jin-Chong Tan
Oxford, UK

Contents

Inorganic Materials Series No. 12
Mechanical Behaviour of Metal–Organic Framework Materials
Edited by Jin-Chong Tan

Published by the Royal Society of Chemistry, www.rsc.org

CHAPTER 1

Fundamentals of MOF Mechanics & Structure–Mechanical Property Relationships

JIN-CHONG TAN*[a]

[a]Multifunctional Materials & Composites (MMC) Laboratory, Department of Engineering Science, University of Oxford, Parks Road, Oxford OX1 3PJ, UK
*E-mail: jin-chong.tan@eng.ox.ac.uk

1.1 An Overview of Metal–Organic Framework Materials

Metal–organic frameworks (MOFs) and porous coordination polymers (PCPs) are terminologies[1] used to describe a vast and continuously expanding family of nanoporous 'hybrid' materials.[2] They are inorganic–organic compounds,[3] constructed by the molecular self-assembly of metal nodes (ions or clusters) and multitopic organic linkers,[4] resulting in the formation of a plethora of network topologies and chemical structures.[5] This bottom-up methodology when coupled with rational design could be harnessed to engineer three-dimensional (3-D) extended frameworks[6,7] as well as two-dimensional (2-D) layered structures,[8] held together by coordination bonds oriented in specific directions. Figure 1.1 shows a few exemplars of MOF structures. Notably, the open framework structures of MOFs afford a

Inorganic Materials Series No. 12
Mechanical Behaviour of Metal–Organic Framework Materials
Edited by Jin-Chong Tan

Published by the Royal Society of Chemistry, www.rsc.org

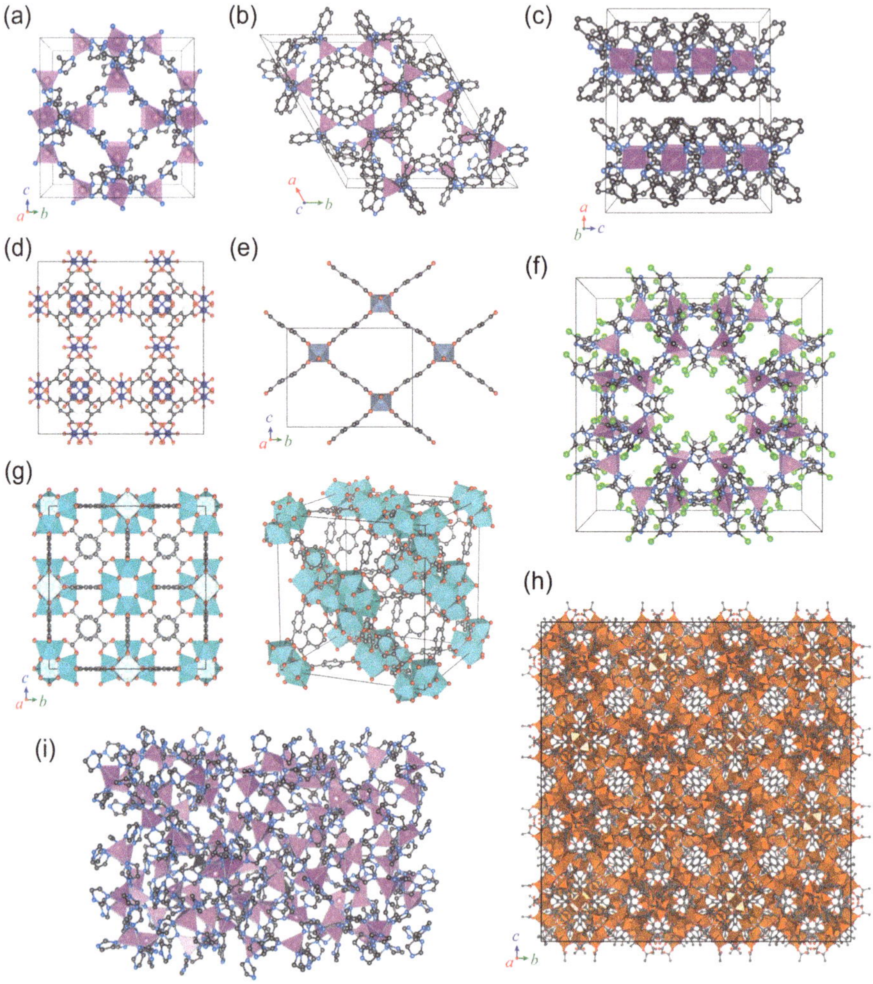

Figure 1.1 Some examples of topical MOF structures. Unit cells of 3D frameworks of (a) ZIF-8 viewed down the cubic *a*-axis, (b) ZIF-7-I (phase I) viewed down the rhombohedral *c*-axis, (c) the 2-D framework of ZIF-7-III (phase III) showing a layered architecture, with van der Waals interactions between adjacent layers, (d) HKUST-1 or Cu_3BTC_2 (BTC = benzene-1,3,5-tricarboxylic acid), viewed down the cubic *a*-axis, (e) MIL-53(Al) viewed down the orthorhombic *a*-axis, (f) ZIF-71 viewed down the cubic *a*-axis, (g) UiO-66(Zr) viewed down the *a*-axis (left) and isometric view of the cubic unit cell (right). (h) A very large cubic unit cell of MIL-100(Fe) viewed down the *a*-axis, comprising over 10 000 atoms. (i) An amorphous *a*-ZIF-4 structure with short-range order but no long-range order, thus with no identifiable unit cell.

precise cavity size and well-defined channel architecture, with pore dimensions ranging from ~2 Å to >1 nm. The most intensively studied MOF/PCP materials are intrinsically crystalline – they possess long-range order and many are highly porous – exhibiting an internal surface area typically of the order of ~1000 $m^2 g^{-1}$.[7] Increasingly, research in the field of MOF/PCP materials has expanded beyond the primary domain of ordered crystalline phases[9] to further encompass frameworks containing topological disorder and amorphous structures, such as MOF monoliths and hybrid glasses.[10] Further still, the porous MOF structure can also serve as a 'host' structure to accommodate and protect luminescent 'guest' molecules confined in its pore cavities, yielding a Guest@MOF 'composite' system with tuneable photophysical and photochemical properties.[11]

By virtue of the many pathways available for combining a multiplicity of organic and inorganic building units, the resultant chemical structures and functionalities of MOFs are particularly varied.[12,13] In principle, one could design, tune, and engineer an unusual combination of physical and chemical properties, which cannot be achieved in purely organic and inorganic materials alone. Indeed, a growing number of potential technological applications have been proposed. Many MOF structures are substantially more porous than any commercially available nanoporous sorbent materials, such as zeolites, silica gels, and activated carbons. Unsurprisingly, the first practical uses identified for MOFs thus encompassed gas separations and storage,[14] catalysis,[15] and CO_2 capture,[16] which may be perceived as 'classical' applications for the more conventional porous materials quoted above. In contrast, more innovative applications have been proposed where MOF serves as the 'active material' for integration into electroluminescent devices[17] and optoelectronics,[18] smart sensors,[19,20] dielectrics,[21] and for accomplishing functions linked to energy harvesting, conversion and transfer.[22,23] Encouragingly, a broad range of promising MOF applications visible on the horizon has attracted the attention of scientists, engineers, and technologists from a wide spectrum of disciplines, with a strong motivation to bridge the gap between fundamental research and real-world applications.

1.2 MOF Mechanics

Much of the earlier research efforts on MOF/PCP and hybrid framework materials are focussed on synthesis, chemical characterisation, and adsorption related properties. However, practical applications require an in-depth understanding of the basic mechanical behaviour, not only of single crystals and microcrystalline powders, but also of polycrystalline thin films and coatings, bulk monoliths and pellets, mixed-matrix membranes and nanocomposites, including bespoke Guest@MOF systems.[24–27] For instance, the sensitivity of a reusable mechanochemical sensor comprising a MOF film adhered to the surface of a silicon microcantilever (substrate) scales with the elastic coupling of stress-and-strain in a MOF structure upon analyte sorption/release, whilst its multicyclic operation depends on the mechanical resilience of the film-to-substrate adhesion.[28] A second exemplar involves the deployment of

MOFs as catalysts in reactors, where the thermomechanical stability of the porous frameworks subject to pressure, temperature, and humid conditions is vital for circumventing excessive framework deformation or stress-induced structural collapse that will gradually degrade performance over time. In yet another scenario, the fluorescence response of mechanochromic Guest@MOF materials[11] is heavily dependent on the interaction of nanoconfined fluorophores with imposed stresses and strains, for accurate calibration to allow force monitoring and optical stress sensing.[29]

In essence, elementary knowledge and precise control of structure–mechanical property relationships are central to the fabrication of advanced devices, for enabling component manufacturing and pellet shaping in industry, while ensuring mechanical durability for a sustained long-term performance in a multitude of commercial and consumer settings.

Underpinning all of these fundamental questions and practical challenges is a rapidly expanding field of research, aptly termed 'MOF Mechanics', which is concerned with an array of topical problems on mechanical phenomena encompassing: elasticity, structural anisotropy and stability, yielding and plastic deformation, time-dependent viscous effects, high-pressure response, interfacial cohesion and debonding, framework dynamics and mechanical dissipation, crack initiation and propagation leading to fracture (rupture of chemical bonds). This book will present to the reader the experimental and theoretical studies in the *New Science of MOF Mechanics*, exemplified by key topics and cutting-edge research addressing the broad range of mechanical behaviour inherent in hybrid framework materials. To this end, the chapters in the book are organised as follows.

- Experimental studies to measure the Young's modulus, hardness, mechanical anisotropy, interfacial adhesion and fracture toughness of crystals, monoliths, films and membranes by means of nanoindentation techniques are presented in Chapter 1 (Tan). This discussion is extended further to cover terahertz framework dynamics, soft modes, and shear deformation mechanisms that could destabilise the porous framework structures. Ashby-style materials selection charts are constructed using the latest available data to reveal general trends in the mechanical properties of MOF materials.
- In Chapter 2 (Marmier), the systematic characterisation of single-crystal elastic constants and the resultant anisotropic mechanical response are explored in detail, drawing from the theoretical and experimental data available to date. Special emphasis is given to highlight unusual framework mechanics responsible for anomalous physical phenomena, such as the negative Poisson's ratio, negative linear compressibility, and negative thermal expansion. This chapter concludes with a critical treatment on the pertinent questions of framework 'flexibility' and proposed mechanisms.
- Computational modelling of MOF mechanics is covered in Chapter 3 (Rogge), where the readers will be exposed to state-of-the-art methodologies for constructing an atomistic model of an extended hybrid framework, and subsequently for simulating its structural response subject

to directional stresses or hydrostatic pressure. Theoretical studies give new insights into complex mechanisms surrounding the elastic, plastic, and phase transformations of framework structures, which cannot be obtained by experiments alone.

- Chapter 4 (Moggach and Turner) focusses on the high-pressure deformation of MOF structures under hydrostatic compression *via* diamond anvil cells. Uniquely, the application of high-pressure X-ray crystallography reveals the evolution of pressure-induced structural deformations, phase transformations, and guest-mediated phenomena attributed to flexible frameworks.
- Rate effects and absorption of mechanical energy by flexible MOF structures are presented in Chapter 5 (Sun and Jiang). The pressure-stimulated liquid intrusion mechanism of hydrophobic MOFs is discussed through a combination of experimental and theoretical studies, elucidating the effect that deformation strain rate (quasistatic, medium, high) has on framework materials performance. Finally, mechanical energy dissipation, by means of non-intrusion mechanisms such as phase transition and pore collapse, is considered.

1.3 Central Concepts in the Study of Mechanical Properties of Solids

When a solid material is subjected to a small mechanical force, the relationship between stress ($\sigma = P/A$, applied force P divided by area A subject to the force) and strain ($\varepsilon = \Delta L/L_0$, change in length ΔL divided by its initial length L_0) is *linear* — this purely elastic behaviour obeys Hooke's law, such that $\sigma \propto \varepsilon$. The material is in equilibrium, where any mechanical deformation or shape change experienced by the solid is reversible upon removal of the applied force. Crucially, in this linear elastic regime, the stress level must not exceed the yield strength (σ_Y) of the material, $\sigma < \sigma_Y$, beyond which its mechanical response will become irreversible and may turn nonlinear, as illustrated in Figure 1.2(a). This important concept of yield strength is related to the hardness (H), which is discussed in Sections 1.4 and 1.8.1.

Within linear elasticity, the mechanical properties of materials can be described by the Young's modulus, Poisson's ratio, shear modulus, and bulk modulus. The Young's modulus or elastic modulus, E, is defined as the ratio of stress to strain, as follows:

$$E = \frac{\sigma}{\varepsilon} \tag{1.1}$$

hence, $\sigma = E\varepsilon$, strictly for unidirectional loading conditions only (*i.e.*, uniaxial tension or uniaxial compression). E is a measure of the mechanical 'stiffness' of the framework structure, this corresponds to the slope of the stress *vs.* strain curve located below the yield point (Figure 1.2(a)). The elastic moduli are typically expressed in units of N m^{-2} or Pascal (Pa). Conversely, the reciprocal of the stiffness property is called the mechanical 'compliance', $S = 1/E$ or E^{-1}, with units of Pa^{-1}.

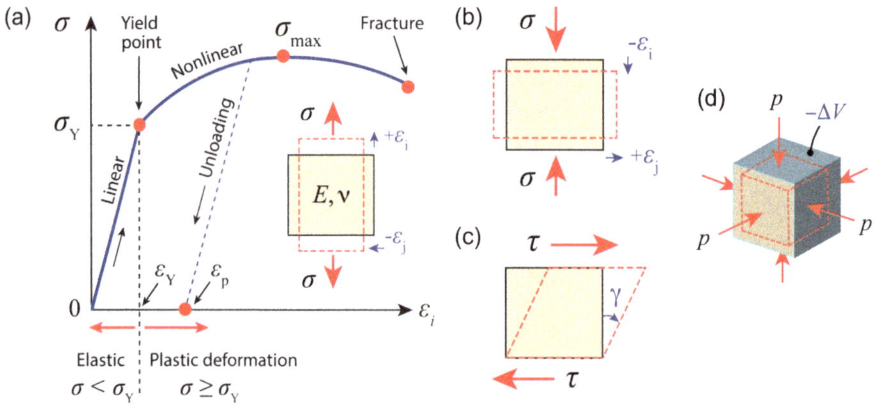

Figure 1.2 (a) Stress *versus* strain (σ–ε) curve under uniaxial tension (inset) for a hypothetical solid material exhibiting nonlinear strain hardening behaviour beyond the yield point, where σ_Y, ε_Y, and ε_p are the yield strength, yield strain, and plastic strain, respectively. The maximum stress (or ultimate strength) is denoted by σ_{max}. E and ν are the Young's modulus and the Poisson's ratio of the isotropic solid, respectively. (b) Uniaxial loading where the applied stress σ is compressive. Subscripts i and j of the resultant strains ε denote the axial and transverse (lateral) directions, respectively. (c) Shear deformation due to application of an external shear stress τ causing a shear strain γ by angular distortion. (d) Hydrostatic pressure p causing a change in volume ΔV (negative sign denotes shrinkage), but with no change to the shape of the cube.

As shown in Figure 1.2(a and b), when the elastic solid is *uniaxially* deformed in the 'longitudinal' direction (ε_i), there will be a resultant *lateral* strain generated in the 'transverse' direction (ε_j). The Poisson's ratio, ν, can be determined from the ratio of lateral strain to axial strain:

$$\nu_{ij} = -\frac{\varepsilon_j}{\varepsilon_i} = -\frac{\varepsilon_{transverse}}{\varepsilon_{longitudinal}} \tag{1.2}$$

The negative sign ensures that the value of ν is always positive for a conventional material: the upper bound is 0.5 for an incompressible solid such as a rubbery polymer, while the lower bound is 0 for a fully compressible material like a foam. However, framework materials can exhibit a negative Poisson's ratio (termed 'auxetic'), with this *anomalous* mechanical phenomenon considered in detail in Chapter 2.

The shear modulus, G, is a measure of the framework's 'rigidity' or torsional stiffness of material subject to an angular distortion, as depicted in Figure 1.2(c). It is defined as the ratio of shear stress to shear strain:

$$G = \frac{\tau}{\gamma} \tag{1.3}$$

The bulk modulus, K, is a measure of the 'volumetric stiffness' (*i.e.*, volumetric stress divided by volumetric strain, $\Delta V/V_0$) of the material subject to a hydrostatic pressure, p, as illustrated in Figure 1.2(d). In this definition, the negative sign accounts for the volumetric contraction experienced in compression, so that the value of K remains positive:

$$K = -\frac{p}{\left(\Delta V/V_0\right)} \tag{1.4}$$

The inverse of bulk modulus is called the compressibility, $\beta = K^{-1}$. Chapter 4 is dedicated to the study of the mechanical behaviour of MOF materials under hydrostatic compression, and it will give a rigorous treatment to this subject matter, including characterisation of framework structures with negative compressibility.

For an *isotropic* solid material, whose mechanical properties are not changing with direction, there are only two independent elastic constants to be established because the elastic properties (E, ν, G, K) are interrelated as follows:

$$G = \frac{E}{2(1+\nu)} \tag{1.5}$$

$$K = \frac{E}{3(1-2\nu)} \tag{1.6}$$

$$\nu = \frac{(3K-2G)}{2(3K+G)} \tag{1.7}$$

For an *anisotropic* solid, however, the mechanical properties determined will be varying with direction of the applied loading. In this situation, the generalised Hooke's law in 3-D can be applied to determine the strains developed along the three orthonormal axes of Figure 1.3(c), when the anisotropic solid is subject to a *triaxial* stress state (σ_1, σ_2, σ_3):

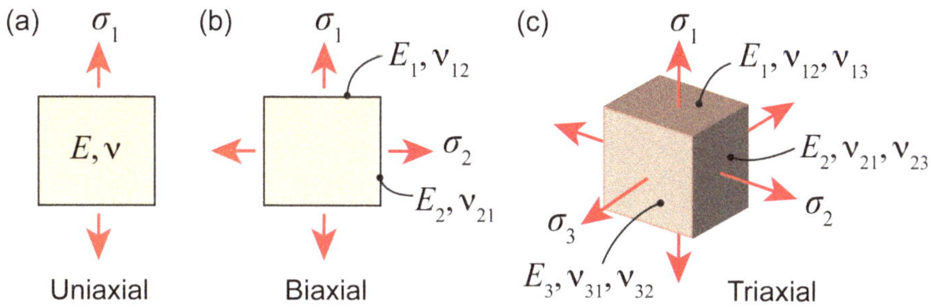

Figure 1.3 (a) Uniaxial, (b) biaxial, and (c) triaxial stress states acting on a solid, where subscripts 1, 2, 3 denote the three orthonormal directions. For material (a), E and ν are the Young's modulus and Poisson's ratio of the isotropic solid. For (b) and (c) the material is anisotropic, hence the Young's moduli ($E_1 \neq E_2 \neq E_3$) and the Poisson's ratios ($\nu_{ij} \neq \nu_{ji}$) are directionally dependent. In the context of a cubic MOF crystal, each direction corresponds to a crystallographic axis oriented normal to the crystal facet.

$$\begin{aligned}\varepsilon_1 &= \frac{\sigma_1}{E_1} - \frac{\nu_{21}\sigma_2}{E_2} - \frac{\nu_{31}\sigma_3}{E_3} \\ \varepsilon_2 &= \frac{\sigma_2}{E_2} - \frac{\nu_{12}\sigma_1}{E_1} - \frac{\nu_{32}\sigma_3}{E_3} \\ \varepsilon_3 &= \frac{\sigma_3}{E_3} - \frac{\nu_{13}\sigma_1}{E_1} - \frac{\nu_{23}\sigma_2}{E_2}\end{aligned} \tag{1.8}$$

For the case of a *biaxial* stress state depicted in Figure 1.3(b), eqn (1.8) can be simplified by letting $\sigma_3 = 0$, thereby corresponding to a 2-D plane stress.

1.4 Nanoindentation of MOF Materials

1.4.1 General Principles of Nanoindentation

Some mechanical properties of MOFs have been studied by employing the instrumented indentation testing (IIT) method, which is commonly called 'nanoindentation'.[30] Nanoindentation is normally performed using a Berkovich diamond indenter tip (three-sided pyramidal probe), with which the Young's modulus (E) and hardness (H) of a small number of MOF materials in the form of single crystals,[31–33] monoliths,[34,35] thin films,[36,37] amorphous particles and glasses[38,39] have been characterised to date.

The different stages of nanoindentation testing are depicted in Figure 1.4. Initially, at loading stage (1), the diamond indenter tip slowly pushes into the sample surface, typically to reach a maximum surface penetration depth of

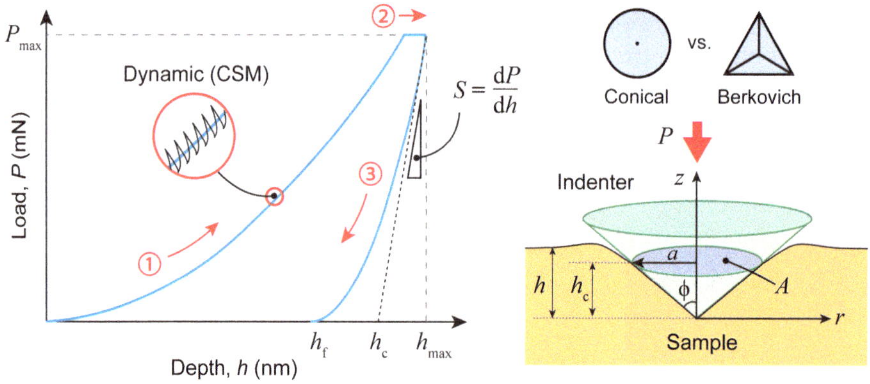

Figure 1.4 A typical load–depth (P–h) curve obtained from nanoindentation testing (right) using a conical indenter tip. Three main test segments comprise: (1) indenter loading, (2) holding at maximum load P_{max}, and (3) indenter unloading. The contact stiffness (S) can be determined from the dynamic continuous stiffness measurement (CSM) and from the slope of the unloading curve. The contact area A is the projected contact area under load, for a conical indenter this is given by $A = \pi a^2 = \pi h_c^2 \tan^2\phi$. The area function for an ideal Berkovich indenter is $A = 24.5h_c^2$, determined using an equivalent conical angle of $\phi = 70.3°$.

around 1–2 μm. Shallower indents of a depth of several hundred nanometres, or less, can be achieved, but this will require a sharp indenter tip (end radius $\lesssim$ 50 nm). Subsequently, the compressed sample is held at a constant maximum load (2), to overcome time-dependent effects (*e.g.*, creep). Finally, the tip is slowly withdrawn from the deformed region, yielding the unloading test segment (3). The changing indenter load (P) and vertical displacement (h) data are continuously recorded during nanoindentation testing, to yield an indentation load–depth (P–h) curve. This P–h curve is subsequently used to compute the values of E and H by employing the Oliver and Pharr (O&P) method,[40] outlined below. For a detailed treatment of basic nanoindentation theory and further data analysis techniques, the reader may consult critical reviews available in the literature.[30,41–44]

Nanoindentation experiments may be performed under either load-controlled (quasi-static) or displacement-controlled (dynamic) modes, the latter is known also as the 'continuous stiffness measurement' (CSM).[41] These techniques differ by the way the elastic contact stiffness, S, is derived. Under quasi-static testing, the elastic contact stiffness is calculated from the slope of the unloading segment in the P–h curve, thus $S = \mathrm{d}P/\mathrm{d}h$ (Figure 1.4). As such, the E and H values are obtained only at the maximum indentation depth, h_{max}. In CSM testing, however, the E and H values can be determined continuously, as a function of the surface penetration depth during the loading segment. This is made possible by superimposing a small sinusoidal displacement at a specific excitation frequency (*e.g.*, 2 nm at 45 Hz)[41] onto the primary loading signal (Figure 1.4 inset), and the dynamic response of the system is utilised to compute the changing magnitude of S with indentation depth, h.

The elastic contact stiffness (S) is later used to calculate the reduced modulus, E_{r}:[45]

$$E_{\mathrm{r}} = \frac{\sqrt{\pi}}{2\beta}\frac{S}{\sqrt{A}} \tag{1.9}$$

where the constant β varies with the geometry of the indenter tip: for example, $\beta = 1$ for a spherical tip, and $\beta = 1.034$ for a Berkovich tip. The contact area established under load, A, is a function of the contact depth, h_{c}. This is given by a tip area function $A(h_{\mathrm{c}})$ as exemplified by the five-term polynomial in eqn (1.10), where C_n are constants obtained by curve fitting.[40] The calibration procedure involves indentation of an isotropic material of known Young's modulus, which is typically a polished sample of fused silica (E = 72 GPa).

$$A(h_{\mathrm{c}}) = C_0h^2 + C_1h + C_2h^{1/2} + C_3h^{1/4} + C_4h^{1/8} \tag{1.10}$$

The reduced modulus, E_{r}, is a function of the Young's moduli and Poisson's ratios of the sample (E_{s}, v_{s}) and the indenter (E_{i}, v_{i}):

$$\frac{1}{E_{\mathrm{r}}} = \left(\frac{1 - v_{\mathrm{s}}^2}{E_{\mathrm{s}}}\right) + \left(\frac{1 - v_{\mathrm{i}}^2}{E_{\mathrm{i}}}\right) \tag{1.11}$$

It follows that the Young's modulus of the sample, E_s, can be determined using the following expression:

$$E_s = \left(1 - \nu_s^2\right)\left[\frac{1}{E_r} - \frac{1 - \nu_i^2}{E_i}\right]^{-1} \tag{1.12}$$

For a diamond indenter probe, E_i = 1141 GPa and ν_i = 0.07. When the stiffness of the sample is significantly lower than the indenter stiffness, $E_s \ll E_i$, eqn (1.12) can be approximated by $E_s = E_r(1 - \nu_s^2)$. The calculation of the Young's modulus of the test sample (E_s) therefore will require knowledge of its Poisson's ratio (ν_s), when the latter is an unknown a value may be chosen based on these guidelines: glasses and ceramics ($\nu_s \sim 0.2$), metals (~0.3), polymers (~0.45), rubbery elastomers (~0.5). Although for stiffer materials like metals, the sensitivity of calculated E_s to the input value of ν_s is weak,[30] for relatively low stiffness MOF-type materials ($E \lesssim 10$ GPa),[24] the error does become more significant.[46] The situation is further complicated by the fact that the Poisson's ratio of MOFs can span a wide range of values (Chapter 2) and, may vary due to their mechanical anisotropy, as described below (Section 1.4).

Indentation hardness, H, also termed 'nanohardness', quantifies the plastic deformation of the material beyond the yield point (further details in Section 1.8.1). Based on the O&P method, H is calculated by dividing the applied load by the projected area of contact:

$$H = \frac{P}{A} \tag{1.13}$$

Therefore, the hardness determined from the maximum indentation depth is given by $P_{max}/A(h_{max})$. Akin to stress, the unit for nanohardness is Pascal (Pa).

The O&P method[40] was derived on the basis that the test sample is homogeneous and elastically isotropic, which is true for many polycrystalline materials that exhibit an approximately isotropic mechanical response. However, this is clearly not the case even for most single crystals with a cubic symmetry.[47] Discrepancies for cubic (Al, Cu, β-brass) and hexagonal single crystals (Zn) of metals caused by elastic anisotropy have been studied in 1994 by Vlassak and Nix,[48] who demonstrated that eqn (1.11) still holds if the elastic modulus terms are redefined as the 'indentation modulus', M, as follows:

$$\frac{1}{M_r} = \left(\frac{1}{M_s}\right) + \left(\frac{1 - \nu_i^2}{E_i}\right)$$
$$M_{\{hkl\}} = \left[\frac{1}{M_r} - \frac{1 - \nu_i^2}{E_i}\right]^{-1} \tag{1.14}$$

where $M_s \equiv M_{\{hkl\}} = E_s/(1 - \nu_s^2)$, designating the indentation modulus oriented normal to the $\{hkl\}$ facet of a single crystal.

In 1998, Hay *et al.*[49] adopted the indentation modulus methodology to probe hexagonal single crystals of β-silicon nitride, which are highly anisotropic, and found that the Young's modulus was underestimated by ~20% in the stiffest direction, whereas in the most compliant direction it was overestimated by ~10% if the indentation results were not corrected for anisotropic effects. In 2004, Bei *et al.*[50] applied this approach to measure the anisotropic Young's moduli of Cr_3Si intermetallics, where they reported consistent results from Berkovich nanoindentation *versus* ultrasonic testing. However, it was not until 2009 that Tan *et al.*[32] demonstrated its implementation for studying the anisotropic mechanical behaviour of two polymorphic Cu-based MOFs, both with a dense structure, one 3-D, the other 2-D. The single-crystal nanoindentation results show a strong structure–mechanical property correlation along specific crystallographic orientations. For example, the stiffest crystal facet of the 3-D framework ($M_{\{100\}} \sim 93$ GPa) is oriented normal to the underlying metal–oxygen–metal (M–O–M) chains, thereby conferring a stiff 'backbone structure'. In contrast, it was established that the most compliant facet of the 2-D framework ($M_{\{010\}} \sim 35$ GPa) is oriented normal to the stacking of hydrogen-bonded layers. Furthermore, the elastic anisotropy of these framework crystals was found to be large (stiffness variation >60%) compared to the hardness anisotropy (hardness variation of ~12%).[32] Interestingly, this parallels the findings for metallic single crystals that despite large anisotropy of yield stress show only small hardness anisotropy (*ca.* 13–20%), with this phenomenon being attributed to the complex plastic strain field generated under the indenter.[48] While the dislocation mechanisms responsible for the plastic deformation of metals are well known, the plastic deformation mechanisms for MOFs and hybrid frameworks are currently not well understood (Section 1.8).

To yield reliable and reproducible nanoindentation measurements of MOF materials, it is generally important to pay attention to the following points in relation to sample preparation. First, the sample to be probed must be secured on a much stiffer 'substrate' material, such as an epoxy mount or a metal stub to eliminate any compliance issues associated with underlying substrate. Second, the sample surface must be microscopically flat for accurate contact area determination (accuracy of $A(h_c)$ has a major effect on E and H calculations) and to minimise the overall experimental scatter. A smooth surface with a mean roughness of ~10 nm can be achieved by cold mounting the crystals or monoliths in an epoxy resin, followed by careful grinding using a non-penetrating lubricant (*e.g.*, water for hydrophobic MOFs or glycerol for water-sensitive samples), and then by polishing with an increasingly finer grade of diamond suspensions to yield the final smooth surface. It is vital to ensure that the chosen surface preparation steps do not significantly alter the properties of the sample surface, which may be caused by several factors, such as surface contamination, guest infiltration into MOF pores, chemical degradation, residual stress, and subsurface cracking.

Two representative studies, where the structure–mechanical property relationships of MOFs have been established with these nanoindentation techniques are discussed in more detail in the following sections.

1.4.2 Nanoindentation of ZIFs

The elastic moduli of the single crystals of a family of zeolitic imidazolate framework (ZIF) materials, encompassing ZIF-4, ZIF-7, ZIF-8, ZIF-9, ZIF-20, ZIF-68, and ZIF-zni, have been systematically characterised by single-crystal nanoindentation.[51] As shown in Figure 1.5(a), the chemical structure, crystal symmetry and network topology of the samples are distinctively different, so are their framework density, porosity and solvent accessible volume (SAV). The *P*–*h* curves of the seven ZIF structures (Figure 1.5(b)), when indented to a maximum depth of 1 μm, revealed a diverse mechanical response in terms of the attained maximum load, slope at initial unloading, and the extent of elastic recovery indicated by the residual depth (h_f) after complete unload. Because the nanoindentation measurements were conducted in CSM mode, the elastic modulus (*E*) can be determined as a function of indentation depth (Figure 1.5(d)). It can be seen that the *E* values are unique to each framework structure, and they remain relatively constant beyond the initial contact ($h >$ 100 nm). For example, ZIF-8, ZIF-20, and ZIF-68, which have a more porous structure (SAV ~ 50%) exhibit a lower structural stiffness in the range of E = 3–4 GPa. In contrast, the densest framework, ZIF-zni (SAV ~ 12%), has elastic

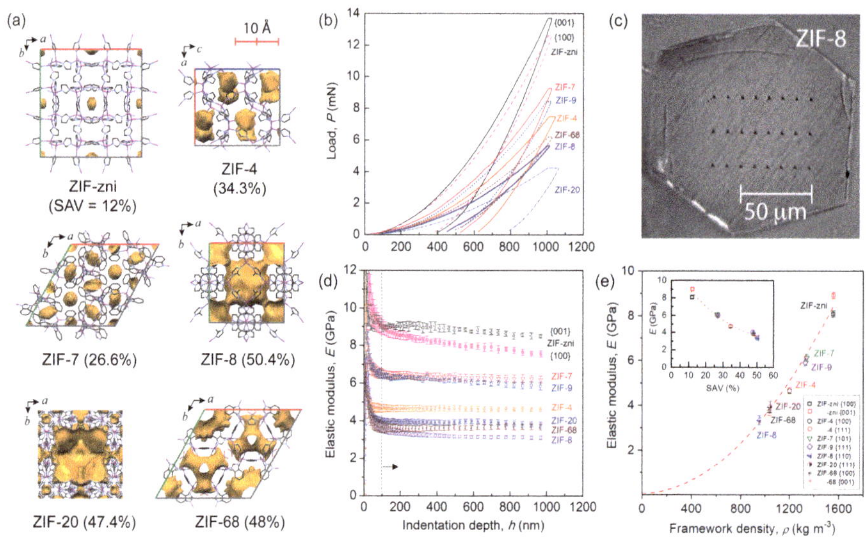

Figure 1.5 (a) ZIF structures with yellow surfaces denoting the solvent accessible volume (SAV, calculated with a probe size of 1.2 Å). Their network topologies are: **zni** for ZIF-zni; **cag** for ZIF-4; **sod** (sodalite) for ZIF-7, ZIF-8, and ZIF-9; **lta** (Linde type A) for ZIF-20; **gme** (gmelinite) for ZIF-68. (b) Load–depth curves from the nanoindentation of ZIF single crystals, like the example of ZIF-8 depicted in (c), showing that 24 residual indents remained on the sample surface after complete unload. (d) Depth-dependent CSM data calculated from the *P*–*h* curves in (b). (e) Elastic moduli of ZIFs as a function of framework density showing a quadratic relationship, and (inset) an inverse correlation with accessible porosity. Adapted from ref. 51 with permission from National Academy of Sciences.

moduli of *ca.* 8–9 GPa, further revealing the mechanical anisotropy associated with its two different crystal facets, where $E_{(001)}$ is higher than $E_{(100)}$.

The nanoindentation results of ZIFs show that the underpinning framework architecture constructed from a varied combination of organic and inorganic blocks has a major influence on the mechanical behaviour of the resultant porous frameworks. Not only the network topology plays a role, but the stereochemistry of the imidazolate-type linkers (*e.g.*, bulkiness and stiffness) and porosity are equally important for determining the mechanics of the resultant frameworks. For instance, while ZIF-7, ZIF-8, and ZIF-9 all have identical sodalite (**sod**) topology, the sterically bulky benzimidazolate (bIm) ligands in ZIF-7 and ZIF-9 confer a greater stiffness value of $E \sim 6$ GPa, compared with ZIF-8, whose $E \sim 3$ GPa due to its less bulky 2-methylimidazolate (mIm) ligands. Moreover, as shown in Figure 1.5(e), an elastic modulus *versus* framework density correlation of the form of $E \propto \rho^2$ has been proposed to describe the elasticity trend for the ZIF family of materials. Likewise, an elastic modulus *versus* SAV relationship of ZIFs has been established, as presented in the Figure 1.5(e) inset, showing an inverse correlation of $E \propto \frac{1}{\text{SAV}}$ exists between structural stiffness and porosity.

1.4.3 Nanoindentation of Perovskite MOFs

Single-crystal nanoindentation has been employed to measure the elastic properties of a family of multiferroic MOFs adopting the ABX_3 perovskite topology, see Figure 1.6. Four isostructural frameworks of dimethylammonium metal formate:[53,54] $[(CH_3)_2NH_2][M(HCOO)_3]$ where $A = [(CH_3)_2NH_2]^+$ and B is M^{2+} (= Ni^{2+}, Co^{2+}, Zn^{2+}, or Mn^{2+}), have been systematically studied. The material is a 'dense' framework structure because the dimethylammonium cation occupies the A site, so there is no porosity left in the otherwise open framework to accommodate additional guest molecules. Nanoindentation measurements were performed on the {012}-oriented facets of the pseudo-cubic crystals (Figure 1.6(a)), where the *P*–*h* curves and Young's moduli of the four isostructural frameworks as a function of indentation depth are shown in Figure 1.6(b).[52] The determined moduli of the dense MOF perovskites lie in the range of $E = 19$–25 GPa comparable to the Young's modulus of a metal-free hybrid perovskite ($E \sim 15$ GPa)[55] and the moduli of several inorganic–organic halide perovskites ($E \sim 12$–19 GPa for $APbX_3$),[56] but they are around one order of magnitude higher than the nanoporous ZIF structures[51] discussed in Section 1.4.2.

The stiffness of the four hybrid perovskite frameworks rises in accordance with the sequence $Mn^{2+} \approx Zn^{2+} < Co^{2+} < Ni^{2+}$. As the compounds are isostructural, the differing mechanical properties can be linked to the different divalent metal cations (M^{2+}) forming the MO_6 octahedral sites. Figure 1.6(c) shows the approximately linear correlation between the Young's moduli and the ligand field stabilisation energy (LFSE) connecting the four cations. However, a straightforward correlation to the cation radius was not found, albeit a general trend of the variation in the elastic moduli with octahedral bond distance can be seen in Figure 1.6(d). Consequently, it was proposed

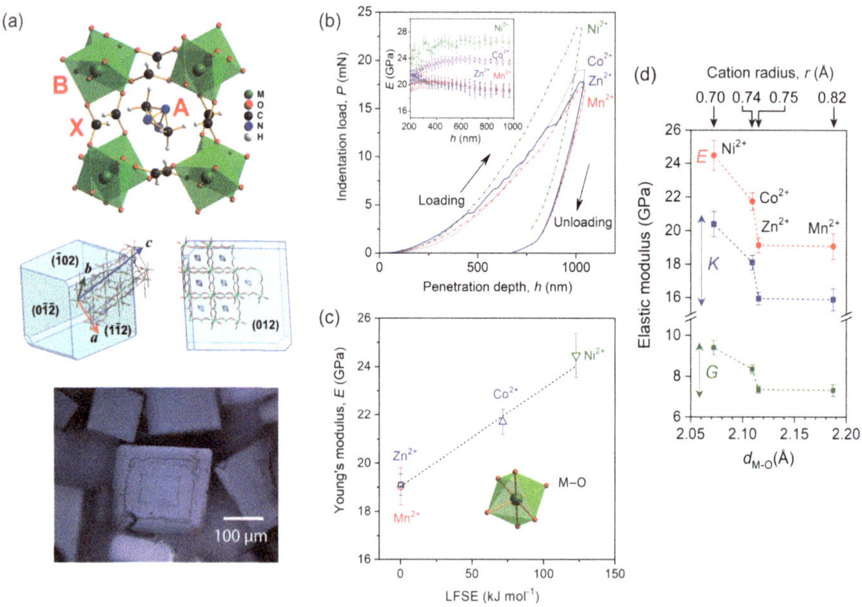

Figure 1.6 (a) Rhombohedral unit cell (top) and pseudo-cubic morphology of dimethylammonium metal formate crystals (metal = Ni, Co, Zn, Mn), showing ABX_3 perovskite architecture and a dense framework. (b) Nanoindentation load–depth curves and CSM data up to 1000 nm for the four isostructural frameworks, the Young's moduli (E) calculated by taking $v_s = 0.3$. (c) Correlation of E to ligand field stabilisation energy (LFSE); the inset shows the MO_6 octahedral site. (d) Trends in the variation of elastic moduli as a function of the octahedral bond distance, d_{M-O}. The values of the shear (G) and bulk (K) moduli were estimated by assuming an isotropic response in accordance with eqn (1.5) and (1.6); note that the dotted lines serve as guides for the eye. Adapted from ref. 52 with permission from the Royal Society of Chemistry.

from a mechanical stability standpoint that, a higher LFSE in the octahedral environment bestows a greater resistance to mechanical deformation at the MO_6 sites (metal nodes), thereby increasing the elastic modulus of the overall framework structure.[52] A follow-on nanoindentation study reported by Li *et al.*[57] on two analogous MOF perovskites further revealed that the Young's moduli increase with an increase in the number of hydrogen bonding interactions established between the A-site molecular cation and the negatively charged framework.

1.4.4 Young's Modulus *vs.* Hardness (*E*–*H*) Materials Property Chart

It is important to be able to see the bigger picture and visualise how the mechanical properties of MOFs and other hybrid framework materials are compared with *conventional* engineering materials, such as metals, polymers, and ceramics. To this end, an Ashby-style plot (materials property chart) may be constructed like the one presented in Figure 1.7, obtained by curating the latest Young's modulus (E) *vs.* hardness (H) datasets from the literature. Chiefly, the E–H domain associated with the hybrid framework materials straddles the borders between the metallic, polymeric (organic), and ceramic (inorganic) materials. Within the hybrid framework domain itself, it can be seen that the upper bound is populated by 'dense' frameworks that are mechanically stiffer and harder (higher E and H), whilst the lower bound is occupied by porous MOFs and open framework structures which are structurally more compliant and softer (lower E and H). There is an 'intermediate' zone, populated by a family of hybrid inorganic–organic perovskites (HOIPs), particularly the multiferroic MOFs with ABX_3 topology and halide perovskites $APbX_3$ described above. Overall, this chart sheds light on the immense tuneability of the mechanical properties of hybrid frameworks as a whole. In fact, this outcome is unsurprising in light of the vast scope to combine a multitude of organic and inorganic moieties to yield products of different architectures.

1.5 Elastic Anisotropy of MOF Single Crystals

The elastic properties of a MOF single crystal, even that with a cubic symmetry, are inherently anisotropic. Accordingly, the magnitude of the elastic constants (E, G, and ν) are orientationally dependent for a single crystal, and thus its mechanical response will change with the direction of loading (Figure 1.3(c)). MOF crystals with a lower crystal symmetry are well known for their extreme elastic anisotropy, which may give rise to anomalous mechanical phenomena, such as a negative Poisson's ratio (auxeticity), negative thermal expansion (NTE), and negative linear compressibility (NLC); these topics are covered in depth in Chapters 2 and 4. Hitherto, the majority of results on the elastic anisotropy of MOFs have been derived from computational modelling studies, such as density functional theory (DFT)[69–71] and molecular dynamics (MD)[72,73] simulations (Chapter 3), complemented by a more limited set of experiments that have precisely measured the elastic stiffness tensor[46] and NLC of MOF crystals.[74,75] This section details a few exemplars in the field.

1.5.1 ZIF-8: Experimental and Theoretical Determination of Elastic Constants

The elastic stiffness tensor (C_{ij}, see definitions in Chapter 2 with the Voigt notation) of a single crystal of a prototypical ZIF material with sodalite topology, termed ZIF-8, has been measured by means of laser Brillouin scattering experiments.[46] Due to its cubic symmetry, ZIF-8 possesses three independent elastic coefficients, the values of which are C_{11} = 9.52 GPa, C_{12} = 6.86 GPa, and C_{44} = 0.97 GPa. The stiffness tensor for a cubic ZIF-8 crystal can be cast into a 6 × 6 matrix:

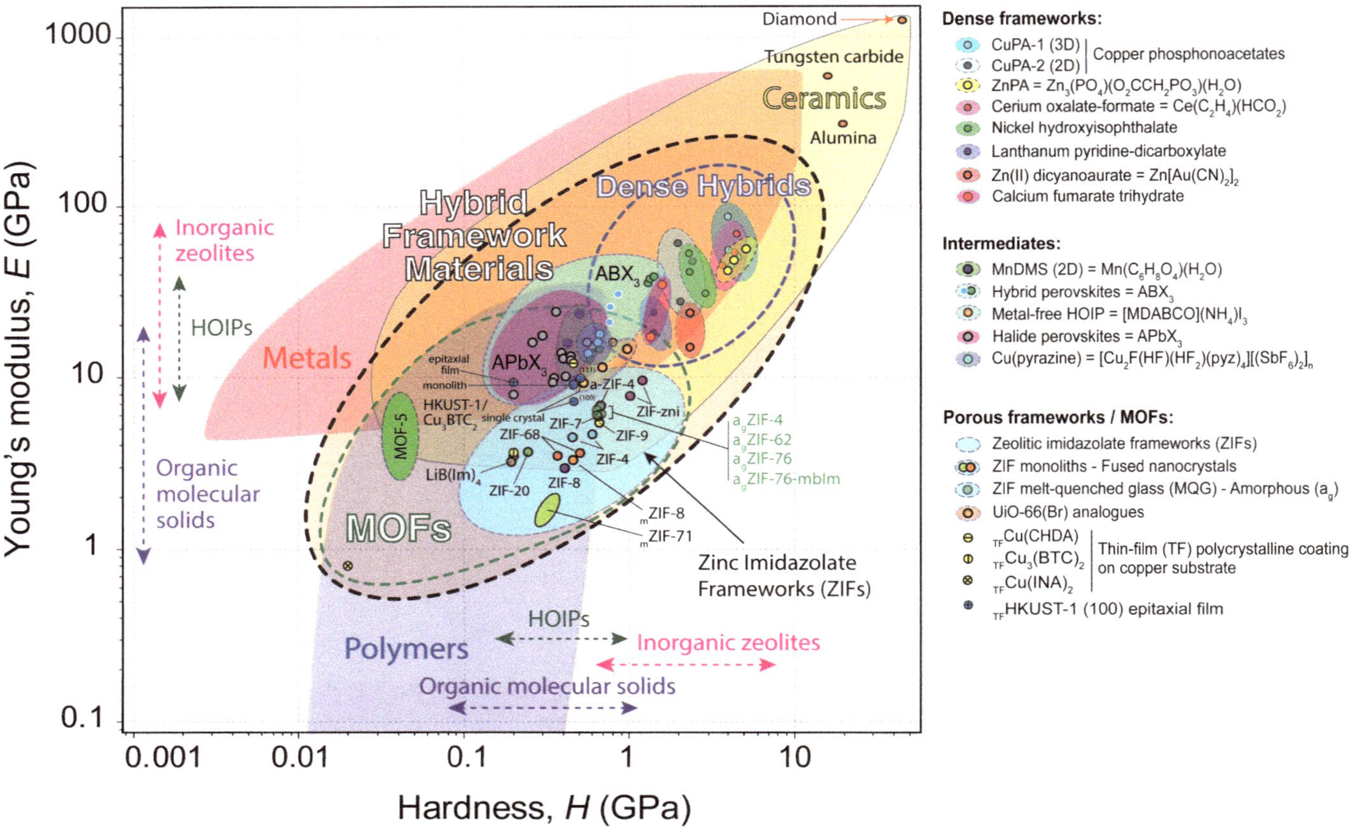

Figure 1.7 Young's modulus (E) plotted against the hardness (H) of MOFs and the wider families of materials. Adapted from ref. 24 and augmented with the latest (E, H) datasets (published up to May 2022) determined mostly by nanoindentation measurements. Exemplars of dense hybrid frameworks include: copper phosphonoacetate (CuPA) polymorphs,[32] zinc phosphate phosphonoacetate hydrate (ZnPA),[58] cerium oxalate–formate,[59] zinc(II) dicyanoaurate,[60] and calcium fumarate trihydrate.[61] Multiple data points for each material bubble signify mechanical anisotropy. Representative MOFs and porous frameworks

$$C_{ij} = \begin{pmatrix} C_{11} & C_{12} & C_{13} & 0 & 0 & 0 \\ C_{21} & C_{22} & C_{23} & 0 & 0 & 0 \\ C_{31} & C_{32} & C_{33} & 0 & 0 & 0 \\ & & & C_{44} & 0 & 0 \\ \vdots & \iddots & & & C_{55} & 0 \\ \text{symm.} & \cdots & & & & C_{66} \end{pmatrix} = \begin{pmatrix} 9.52 & 6.86 & 6.86 & 0 & 0 & 0 \\ 6.86 & 9.52 & 6.86 & 0 & 0 & 0 \\ 6.86 & 6.86 & 9.52 & 0 & 0 & 0 \\ 0\ 0\ 0 & & & 0.97 & 0 & 0 \\ 0\ 0\ 0 & & & 0 & 0.97 & 0 \\ 0\ 0\ 0 & & & 0 & 0 & 0.97 \end{pmatrix} \text{GPa} \tag{1.15}$$

Herewith, the coefficient C_{11} (= C_{22} = C_{33} because of cubic symmetry) designates the stiffness along the orthonormal a, b, and c principal crystal axes, respectively, when subject to uniaxial strain. The shear coefficient C_{44} (= C_{55} = C_{66}) is the stiffness against an angular distortion, when subject to a shear strain. Finally, the stiffness coefficient C_{12} (= C_{13} = C_{23} = C_{21} = C_{31} = C_{32}) corresponds to tensile–tensile coupling between any two orthonormal axes. It follows that the C_{ij} values of ZIF-8 obey the fundamental stability criteria of a cubic crystal: $C_{11} > |C_{12}|$, $C_{11} + 2C_{12} > 0$, and $C_{44} > 0$.[47] Subsequently, the degree of elastic anisotropy for a cubic crystal can be characterised by the Zener ratio A,[76] where $A = \dfrac{C_{44}}{1/2(C_{11} - C_{12})}$. For ZIF-8, A = 0.73, thus it is moderately anisotropic, noting that A is unity for an isotropic material.[46]

Tensorial analysis of C_{ij} (see Section 2.2 of Chapter 2) is useful for analysing the direction-dependent E and ν, with the complete picture revealing the elastic behaviour of a single crystal of ZIF-8 illustrated in Figure 1.8. In terms of Young's modulus, the representation surface of E exhibits protuberances along the ⟨100⟩ cube axes, corresponding to the highest stiffness of $E_{\max(100)}$ ~ 3.8 GPa oriented normal to the 4-membered rings (4MRs) of the sodalite framework. Moreover, its lowest stiffness is $E_{\min(111)}$ ~ 2.8 GPa, oriented normal to the 6-membered rings (6MRs). It follows that the Young's modulus anisotropy of ZIF-8 is given by $E_{\max}/E_{\min}$ = 1.35. Furthermore, an intermediate stiffness of $E_{(110)}$ ~ 3 GPa is present along the ⟨110⟩ face diagonal of the cubic unit cell. Nanoindentation experiments conducted on the above three selected crystal facets are consistent with the results obtained from Brillouin scattering, see Table 1.1. Additionally, theoretical DFT calculations

include ZIFs (single crystals,[51] nanocrystalline monoliths,[35] a-ZIF-4 and recrystallised ZIF-zni),[38] lithium–boron analogue of ZIF [LiB(Im)$_4$],[62] melt-quenched MOF glasses,[39] HKUST-1 (single crystals,[63] nanocrystalline monolith,[34] epitaxial film),[36] MOF-5,[31] UiO-66(Br) analogues,[64] and Cu-MOF polycrystalline films.[37] Intermediates bridging the porous and dense framework regimes, encompassing hybrid organic–inorganic perovskites (HOIPs) such as halide perovskites (APbX$_3$),[65,66] MOFs with perovskite ABX$_3$ topology,[52] and metal-free HOIP.[55] Other intermediates include Mn 2,2-dimethylsuccinate (2-D layered structure of MnDMS)[67] and a copper pyrazine framework.[68] Adapted from ref. 24 with permission from the Royal Society of Chemistry.

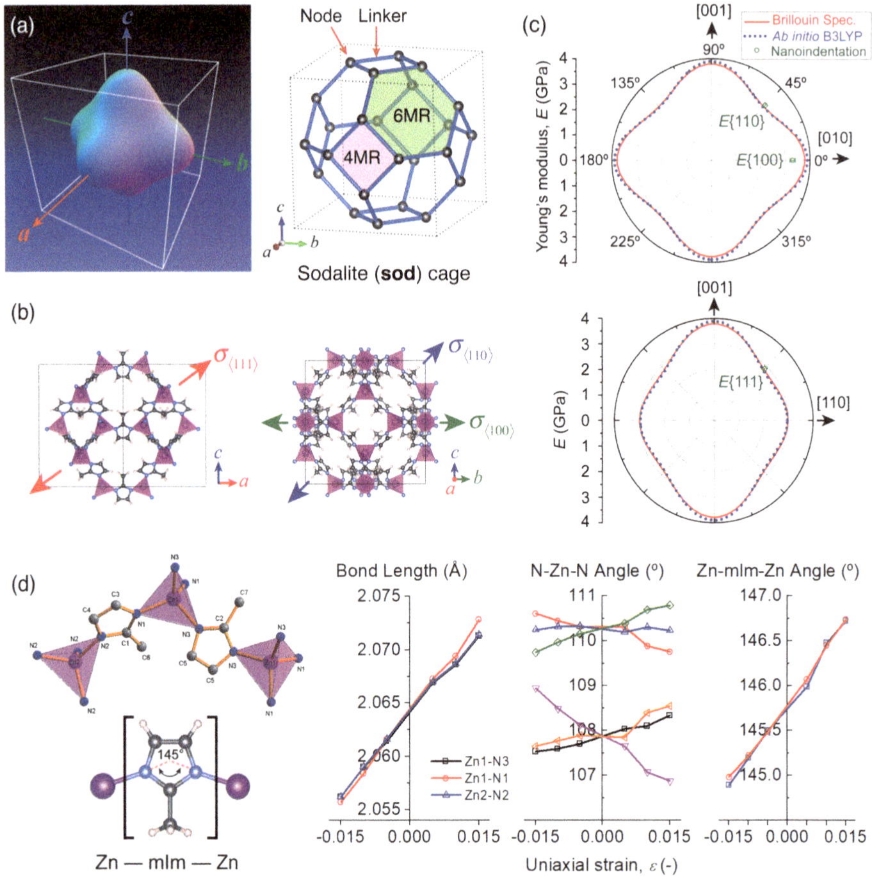

Figure 1.8 (a) 3-D representation surfaces of the anisotropic Young's modulus (E) of ZIF-8. (Right) A sodalite topology highlighting the four- and six-membered rings (4MR and 6MR). (b) Uniaxial stresses applied in the $\langle uvw \rangle$ axes of the cubic unit cell of ZIF-8, resulting in the maximum, intermediate, and minimum values of E. (c) Polar plots of E projected onto the (100) and $(1\bar{1}0)$ planes, respectively. (d) DFT predictions of the evolution of bond length and bond angles as a function of the imposed uniaxial strain of $\varepsilon = \pm 1.5\%$. ZnN_4 tetrahedra are shown in pink, carbon in grey, and hydrogen atoms are omitted for clarity. Adapted from ref. 46 with permission from American Physical Society, Copyright 2012.

also produced a reasonably good agreement to the experiments (Table 1.1). Crucially, the theoretical results revealed that the underpinning deformation mechanism during uniaxial loading is controlled by the tension/compression of the Zn–N chemical bonds (while imidazolate rings remain rigid), accommodated by bending of the bond angles in the N–Zn–N tetrahedra and Zn–mIm–Zn bridging linkages (Figure 1.8(d)).

The complete anisotropic shear response of ZIF-8 is shown in Figure 1.9 for the 3-D representation surfaces of both the minimum and the maximum

Table 1.1 Elastic properties of a ZIF-8 single crystal measured by Brillouin scattering, compared with nanoindentation experiments and theoretical DFT predictions of an idealised defect-free crystal. The elastic compliance coefficients S_{ij} are the inversions of C_{ij}, where the very large magnitude of S_{44} signifies an exceedingly low resistance against shear deformation. Reproduced from ref. 46 with permission from American Physical Society, Copyright 2012.

Elastic Properties		Experimental Data (295 K)		*Ab initio* DFT calculations, B3LYP (0 K)
		Brillouin scattering	Nanoindentation	
Stiffness coefficient, C_{ij} (GPa)	C_{11}	9.5226 ± 0.0066	—	11.038
	C_{12}	6.8649 ± 0.0144		8.325
	C_{44}	0.9667 ± 0.0044		0.943
Compliance coefficient, S_{ij} (GPa^{-1})	S_{11}	0.2652	—	0.2578
	S_{12}	−0.1111		−0.1108
	S_{44}	1.0345		1.0605
Acoustic wave velocities, V ($km\ s^{-1}$)	Longitudinal (max. and min.)	3.17 and 3.08	—	3.41 and 3.32
	Transverse (max. and min.)	1.18 and 1.01		1.19 and 1.00
Young's modulus, E (GPa)	$E_{max} = E_{\{100\}}$	3.77 ± 0.01	3.29 ± 0.11	3.879
	$E_{\{110\}}$	2.98 ± 0.01	3.07 ± 0.07	2.953
	$E_{min} = E_{\{111\}}$	2.78 ± 0.01	2.87 ± 0.09	2.736
Shear modulus, G (GPa)	$G_{max} = ½ (C_{11}-C_{12})$	1.329 ± 0.005	—	1.36
	$G_{min} = C_{44}$	0.967 ± 0.005		0.94
Poisson's ratio, ν (–)	$\nu_{max} = \nu\langle 110, 1\bar{1}0\rangle$	0.54	—	0.57
	$\nu_{min} = \nu\langle 110, 001\rangle$	0.33		0.33
Anisotropy measure	Zener, A (=1 if isotropic)	0.73	—	0.70
	E_{max}/E_{min}	1.35	1.22	1.42

shear moduli. Noteworthy is the exceedingly low shear modulus of ZIF-8, where G_{min} (= C_{44}) $\lesssim$ 1 GPa, suggesting a low framework rigidity against structural distortion when subject to shear stresses applied in antiparallel directions of the plane of the 4MRs. In fact, this configuration corresponds to a structurally compliant 4-noded framework, which is susceptible to collapse under shear deformation. DFT reveals that the low shear resistance is due to the pliant ZnN_4 tetrahedra, allowing the framework to distort through the bending of the N–Zn–N and Zn–mIm–Zn bond angles (Figure 1.9(c)). In contrast, the maximum shear modulus is G_{max} = 1.3 GPa and lies on the planes of the 6MRs, giving better resistance to angular distortion subject to shear stress.

Turning to the Poisson's ratio, the 3-D representation surfaces of ν_{max} and ν_{min} are depicted in Figure 1.10. More specifically, the lower and upper

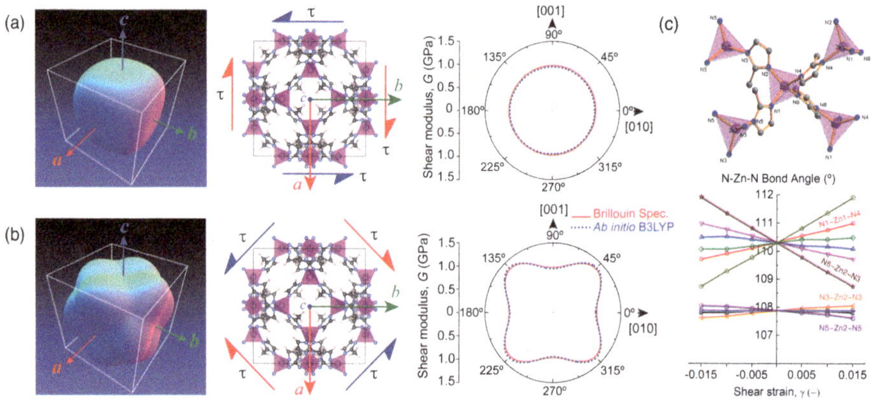

Figure 1.9 Shear modulus G representation surfaces of ZIF-8 for (a) G_{min} and (b) G_{max}, derived from Brillouin spectroscopic measurements. Unit cells show the directions of the opposing pairs of shear stresses τ yielding the minimum and maximum shear deformations, while the polar plots compare values obtained from the experiments and DFT calculations. (c) Variation of bond angles subject to a shear strain of $\gamma = \pm 1.5\%$, determined from DFT. The ZnN_4 tetrahedra are shown in pink, carbon in grey, and hydrogen atoms are omitted for clarity. Adapted from ref. 46 with permission from American Physical Society, Copyright 2012.

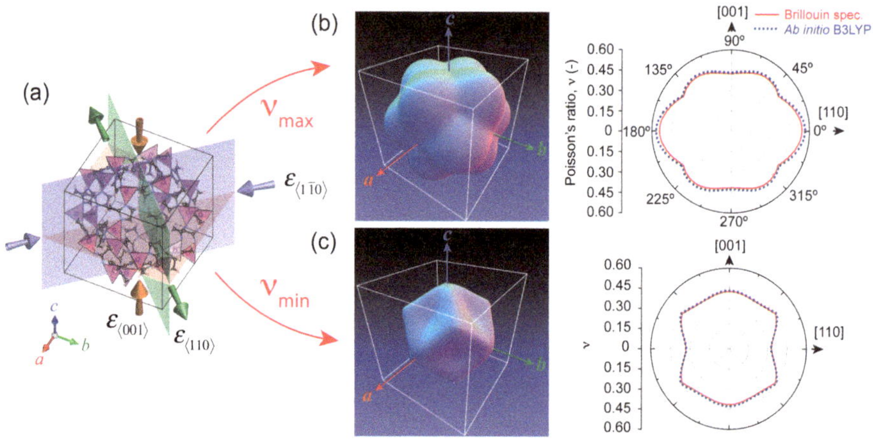

Figure 1.10 Poisson's ratio 3-D representation surfaces of ZIF-8 and their projections onto 2-D polar plots (experiments *vs.* DFT predictions). (a) Axial and lateral strains acting on a unit cell of ZIF-8 that result in the (b) maximum Poisson's ratio, $\nu_{max} = \nu\langle 110, 1\bar{1}0\rangle$, and (c) minimum Poisson's ratio, $\nu_{min} = \nu\langle 110, 001\rangle$. Adapted from ref. 46 with permission from American Physical Society, Copyright 2012.

Table 1.2 Isotropic aggregate properties for ZIF-8 and HKUST-1 determined from the Voigt–Reuss–Hill (VRH) averaging scheme. All units are in GPa apart from ν (dimensionless). Data from references indicated.

MOF	Method	Isotropic elastic properties				Reference
		E_{VRH}	G_{VRH}	K	ν_{VRH}	
ZIF-8	Brillouin scattering experiments (ambient temperature)	3.145 ± 0.013	1.095 ± 0.005	7.751 ± 0.011	0.43	46
	DFT, B3LYP (0 K)	3.15	1.09	9.23	0.44	
ZIF-8	DFT, PBEsol0-3c (0 K)					
i. Defect-free		3.33	1.15	10.52	0.45	79
ii. Missing Zn		2.70	0.95	5.60	0.42	
iii. Missing mIm linker		3.36	1.18	7.01	0.42	
HKUST-1 (Cu_3BTC_2)	DFT, B3LYP (0 K)	8.10	2.80	26.39	0.45	81

bounds are $0.33 \leq \nu \leq 0.54$. Interestingly, the exceptionally low shear modulus of ZIF-8 is accompanied by a high value of Poisson's ratio, which surpasses the normal maximum of 0.5 for incompressible rubbery solids (isotropic).

Further to knowing the single-crystal properties, for practical reasons generally it would be of interest to determine the isotropic aggregate properties of a 'bulk' polycrystalline solid. To this end, averaging methods such as the Voigt–Reuss–Hill (VRH) scheme can be employed to estimate the mechanical properties of crystal aggregates based on the stiffness and compliance tensors (see Section 2.2.1.2 of Chapter 2). This is exemplified in Table 1.2 for ZIF-8. The obtained values correspond to that of a textureless polycrystalline material; this hypothetical solid comprising ZIF-8 crystals will exhibit isotropic elastic properties.

1.5.2 ZIF-8: Impact of Structural Defects on Elasticity

Research about the role of defects on the function of MOFs is becoming increasingly relevant in the field.[77,78] In this context, it is important to understand how structural defects may impact anisotropic mechanical properties. For example, Möslein *et al.*[79] demonstrated the use of tip force microscopy (TFM) to map the local stiffnesses of ZIF-8 nanocrystals harvested from different growth times (see Figure 1.11(a)), where a lower mean stiffness and higher anisotropy in Young's modulus distribution were observed for the defective single crystals. Further insights can be obtained from DFT simulations of ZIF-8 with structural defects, by comparing the elastic constants (C_{ij}) of ideal frameworks with those containing 'missing zinc' and 'missing mIm linker' defects. It was found that the defective structures give rise to a higher degree of mechanical anisotropy, because the maximum and minimum values of Young's moduli are greatly modified by structural imperfection, as

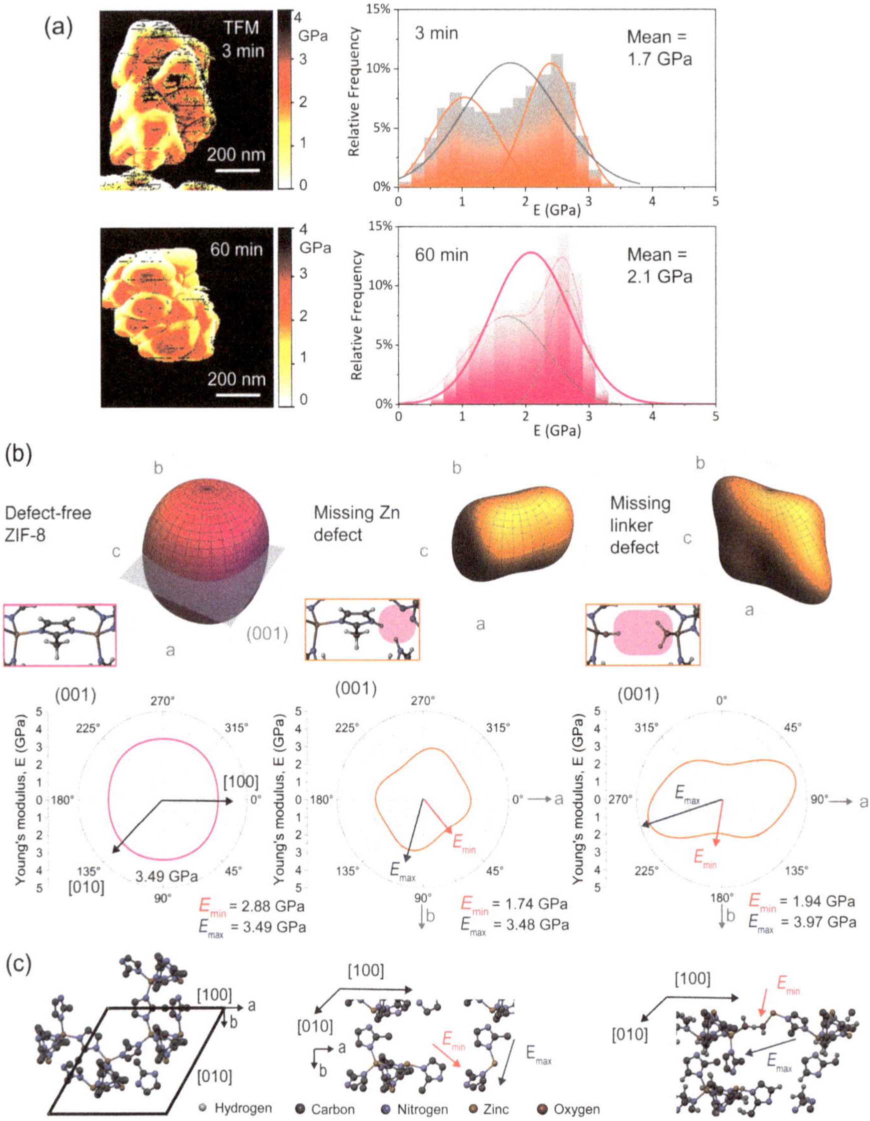

Figure 1.11 (a) Tip force microscopy (TFM) characterisation of the Young's modulus of ZIF-8 nanocrystals. The histograms on the right show stiffness distributions of nanocrystals with 3 and 6 min growth times. (b) 3-D representation surfaces of the Young's moduli (E) of the defect-free ZIF-8, compared with defective structures with missing Zn and missing mIm linkers. The bottom panels show a selected section through the 3-D surface to illustrate the maximum and minimum E values. (c) Crystallographic orientations corresponding to the maximal and minimal moduli in defect-free and defective ZIF-8 structures. Adapted from ref. 79, https://doi.org/10.1021/acsanm.2c00493, under the terms of the CC BY 4.0 license https://creativecommons.org/licenses/by/4.0/.

depicted in Figure 1.11(b). The theoretical models allow one to pinpoint the structure–property relationships of a defective structure, where the organic and inorganic connectivity, or indeed the lack of it (see Figure 1.11(c)), alters the mechanical response in very specific directions. Equally, averaged polycrystalline properties from the VRH scheme suggest there will be distinct and sometimes unpredictable changes to the isotropic behaviour stemming from framework defects (Table 1.2). Another DFT study considers the role of missing clusters on the mechanical anisotropy of UiO-66,[80] the findings of which are presented in Section 3.3.1 of Chapter 3.

1.5.3 HKUST-1: Elastic Anisotropy From Density Functional Theory (DFT)

Theoretical DFT calculations of the elastic behaviour of a HKUST-1 (Cu_3BTC_2, BTC = benzene-1,3,5-tricarboxylate) single crystal reported by Ryder *et al.*[81] demonstrated a highly anisotropic mechanical response, as depicted in Figure 1.12. Because of its cubic symmetry, HKUST-1 has three independent

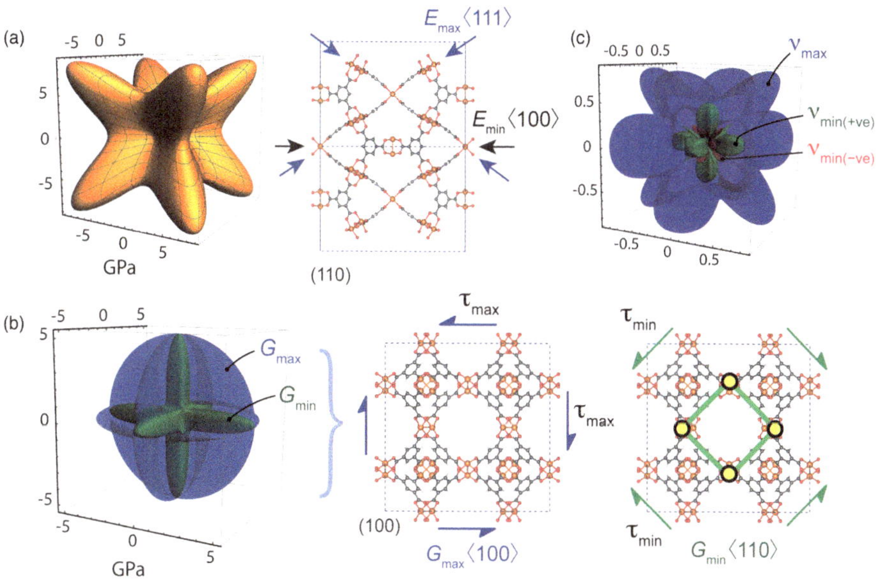

Figure 1.12 Anisotropic elastic properties of a single crystal of HKUST-1 computed using the C_{ij} elastic tensors derived from DFT calculations. (a) Young's modulus representation surface and the structural origins of the maximum and minimum magnitudes. (b) Shear modulus surfaces and the mechanisms giving the extremum shear stresses τ_{max} and τ_{min}. The four-noded frame (green square) is susceptible to shear-induced structural collapse. (c) Anisotropic Poisson's ratios, where the blue surfaces are the maximum values, while green and red correspond to the positive and negative minimum values, respectively. Adapted from ref. 81 with permission from the Royal Society of Chemistry.

elastic constants: C_{11} = 27.7 GPa, C_{12} = 25.7 GPa, and C_{44} = 5.4 GPa, where the C_{ij} values were computed using a DFT B3LYP functional. The 3-D representation surfaces of E, G, and ν all show large elastic anisotropies. We first consider its Young's modulus, where the extremum values are: $E_{max}\langle 111 \rangle$ ~ 15 GPa oriented along the cube body diagonals, while $E_{min}\langle 100 \rangle$ ~ 3 GPa directed along the principal cube axes (Figure 1.12(a)). Therefore, the anisotropic ratio for the Young's modulus is E_{max}/E_{min} = 5, which is notably greater than that of ZIF-8 discussed above. Turning to the shear modulus, the DFT predicted values are $G_{max}\langle 100 \rangle$ = 5.4 GPa and $G_{min}\langle 100 \rangle$ = 1 GPa. The maximum and minimum resistance against shear-induced structural distortion can thus be understood by examining how the shear deformation acts on the arrangement of nodal 'hinges' (Cu paddlewheel clusters) connecting the 'rigid' BTC linkers, see Figure 1.12(b).

The theoretical Poisson's ratios of HKUST-1 exhibit anomalous values of ν_{max} = 1.2 and ν_{min} = −0.3 (Figure 1.12(c)), which can be attributed to its high elastic anisotropy (A = 5.4). The negative Poisson's ratio suggests that the single crystal is auxetic when loaded in specific directions (further analysis in Section 2.3.3 of Chapter 2), so the mechanism could be linked to the dynamics of the node-linkages that uncoil when the framework is stretched along the $\langle 110 \rangle$ axis, thereby resulting in expansion in the transverse $\langle 1\bar{1}0 \rangle$ direction. Furthermore, it is worth considering the (bulk) polycrystalline mechanical properties of HKUST-1, see Table 1.2, derived from the VRH averaging scheme. The Young's modulus of a 'bulk' sample was found to be E_{VRH} ~ 8 GPa, which is in close agreement to that determined from nanoindentation for isotropic monolithic HKUST-1 (E_{mono}HKUST-1 ~9 GPa).[34] For the Poisson's ratio of a bulk HKUST-1 sample, it is intriguing to see that it is positive (ν_{VRH} = 0.45) even though it can be auxetic as a single crystal (ν_{min} = −0.3). The findings thus reveal the importance of controlling the precise crystallographic orientation of the sample or polycrystalline film in order to achieve auxeticity, since a random distribution of grains will diminish this effect. However, the epitaxial growth of accurately oriented MOF films and coatings remains a challenging task, and this is still limited only to a few examples of MOFs.[36,82,83]

1.6 Mechanical Properties of Monoliths, Glasses, Nanocrystals and Thin Films

1.6.1 MOF Monoliths: HKUST-1

The mechanical behaviour of porous monolithic MOFs has been little explored to date. Tian *et al.* reported the enhanced methane adsorption capacity of a monolithic HKUST-1 (Cu_3BTC_2) MOF derived from a sol–gel synthesis route.[34] The monolith in fact comprises a polycrystalline aggregate of nanosized HKUST-1 crystals, with the efficient packing and densification route resulting in the formation of a porous solid with a record-breaking methane storage capacity of 259 cm^3 (STP) cm^{-3} (*versus* the US Department of Energy (DOE) target of 263 cm^3 (STP) cm^{-3}). Figure 1.13 shows the

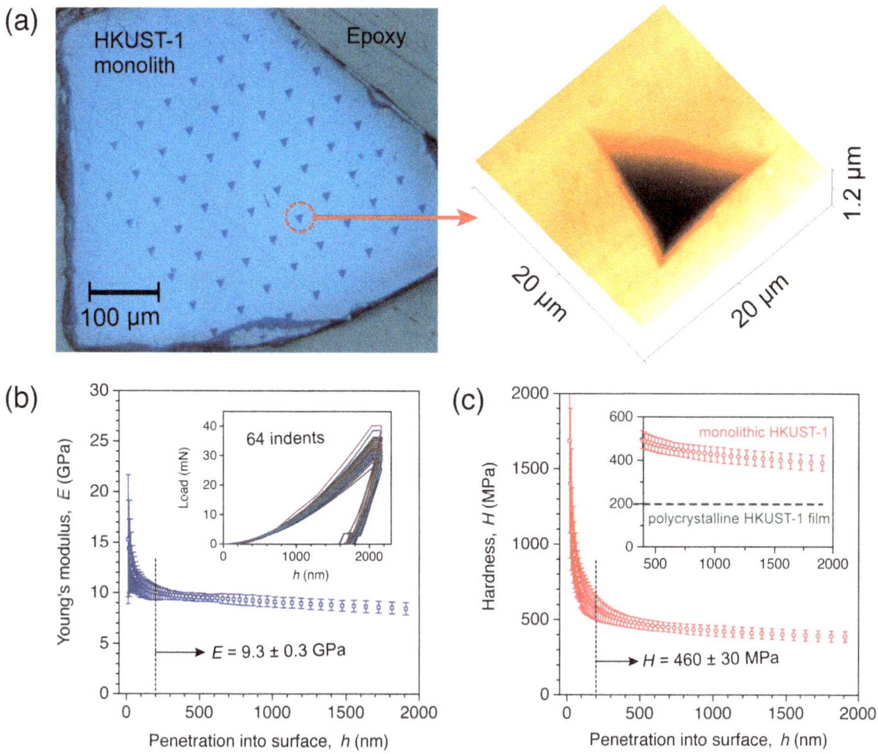

Figure 1.13 (a) Nanoindentation of a HKUST-1 (Cu_3BTC_2) monolith mounted on epoxy substrate, showing the array of residual indents and (right) AFM height profile with no sign of surface cracking. (b) Young's modulus and (c) hardness as a function of indentation depth from the CSM method, and their averaged values derived from a surface penetration depth of 200–2000 nm. Adapted from ref. 34 with permission from Springer Nature, Copyright 2018.

nanoindentation results determined from a sample of HKUST-1 monolith (termed $_{\text{mono}}$HKUST-1) mounted on an epoxy resin, where 60 residual indents after unloading from 2 μm deep indentations can be observed. The Young's modulus of $_{\text{mono}}$HKUST-1 was found to be $E \sim 9.3$ GPa (taking $\nu = 0.433$),[36] which is comparable with the stiffness value predicted from theoretical DFT ($E = 8.1$ GPa, see Table 1.2)[81] that assumes a combined elastic response from a polycrystalline aggregate of HKUST-1 crystals. While the modulus of $_{\text{mono}}$HKUST-1 matches the conventional HKUST-1, the monolith hardness ($H \sim 460$ MPa) is more than 130% higher than its conventional counterpart ($H \sim 200$ MPa),[36] see the inset in Figure 1.13(c). The results suggest that the high bulk density of $_{\text{mono}}$HKUST-1 ($\rho_{\text{monolith}} = 1.06$ g cm^{-3} *vs.* $\rho_{\text{crystal}} = 0.883$ g cm^{-3}) not only improves the volumetric adsorption capacity, but also enhances the mechanical durability of the densified MOF monolith. Of note, there is no sign of surface cracking detected in the vicinity of the residual indents, as evidenced from the AFM topographic image in Figure 1.13(a), indicating the good mechanical resilience of densified $_{mono}$HKUST-1. The findings also suggest that the sol–gel method could potentially be deployed to fabricate mechanically tough monolithic MOF materials aimed at practical engineering use.

1.6.2 Amorphous Phases: ZIFs

Crystalline MOFs can experience structural collapse and framework densification to yield an amorphous phase when subject to an external stimulus, such as high temperature, hydrostatic pressure, or shear stress. Figure 1.14 depicts the crystal–amorphous transition of a ZIF-4 crystal that transforms into amorphous *a*-ZIF upon heating to 300 °C, and its subsequent recrystallisation to form crystalline ZIF-zni by heating to 400 °C. The recovered monolith shown in Figure 1.14(c) reveals two disparate phases (designated as 'dark' and 'bright'), where nanoindentation was applied for differentiating between *a*-ZIF and ZIF-zni based on their mechanical properties. It can be seen in Figure 1.14(d) that, the Young's modulus and hardness of the 'dark' phase ($E \sim 6.5$ GPa, $H \sim 0.65$ GPa) perfectly match those of a pristine monolithic a-ZIF material that is isotropic.[38] The disordered network topology of the amorphous phase implies mechanical isotropy, more precisely, *a*-ZIF has

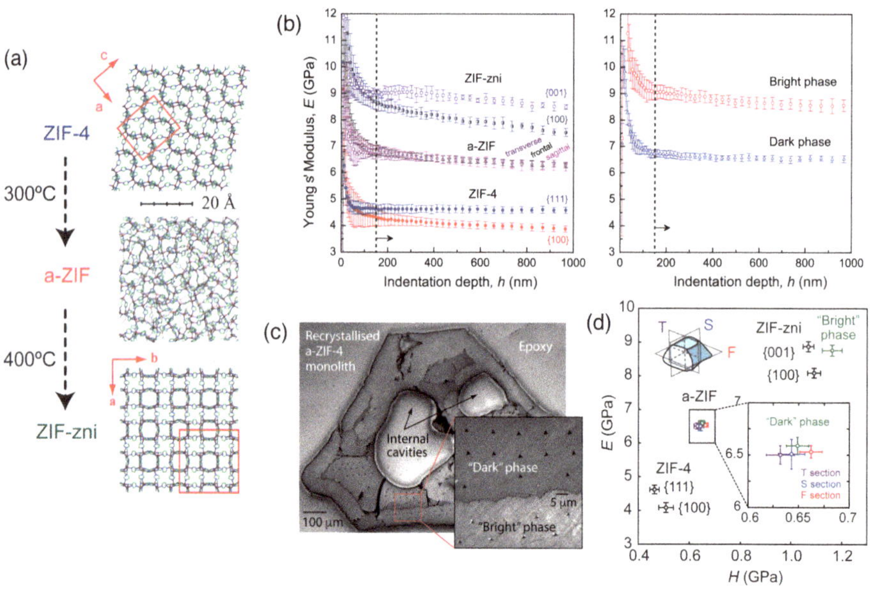

Figure 1.14 (a) Phase transition of ZIF-4 → *a*-ZIF → ZIF-zni during heating. (b) CSM nanoindentation measurements of the different phases depicted in (c) the optical micrograph of a partially recrystallised *a*-ZIF-4 monolith. (d) Summary of the Young's modulus *versus* hardness data of the three distinct phases. Adapted from ref. 38 with permission from American Physical Society, Copyright 2010.

lost the elastic anisotropy of its ZIF-4 parent phase (where $E_{max}/E_{min} \sim 1.6$ for a ZIF-4 crystal).[84] Remarkably, *a*-ZIF exhibits glass-like behaviour evident from the curved external surfaces and internal cavities present on the monolith (Figure 1.14(c)), indicating that the amorphisation process involves viscous flow during formation of *a*-ZIF at 300 °C. Subsequently, systematic nanoindentation studies on a family of ZIF melt-quenched glasses comprising a_gZIF-4, a_gZIF-62, a_gZIF-76, and a_gZIF-76-mbIm, have yielded Young's moduli (in GPa) of 6.9, 6.6, 6.3, and 6.1, respectively. Likewise, the hardness values of these ZIF glasses are similar to that of *a*-ZIF, all of which are distributed in a narrow range of $H = 0.66–0.68$ GPa;[39] their properties can best be contrasted against other crystalline open frameworks shown in the Ashby-style plot in Figure 1.7.

Turning to the nanoindentation of the 'bright' phase that can be observed in Figure 1.14(c), the moduli and hardness values ($E \sim 9$ GPa, $H \sim 1.2$ GPa) were found to be in the range of the ZIF-zni crystal, which is anisotropic (Figure 1.14(b)). In fact, the recrystallised ZIF-zni is a polycrystalline monolith, comprising submicron grains of random orientation, as confirmed from its electron diffraction pattern. One can therefore expect that the nanoindentation results of the above specimen will lie within the upper and lower bounds of a single crystal of ZIF-zni (DFT predictions: $E_{max} \sim 12.3$ GPa and $E_{min} \sim 4.7$ GPa).[84]

1.6.3 Nanocrystals: ZIF-8

Precise measurement of the mechanical properties of nanocrystals (~100 s of nm) and isolated micron-sized crystals (a few microns) cannot be accomplished using a conventional instrumented indentation technique because of the extremely small volume of samples involved. The latter technique requires a sample size of at least 50–100 μm[59] for it to be securely mounted in epoxy and surface polished. The nanomechanical characterisation of individual nanocrystals can be achieved, for example, through the application of atomic force microscopy (AFM)-based nanoindentation techniques.

Zeng and Tan[85] demonstrated the application of a diamond-tipped stainless steel cantilever probe (with calibrated cube-corner indenter) for quantitative characterisation of fine-scale crystals of ZIF-8. Figure 1.15(a) shows the AFM topographic scan of the ZIF-8 nanocrystals with a size of around 500 nm, prepared by drop casting of its suspension onto a glass substrate. Representative *P–h* curves for a set of very shallow indents ($h_{max} \sim 10–30$ nm) and another set of deeper indents ($h_{max} < 80$ nm) are shown in Figure 1.15(b and c), respectively, corresponding to applied loads in the range of *ca.* 2–12 μN. Large elastic recovery upon tip unloading is evident, especially for the set of shallow indents in contrast to the deeper ones that exhibit more permanent deformation after indenter unload. It was demonstrated that reliable nanoindentation curves in the few μN load range (for surface penetration depth of tens of nm) can be better obtained through a high unloading strain rate of $\dot{\varepsilon} > 60$ s^{-1}, see Figure 1.15(d). It can be seen that a Young's modulus of ZIF-8 in

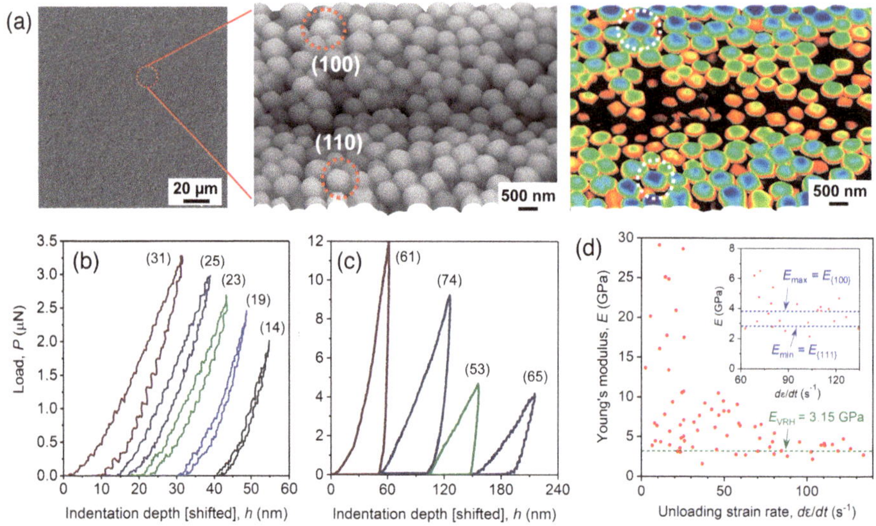

Figure 1.15 (a) Nanocrystals of ZIF-8 deposited on a glass substrate for AFM nanoindentation study. (Right) AFM height profiles of the thin-film polycrystalline coating. (b, c) Load–depth curves measured using a cube-corner diamond indenter mounted at the end of a stainless-steel AFM cantilever probe. (d) Young's modulus as a function of the unloading strain rate of an AFM tip in quasi-static indentation testing. Adapted from ref. 85 with permission from American Chemical Society, Copyright 2017.

the range of $E \sim 3$–4 GPa can be determined, in agreement with conventional instrumented nanoindentation[51] but the crystal size concerned is around 1000 times smaller by employing the AFM nanoindentation approach.[85] Whilst the results are promising, the wider applicability of this AFM-based indentation technique to probe a wide range of MOF nanocrystals should be systematically investigated.

1.6.4 Thin Films: HKUST-1

The mechanical properties of polycrystalline thin films and surface coatings of MOF materials have also been studied by nanoindentation, the available data of which are summarised in Figure 1.7. Bundschuh *et al.*[36] applied a liquid epitaxy method to fabricate a {100}-oriented film comprising HKUST-1 crystals with a thickness of *ca.* 1 μm, grown on a gold-coated silicon substrate. The epitaxial film has a surface roughness of *ca.* 10–15 nm (Figure 1.16(a)), deemed to be suitable for nanoindentation study without additional surface preparation steps. However, the *P*–*h* curve of the film sample shown in Figure 1.16(b) does not conform to the standard parabolic response because of the strong influence of the stiffer and harder substrate located beneath the relatively thin sample of HKUST-1. The substrate effect becomes

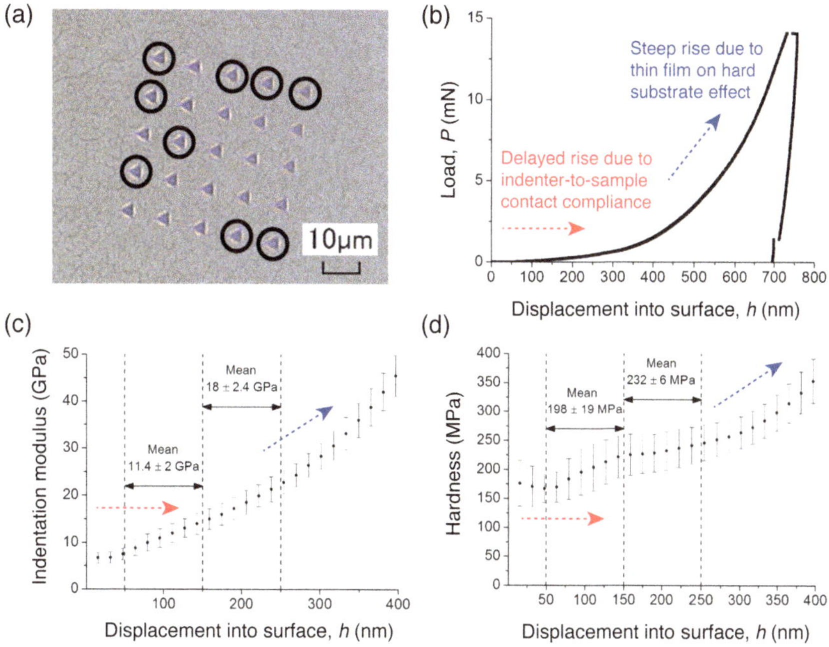

Figure 1.16 (a) Optical micrograph of the polycrystalline thin film of HKUST-1 epitaxially grown on a silicon substrate. A 5 × 5 array of residual indents, where the circled indents are invalid and were omitted in subsequent analysis. (b) A load–depth indentation curve, highlighting the atypical response of indenting a 'soft' thin film on a 'hard' substrate. (c) Indentation modulus and (d) hardness values as a function of indentation depth, the mean values were derived from 16 indents depicted in (a) and the error bars are standard deviations. Adapted from ref. 36 with permission from American Institute of Physics Publishing, Copyright 2012.

prevalent in Figure 1.16(c and d), revealing how the indentation modulus rises sharply as a function of indentation depth (in contrast to a relatively constant moduli when probing a single crystal or monolith, see Figure 1.14). A 10% value of the surface penetration depth was typically applied to minimise the substrate contribution,[42] corresponding to an indentation depth of ~100 nm, where the indentation modulus was determined to be 11.4 ± 2 GPa. Subsequently, assuming isotropy and by taking a theoretical Poisson's ratio of $v = 0.433$,[86] the Young's modulus of the HKUST-1 film was estimated as $E_{(100)\ \text{epitaxial film}} = 9.3$ GPa. Note that this is in contradiction to a theoretical DFT study that found that the $E_{(100)}$ of a single crystal of HKUST-1 is minimum, and its magnitude is three times lower (Figure 1.12(a)).[81] The contributions from grain boundaries in epitaxial film combined with the hard substrate effect may explain the observed discrepancy, and indeed the wider spread of data depicted in Figure 1.7 for a number of MOF thin film samples indented to date. Interestingly, the magnitude of $E_{(100)\ \text{epitaxial film}}$ matches that of a monolithic HKUST-1 sample comprising a random aggregate of nanocrystals: E_{mono}HKUST-1 = 9.3 ± 0.3 GPa,[34] which represents an isotropic response determined from a more substantial indentation depth of 2000 nm (see Figure 1.13).

This study highlights some outstanding challenges surrounding the reliable nanoindentation measurements of thin-film MOF samples, especially those possessing a thickness of $\lesssim$1 μm; future efforts are warranted in this area to explore the efficacy of AFM nanoindentation for characterising MOF thin films. Further discussions on the mechanics of MOF films and polycrystalline coatings are given in Section 1.8.3.

1.7 Framework Lattice Dynamics Dictating MOF Mechanics

1.7.1 Terahertz Modes

Many of the unusual mechanical features of MOFs and their unique functions observed on the macroscopic and microscopic scales can be explained by examining the terahertz (THz) vibrations of the underpinning molecular framework. Terahertz vibrational modes are collective lattice dynamics with an oscillation frequency of typically below 20 THz or ~700 cm^{-1} (1 THz = 10^{12} Hz = 4.14 meV ≈ 33.3 cm^{-1}); indeed they are low-energy phonons that stem from the structural 'flexibility' of the MOF framework. Experimentally, the THz vibrations of MOFs have been measured by high-resolution inelastic neutron scattering (INS),[87–89] synchrotron radiation far-infrared (SR-FIR) spectroscopy,[87] Raman spectroscopy,[81,90] and terahertz-time domain spectroscopy (THz-TDS)[91] techniques. Identification of the nature of the collective vibrational modes is challenging, because these low-energy vibrations belong to the collective dynamics of the entire framework. There are hence no characteristic frequencies associated with the standard functional groups that are commonplace to, for example, mid-infrared (MIR) spectroscopy, where the fingerprinting of diverse chemical moieties exists. Instead, the precise identification of the THz modes of MOF structures requires the application of theoretical simulations, such as quantum mechanical DFT calculations.[87,91,92] A combined experimental and theoretical methodology can offer rich insights into the THz modes of nanoporous frameworks. This approach sheds new light on how low-energy phonons may affect the thermal and mechanical stability of extended framework structures,[89,93,94] helping to reveal complex physicochemical phenomena through soft phonon modes[95] and rotor dynamics,[96] which may affect the performance of nanoconfined guest–host systems.[29,97]

1.7.2 THz Dynamics of ZIFs

The THz dynamics of ZIF structures have been relatively well researched compared with other families of MOF materials. Using SR-FIR experiments coupled with DFT calculations to investigate the far-infrared (FIR) modes of

ZIF-4, ZIF-7, and ZIF-8, Ryder *et al.*[87] established that ZIFs exhibit two commonly shared collective dynamics under ~21 THz, as shown in Figure 1.17(a). First, the vibrations in the range of 8–10 THz (265–325 cm^{-1}) arise from flexible ZnN_4 tetrahedra. Second, at around twice greater than the energy of the former band, the vibrational modes at 18–21 THz (600–700 cm^{-1}) originate from ring deformations of the imidazolate-type linkers. FIR absorption spectroscopy is a highly sensitivity probe for characterising vibrational modes linked to C, N, and H that prevail in ZIFs. When combined with a synchrotron light source, SR-FIR enables the collection of THz signals with a high signal-to-noise ratio down to 0.6 THz (~20 cm^{-1}). Theoretically, DFT vibrational calculations suggest that notable lattice dynamics of porous frameworks are ubiquitous in the region below 3 THz ($\lesssim$100 cm^{-1}). In practice, however, it can be seen that the intensity of FIR modes is weaker in this low wavenumber region, in contrast to the more intense energy loss signals measured *via* inelastic neutron scattering (INS).

Unlike optical techniques (*e.g.*, IR and Raman), neutrons have no optical selection rules, meaning that in principle all transitions are active.[98] Furthermore, it is advantageous to use INS to study ZIFs (and MOFs in general)[89,97,99,100] due to its high sensitivity for detecting hydrogen modes, simply by leveraging the large incoherent neutron cross section of hydrogen. For the low wavenumber region below 100 cm^{-1} a precise match between experiment and theory remains challenging for ZIFs, though a reasonable agreement between them can be observed (see Figure 1.17(b)). Improved peak-to-peak

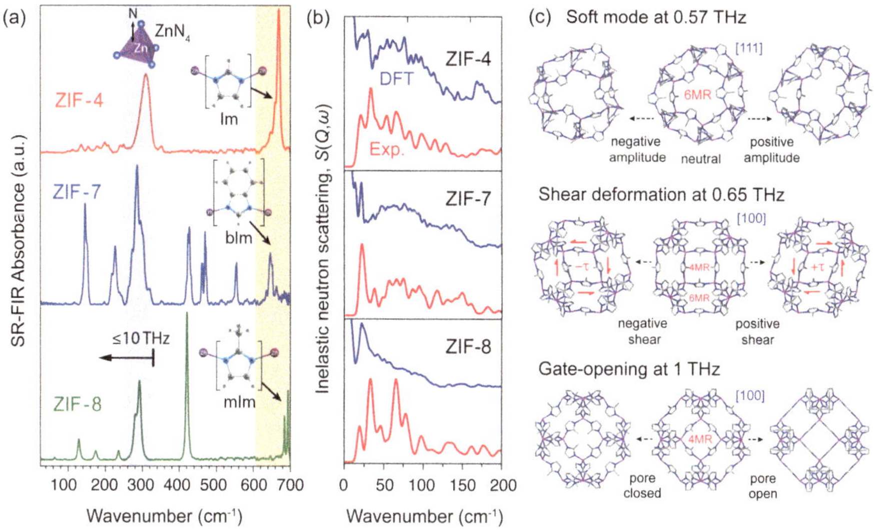

Figure 1.17 (a) Synchrotron far-infrared spectra of ZIFs in the region of <20 THz. (b) INS spectra in the region of below <6 THz, comparing the theoretical DFT and neutron experimental data. (c) Low-energy lattice dynamics of ZIF-8 illustrating the notable THz modes. Adapted from ref. 87 with permission from American Physical Society, Copyright 2014.

assignment between the INS and DFT spectra has been demonstrated in another study on a Zr-based MOF, termed MIL-140A, within the low-energy THz region of 0–250 cm^{-1}.[96] In the following section we shall consider lattice dynamics determined from DFT simulations for exemplifying several interesting mechanical and physical phenomena underpinning the functions of porous ZIF structures.

In the case of ZIF-8 as depicted in Figure 1.17(c), it is interesting to consider the collective lattice dynamics of a 'soft mode' at 0.57 THz (~19 cm^{-1}) and a shear-induced deformation mode at 0.65 THz (~22 cm^{-1}).[87] Importantly, these two sub-THz vibrations influence the elasticity of the porous network and potentially trigger structural destabilisation of the sodalite cage. The presence of a soft phonon mode may explain the occurrence of a mechanically-triggered phase transition under pressure[101] and guest-induced structural response[102] evidenced in ZIF-8. When the framework is directionally stressed, the six-membered rings (6MR) distort to accommodate tensile and compressive strains propagated through the interconnected mIm–Zn–mIm linkages, permitting framework flexibility in accordance with the dynamics of the soft mode. Turning to the shearing mode, this involves the angular distortion of the four-membered rings (4MR) caused by a pair of antiparallel shear forces. The four-node geometrical configuration of the 4MR is mechanically unstable, making intrinsically susceptible to undergo a structural collapse under shear deformation.

Likewise, the porous structures of ZIF-4 and ZIF-7 can experience lattice distortions driven by the shearing dynamics in the low-frequency THz regime ($\lesssim$1 THz), attributed to the flexibility of the 4MRs. Crucially, the crystalline structures of ZIF-4, ZIF-7, and ZIF-8 can be made amorphous *via* ball milling,[103,104] where the shear stress induced by the sliding motion of the impacting balls induces structural collapse and densification of the porous frameworks. It has been postulated that the stress-induced amorphisation of ZIFs, and more generally the structural destabilisation of MOFs, can be triggered by THz shear modes inherent in porous solids. Moreover, research has found that when the shear modes are not causing framework amorphisation, they play a vital role by facilitating lattice modes that alter specific functions. For instance, the soft modes at sub-THz frequencies in ZIF-4 enable the stretching and compression of 6MR apertures, which are accommodated by shearing of 4MR in the perpendicular orientation (Figure 1.18(a)). This mechanism offers a pathway for phase transitions observed in desolvated ZIF-4,[105] further modifying the accessible pore volume for enhanced gas adsorption.[106] Another example involves the sub-THz 'breathing' mode of ZIF-7 (Figure 1.18(b)), where the shearing of 4MR enables the twisting spiral motion of the bulky benzimidazolate (bIm) linkers surrounding the 6MR aperture. This kind of pore opening mechanism, akin to 'gate-opening' can be leveraged to regulate the level of CO_2 adsorption by ZIF-7.[107] In particular, the collective vibrational mode of the order of ~1 THz associated with the dynamics of gate-opening/-closing of ZIF-8 (Figure 1.17(c)), is the key to understanding its unusual guest uptake characteristics,[101,108] chemical

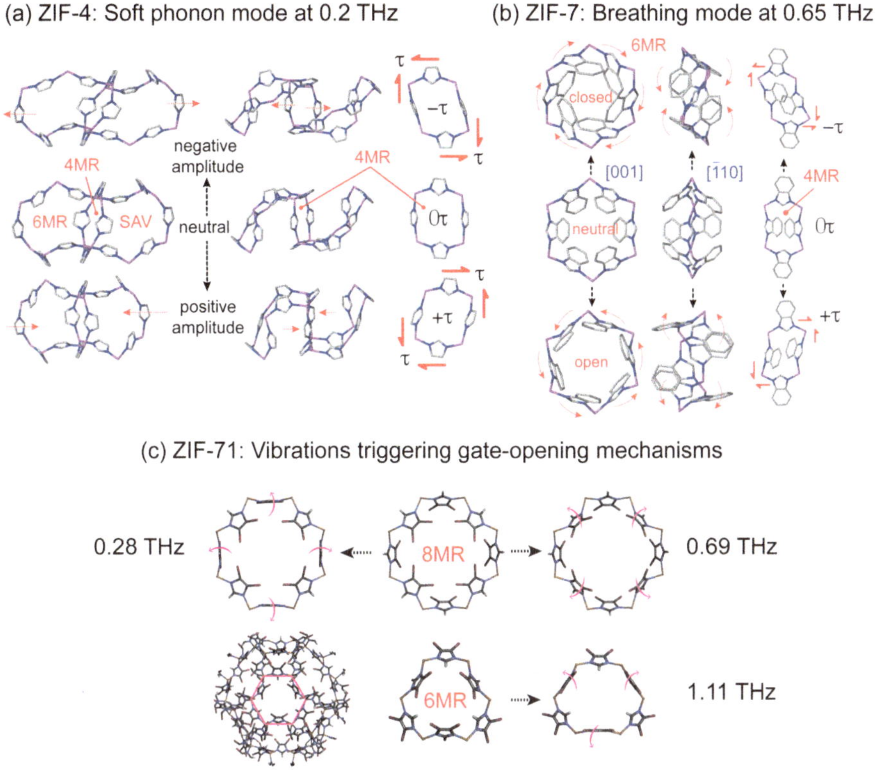

Figure 1.18 Sub-terahertz collective lattice dynamics of (a) ZIF-4 and (b) ZIF-7, highlighting the shear (τ) induced deformations of four-membered rings (4MRs) that are responsible for modifying the geometry of the pore apertures comprising six-membered rings (6MRs). Adapted from ref. 87 with permission from American Physical Society, Copyright 2014. (c) THz and sub-THz modes of ZIF-71 that trigger gate openings of 8MR and 6MR in RHO topology. Adapted from ref. 110, https://doi.org/10.1021/acs.jpclett.2c00081, under the terms of the CC BY 4.0 license https://creativecommons.org/licenses/by/4.0/.

selectivity,[102] and basic mechanics[92,109] in the wider context. A recent study by Möslein *et al.* on large-cage ZIF-71 with the RHO topology,[110] further elucidated that low-frequency THz dynamics (Figure 1.18(c)) play a key role in influencing the mechanical properties and various physical phenomena that control the functions of porous frameworks.

1.7.3 THz Dynamics of HKUST-1

The HKUST-1 framework has a few distinct classes of low-energy lattice dynamics in the THz and sub-THz regimes, which are linked to the elastic behaviour of the porous structure.[81] Figure 1.19(a) illustrates the 'trampoline-like' deformation modes viewed down the ⟨111⟩ and ⟨110⟩

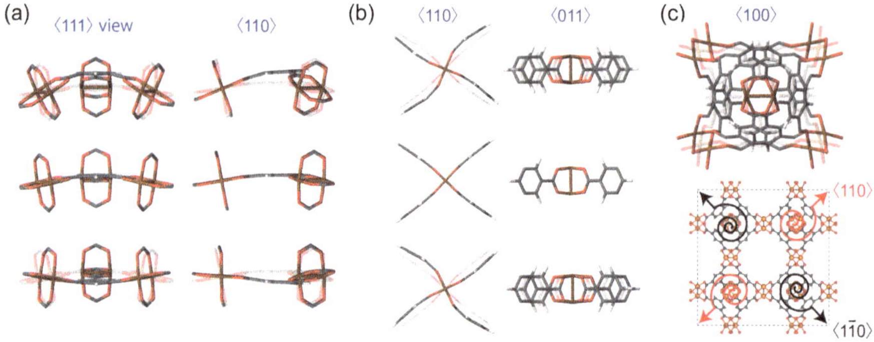

Figure 1.19 Terahertz vibrations of HKUST-1 determined from DFT calculations. (a) Trampoline-like motion at 2.4 THz, viewed down the ⟨111⟩ and ⟨110⟩ crystallographic axes. (b) Rotor dynamics of the copper paddlewheel at 1.7 THz. (c) 0.5 THz collective vibrations with a coupled cluster rotation mechanism (top), which is a source of auxetic deformation (bottom). Adapted from ref. 81 with permission from the Royal Society of Chemistry.

crystallographic directions. There are four such vibrational motions located in the range of 1.7–3 THz, which are either IR, non-optically active, or Raman modes. The first two modes comprise the out-of-plane bending deformation of the benzene-1,3,5-tricarboxylate (BTC) organic linker that mimics the oscillatory dynamics of a trampoline. The lower-energy Raman-active mode at 1.7 THz is unique in the sense that it is accompanied by the simultaneous rotational motion of the Cu paddlewheel, see Figure 1.19(b). Significantly, such combined trampolining and molecular rotor dynamics offer the mechanism needed to afford negative thermal expansion (NTE), as experimentally observed in HKUST-1.[111] More precisely, in the temperature range of *ca.* 100–400 K the HKUST-1 framework experiences shrinkage (instead of expansion) when heated to yield a smaller unit cell volume. In the same way, similar lattice dynamics could be the source of NTE phenomenon in MOF-5 [93,95] and other MOF structures constructed from pliant linkages.[112]

Now, we consider a sub-THz vibration of HKUST-1 with synchronous cluster dynamics, where this mode is postulated to be the origin of the negative Poisson's ratio (NPR) or auxetic phenomenon predicted by DFT. Specifically, the collective dynamics at 0.5 THz couple the rocking and translational motions, thereby compensating for the coordinated rotation of the linker–paddle-wheel clusters, as depicted in Figure 1.19(c). From a mechanical standpoint, this mechanism enables the organic–inorganic clusters to rotate in a spiral fashion upon uniaxial stretching in the axial direction, producing another elongation (instead of contraction) in the lateral direction, and thus exhibiting a auxetic response where $\nu_{min} = -0.3$ for HKUST-1.[81]

1.7.4 THz Dynamics of DUT and MIL Structures

Further to the dynamics of ZIFs and HKUST-1 materials, the study of THz vibrations has been extended to a family of structurally more flexible MOFs, such as DUT-8(Ni) and MIL-53(Al). Krylov *et al.*[90] employed Raman spectroscopy to probe the collective vibrational modes of the 'rigid' *vs.* 'flexible' forms of the pillared-layer DUT-8(Ni) framework, focussing on the low wavenumber region below 300 cm^{-1} ($\lesssim$10 THz). While the rigid DUT-8(Ni) nanocrystals can be desolvated without undergoing a phase transformation, the flexible DUT-8(Ni) macro crystals reversibly transform between a desolvated closed-pore (CP) phase and a large-pore (LP) phase upon adsorption of gas or liquid guest molecules. Figure 1.20(a) shows that the rigid framework of DUT-8(Ni) exhibits a persistent sub-THz band at 23 cm^{-1} (0.69 THz), even after desolvation, corresponding to the LP phase. In contrast, the CP phase of flexible DUT-8(Ni) after desolvation exhibits a higher characteristic band at 59 cm^{-1}, although this 1.8 THz mode is absent in the as-synthesised version of flexible DUT-8(Ni).

Notable also is the width of the 59 cm^{-1} Raman band in the CP form, which is significantly broader than the 23 cm^{-1} band in the LP form, indicating an increase in the intermolecular π–π interactions within the crystal when it adopts a closed pore configuration (Figure 1.20(b)). In terms of its elastic properties, a high degree of mechanical anisotropy is expected for DUT-8(Ni) akin to its isostructural analogue termed DMOF-1, where the 3-D representation surface of the Young's modulus of the latter is depicted in Figure 1.20(c).

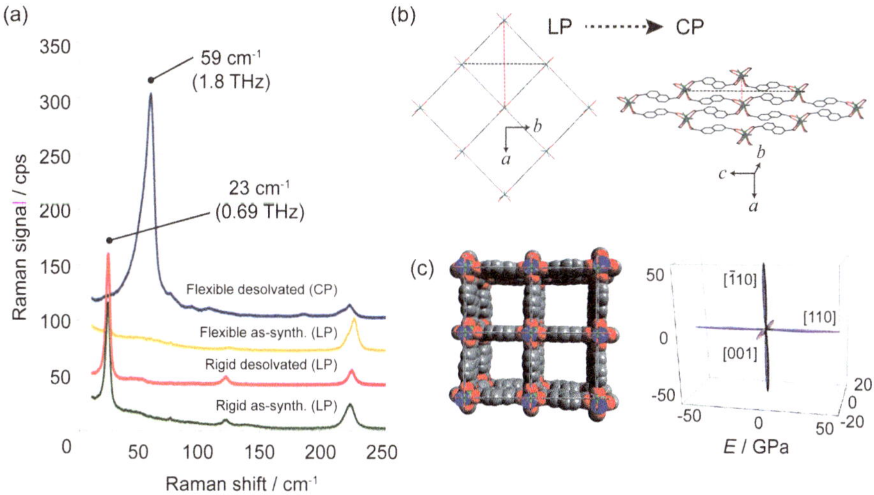

Figure 1.20 (a) Raman spectra of DUT-8(Ni) in the THz region. (b) Transformation of the crystal structure from the large-pore (LP or open pore) to the closed-pore (CP) phase. The dotted lines show the mechanically pliant directions. (c) Anisotropic Young's modulus of DMOF-1. Adapted from ref. 90 with permission from the Royal Society of Chemistry.

On the basis of the LP structure of DUT-8(Ni), it is anticipated that the mechanically stiff directions are oriented along the naphthalene dicarboxylate linkers pointing towards the {110} and {$\bar{1}$00} axes, and in the {001} out-of-plane direction where the Ni_2(dabco) (dabco = 1,4-diazabicyclo[2.2.2]octane) chains are present. Clearly the mechanically pliant directions correspond to the *a*- and *b*-axes of the porous channels marked in Figure 1.20(b), where the soft phonon mode responsible for the breathing mechanism of DUT-8(Ni) LP is operational and has a major impact on the mechanical behaviour of the entire framework.

Hoffman *et al.*[113] investigated the role of THz vibrations on the mechanics of the breathing behaviour and thermodynamic properties of flexible MIL-53(Al) using periodic DFT calculations. Like DUT-8(Ni) elucidated above, the structure of MIL-53(Al) can switch between the LP and CP phases upon guest adsorption or under mechanical stress.[114,115] The theoretical results reveal that the soft modes are vital as they help to stabilise the CP structure at low temperatures, and they could also incite LP-to-CP phase transformation. Figure 1.21 shows the volume–frequency characteristics of MIL-53(Al) in the low-frequency THz region, in which some collective motions are observed to undergo relatively large frequency variations ascribed to either molecular repulsion or internal strain effects. It is conceivable that such vibrational fluctuations could trigger phase transformation. Figure 1.21 shows the trampoline-like deformation motions of the linker fluctuate in the range of *ca.* 2–4 THz as the cell volume evolves from the LP-to-CP phase. The findings

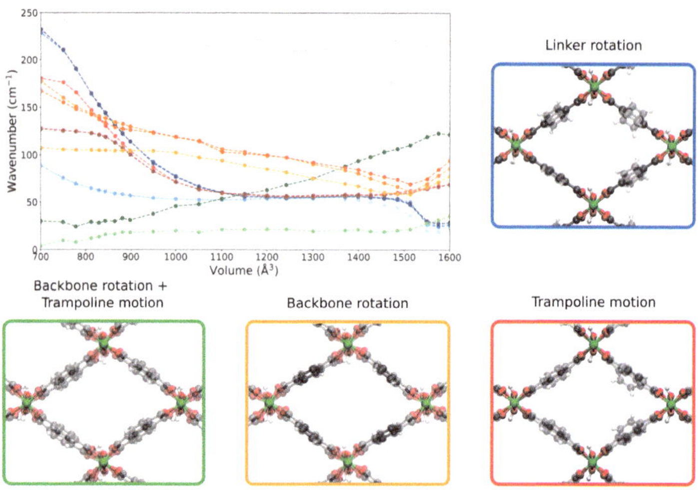

Figure 1.21 Volume–frequency relationships of the THz vibrations in MIL-53(Al) predicted from DFT calculations. The highlighted low-frequency modes are linker rotations, trampoline motions, and soft modes comprising the collective dynamics of the aluminium oxide backbone and bridging linkers. Reproduced from ref. 113 with permission from Walter de Gruyter and Company, Copyright 2019.

thus suggest that the THz mode concerned may be responsible for stimulating the breathing mechanism observed in MIL-53(Al). Furthermore, the theoretical results suggest that the THz vibrations have a pronounced effect on the thermodynamic properties (*e.g.*, specific heat capacity, bulk modulus, thermal expansion coefficient), notable in such a way that even a small deviation in the predicted frequencies results in a large change in the resultant calculated properties. This study demonstrates that computational simulations are thus a powerful tool to use to gain insights into the detailed mechanical mechanisms of highly flexible MOF structures (Chapter 3), which might be intractable from experiments alone.

Related to the discussion above, a theoretical DFT study by Wang *et al.*[116] revealed the dramatic effects that guest molecules have on modifying the elastic properties of a flexible MIL-53(Al) 'host' structure. The hydrated structure of MIL-53(Al) contains guest water molecules that are hydrogen bonded to the porous host, conferring a reinforced wine-rack 'guest@host' assembly, with augmented elastic moduli but reduced mechanical anisotropy. For example, for the LP phase, the vacant structure of MIL-53(Al) was predicted to exhibit $E_{max(LP/empty)}$ = 94.4 GPa and $E_{min(LP/empty)}$ = 0.9 GPa, in stark contrast to the water-occupied MIL-53(Al) structure with $E_{max(LP/water)}$ = 75.4 GPa and $E_{min(LP/water)}$ = 21.6 GPa. It follows that the anisotropy ratio of the Young's moduli significantly fell from E_{max}/E_{min} ~ 105 to just 3.5; with the model assuming that no phase transformation occurs upon water intrusion. Likewise, for the CP phase of MIL-53(Al), it was predicted that the hydrated structure has a maximum Young's modulus of $E_{max(CP/water)}$ = 126 GPa, which represents a major increase from the $E_{max(CP/empty)}$ = 71 GPa of the vacant framework; accordingly the anisotropy ratio fell from around 44 to 28 upon hydration of the CP phase. The simulations reveal that hydrogen bonds from the adsorbed water molecules mechanically reinforce the initially porous framework and by doing so they not only suppress mechanical anisotropy but shift the pliant directions of the original structure.[116] While direct comparison with experiments are currently not available to validate the above elastic calculations of MIL-53(Al), there are reported single-crystal nanoindentation measurements that have been performed on several desolvated *vs.* solvent-containing MOF structures that suggest a highly tuneable mechanical response as a function of the guest pore occupancy.[33,51]

Finally, to conclude this section we shall consider the molecular rotational dynamics of a zirconium-based framework, called MIL-140A [$ZrO(O_2C–C_6H_4–CO_2)$], where the rotor-like twisting motions of the organic linkers are prevalent.[96] Figure 1.22(a) summarises the cooperative THz dynamics of the type-A and type-B rotors observed under 3 THz (~100 cm^{-1}), in which the hindered rotations of the C_6H_4 aromatic rings within the 1,4-benzenedicarboxylate (BDC) linkers can be classified into either the symmetric or asymmetric modes. Of great importance is the consequence that rotor motions have on the dynamic evolution of the SAV, which quantifies the accessible voids present in the framework. The rotational energy barriers have been estimated by DFT calculations for the $\varphi = 0° \rightarrow 180°$ twist, revealing that the

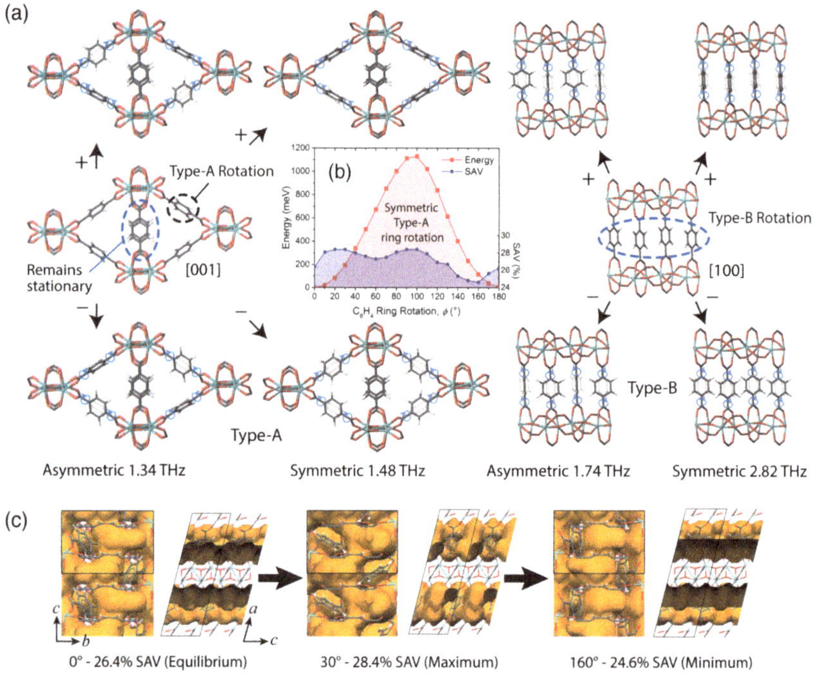

Figure 1.22 (a, b) Rotor dynamics of MIL-140A. (c) Evolution of the solvent accessible volume (SAV using a probe size of 1.2 Å) according to the twist angle of the phenyl ring (φ). Adapted from ref. 96 with permission from American Physical Society, Copyright 2017.

type-A rotational barrier is ~283 meV per aromatic ring (Figure 1.22(b)). Moreover, the barrier for type-B rotors cannot be reliably obtained because of the steric hindrance of the overlapping linkers. As depicted in Figure 1.22(c), it is striking to see that the morphology of the SAV evolves continuously with the twist angle φ, oscillating between ~25% and 28% ($\varphi_{\text{min/max}}$ respectively), whereby the SAV equilibrium is ~26% when $\varphi = 0°$. When $\varphi_{\text{min}} = 160°$, this angular motion confers a fully 'gate-open' geometry when the adjacent pores coalesce to form continuous 1-D channels along the *c*-axis. More broadly, this exemplar shows how entwined the low-energy THz rotational modes with the MOF mechanics, thereby driving structural transformation and altering the physicochemical response of porous frameworks.

1.8 Beyond Elasticity: Inelastic Mechanical Behaviour and Structural Failure

1.8.1 Hardness, Yield Strength, and Plasticity

The 'hardness' of a material is a measure of its mechanical resistance against localised *plastic* deformation. Generally, plasticity occurs (in *metals*) when the applied stress exceeds the yield strength of the material, when $\sigma_{\text{applied}} \geq \sigma_{\text{Y}}$.

As illustrated in Figure 1.2(a), plastic deformation produces a permanent plastic strain (ε_p), which is an *irreversible* deformation on a material that is sustained after complete removal of the applied force.

Typically, the hardness value is characterised by indentation-based techniques (*e.g.*, Mohs, Vickers, Knoop, Shore, Rockwell, and Brinell hardness tests), or by employing a scratch test to assess the materials surface resistance towards abrasion. Mohs hardness is the earliest hardness scale that was introduced, comprising values of 1 to 10 established by ranking a series of progressively harder materials to resist scratching, namely (softest → hardest): talc, gypsum, calcite, fluorite, apatite, feldspar, quartz, topaz, corundum, and diamond.[117] Of course, this is a crude estimate of the different degrees of hardness, expressed in a qualitative manner. A more quantitative approach, such as the Vickers 'microhardness' test, involves the use of a Vickers diamond indenter (four-sided pyramid) and indentation loads ranging from 100 mN to 10 N. This technique is commonly used for measuring the hardness of (bulk) engineering materials, such as metals, ceramics, polymers, and composites. The measured hardness is expressed in units of HV.

Because of the relatively small sample volumes involved compared with conventional solids, the characterisation of the hardness of MOF materials is often achieved by instrumented nanoindentation (Section 1.4). Using a Berkovich diamond tip (three-sided pyramid) and performed under a maximum load of just tens of mN or less.[33,35,57,59,64] The value of hardness derived from nanoindentation, also termed 'nanohardness', can be calculated from eqn (1.13): $H = P/A$, where P is the indentation load and A is the projected contact area under load (see Figure 1.4). The nanohardness is expressed in units of Pascal (Pa).

From the foregoing discussion, it should be clear that the determined hardness value is not a unique materials property (like the Young's and shear moduli), but a function of the chosen test method and its accompanying test parameters, including indenter geometry, applied load, indentation depth, and formula/model used to compute the magnitude of hardness. It is therefore important to note that the H values generated from different techniques are not necessarily comparable, unless the main features of the tests are similar.[24] In this chapter, only the hardness measurements of MOFs derived from nanoindentation are considered. From the E–H materials selection chart (Figure 1.7), it can be seen that the hardness values of porous ZIFs, boron imidazolate framework (BIF), HKUST-1, *a*-ZIF, MOF glasses, and dense ZIF-zni typically lie between several hundreds of MPa and ~1 GPa. In contrast, the hardness values of other dense hybrid frameworks (comprising inorganic and organic building blocks) are at least one order of magnitude higher, thereby lying mainly in the range of about 1–10 GPa.

More specifically, in the family of ZIF structures there is an inverse correlation between the hardness of single crystals and the SAV, as elucidated below.[51] What can be considered as 'soft porous crystals' are the highly porous framework structures (SAV ~50%) of ZIF-20, ZIF-68, and ZIF-8, the hardness values of which are: H = ~250 MPa, ~300–500 MPa, and ~550 MPa, respectively. Lesser porous frameworks, such as ZIF-7 and ZIF-9 (both SAV ~ 26%), while having the same sodalite (**sod**) topology as ZIF-8, have a greater hardness of H = ~650–700 MPa due to the sterically bulkier bIm linkers of ZIF-7 and ZIF-9 *vs.* the mIm linkers in ZIF-8. The least porous ZIF is ZIF-zni (SAV = 12.2%), which to date remains the hardest compound in the ZIF family with H = ~1.1 GPa. Surprisingly, the isostructural BIF-1-Li crystal ($LiB(Im)_2$, SAV = 5.3%) has a substantially lower hardness of ~0.15 GPa, which is an order of magnitude lower than that of ZIF-zni.[62] From the relative deformation data of the different tetrahedra (LiN_4 > ZnN_4 > BN_4), it can be reasoned that the lithium coordination environment and adjoining Li–Im–B linkages are more susceptible to mechanical deformation compared with the zinc counterpart.

Hardness (H) and yield strength (σ_Y) of *conventional* materials are correlated through the Tabor relationship:[119]

$$H = C \cdot \sigma_Y \tag{1.16}$$

where C is termed the 'constraint factor'. The value of C is materials dependent, ranging from a value of ~1.5 for brittle solids such as glasses (small E/σ_Y ratio) to ~3 for ductile materials such as metals (large E/σ_Y ratio).[120] In the context of nanoindentation with a spherical indenter tip, it has been proposed that an identical relationship exists between hardness and the yield pressure (P_Y), such that $H = C \cdot P_Y$. Unlike a Berkovich tip, the spherical geometry (typical tip radius in the order of 1 μm) generates a much greater contact area at the same penetration depth, thereby delaying the onset of plasticity as shown in Figure 1.23(a) for the comparative indentation tests conducted on a single crystal of ZIF-8. The 'indentation stress–strain' curve (Figure 1.23(b)), in the form of contact pressure (P_m) *versus* contact strain

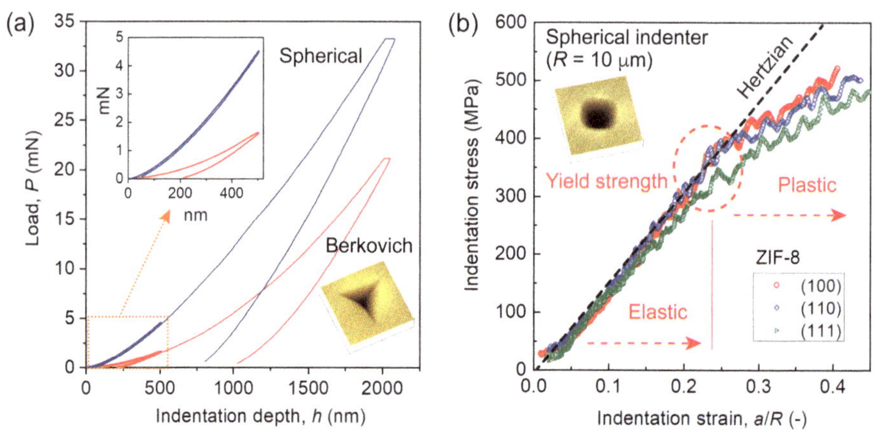

Figure 1.23 (a) Nanoindentation load–depth curves obtained from a Berkovich indenter *versus* the spherical probe (radius = 10 μm) on the (100) facet of a ZIF-8 single crystal. The inset shows the shallower indents to 500 nm. (b) Estimated yield strengths of ZIF-8 crystals and plastic anisotropy of the (100), (110), and (111) facets.[118]

(a/R) can subsequently be derived from a spherical indentation experiment (for method, see ref. 121). The point of deviation from the Hertzian linear response indicates the yield pressure, P_Y. For ZIF-8, using this methodology it can be estimated that the magnitudes of P_Y lie in the range of 300–350 MPa and there is also sign of plastic anisotropy associated with the (100), (110), and (111) facets. Beyond the yield point, the indentation stress–strain curves appear to show a power-law hardening behaviour, given by $P_m = (a/R)^n$, where n is the strain hardening exponent. While this kind of hardening curve is typically linked to dislocation entanglement prevalent in metallic materials, the underlying mechanism is currently unknown for inorganic–organic (hybrid) solids constructed from strongly directional coordination/covalent bonds.

Spherical indentation studies on dense 3-D inorganic–organic frameworks have shown considerably higher yield pressures of up to $P_Y \sim 2.3$ GPa, and with constraint factors of $C \sim 2$ and 2.4 depending on the indented crystal orientation.[32] Imaging techniques, such as cross-sectional transmission electron microscopy (TEM) revealed whether dislocation glides are present beneath the indented zone, which may accommodate plastic flows observed in a number of inorganic crystals that are intrinsically brittle in nature.[122,123] On the theoretical front, Banlusan *et al.*[124] have employed large-scale MD simulations to investigate the plastic deformation behaviour of a cubic MOF-5 crystal subject to uniaxial compression, see Figure 1.24. They show that irreversible deformation mechanism in the crystal is governed by slip of dislocations driven by compressive and shear stresses, leading to the formation and propagation of shear collapse bands shown in Figure 1.24(e–h). This theoretical study demonstrates that the activation of the multiple ⟨001⟩ {100} slip systems controls the plastic deformation of a cubic framework crystal, where the organic–inorganic linkages are all oriented along the cubic axes. The flexibility of the metal clusters facilitates the rotation of the organic linkers to initiate structural yielding under a shear strain. It can be seen that

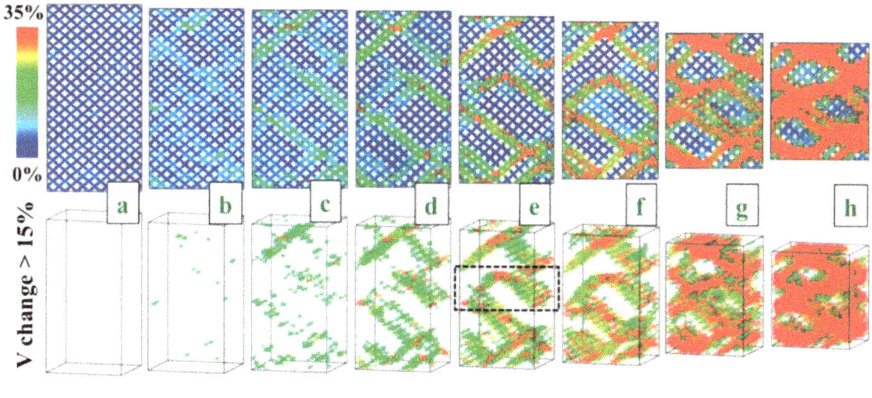

Figure 1.24 MD simulations of the plastic deformation of MOF-5 by uniaxial compression along the [101] crystal axis. Panels (a) to (h) correspond to the incremental strains from *ca.* −10% to −45%. Reproduced from ref. 124 with permission from American Chemical Society, Copyright 2015.

the amount of pore collapse and structural densification caused by the application of compressive stress is very substantial beyond the yield point. The theoretical findings are very interesting, but there is still no direct experimental evidence of dislocation activity in MOFs for validating the large-strain predictions described above.

Recent developments on the study of the yield stress and hardening behaviour of ZIF monoliths show that by using a combination of Berkovich/cube-corner indentation and finite-element (FE) modelling (Figure 1.25(a)) it is possible to determine an improved value of yield strengths, where $\sigma_Y = 200$ MPa for a ZIF-8 monolith and $\sigma_Y = 130$ MPa for a ZIF-71 monolith.[35] The constraint factor was found to be $C = 2.1$, which is an intermediate value for a brittle and a ductile material. Beyond the elastic limit, FE simulations show that the monoliths experience very limited plastic hardening, supported by the lack of materials pile-up characterised by AFM in the vicinity of the residual indents (Figure 1.25(d)). Therefore, the use of an elastic-perfectly plastic constitutive model is sufficient to simulate the experimental indentation curves of the ZIF monoliths considered in this work. The nanograined microstructure of the ZIF monoliths was revealed by tip force microscopy

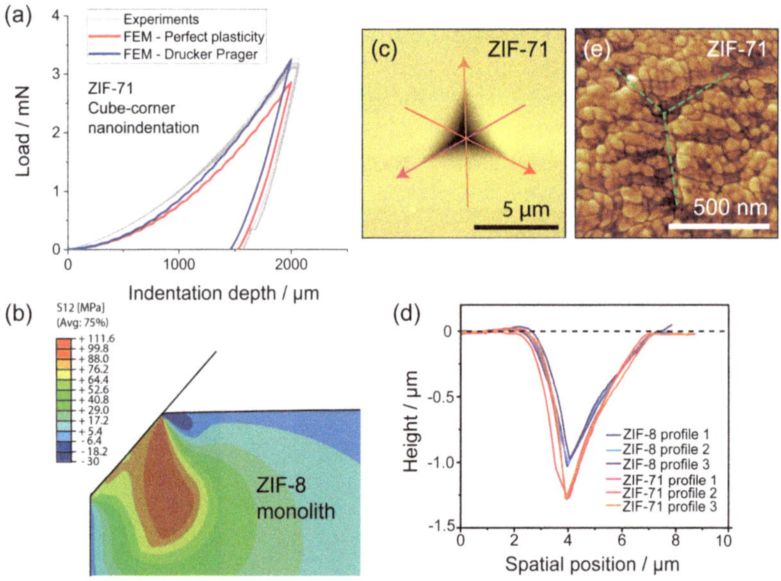

Figure 1.25 Nanomechanical characterisation of ZIF-8 and ZIF-71 monoliths. (a) FE simulations using a cube-corner indenter and experimental *P–h* curves of the ZIF-71 monolith. (b) Shear stress contours of the ZIF-8 monolith from FE recorded at a maximum surface penetration depth of 2000 nm. (c, d) AFM height profiles of residual indents on the polished monolith surface, showing negligible pile ups. (e) TFM stiffness map of a shallow residual indent on the unpolished monolith surface, showing the nanostructured grains. Adapted from ref. 35 with permission from Elsevier, Copyright 2022.

(TFM), suggesting that grain boundary sliding is likely operational under stress (Figure 1.25(e)); this mechanism is also consistent with the absence of surface cracking when tested under a Berkovich indenter (see Section 1.8.4 on fracture toughness). This study also demonstrates the novel application of near-field infrared nanospectroscopy (nanoFTIR)[125] to characterise the local deformations across a residual indent of MOF at ~20 nm spatial resolution, revealing the effect of stress concentration on the framework distortion/collapse and breakage of chemical bonds.[35]

1.8.2 Nanosheets of 2-D MOFs: Pop-ins and Shear-induced Failures

This section considers the plastic deformation behaviour of layered 2-D metal–organic nanosheets (MONs) when stressed beyond the elastic limit. The earliest example concerns the 'pop-in' phenomenon (defined as a displacement burst under a constant load)[126,127] observed during nanoindentation of a copper phosphonoacetate (CuPA) single crystal with a (dense) layered inorganic–organic framework architecture.[32] The pop-ins detected in the *P*–*h* curves arise from the breakage of the hydrogen bonds binding the adjacent 2-D layers together, but gave way under shear stresses acting on the 2-D planes. The AFM topography height image obtained from the plastically deformed region revealed the formation of shear bands, where the height of the slip steps was in the range of 10–50 nm.

Another exemplar concerns a dense 2-D nanosheet crystal constructed from a Mn 2,2-dimethylsuccinate (MnDMS) framework (Figure 1.26(a)), where the thickness of the monolayer is around 1 nm and the 2-D stack is held together by van der Waals interactions.[67] Spherical nanoindentation

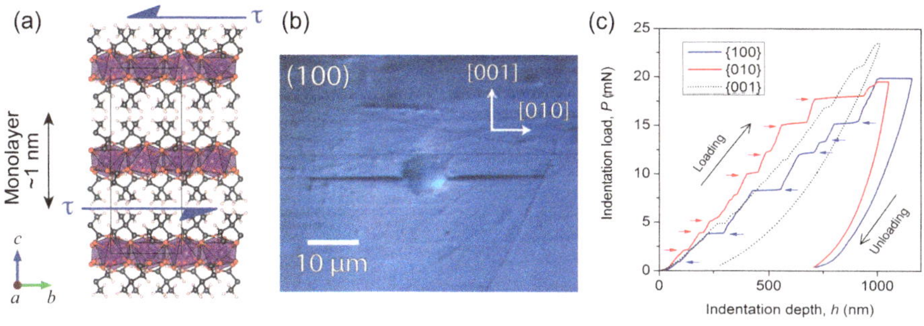

Figure 1.26 (a) 2-D layered architecture of a dense framework of Mn 2,2-dimethylsuccinate (MnDMS) with an orthorhombic unit cell. (b) Optical micrograph of the delamination of van der Waals layers under spherical indentation to a maximum penetration depth of 1000 nm. (c) Indentation load–depth curves showing pop-ins (marked by horizontal arrows) and the degree of mechanical anisotropy associated with three orthonormal crystal facets. Adapted from ref. 67 with permission from American Chemical Society, Copyright 2012.

performed normal to the (100) and (010) crystal facets yielded substantial pop-in displacements caused by the shear-induced delamination of the weakly bonded 2-D layers, see Figure 1.26(b and c). Using the knowledge of critical load (P^*) measured from the first pop-in event (*i.e.*, the onset of plasticity), and by applying the Hertzian elastic contact theory (eqn (1.17)),[128] it is possible to estimate the magnitude of the critical resolved shear stress (τ_{crit}) to initiate splitting of the two adjacent layers in the 2-D framework, given by:

$$\tau_{crit} = 0.31\left(\frac{6P^* E_r^{\,2}}{\pi^3 R^2}\right)^{\frac{1}{3}} \tag{1.17}$$

where E_r is the reduced modulus and R is the radius of the spherical indenter tip. It was reported that the values of critical resolved shear stress are as low as τ_{crit} = 0.24–0.39 GPa for the in-plane directions of the layered crystals of MnDMS.[67] In contrast, the 2-D layers of CuPA held together by hydrogen bonds have a distinctively higher value of τ_{crit} ~ 1 GPa.[32] The above findings explain why micromechanical exfoliation by ultrasonication and mechanical shearing present an effective pathway by which to produce thin nanostructures of 2-D MOFs held together by weak interactions.

The failure mechanisms of a 2-D MOF nanosheet with atomic-sized pores, comprising a copper 1,4-benzenedicarboxylate (CuBDC) framework, have been studied by Zeng *et al.* utilising AFM-based nanoindentation experiments coupled with FE modelling.[129] In terms of its elastic–plastic properties in the through-thickness directions, the Young's modulus is E ~ 23 GPa, and the yield strength is estimated to be σ_Y ~450 MPa. In this study, a power-law hardening behaviour was proposed beyond the yield point to simulate the *P–h* curves observed in AFM indentation (Figure 1.27). Of note, this study shows that it is possible to differentiate between the modes of mechanical failure that occur during nanoindentation, by carefully examining the distorted shapes of the indentation *P–h* curves. For CuBDC nanosheets, three distinctive plastic deformation modes have been proposed: mode I – interfacial slippage between nanosheets, mode II – fracture of the nanosheets, and mode III – interfacial delamination of the nanosheets. The failure mechanisms of the CuBDC nanosheets are summarised in Figure 1.27(c and d), where these mechanisms are potentially applicable to a broader family of van der Waals solids[130,131] and layered 2-D compounds such as covalent organic frameworks (COFs).[132,133]

Under the influence of mode I, the coplanar layers are pushed apart sideways by the nanoindenter as it penetrates deeper into the 2-D stack, where the van der Waals interactions are ruptured predominantly by shear-induced deformation. This kind of slippage failure results in the occurrence of a 'pop-in' phenomenon during the loading stage. In the case of mode II, the failure mechanism can be attributed to a stress build-up at the indenter tip prior to fracture, identifiable by the formation of anomalous 'humps' evident in the *P–h* curves (identifiable by a distinct rise in load, prior to its decline). For mode III, the structural bending mechanism causes the interlayer

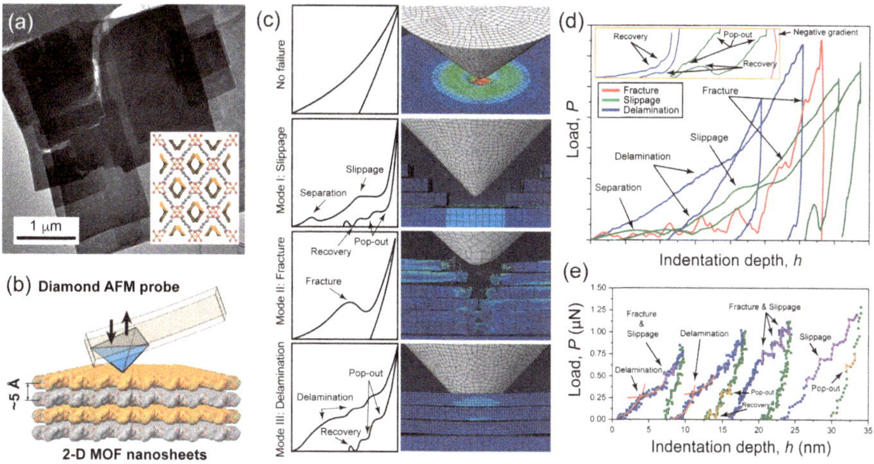

Figure 1.27 (a) Transmission electron microscopy (TEM) image of copper 1,4-benzenedicarboxylate (CuBDC) nanosheets. The inset shows the porous nanosized channels down the [001] crystal axis of the layered 2-D framework. (b) Schematic of AFM nanoindentation with a cube-corner diamond indenter mounted on a stainless-steel cantilever probe. (c) FE modelling of failure modes and (d) the corresponding predicted distorted *P*–*h* curves. (e) Experimental data of the various modes of nanosheet failure. Adapted from ref. 129 with permission from the Royal Society of Chemistry.

delamination of the 2-D stack during indenter loading, followed by the occurrence of the 'pop-out' and recovery phenomena during the unloading stage. All three mechanisms are operational at different stages of the indentation experiment, as evidenced from the AFM nanoindentation data shown in Figure 1.27(e). A vital implication of local nanoscale failure on the overall mechanical properties of the nanosheets can be recognised through the substantial decline of their Young's moduli: from $E \sim 23$ GPa (in the normal *P*–*h* curves), falling to ~12 GPa as a result of delamination failure, and the stiffness further decreases to ~4 GPa due to the slippage failure and fracture of the 2-D layers.

1.8.3 Interfacial Strengths of MOF Polycrystalline Films and Coatings

For device applications in the dielectrics, sensors, lighting and optoelectronics sectors, it is important to fabricate MOF materials in the form of high-quality thin films or well-adhered coatings onto a range of engineering substrates and components.[134] These are in fact polycrystalline films encompassing fine-scale MOF crystals prepared through a range of methods,[135] such as dip coating in a colloidal dispersion of preformed nanocrystals, seeded growth, layer-by-layer deposition (Figure 1.16), electrochemical methods, or more recently by employing more sophisticated techniques such as chemical vapour deposition (CVD) and lithography.[136] Mechanical characterisation of the properties of the polycrystalline thin film (apart from the E and H values) is not commonly reported in the literature, although the knowledge of surface adhesion (against delamination failure), interfacial behaviour, and fracture strength are central to the practical applications of technological devices.

Figure 1.28(a) shows the results of a nanomechanical study performed on a polycrystalline film comprising submicron crystals of ZIF-8, fabricated *via* dip coating on a glass substrate.[85] A cube-corner AFM diamond probe was used to generate a significantly deeper indentation up to $h \sim 200$ nm (*vs.* typically tens of nm for AFM nanoindentation), with which the collective mechanical response of a polycrystalline film can be measured. By comparing the AFM height topographic images of the polycrystalline surface, obtained before and after indentation testing, these reveal a sliding mechanism accompanied by pile-up/sink-in from the interparticle slippage of adjacent nanocrystals caused by shear-induced glide. The P–h curves exhibit load build-up, pop-in, and subsequent load-drop phenomena that can be explained by the grain boundary sliding mechanism that occurs at the interfaces of the polycrystals. The data reveal that for a thin-film MOF coating prepared by dip coating/drop casting the sliding deformation or glide of the neighbouring nanocrystals could be triggered by a relatively small external load, on the order of just several micro-Newtons (μN).

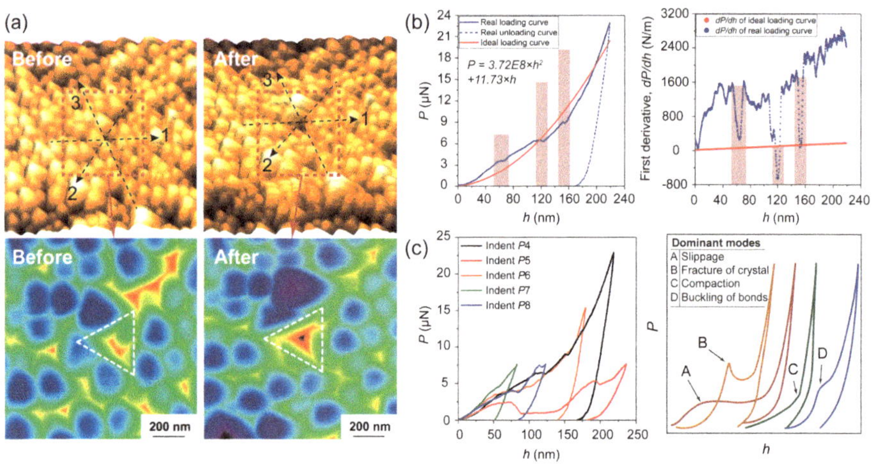

Figure 1.28 Failure modes of a polycrystalline thin film comprising ZIF-8 nanocrystals. (a) AFM scans before and after indentation tests to study the gliding mechanism. (b) Analysis of the distorted load–depth indentation curve (left) to identify failure points during loading, marked by red bands (right). (c) Experimental data (left) of the failure modes A–D illustrated in the right panel. Adapted from ref. 85 with permission from American Chemical Society, Copyright 2017.

The critical depth (h^*) at when the first grain-boundary slippage occurs can be established from the experimental *P–h* curves plotted against an ideal function (assuming no interfacial sliding), as shown in Figure 1.28(b). Of note, it was found that an indentation depth of no greater than ~1/3 of the size of the smallest individual nanocrystal (~300 nm) is all that is required to trigger the first slippage for this thin-film sample. It is expected that this magnitude is dependent upon the fabrication routes employed as the interfacial response of grain boundaries should vary with the adhesion strength of the polycrystals, local packing pattern, and the type of the underlying substrate on which the crystals are deposited. Four different failure modes of the polycrystalline film, denoted as A to D are shown in Figure 1.28(c), from which the critical stresses corresponding to each mechanism can be estimated. Generally, mode A is attributed to grain boundary slippage, which occurs at a shallow deformation of ~10 nm. Mode B is associated with polycrystalline fracture, where a failure strength of up to 1 GPa has been detected. Mode C is due to the accumulated compaction of the porous framework under compression, while mode D might be linked to the buckling of chemical bonds in the direction of loading.

Nanoscratch measurements under an instrumented nanoindenter have been applied by Buchan *et al.* on a polycrystalline HKUST-1 (Cu_3BTC_2) coating (Figure 1.29(a)), fabricated by electrochemical reaction on a pure

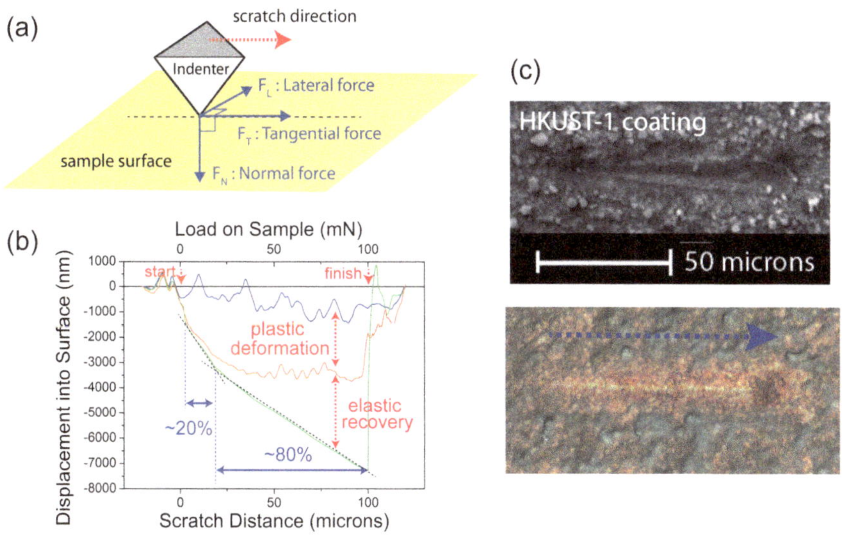

Figure 1.29 (a) Schematic of nanoscratch testing utilising a Berkovich indenter probe, showing the components of the normal (F_N), tangential (F_T), and lateral (F_L) forces acting on the indenter tip. (b) Ramp-load measurement of the HKUST-1 coating over a scratch distance of 100 μm, subject to a monotonically increasing normal load from 0 to 100 mN. (c) SEM and optical profilometry images of the failed coating with exposed copper substrate. Adapted from ref. 137 with permission from American Chemical Society, Copyright 2015.

copper substrate to study the film-to-substrate adhesion properties and failure mode.[137] Although increasing the reaction time of the electrochemical process increases the overall film thickness and coverage, it has the less desired effect of generating a higher surface roughness (on the order of a few μm) due to a larger average crystal size accompanied by secondary growth. The results obtained from nanoscratch tests are usually semi-quantitative as these measurements are affected by a combination of materials factors and test parameters chosen for a specific study. The 'ramp-load' test may be employed to identify the critical force needed to generate surface failure. An example is shown in Figure 1.29(b) for a film with a mean thickness of 4 μm, which resulted in an exposed substrate during a 100 μm scratch test ($F_{N(max)}$ = 100 mN) using a Berkovich indenter tip. The first 20% of the scratch distance yielded a steeper surface penetration curve (green), but reduced in its gradient with distance. For the remaining 80% scratch distance, however, a linear displacement curve with a constant slope was obtained. After unloading, scanning of the film surface revealed the extent of elastic recovery and plastic deformation of the underlying copper substrate (orange curve). In this example, the film could survive only the initial 20% scratch distance corresponding to a critical normal load of 20 mN, as corroborated by the electron microscopy and optical profilometry images shown in Figure 1.29(c).

Apart from polycrystalline MOF films, the ramp-load nanoscratch methodology described above has also been implemented by Li *et al.*[39] to study the scratch resistance of a bulk MOF glass sample prepared using a melt-quenched ZIF material, termed as a_gZIF-62. The experiments used a conical diamond indenter with a spherical tip radius of ~5 μm, traversing over a distance of 500 μm to produce a monotonically increasing normal load reaching a maximum value of $F_{N(max)}$ = 50 mN. The onset of yielding was detected by comparing the surface profiles before and after scratch testing, see Figure 1.30(a). However, there was no sudden jump observed in the frictional coefficient ($\mu = F_T/F_N$) curve of a_gZIF-62 as a function of scratch distance (Figure 1.30(b)), suggesting that there was no ductile fracture in the limit of the spherical scratch regime.[138] This is an interesting finding, nonetheless, the authors did not report images of the surface topography in the region of the scratch to corroborate the absence of any form of surface or subsurface fractures, thus presenting opportunities for future studies.

Another approach for nanoscratch testing is using a 'pass-and-return' wear test method, this is typically conducted at a lower load so that the film underneath the probe can survive a higher number of passes before delamination failure sets in. Figure 1.31 shows the results of such a cyclic wear test (constant F_N = 20 mN for 100 μm, 10 cycles) performed on a polycrystalline film of a Zr-based MOF called UiO-66 (for its crystal structure see Figure 1.1(g)), elucidating the differential surface damage experienced by the anodically- *versus* cathodically-deposited films grown on a zirconium substrate (foil) *via* an electrochemical method.[139] Stassen *et al.* demonstrated that while the

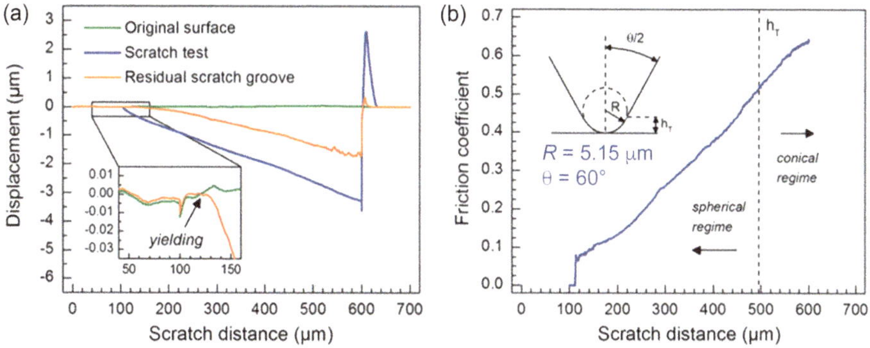

Figure 1.30 (a) Ramp-load scratch testing of a_gZIF-62 using a conospherical diamond indenter (inset in b), traversing at a constant speed of 50 μm s^{-1} in the tangential direction. (b) Evolution of the coefficient of friction during the scratch testing. Reproduced from ref. 39 with permission from American Chemical Society, Copyright 2019.

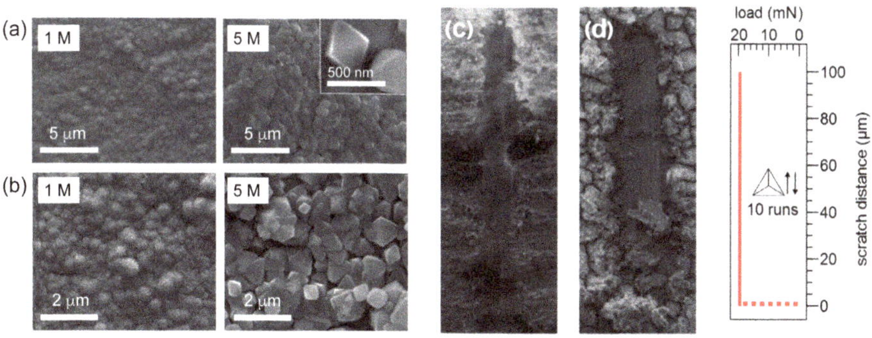

Figure 1.31 SEM images of the electrochemically grown UiO-66 films on a Zr foil substrate by means of (a) anodic and (b) cathodic depositions (for 30 min) in acetic acid concentrations of 1 M and 5 M, respectively. (c, d) Different levels of film damage caused by a ten-cycle pass-and-return wear test using a Berkovich diamond indenter, the test profile of which is depicted in the panel to the right of (d). Reproduced from ref. 139 with permission from American Chemical Society, Copyright 2015.

cathodically-deposited film detaches completely from the substrate due to its poor cohesion strength, it can be seen that the anodically-deposited film has been compressed but still remains adhered to the substrate. The enhanced interfacial adhesion strength of the anodically-deposited film is attributed to the zirconium oxide film acting as a bridging layer between the UiO-66 nanocrystals and the metallic substrate. From a mechanical standpoint, this is an important finding as it holds the key to the fabrication of damage-tolerant MOF films required for practical applications.

1.8.4 Fracture Toughness, Bond Breakage and Cracking

Fracture toughness (K_c) is a measure of materials resistance to crack propagation starting from some pre-existing microscopic flaws, which act as stress raisers. This is a crucial materials property, especially for materials of limited ductility, as catastrophic failures occur at a stress level that is well below the yield strength of the material.

In the discipline of fracture mechanics, K_{Ic} denotes the mode-I fracture toughness value, determined under the 'crack opening' or tensile configuration in plane strain. Mode I is the most common fracture mode in practice. Modes-II and -III correspond to the sliding and shear loading configurations, respectively.[140] The SI unit for K_c is MPa $m^{1/2}$.

Only a few studies thus far have attempted to characterise either K_{Ic} or the associated fracture surface energy (γ_s) of MOFs and inorganic–organic framework materials, as can be seen in Table 1.3. The available K_{Ic} *vs.* E data are presented as an Ashby-style plot in Figure 1.32, showing the projected upper and lower bounds in relation to conventional engineering materials. The existing data suggest a relatively low fracture toughness value for MOFs and dense framework materials on the order of 0.1 MPa $m^{1/2}$, much below the magnitudes expected for brittle ceramics and glassy polymers. It is therefore imperative to study the toughness properties of MOF materials. On the one hand, it is critical to understand the mechanisms involved and on the other hand to be able to design framework materials with adequate robustness to withstand the rigour of practical applications.

For two polymorphic crystals of copper phosphonoacetate (CuPA) with a dense framework,[32] the Berkovich nanoindenter was employed to generate radial cracks under an applied load of 50–100 mN. The values of K_{Ic} were calculated in accordance with the Laugier's expression given in eqn (1.18), which assumes a Palmqvist crack configuration:[141]

Table 1.3 Fracture toughness and surface energy of inorganic–organic framework materials, determined from indentation toughness experiments and computational modelling.

MOF-type materials	K_{Ic}/MPa $m^{1/2}$	G_c/kJ m^{-2}	γ_s/J m^{-2}	Reference
ZIF-8 monolith	0.074 ± 0.023	0.0017	—	35
ZIF-71 monolith	0.145 ± 0.050	0.0126	—	
ZIF-8 crystal (DFT), defect-free				144
{110} facets	—	—	0.43	
{100} facets			0.72	
ZIF-62 glass	0.104 ± 0.02	0.00104	0.82 ± 0.31	145
ZIF-62 glass (MD), pre-cracked	0.097 ± 0.009	0.00115	0.98	
Dense (3D) CuPA-1 single crystal	~0.10–0.33	<0.0012	—	32
Layered (2D) CuPA-2 single crystal	~0.08–0.12	<0.0004	—	

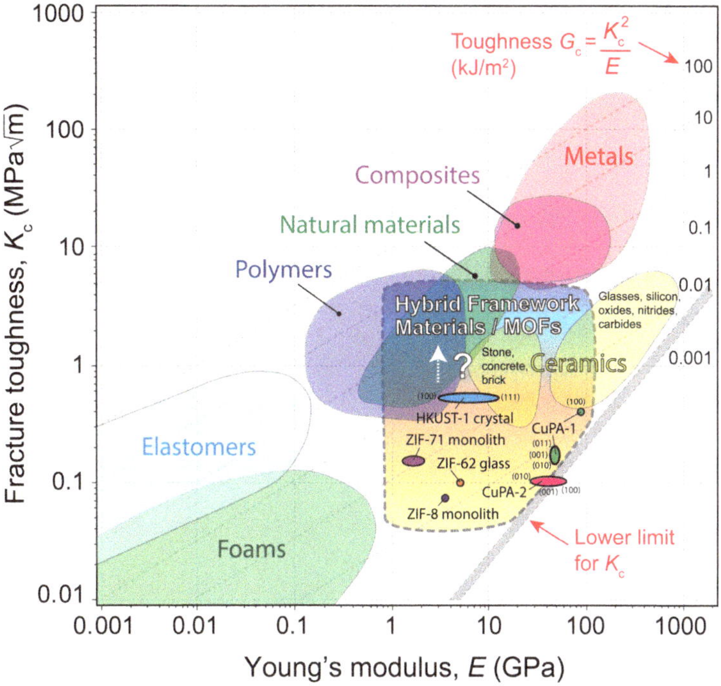

Figure 1.32 Ashby plot of fracture toughness (K_c) *versus* the Young's modulus (E) of common engineering materials (including natural materials) compared with the projected toughness of MOFs and inorganic–organic framework materials. The experimental data of MOFs and hybrid frameworks are sourced from Table 1.3. The recent fracture toughness data of HKUST-1 single crystals were measured using micropillar compression on the {100}- and {111}-oriented facets.[146]

$$K_{Ic} = k\left(\frac{a}{l}\right)^{1/2}\left(\frac{E}{H}\right)^{2/3}\left(\frac{P}{c^{3/2}}\right) \tag{1.18}$$

where P is the maximum indentation load, k is an empirical constant of the indenter shape ($k = 0.016$ for Berkovich;[142] $k = 0.057$ for cube corner),[143] a is the half-diagonal length, l is the crack length and $c = a + l$ (see Figure 1.33). From Figure 1.32, it can be seen that CuPA-1, which has an extended 3-D structure, exhibits a more pronounced toughness anisotropy compared with the layered architecture of CuPA-2. Moreover, it is striking to see that these two dense frameworks are located in the region near the lower limit for K_c, and both of which possess a toughness value (eqn. (1.19)) of $G_c \lesssim 0.001$ kJ m^{-2}. Brittle solids such as glasses, silicon, oxides and carbides are also located close to this lower limit for K_{Ic}, but their toughness values are notably higher, with $0.002 \lesssim G_c \lesssim 0.2$ kJ m^{-2}.

$$G_c = \frac{K_{Ic}^{\ 2}}{E} \tag{1.19}$$

On the nanoporous ZIF monoliths studied by Tricarico *et al.*,[35] it was not possible to generate cracks using a Berkovich indenter due to the large included angle of ~142° (face angle = 65.3°). Likewise, no cracks were observed for the HKUST-1 monolith tested using the Berkovich indenter,[34] as shown in Figure 1.13(a). For the ZIF-8 and ZIF-71 monoliths, the radial cracks required for the K_{Ic} analysis of eqn (1.18) can only be attained using a cube-corner indenter tip (face angle = 35.3°) to induce higher stresses for crack initiation/propagation. The different geometries of the indenter tips are compared in Figure 1.33(a). Between the two ZIF monoliths processed using a sol–gel method, Tricarico *et al.*[35] demonstrated that the ZIF-8 monolith is relatively easier to fracture. As shown in Figure 1.33(b), radial cracks can be seen propagating from all three corners of the cube-corner indent when subjected to a maximum indentation depth of 2 μm (P_{max} ~ 5 mN). In contrast, under an identical test setup, a maximum depth of 5 μm was necessary to induce cracking on the ZIF-71 monolith. Notably,

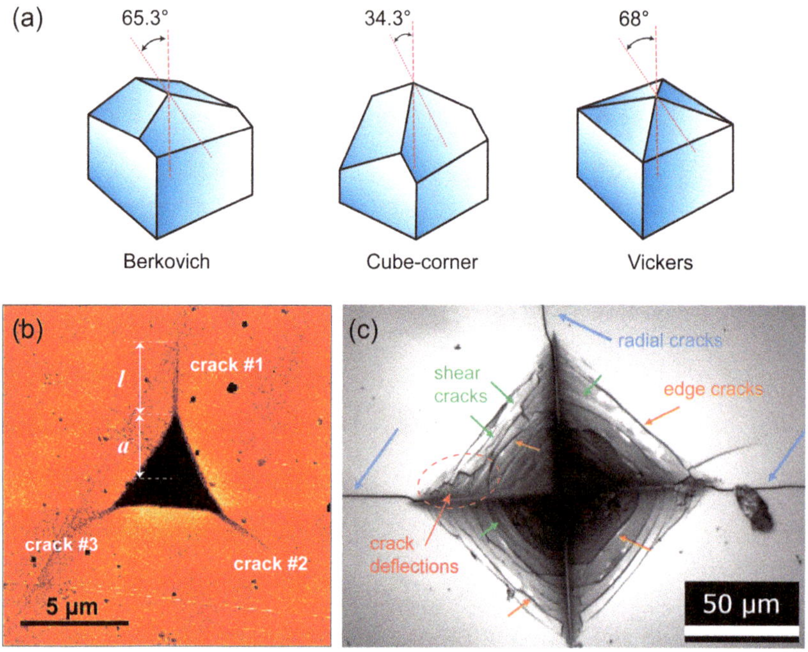

Figure 1.33 (a) Geometry of the Berkovich, cube-corner, and Vickers indenter tips and their corresponding centreline-to-face angles. (b) Cube-corner nanoindentation cracking on a ZIF-8 monolith (sol–gel method). The micron-sized radial cracks become visibly clear *via* nearfield infrared microscopy (s-SNOM). Adapted from ref. 35 with permission from Elsevier, Copyright 2022. (c) Vickers microindentation cracking of the ZIF-62 glass (melt–quench method). Different crack patterns of tens of microns in length are highlighted. Adapted from ref. 148 with permission from National Academy of Sciences.

the fracture toughness value of the ZIF-71 monolith was determined to be K_{Ic} = 0.145 MPa $m^{1/2}$, which is almost twice as high as that of the ZIF-8 monolith (Table 1.3). Because the Young's modulus of the ZIF-71 monolith (E = 1.67 GPa) is relatively lower than the monolith of ZIF-8 (3.18 GPa), the toughness of ZIF-71 (G_c ~ 0.013 kJ m^{-2}) is appreciably higher that of ZIF-8 (G_c ~ 0.002 kJ m^{-2}), see Figure 1.32. The fine-grained nanostructure of the ZIF-71 monolith capable of grain boundary sliding (Figure 1.25(e)) may contribute to improved toughness,[35] compared with a monolith of a melt-quenched a_gZIF-62 glass with a lower fracture toughness of 0.104 MPa $m^{1/2}$ and toughness of ~0.001 kJ m^{-2}.[145] Noteworthy, the mode-I fracture toughness of glassy ZIF-62 was measured using the single-edge precracked beam (SEPB) method,[147] enabled by the availability of a millimetre-sized specimen fabricated *via* the melt-quenching and shaping of a ZIF-62 powder.

Figure 1.33(c) shows the varied fracture patterns evidenced in the ZIF-62 glass, where the sample was tested by Vickers microindentation employing a maximum load of 5 N. Stepniewska *et al.*[148] attributed the formation of the shear cracks to the weaker Zn–N coordinative bonding of the ZIF structure compared with other glass families containing stronger covalent, ionic, or metallic bonds. The crack deflection in the indent leading to the formation of large radial cracks was observed in the ZIF-62 glass. Intriguingly, in 1983 Lawn *et al.*[149] reported a similar shear-induced cracking pattern *via* the Vickers indentation of a soda-lime glass. In particular, the 'shear faults' (shear cracks) they reported are reminiscent of the patterns observed in Figure 1.33(c), emphasising that the narrow shear faults are not 'slip bands' characteristic of dislocation multiplications prevalent in metals. Of course, the latter mechanism is unlikely in ZIF-62 as it is amorphous. Instead, the origin of the shear cracks may be linked to the maximum shear trajectory surfaces that develop during the plastic indentation of brittle materials.[150] In depth research on the crack morphologies and failure mechanisms of MOF and framework materials is thus warranted to shed new light on this largely unexplored topic.

Reactive MD simulations by To *et al.*[145] revealed the breakage of the Zn–N bonds in ZIF-62 glass under uniaxial deformation, see Figure 1.34. The model introduced a precrack at the atomic scale (a flaw for stress concentration) prior to application of a tensile strain (ε) to propagate the crack. With increasing strain the crack propagates transversely to the direction of the axial loading, with the simulation terminated when all stretched bonds were ruptured across the crack to form two new macroscopic surfaces. The MD simulations yielded a fracture toughness of 0.097 MPa $m^{1/2}$, which is similar to that of the experimental value (~0.1 MPa $m^{1/2}$). Subsequently, the surface energy for melt-quenched ZIF-62 glass was calculated from the Irwin formula in eqn (1.20), assuming that the material is in plane strain.

$$\gamma_s = \frac{K_{Ic}^2}{2}\left(\frac{1-\nu^2}{E}\right) \tag{1.20}$$

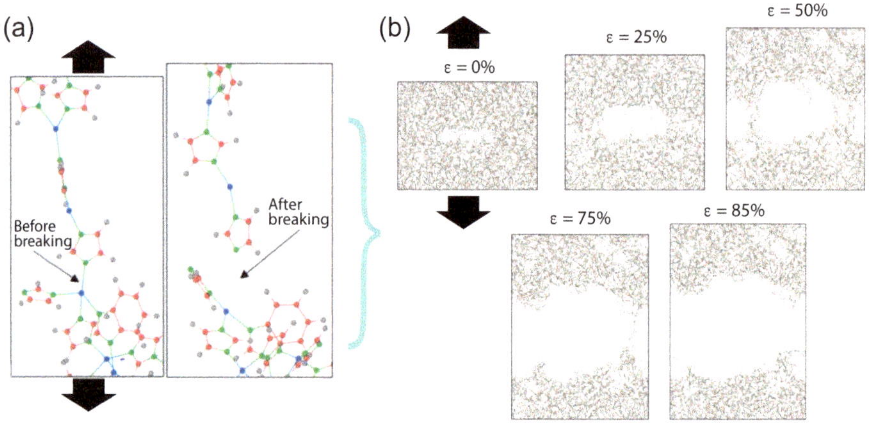

Figure 1.34 MD used to simulate the fracture mechanism of ZIF-62 glass. (a) Breakage of the Zn–N bond in uniaxial tension, designated by the pair of arrows. (b) Crack growth starting from a precrack at 0% strain, propagating with increasing tensile strain to complete rupture marking a failure strain of 85%. Adapted from ref. 145, https://doi.org/10.1038/s41467-020-16382-7, under the terms of the CC BY 4.0 license https://creativecommons.org/licenses/by/4.0/.

The theoretical value was found to be $\gamma_s = 0.98$ J m^{-2}, taking $E = 4.1$ GPa and $\nu = 0.395$ derived from MD simulations. Using the experimental values ($E = 5.2$ GPa and $\nu = 0.343$), the fracture surface energy was determined to be 0.82 J m^{-2}. However, the discrepancies between the experiments and simulations are considerable in terms of the magnitudes of strength (σ_{max} = 8 MPa (exp) *vs.* 703 MPa (MD)) and strain at maximum stress ($\varepsilon_{\sigma_{max}} = 0.3\%$ (exp) *vs.* 17.38% (MD)). Such discrepancies can be explained by surface flaws inherent to the physical sample, but are absent in the (idealised) theoretical sample that has only a prescribed flaw (Figure 1.34(b)); there is also a major difference in sample size between the experiments and simulations.[145]

The foregoing discussion highlights the many opportunities as well as challenges faced in this emerging field. The research to date demonstrates that the fracture behaviour of glassy MOFs, bulk monoliths, single crystals, and composites of hybrid frameworks is very much an underexplored research territory. The basic mechanisms responsible for crack initiation, propagation, and catastrophic fracture thus must be systematically investigated to unravel structure–property relationships that will guide future work in improving the damage tolerance of framework materials. For example, Mahdi *et al.* systematically characterised the fracture energy of mixed-matrix membranes (MMMs) incorporating MOF nanoparticles as fillers, where the composite samples were tested under large-strain uniaxial tension until rupture. The results show that for the (glassy) Matrimid/ZIF-8 nanocomposite membrane, a nanoparticle loading of higher than ~10 wt% significantly impaired the toughness properties of the membrane due to embrittlement effects.[151] A similar trend was observed for a PVDF (polyvinyl difluoride)/ZIF-90 nanocomposite membrane.[152] When employing rubbery polyurethane (PU) as a matrix, it was found that the resultant PU/ZIF-8 nanocomposite also suffered from reduced ductility with increased concentration of MOF nanocrystals, but this effect set in only above 30 wt% of MOF filler when the hyperelastic matrix (initially rubbery) began to experience some embrittlement.[153] Such mechanical effects on a much wider family of polymer matrices,[154–156] hitherto, have not been systematically studied by means of experiments or theory.

1.9 Time-dependent Mechanical Behaviour

Finally, we shall briefly touch upon the time-dependent mechanical properties of MOFs and hybrid framework materials, which are connected to phenomena such as viscoelasticity (at a small strain), stress relaxation (at a constant strain), creep (at a constant stress), and viscoplasticity (at a large strain). Research in this topic area is scarce in the context of MOFs, with only a few exemplars known to date. As they are constructed from organic and inorganic building blocks, it is expected that the mechanical response of hybrid framework structures and MOF-derived composites should exhibit some form of rate dependency. A better elucidation of the structure–property relationships underpinning rate-dependent MOF mechanics will be beneficial for real-world scenarios where engineering materials are commonly exposed to a combination of external stimuli in the thermo–mechanical domain (*e.g.*, rate of deformation, temperature gradient, impact, friction, non-hydrostatic pressure). Crucially, combined thermomechanical loading can generate unexpected results, which are hard to predict by studying the effects of the individual components in isolation.

The time-dependent mechanical response and stress relaxation behaviour of MOF crystals[109] associated with the structural 'flexibility' and chemistry of porous framework are treated in greater detail in Section 5.4 of Chapter 5. Here, worth highlighting are the indentation strain–rate effects on the measured hardness of a family of melt–quenched ZIF glasses, reported by Li *et al.*[39] Using an indentation strain–rate jump method,[157] they determined the strain–rate sensitivity values (m, which reflects a material's susceptibility to creep deformation under a constant stress) of glassy ZIFs to lie in the range of $m = 0.0579–0.0757$, akin to glassy polymers ($m = 0.05–0.10$).[39] This is an interesting development, as such a fine-scale indentation approach could readily be applied to probe the time-dependent deformation of a vast range of crystalline phases and non-amorphous MOF monoliths.

A few studies have characterised the dynamic modulus (E^*) of mixed-matrix membranes incorporating MOF as a filler,[151–153] and of polycrystalline powders (*e.g.*, MOF nanosheets)[158] subject to a cyclic loading under a small strain. This can be achieved using dynamic mechanical analysis (DMA) techniques, which are well established for measuring the viscoelasticity of polymers and soft matter across a range of temperatures and frequencies. The measured viscoelastic properties include the storage and loss moduli, E' and E'', respectively, relaxation temperatures ($T_{\alpha,\beta,\gamma}$), and the dissipation factor (loss tangent, $\tan\delta$), defined by:

$$E^* = \frac{\sigma^*}{\varepsilon^*} = E' + \mathrm{i}E''; \quad \tan\delta = \frac{E''}{E'} \tag{1.21}$$

where σ^* and ε^* are the time-varying oscillatory stress and strain (in sinusoidal waveform), respectively, while $\mathrm{i} = \sqrt{-1}$ denotes the imaginary component of E^*. E' corresponds to the amount of elastic energy stored in the viscoelastic media that is recoverable, while E'' corresponds to energy dissipated by inelastic processes such as entropic motions and/or rotations of flexible structures. To illustrate the associated phenomena, let us consider the PU/ZIF-8 nanocomposite system[153] depicted in Figure 1.35. In a rubbery polymer matrix such as PU, above the glass transition temperature ($T > T_g$) a distinct scaling can be observed between the storage moduli and filler content rising systematically from 0 to 40 wt%. Conversely, at T_g the loss moduli systematically decline with increasing filler content, although the value of T_g

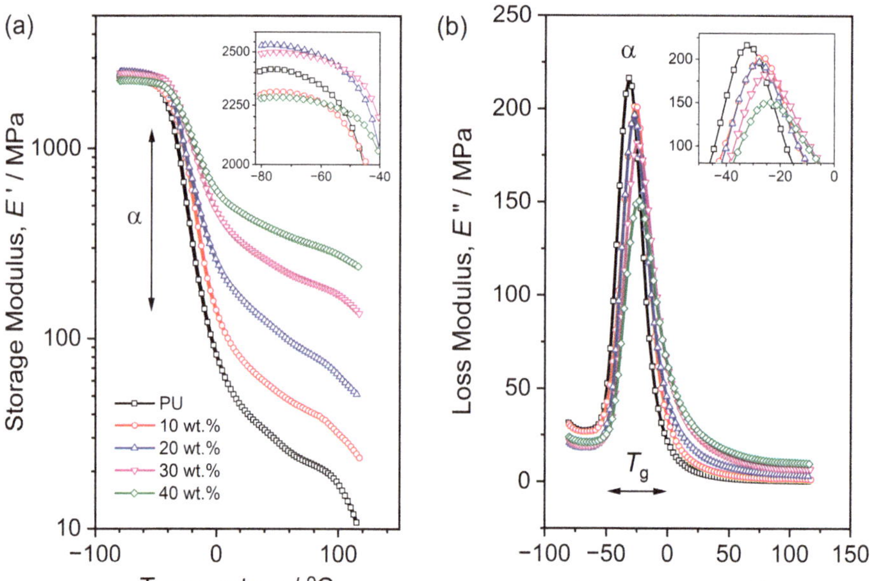

Figure 1.35 (a) Storage and (b) loss moduli of the PU/ZIF-8 nanocomposite membranes tested at an oscillatory frequency of 10 Hz in DMA under a uniaxial tensile mode. The insets show the thermomechanical response at around the glass transition temperature, $T_g \sim -15$ °C. The primary relaxation temperature is designated as α. Adapted from ref. 153 with permission from Elsevier, Copyright 2016.

itself is only marginally affected. The thermomechanical results reveal the intimate interactions between the MOF nanoparticles and their surrounding rubbery matrix; this pinning effect reduces chain mobility and impedes structural relaxation. The highly tuneable viscoelastic response demonstrated here is fascinating, and should be further explored to uncover how such a coupled thermomechanical response may be translated to other combinations of MOF–polymer nanocomposite systems.[26,159,160]

1.10 Summary and Outlook

Chapter 1 introduced the readers to the core principles of 'MOF mechanics'. Using representative examples, we discussed the basic ideas and elucidated the latest methodologies underpinning the study of the elasticity and anisotropy, as well as inelasticity and fracture, of MOFs, including the terahertz dynamics ubiquitous to framework structures. The most up-to-date mechanical properties charts are constructed and presented herein. These charts not only capture the latest findings that have been accomplished since the first review article on the mechanical properties of hybrid frameworks was published over a decade ago,[24] but also reveal directions for possible new discoveries in the field. The foundations gained from this opening chapter prepare the readers for the forthcoming topics, focussing on anomalous elasticity and framework flexibility (Chapter 2), the computational modelling of MOF mechanics (Chapter 3), high-pressure deformations (Chapter 4), rate effects and mechanical energy absorption (Chapter 5).

To conclude, what does the new science of MOF mechanics have to offer? In a nutshell, the research into the mechanical properties of MOFs is *no longer* just about determining the 'mechanical stability' of framework materials. Clearly, it has been proven to be a rich platform for conducting multi-faceted research that appeals to numerous types of scientists, engineers, and technologists. Chapter 1 represents only the tip of the iceberg – the aim is to inspire new research, with the quest for establishing how chemistry governs structure–mechanical property relationships, and to invite readers to explore uncharted territories *via* basic and applied research on MOF mechanics.

List of Abbreviations

DUT	Dresden University of Technology
HKUST	Hong Kong University of Science and Technology
MIL	Materials Institute Lavoisier
MOF	Metal–Organic Framework
PCP	Porous Coordination Polymer
UiO	Universitetet i Oslo
ZIF	Zeolitic Imidazolate Framework

References

1. S. R. Batten, N. R. Champness, X.-M. Chen, J. Garcia-Martinez, S. Kitagawa, L. Öhrström, M. O'Keeffe, M. Paik Suh and J. Reedijk, *Pure Appl. Chem.*, 2013, **85**, 1715.
2. J. C. Tan and B. Civalleri, *CrystEngComm*, 2015, **17**, 197.
3. A. K. Cheetham, C. N. Rao and R. K. Feller, *Chem. Commun.*, 2006, 4780.
4. W. Lu, Z. Wei, Z. Y. Gu, T. F. Liu, J. Park, J. Park, J. Tian, M. Zhang, Q. Zhang, T. Gentle, M. Bosch and H. C. Zhou, *Chem. Soc. Rev.*, 2014, **43**, 5561.
5. M. O'Keeffe and O. M. Yaghi, *Chem. Rev.*, 2012, **112**, 675.
6. S. Kitagawa, R. Kitaura and S. Noro, *Angew. Chem., Int. Ed.*, 2004, **43**, 2334.
7. H. Furukawa, K. E. Cordova, M. O'Keeffe and O. M. Yaghi, *Science*, 2013, **341**, 1230444.
8. D. J. Ashworth and J. A. Foster, *J. Mater. Chem. A*, 2018, **6**, 16292.
9. T. D. Bennett, F. X. Coudert, S. L. James and A. I. Cooper, *Nat. Mater.*, 2021, **20**, 1179.
10. T. D. Bennett, J. C. Tan, Y. Yue, E. Baxter, C. Ducati, N. J. Terrill, H. H. Yeung, Z. Zhou, W. Chen, S. Henke, A. K. Cheetham and G. N. Greaves, *Nat. Commun.*, 2015, **6**, 8079.
11. M. Gutiérrez, Y. Zhang and J. C. Tan, *Chem. Rev.*, 2022, **122**, 10438.
12. N. Stock and S. Biswas, *Chem. Rev.*, 2012, **112**, 933.
13. S. Yuan, L. Feng, K. Wang, J. Pang, M. Bosch, C. Lollar, Y. Sun, J. Qin, X. Yang, P. Zhang, Q. Wang, L. Zou, Y. Zhang, L. Zhang, Y. Fang, J. Li and H. C. Zhou, *Adv. Mater.*, 2018, **30**, e1704303.
14. J. R. Li, R. J. Kuppler and H. C. Zhou, *Chem. Soc. Rev.*, 2009, **38**, 1477.
15. M. Yoon, R. Srirambalaji and K. Kim, *Chem. Rev.*, 2012, **112**, 1196.
16. K. Sumida, D. L. Rogow, J. A. Mason, T. M. McDonald, E. D. Bloch, Z. R. Herm, T. H. Bae and J. R. Long, *Chem. Rev.*, 2012, **112**, 724.
17. M. Gutiérrez, C. Martín, M. Van der Auweraer, J. Hofkens and J. C. Tan, *Adv. Opt. Mater.*, 2020, **8**, 2000670.
18. V. Stavila, A. A. Talin and M. D. Allendorf, *Chem. Soc. Rev.*, 2014, **43**, 5994.
19. F.-Y. Yi, D. Chen, M.-K. Wu, L. Han and H.-L. Jiang, *ChemPlusChem*, 2016, **81**, 675.
20. A. K. Chaudhari, H. J. Kim, I. Han and J. C. Tan, *Adv. Mater.*, 2017, **29**, 1701463.
21. M. R. Ryder, Z. Zeng, K. Titov, Y. Sun, E. M. Mahdi, I. Flyagina, T. D. Bennett, B. Civalleri, C. S. Kelley, M. D. Frogley, G. Cinque and J. C. Tan, *J. Phys. Chem. Lett.*, 2018, **9**, 2678–2684.
22. M. R. Ryder and J. C. Tan, *Mater. Sci. Technol.*, 2014, **30**, 1598.
23. E. A. Dolgopolova and N. B. Shustova, *MRS Bull.*, 2016, **41**, 890.
24. J. C. Tan and A. K. Cheetham, *Chem. Soc. Rev.*, 2011, **40**, 1059.
25. B. B. Shah, T. Kundu and D. Zhao, *Top. Curr. Chem.*, 2019, **377**, 25.

26. E. M. Mahdi and J. C. Tan, in *Comprehensive Composite Materials II*, ed. P. W. R. Beaumont and C. H. Zweben, Academic Press, Oxford, 2018, vol. 4, pp. 525–553.
27. N. C. Burtch, J. Heinen, T. D. Bennett, D. Dubbeldam and M. D. Allendorf, *Adv. Mater.*, 2017, 1704124.
28. A. Venkatasubramanian, J.-H. Lee, V. Stavila, A. Robinson, M. D. Allendorf and P. J. Hesketh, *Sens. Actuators, B*, 2012, **168**, 256.
29. Y. Zhang, T. Xiong, A. F. Möslein, S. Mollick, V. Kachwal, A. S. Babal, N. Amin and J.-C. Tan, *Appl. Mater. Today*, 2022, **27**, 101434.
30. J. Hay, *Exp. Tech.*, 2009, **33**, 66.
31. D. F. Bahr, J. A. Reid, W. M. Mook, C. A. Bauer, R. Stumpf, A. J. Skulan, N. R. Moody, B. A. Simmons, M. M. Shindel and M. D. Allendorf, *Phys. Rev. B*, 2007, **76**, 184106.
32. J. C. Tan, C. A. Merrill, J. B. Orton and A. K. Cheetham, *Acta Mater.*, 2009, **57**, 3481.
33. S. Henke, W. Li and A. K. Cheetham, *Chem. Sci.*, 2014, **5**, 2392.
34. T. Tian, Z. Zeng, D. Vulpe, M. E. Casco, G. Divitini, P. A. Midgley, J. Silvestre-Albero, J. C. Tan, P. Z. Moghadam and D. Fairen-Jimenez, *Nat. Mater.*, 2018, **17**, 174.
35. M. Tricarico and J.-C. Tan, *Mater. Today Nano*, 2022, **17**, 100166.
36. S. Bundschuh, O. Kraft, H. K. Arslan, H. Gliemann, P. G. Weidler and C. Wöll, *Appl. Phys. Lett.*, 2012, **101**, 101910.
37. B. Van de Voorde, R. Ameloot, I. Stassen, M. Everaert, D. De Vos and J. C. Tan, *J. Mater. Chem. B*, 2013, **1**, 7716.
38. T. D. Bennett, A. L. Goodwin, M. T. Dove, D. A. Keen, M. G. Tucker, E. R. Barney, A. K. Soper, E. G. Bithell, J. C. Tan and A. K. Cheetham, *Phys. Rev. Lett.*, 2010, **104**, 115503.
39. S. Li, R. Limbach, L. Longley, A. A. Shirzadi, J. C. Walmsley, D. N. Johnstone, P. A. Midgley, L. Wondraczek and T. D. Bennett, *J. Am. Chem. Soc.*, 2019, **141**, 1027.
40. W. C. Oliver and G. M. Pharr, *J. Mater. Res.*, 2004, **19**, 3.
41. J. Hay, P. Agee and E. Herbert, *Exp. Tech.*, 2010, **34**, 86.
42. A. C. Fischer-Cripps, *Surf. Coat. Technol.*, 2006, **200**, 4153.
43. J. C. Hay, A. Bolshakov and G. M. Pharr, *J. Mater. Res.*, 1999, **14**, 2296.
44. G. M. Pharr and A. Bolshakov, *J. Mater. Res.*, 2002, **17**, 2660.
45. I. N. Sneddon, *Int. J. Eng. Sci.*, 1965, **3**, 47.
46. J. C. Tan, B. Civalleri, C. C. Lin, L. Valenzano, R. Galvelis, P. F. Chen, T. D. Bennett, C. Mellot-Draznieks, C. M. Zicovich-Wilson and A. K. Cheetham, *Phys. Rev. Lett.*, 2012, **108**, 095502.
47. J. F. Nye, *Physical Properties of Crystals*, Clarendon Press, Oxford, 1985.
48. J. J. Vlassak and W. D. Nix, *J. Mech. Phys. Solids*, 1994, **42**, 1223.
49. J. C. Hay, E. Y. Sun, G. M. Pharr, P. F. Becher and K. B. Alexander, *J. Am. Ceram. Soc.*, 1998, **81**, 2661.
50. H. Bei, E. P. George and G. M. Pharr, *Scr. Mater.*, 2004, **51**, 875.

51. J. C. Tan, T. D. Bennett and A. K. Cheetham, *Proc. Natl. Acad. Sci. U. S. A.*, 2010, **107**, 9938.
52. J. C. Tan, P. Jain and A. K. Cheetham, *Dalton Trans.*, 2012, **41**, 3949.
53. P. Jain, N. S. Dalal, B. H. Toby, H. W. Kroto and A. K. Cheetham, *J. Am. Chem. Soc.*, 2008, **130**, 10450.
54. P. Jain, V. Ramachandran, R. J. Clark, H. D. Zhou, B. H. Toby, N. S. Dalal, H. W. Kroto and A. K. Cheetham, *J. Am. Chem. Soc.*, 2009, **131**, 13625.
55. M. G. Ehrenreich, Z. Zeng, S. Burger, M. R. Warren, M. W. Gaultois, J. C. Tan and G. Kieslich, *Chem. Commun.*, 2019, **55**, 3911.
56. L. Ma, W. Li, K. Yang, J. Bi, J. Feng, J. Zhang, Z. Yan, X. Zhou, C. Liu, Y. Ji, J. C. Huang and X. Han, *APL Mater.*, 2021, **9**, 041112.
57. W. Li, A. Thirumurugan, P. T. Barton, Z. Lin, S. Henke, H. H. Yeung, M. T. Wharmby, E. G. Bithell, C. J. Howard and A. K. Cheetham, *J. Am. Chem. Soc.*, 2014, **136**, 7801.
58. M. Kosa, J. C. Tan, C. A. Merrill, M. Krack, A. K. Cheetham and M. Parrinello, *ChemPhysChem*, 2010, **11**, 2332.
59. J. C. Tan, J. D. Furman and A. K. Cheetham, *J. Am. Chem. Soc.*, 2009, **131**, 14252.
60. C. S. Coates, M. R. Ryder, J. A. Hill, J. C. Tan and A. L. Goodwin, *APL Mater.*, 2017, **5**, 066107.
61. S. Sun, S. Henke, M. T. Wharmby, H. H. Yeung, W. Li and A. K. Cheetham, *Inorg. Chem.*, 2015, **54**, 11186.
62. T. D. Bennett, J. C. Tan, S. A. Moggach, R. Galvelis, C. Mellot-Draznieks, B. A. Reisner, A. Thirumurugan, D. R. Allan and A. K. Cheetham, *Chem. – Eur. J.*, 2010, **16**, 10684.
63. Z. Zeng, Quantitative studies of the nanoscale mechanical properties of metal-organic framework materials, DPhil Thesis, University of Oxford, 2018.
64. R. J. Marshall, T. Richards, C. L. Hobday, C. F. Murphie, C. Wilson, S. A. Moggach, T. D. Bennett and R. S. Forgan, *Dalton Trans.*, 2016, **45**, 4132.
65. S. Sun, F. H. Isikgor, Z. Deng, F. Wei, G. Kieslich, P. D. Bristowe, J. Ouyang and A. K. Cheetham, *ChemSusChem*, 2017, **10**, 3740.
66. Y. Rakita, S. R. Cohen, N. K. Kedem, G. Hodes and D. Cahen, *MRS Commun.*, 2015, **5**, 623.
67. J. C. Tan, P. J. Saines, E. G. Bithell and A. K. Cheetham, *ACS Nano*, 2012, **6**, 615.
68. W. Li, M. S. R. N. Kiran, J. L. Manson, J. A. Schlueter, A. Thirumurugan, U. Ramamurty and A. K. Cheetham, *Chem. Commun.*, 2013, **49**, 4471.
69. A. U. Ortiz, A. Boutin, A. H. Fuchs and F.-X. Coudert, *Phys. Rev. Lett.*, 2012, **109**, 195502.
70. M. R. Ryder and J. C. Tan, *Dalton Trans.*, 2016, **45**, 4154.
71. M. R. Ryder, B. Civalleri and J. C. Tan, *Phys. Chem. Chem. Phys.*, 2016, **18**, 9079.
72. S. M. J. Rogge, M. Waroquier and V. Van Speybroeck, *Acc. Chem. Res.*, 2018, **51**, 138.

73. A. U. Ortiz, A. Boutin, A. H. Fuchs and F.-X. Coudert, *J. Phys. Chem. Lett.*, 2013, **4**, 1861.
74. W. Li, M. R. Probert, M. Kosa, T. D. Bennett, A. Thirumurugan, R. P. Burwood, M. Parinello, J. A. K. Howard and A. K. Cheetham, *J. Am. Chem. Soc.*, 2012, **134**, 11940.
75. P. Serra-Crespo, A. Dikhtiarenko, E. Stavitski, J. Juan-Alcaniz, F. Kapteijn, F. X. Coudert and J. Gascon, *CrystEngComm*, 2015, **17**, 276.
76. C. Zener, *Elasticity and Anelasticity of Metals*, University of Chicago Press, Chicago, 1948.
77. Z. L. Fang, B. Bueken, D. E. De Vos and R. A. Fischer, *Angew. Chem., Int. Ed.*, 2015, **54**, 7234.
78. I. E. Khalil, C. Xue, W. J. Liu, X. H. Li, Y. Shen, S. Li, W. N. Zhang and F. W. Huo, *Adv. Funct. Mater.*, 2021, **31**, 2010052.
79. A. F. Möslein, L. Donà, B. Civalleri and J.-C. Tan, *ACS Appl. Nano Mater.*, 2022, **5**, 6398.
80. A. W. Thornton, R. Babarao, A. Jain, F. Trousselet and F. X. Coudert, *Dalton Trans.*, 2016, 4352.
81. M. R. Ryder, B. Civalleri, G. Cinque and J. C. Tan, *CrystEngComm*, 2016, **18**, 4303.
82. K. Otsubo, T. Haraguchi, O. Sakata, A. Fujiwara and H. Kitagawa, *J. Am. Chem. Soc.*, 2012, **134**, 9605.
83. F. I. Pambudi, M. W. Anderson and M. P. Attfield, *Faraday Discuss.*, 2021, **231**, 112.
84. J. C. Tan, B. Civalleri, A. Erba and E. Albanese, *CrystEngComm*, 2015, **17**, 375.
85. Z. Zeng and J. C. Tan, *ACS Appl. Mater. Interfaces*, 2017, **9**, 39839.
86. M. Tafipolsky, S. Amirjalayer and R. Schmid, *J. Phys. Chem. C*, 2010, **114**, 14402.
87. M. R. Ryder, B. Civalleri, T. D. Bennett, S. Henke, S. Rudić, G. Cinque, F. Fernandez-Alonso and J. C. Tan, *Phys. Rev. Lett.*, 2014, **113**, 215502.
88. M. Casco, Y. Cheng, L. L. Daemen, D. Fairen-Jimenez, E. V. Ramos Fernandez, A. J. Ramirez-Cuesta and J. Silvestre-Albero, *Chem. Commun.*, 2016, **52**, 3639.
89. K. T. Butler, P. Vervoorts, M. G. Ehrenreich, J. Armstrong, J. M. Skelton and G. Kieslich, *Chem. Mater.*, 2019, **31**, 8366.
90. A. S. Krylov, A. Vtyurin, P. Petkov, I. Senkovska, M. Maliuta, V. Bon, T. Heine, S. Kaskel and E. A. Slyusareva, *Phys. Chem. Chem. Phys.*, 2017, **19**, 32099.
91. N. Y. Tan, M. T. Ruggiero, C. Orellana-Tavra, T. Tian, A. D. Bond, T. M. Korter, D. Fairen-Jimenez and J. A. Zeitler, *Chem. Commun.*, 2015, **51**, 16037.
92. F. Formalik, B. Mazur, M. Fischer, L. Firlej and B. Kuchta, *J. Phys. Chem. C*, 2021, **125**, 7999.
93. L. H. Rimmer, M. T. Dove, A. L. Goodwin and D. C. Palmer, *Phys. Chem. Chem. Phys.*, 2014, **16**, 21144.
94. M. R. Ryder, T. D. Bennett, C. S. Kelley, M. D. Frogley, G. Cinque and J. C. Tan, *Chem. Commun.*, 2017, **53**, 7041.

95. N. Lock, M. Christensen, Y. Wu, V. K. Peterson, M. K. Thomsen, R. O. Piltz, A. J. Ramirez-Cuesta, G. J. McIntyre, K. Noren, R. Kutteh, C. J. Kepert, G. J. Kearley and B. B. Iversen, *Dalton Trans.*, 2013, **42**, 1996.
96. M. R. Ryder, B. Van de Voorde, B. Civalleri, T. D. Bennett, S. Mukhopadhyay, G. Cinque, F. Fernandez-Alonso, D. De Vos, S. Rudic and J. C. Tan, *Phys. Rev. Lett.*, 2017, **118**, 255502.
97. B. E. Souza, S. Rudic, K. Titov, A. S. Babal, J. D. Taylor and J. C. Tan, *Chem. Commun.*, 2019, **55**, 3868.
98. *Neutron Scattering – Fundamentals, Experimental Methods in the Physical Sciences*, ed. F. Fernandez-Alonso and D. L. Price, Academic Press, New York, 2013.
99. C. Cuadrado-Collados, J. Fernandez-Catala, F. Fauth, Y. Q. Q. Cheng, L. L. Daemen, A. J. Ramirez-Cuesta and J. Silvestre-Albero, *J. Mater. Chem. A*, 2017, **5**, 20938.
100. K. Titov, D. B. Eremin, A. S. Kashin, R. Boada, B. E. Souza, C. S. Kelley, M. D. Frogley, G. Cinque, D. Gianolio, G. Cibin, S. Rudic, V. P. Ananikov and J.-C. Tan, *ACS Sustainable Chem. Eng.*, 2019, **7**, 5875.
101. S. A. Moggach, T. D. Bennett and A. K. Cheetham, *Angew. Chem., Int. Ed.*, 2009, **48**, 7087.
102. C. L. Hobday, C. H. Woodall, M. J. Lennox, M. Frost, K. Kamenev, T. Duren, C. A. Morrison and S. A. Moggach, *Nat. Commun.*, 2018, **9**, 1429.
103. S. Cao, T. D. Bennett, D. A. Keen, A. L. Goodwin and A. K. Cheetham, *Chem. Commun.*, 2012, **48**, 7805.
104. T. D. Bennett and A. K. Cheetham, *Acc. Chem. Res.*, 2014, **47**, 1555.
105. M. T. Wharmby, S. Henke, T. D. Bennett, S. R. Bajpe, I. Schwedler, S. P. Thompson, F. Gozzo, P. Simoncic, C. Mellot-Draznieks, H. Tao, Y. Yue and A. K. Cheetham, *Angew. Chem., Int. Ed.*, 2015, **54**, 6447.
106. S. Henke, M. T. Wharmby, G. Kieslich, I. Hante, A. Schneemann, Y. Wu, D. Daisenberger and A. K. Cheetham, *Chem. Sci.*, 2018, **9**, 1654.
107. S. Aguado, G. Bergeret, M. P. Titus, V. Moizan, C. Nieto-Draghi, N. Bats and D. Farrusseng, *New J. Chem.*, 2011, **35**, 546.
108. D. Fairen-Jimenez, S. A. Moggach, M. T. Wharmby, P. A. Wright, S. Parsons and T. Duren, *J. Am. Chem. Soc.*, 2011, **133**, 8900.
109. Y. Sun, S. M. J. Rogge, A. Lamaire, S. Vandenbrande, J. Wieme, C. R. Siviour, V. Van Speybroeck and J. C. Tan, *Nat. Mater.*, 2021, **20**, 1015.
110. A. F. Möslein and J. C. Tan, *J. Phys. Chem. Lett.*, 2022, **13**, 2838.
111. Y. Wu, A. Kobayashi, G. J. Halder, V. K. Peterson, K. W. Chapman, N. Lock, P. D. Southon and C. J. Kepert, *Angew. Chem., Int. Ed.*, 2008, **47**, 8929.
112. D. Dubbeldam, K. S. Walton, D. E. Ellis and R. Q. Snurr, *Angew. Chem., Int. Ed.*, 2007, **46**, 4496.
113. A. E. J. Hoffman, J. Wieme, S. M. J. Rogge, L. Vanduyfhuys and V. Van Speybroeck, *Z. Kristallogr. – Cryst. Mater.*, 2019, **234**, 529.
114. P. G. Yot, K. Yang, V. Guillerm, F. Ragon, V. Dmitriev, P. Parisiades, E. Elkaïm, T. Devic, P. Horcajada, C. Serre, N. Stock, J. P. S. Mowat, P. A. Wright, G. Férey and G. Maurin, *Eur. J. Inorg. Chem.*, 2016, 4424.

115. K. Titov, Z. Zeng, M. R. Ryder, A. K. Chaudhari, B. Civalleri, C. S. Kelley, M. D. Frogley, G. Cinque and J. C. Tan, *J. Phys. Chem. Lett.*, 2017, **8**, 5035.
116. M. Wang, X. Zhang, Y. Chen and D. Li, *J. Phys. Chem. C*, 2016, **120**, 5059.
117. D. Tabor, *Proc. Phys. Soc. B*, 1954, **67**, 249.
118. J. C. Tan, *Unpublished Work on Spherical Nanoindentation of ZIF-8 Single Crystals*, 2012.
119. D. Tabor, *Philos. Mag. A*, 1996, **74**, 1207.
120. A. C. Fischer-Cripps, *Nanoindentation*, Springer, New York, 2nd edn, 2004.
121. M. V. Swain, *J. Mater. Sci. Eng. A*, 1998, **253**, 160.
122. J. E. Bradby, J. S. Williams, J. Wong-Leung, M. V. Swain and P. Munroe, *Appl. Phys. Lett.*, 2001, **78**, 3235.
123. P. Mogilevsky, *Philos. Mag.*, 2005, **85**, 3511.
124. K. Banlusan, E. Antillon and A. Strachan, *J. Phys. Chem. C*, 2015, **119**, 25845.
125. A. F. Möslein, M. Gutiérrez, B. Cohen and J. C. Tan, *Nano Lett.*, 2020, **20**, 7446.
126. H. Bei, E. P. George, J. L. Hay and G. M. Pharr, *Phys. Rev. Lett.*, 2005, **95**, 045501.
127. D. J. Oliver, J. E. Bradby, J. S. Williams, M. V. Swain and P. Munroe, *J. Appl. Phys.*, 2007, **101**, 043524.
128. K. L. Johnson, *Contact Mechanics*, Cambridge Univ. Press, Cambridge, 1985.
129. Z. Zeng, I. S. Flyagina and J.-C. Tan, *Nanoscale Adv.*, 2020, **2**, 5181.
130. B. Choi, J. Yu, D. W. Paley, M. T. Trinh, M. V. Paley, J. M. Karch, A. C. Crowther, C. H. Lee, R. A. Lalancette, X. Y. Zhu, P. Kim, M. L. Steigerwald, C. Nuckolls and X. Roy, *Nano Lett.*, 2016, **16**, 1445.
131. D. N. Basov, M. M. Fogler and F. J. Garcia de Abajo, *Science*, 2016, **354**, aag1992.
132. A. Giri, A. M. Evans, M. A. Rahman, A. J. H. McGaughey and P. E. Hopkins, *ACS Nano*, 2022, **16**, 2843.
133. J. H. Dou, M. Q. Arguilla, Y. Luo, J. Li, W. Zhang, L. Sun, J. L. Mancuso, L. Yang, T. Chen, L. R. Parent, G. Skorupskii, N. J. Libretto, C. Sun, M. C. Yang, P. V. Dip, E. J. Brignole, J. T. Miller, J. Kong, C. H. Hendon, J. Sun and M. Dinca, *Nat. Mater.*, 2021, **20**, 222.
134. J. Liu and C. Woll, *Chem. Soc. Rev.*, 2017, **46**, 5730.
135. A. Betard and R. A. Fischer, *Chem. Rev.*, 2012, **112**, 1055.
136. M. Tu, B. Xia, D. E. Kravchenko, M. L. Tietze, A. J. Cruz, I. Stassen, T. Hauffman, J. Teyssandier, S. De Feyter, Z. Wang, R. A. Fischer, B. Marmiroli, H. Amenitsch, A. Torvisco, M. J. Velasquez-Hernandez, P. Falcaro and R. Ameloot, *Nat. Mater.*, 2021, **20**, 93.
137. I. Buchan, M. R. Ryder and J. C. Tan, *Cryst. Growth Des.*, 2015, **15**, 1991.
138. K. Lee, K. P. Marimuthu, C.-L. Kim and H. Lee, *Tribol. Int.*, 2018, **120**, 398.
139. I. Stassen, M. Styles, T. Van Assche, N. Campagnol, J. Fransaer, J. Denayer, J. C. Tan, P. Falcaro, D. De Vos and R. Ameloot, *Chem. Mater.*, 2015, **27**, 1801.

140. W. D. Callister, *Materials Science and Engineering: An Introduction*, John Wiley & Sons, Inc, USA, 4th edn, 1997.
141. M. T. Laugier, *J. Mater. Sci. Lett.*, 1987, **6**, 355.
142. R. D. Dukino and M. V. Swain, *J. Am. Ceram. Soc.*, 1992, **75**, 3299.
143. N. Cuadrado, D. Casellas, M. Anglada and E. Jiménez-Piqué, *Scr. Mater.*, 2012, **66**, 670.
144. S. H. Pang, C. Han, D. S. Sholl, C. W. Jones and R. P. Lively, *Chem. Mater.*, 2016, **28**, 6960.
145. T. To, S. S. Sorensen, M. Stepniewska, A. Qiao, L. R. Jensen, M. Bauchy, Y. Yue and M. M. Smedskjaer, *Nat. Commun.*, 2020, **11**, 2593.
146. Z. Zeng, Y. Xiao, J. M. Wheeler and J. C. Tan, *arXiv preprint*, https://arxiv.org/abs/2210.12489.
147. T. Rouxel and S. Yoshida, *J. Am. Ceram. Soc.*, 2017, **100**, 4374.
148. M. Stepniewska, K. Januchta, C. Zhou, A. Qiao, M. M. Smedskjaer and Y. Yue, *Proc. Natl. Acad. Sci. U. S. A.*, 2020, **117**, 10149.
149. B. R. Lawn, T. P. Dabbs and C. J. Fairbanks, *J. Mater. Sci.*, 1983, **18**, 2785.
150. J. T. Hagan and M. V. Swain, *J. Phys. D: Appl. Phys.*, 1978, **11**, 2091.
151. E. M. Mahdi and J. C. Tan, *J. Membr. Sci.*, 2016, **498**, 276.
152. I. S. Flyagina, E. M. Mahdi, K. Titov and J. C. Tan, *APL Mater.*, 2017, **5**, 086104.
153. E. M. Mahdi and J. C. Tan, *Polymer*, 2016, **97**, 31.
154. S. J. D. Smith, C. H. Lau, J. I. Mardel, M. Kitchin, K. Konstas, B. P. Ladewig and M. R. Hill, *J. Mater. Chem. A*, 2016, **4**, 10627.
155. M. Kalaj, M. S. Denny Jr, K. C. Bentz, J. M. Palomba and S. M. Cohen, *Angew. Chem., Int. Ed.*, 2019, **58**, 2336.
156. S. M. Vornholt, M. J. Duncan, S. J. Warrender, R. Semino, N. A. Ramsahye, G. Maurin, M. W. Smith, J. C. Tan, D. N. Miller and R. E. Morris, *ACS Appl. Mater. Interfaces*, 2020, **12**, 58263.
157. V. Maier, K. Durst, J. Mueller, B. Backes, H. W. Höppel and M. Göken, *J. Mater. Res.*, 2011, **26**, 1421.
158. Z. G. Hu, E. M. Mahdi, Y. W. Peng, Y. H. Qian, B. Zhang, N. Yan, D. Q. Yuan, J. C. Tan and D. Zhao, *J. Mater. Chem. A*, 2017, **5**, 8954.
159. Y. Cheng, Y. Ying, S. Japip, S. D. Jiang, T. S. Chung, S. Zhang and D. Zhao, *Adv. Mater.*, 2018, **30**, e1802401.
160. B. Seoane, J. Coronas, I. Gascon, M. E. Benavides, O. Karvan, J. Caro, F. Kapteijn and J. Gascon, *Chem. Soc. Rev.*, 2015, **44**, 2421.

CHAPTER 2

Anomalous Mechanical Behaviour Arising From Framework Flexibility

ARNAUD MARMIER*[a]

[a]Department of Engineering, Design and Mathematics, University of the West of England, Bristol, UK
*E-mail: Arnaud.marmier@uwe.ac.uk, a.s.h.marmier@gmail.com

2.1 Introduction

For a while, the authors of articles on the mechanical and thermo-mechanical behaviour of framework materials were caught in a frenzy of overexcitement as their titles were advertising *exceptionally large*, *giant* or even, my personal favourite, *colossal* exotic properties. There is no doubt that a lot of this was initially tongue in cheek, but the damage to our collective psyche has been done.

Jest aside, it is true that many instances of anomalous elastic and thermo-elastic behaviour seem to occur in MOFs. Negative linear or area compressibility (NLC and NAC), negative Poisson's ratio (NPR) and negative thermal expansion (NTE) are not rare in MOFs. In fact, some properties seem to be so frequent, intrinsic even, in some MOF families that they should not be considered anomalous at all, but perhaps just slightly unusual.

Inorganic Materials Series No. 12
Mechanical Behaviour of Metal–Organic Framework Materials
Edited by Jin-Chong Tan

Published by the Royal Society of Chemistry, www.rsc.org

MOFs are already promising to become extraordinarily useful for applications where adsorption is an essential mechanism, for instance gas separation, fuel storage, sensing or catalysis. Their more unusual properties could allow other potential applications:

- NLC materials have been proposed as artificial muscles and amplificators of piezoelectric response for sensors and actuators.
- NTE compounds have been suggested as fillers in composites with tailored thermal expansion, for instance in dental fillers that match tooth enamel.
- NPR materials tend to have improved mechanical impact absorption, and if they exhibit good pore size tunability they have been proposed for use in shock absorption or filtration.

It should be noted that, in general, MOFs suffer from poor fundamental elastic properties compared with conventional engineering materials, with typically low Young's modulus, strength and toughness values (see Section 1.8.4 of Chapter 1). In addition, MOFs might be prohibitively expensive for applications that require a large amount of material. These two factors might preclude some of the possible applications otherwise reserved to more robust and abundant materials; it is difficult for instance to imagine MOFs being used as body armour, a mooted application of NPR materials.

This chapter is organised as follows. In order to establish a theoretical basis and define the various properties of interests, Section 2.2 recalls the basis of the theory of elasticity and summarises the main elements of thermo-elasticity. This section also consolidates as many published elastic tensors of MOFs as the author could collect. Section 2.3 describes in more detail the main anomalous elastic and thermoelastic properties, presents their main mechanisms, and systematically classifies their occurrences in MOF structures. Section 2.4 explores the somewhat ill-defined concept of 'flexibility', and contrasts several of the competing definitions. The chapter concludes with Section 2.5 that also considers future developments in the field.

2.2 Anisotropic Elasticity

This section details the basic theoretical framework of elasticity, with an emphasis on fundamental properties like the Young's, shear and bulk moduli and Poisson's ratio, and slightly more niche ones such as linear compressibility. Isotropic elasticity with only two independent parameters is relatively straightforward. The jump in complexity provided by extra parameters beyond the basic two (from three for cubic crystal systems to 26 for triclinic crystal systems) is considerable, and it is helpful to start organising materials following a systematic measure of elastic anisotropy. This concept is not unique however, and several metrics are discussed.

2.2.1 Elasticity Theory

In essence, the theory of elasticity explains how bodies deform under loads that are small enough to leave the initial body unchanged once the load is released. This theory has been accepted for a long time, works very well, and is really not very complicated, mainly because the relationships are linear. The main challenge concerns the fact that some of its objects are four-dimensional and as such can be difficult to visualise. However, elasticity is often used to introduce interesting mathematical concepts such as tensors to undergraduates in physical sciences. This section introduces the concepts and quantities, but in a broad manner, which will benefit all readers regardless of their background. Many excellent textbooks, for instance ref. 1 and 2, provide much more detailed and systematic expositions.

Before getting into the details of the theory, it is useful to recall the atomistic origin of elasticity, as illustrated in Figure 2.1. The standard picture is of crystals comprising molecular bonds that act like 'springs'. This model is obviously very simplified, and it does not really hold even for covalent systems where torsion (angular) springs would be needed. Conceptually, it is even worse when bonds are less clearly defined as in metallic systems, but the idea of treating bonds as springs still reflects well the fact that the potential energy well at equilibrium is quadratic in nature.

In practice, the theory of elasticity linearly relates stresses to strains, and defining these two quantities is a good way to start.

The stress describes the surface forces acting on a unit volume element in a continuum. It can be represented by a symmetric 2nd order stress tensor, $\boldsymbol{\sigma} = \sigma_{ij}$, where subscripts $i, j = x, y, z$ (or 1, 2, 3). Nominally, $\boldsymbol{\sigma}$ has nine components but with only six independent components: three normal stresses

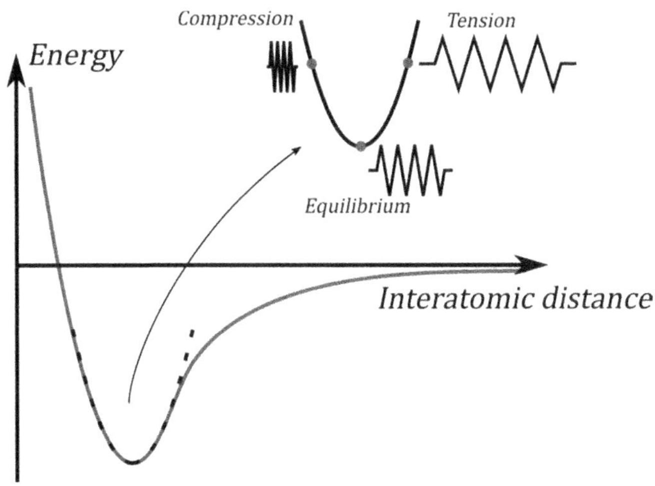

Figure 2.1 Atomistic origin of elasticity. At near equilibrium position, chemical bonds behave like springs as depicted in the insets.

acting perpendicular to the faces (diagonal terms, note $\sigma_{ii} \equiv \sigma_i$) and three shear stresses acting parallel to the faces (off-diagonal terms, note $\tau_{ij} = \tau_{ji}$),

$$\boldsymbol{\sigma} = \begin{pmatrix} \sigma_x & \tau_{xy} & \tau_{xz} \\ \tau_{yx} & \sigma_y & \tau_{yz} \\ \tau_{zx} & \tau_{zy} & \sigma_z \end{pmatrix}. \tag{2.1}$$

The cube in Figure 2.2 is a canonical representation of the stress components as unit surface forces.

The strain describes the state of deformation of a solid body. It can also be represented by a 2nd order strain tensor, $\boldsymbol{\varepsilon} = \varepsilon_{ij}$, with six independent components: three normal strains that affect the lengths (diagonal terms, $\varepsilon_{ii} \equiv \varepsilon_i$) and three shear strains (off-diagonal terms, $\gamma_{ij} = \gamma_{ji}$) that affect the angles,

$$\boldsymbol{\varepsilon} = \begin{pmatrix} \varepsilon_x & \gamma_{xy} & \gamma_{xz} \\ \gamma_{yx} & \varepsilon_y & \gamma_{yz} \\ \gamma_{zx} & \gamma_{zy} & \varepsilon_z \end{pmatrix}. \tag{2.2}$$

The deformed parallelepiped and parallelogram in Figure 2.2 depict the three strain components.

The stiffness tensor, C_{ijkl}, expresses the stress tensor in terms of the strain tensor, using Einstein's summation rules,

$$\sigma_{ij} = C_{ijkl}\varepsilon_{kl}. \tag{2.3}$$

It is a property of the crystal, a tensor of 4th order, and its components depend on the choice of axis. Eqn (2.3) is a generalised version of Hooke's law.

The compliance tensor, S_{ijkl}, is the inverse of the stiffness tensor and expresses the strain tensor in terms of the stress tensor,

$$\varepsilon_{ij} = S_{ijkl}\sigma_{kl}. \tag{2.4}$$

The Young's modulus (E), or modulus of elasticity, is defined as the ratio of normal stress to linear normal strain (both in the direction of applied load).

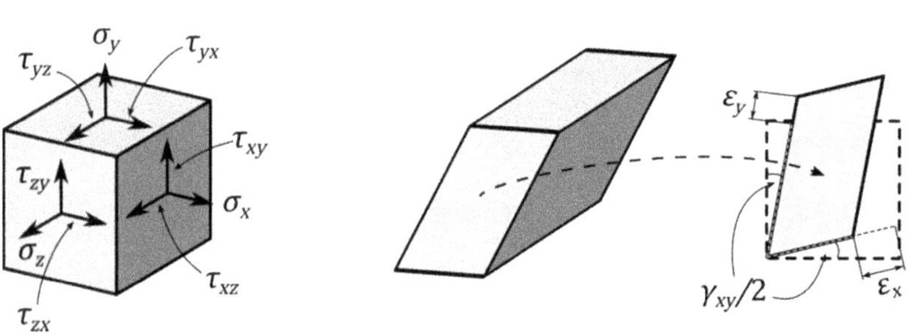

Figure 2.2 Pictorial representations of the stresses and strains acting on an infinitesimal cube, comprising both the normal and shear components. For equilibrium, note that $\tau_{xy} = \tau_{yx}$, $\gamma_{xy} = \gamma_{yx}$, *etc.*

The shear modulus (G), or modulus of rigidity, is defined as the ratio of shear stress to linear shear strain. The Poisson's ratio (ν) is defined as the ratio of transverse strain (normal to the applied load), to axial strain (in the direction of the applied load). These basic concepts are presented in Section 1.3 of Chapter 1. The linear compressibility is the ratio of the induced strain along a given direction and of the hydrostatic pressure, and except in crystals of cubic symmetry, where compressibility is isotropic, it is dependent on direction.

Due to translational and rotational symmetries, the number of independent components of the 4th order tensor reduces to 21 – from 81 – for the least symmetric case.[1] Crystals (and more generally 3D periodic structures or space groups) can be grouped into seven crystal systems, and the associated symmetries reduce the number of independent components for the 4th order tensors further: triclinic (21), monoclinic (15), orthorhombic (9), trigonal (7), tetragonal (5), hexagonal (5) and cubic (3).

Six components are sufficient to describe stress and strain. The Voigt[3] scheme uses this fact and replaces the cumbersome 2nd and 4th order tensors in a three-dimension vector space by vectors and matrices in a six-dimension vector space.

Tensor notation	11	22	33	23,32	31,13	12,21
Voigt's notation	1	2	3	4	5	6

These transformation rules apply directly for stress and stiffness, but the use of a corrective coefficient is required for strain and compliances:

$$\begin{array}{ll} \varepsilon_{ii} = \varepsilon_p & S_{ijkl} = S_{pq} \text{ if } p \text{ and } q \text{ are 1, 2, 3 only} \\ \varepsilon_{ij} = \frac{1}{2}\varepsilon_p \text{ if } i \text{ and } j \text{ are different} & S_{ijkl} = \frac{1}{2}S_{pq} \text{ if either } p \text{ or } q \text{ are 4, 5, 6} \\ & \text{(and the other is 1, 2 or 3)} \\ & S_{ijkl} = \frac{1}{4}S_{pq} \text{ if } p \text{ and } q \text{ are 4, 5, 6 only} \end{array} \tag{2.5}$$

Another more symmetric scheme, where the stress and strain vectors, and the stiffness and compliance matrices, are treated similarly, with $1/\sqrt{2}$ correction factors for the off-diagonal terms, has been proposed by Mandel,[4] but it has unfortunately not displaced Voigt's notation.

2.2.1.1 *Tensor Transformations and Euler Angles*

A second order tensor transforms into a new basis set following the rule:

$$\text{'new' in terms of 'old'}, \; T'_{\alpha\beta} = r_{\alpha i} r_{\beta j} T_{ij}, \tag{2.6}$$

where Einstein's summation rule is used and, where $r_{\alpha i}$ and $r_{\beta j}$ are the components of the rotation matrix $\mathbf{r}$ (or direction cosines).

A fourth order tensor transforms into a new basis set following the rule

$$T'_{\alpha\beta\gamma\delta} = r_{\alpha i} r_{\beta j} r_{\gamma k} r_{\delta l} T_{ijkl}. \tag{2.7}$$

For more details, the reader is invited to consult Nye's 'Physical Properties of Crystals'.[1]

A direction in Cartesian space, corresponding to an elastically significant distortion, for instance uniaxial stress or response to isotropic pressure, can be represented as a point on the unit sphere (unit vector), and advantageously by the first two Euler angles. This vector **a** is chosen as the first unit vector of a new basis set. It is fully characterised by the angles θ (0, π) and φ (0, 2π), as illustrated in Figure 2.3. The determination of some elastic properties (shear modulus, Poisson's ratio) requires another, perpendicular, direction. This is defined by another unit vector, **b**, perpendicular to the first, and characterised by the third Euler angle χ (0, 2π). The coordinates of the two vectors can be expressed as

$$\begin{aligned} \boldsymbol{a} &= \begin{pmatrix} r_{11} \\ r_{12} \\ r_{13} \end{pmatrix} = \begin{pmatrix} \sin\theta\cos\varphi \\ \sin\theta\sin\varphi \\ \cos\theta \end{pmatrix} \\ \boldsymbol{b} &= \begin{pmatrix} r_{21} \\ r_{22} \\ r_{23} \end{pmatrix} = \begin{pmatrix} \cos\theta\cos\varphi\cos\chi + \sin\varphi\sin\chi \\ \cos\theta\sin\varphi\cos\chi - \cos\varphi\sin\chi \\ -\sin\theta\cos\chi \end{pmatrix}. \end{aligned} \tag{2.8}$$

By definition, the components of the first two columns of the rotation matrix **r** are the coordinates of **a** and **b**. These are enough to obtain all the components of the fourth order tensors in the vector subspace defined by directions 1 and 2, for instance:

$$S'_{12} = S'_{1122} = r_{1i} r_{1j} r_{2k} r_{2l} S_{ijkl} = a_i a_j b_k b_l S_{ijkl}, \text{ and}$$
$$S'_{66} = S'_{1212} = r_{1i} r_{2j} r_{1k} r_{2l} S_{ijkl} = a_i b_j a_k b_l S_{ijkl},$$

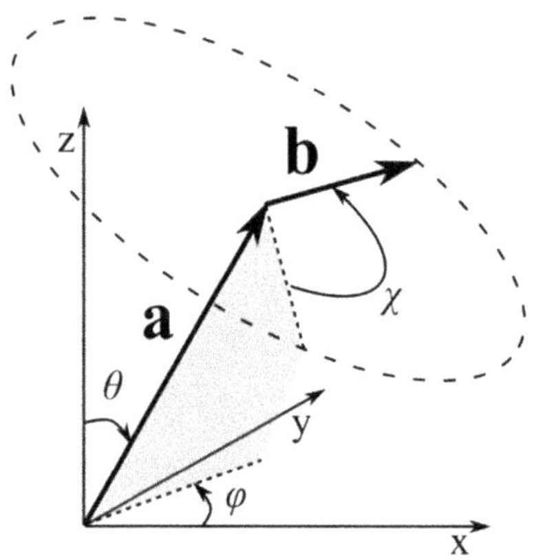

Figure 2.3 Euler angles and vectors describing the longitudinal, **a**, and transverse, **b**, directions.

where the primed components correspond to the new basis set.

The angles θ, φ, and χ can be scanned to access all the components in the new basis set without requiring the third unit vector of the new basis set.

2.2.1.2 *Averaging Schemes*

The elastic properties of an anisotropic material are often replaced by those of an 'equivalent' isotropic material. These processes of averaging are especially important for materials consisting of crystalline grains of random orientation. There are four main schemes: Voigt,[3] Reuss,[5] Hill,[6] and direct.

The Voigt averaging scheme is based on the stiffness matrix (it assumes a given uniform strain) and the bulk modulus K and the shear modulus G are given by

$$K_V = \frac{A+2B}{3} \text{ and } G_V = \frac{A-B+3C}{5},$$

where

$$A = \frac{C_{11}+C_{22}+C_{33}}{3},\ B = \frac{C_{23}+C_{13}+C_{12}}{3},\ C = \frac{C_{44}+C_{55}+C_{66}}{3}. \tag{2.9}$$

Conversely, the Reuss averaging scheme is based on the compliance matrix (it assumes a given uniform stress), resulting in:

$$K_R = \frac{1}{3a+6b} \text{ and } G_R = \frac{5}{4a-4b+3c},$$

where

$$a = \frac{S_{11}+S_{22}+S_{33}}{3},\ b = \frac{S_{23}+S_{13}+S_{12}}{3},\ c = \frac{S_{44}+S_{55}+S_{66}}{3}. \tag{2.10}$$

In both cases, the Young's modulus E and the Poisson's ratio ν are given by

$$E = \left(\frac{1}{3G}+\frac{1}{9K}\right)^{-1} \text{ and } \nu = \frac{1}{2}\left(1-\frac{3G}{3K+G}\right). \tag{2.11}$$

The Hill average (also known as Voigt–Reuss–Hill (VRH)) is the arithmetic average of the Voigt and Reuss values.

A non-analytical direct averaging scheme can also be obtained, based on a numerical average over a sample of all possible combinations of directions.

2.2.1.3 *One-direction Properties: Young's Modulus and Linear Compressibility*

Some properties can be simply expressed in terms of the compliance tensor **S**.

The Young's modulus can be obtained using the purely normal stress in eqn (2.4) in its vector form and is given by

$$E(\theta,\varphi) = \frac{1}{S'_{11}(\theta,\varphi)} = \frac{1}{a_i a_j a_k a_l S_{ijkl}}. \tag{2.12}$$

The linear compressibility follows a slightly different scheme but is even simpler to compute. It is obtained by applying an isotropic stress corresponding to a pressure p in tensor form $\sigma_{kl} = -p\delta_{kl}$, ($\delta_{kl}$ is the Kronecker delta, or a diagonal unit matrix). Hooke's law thus becomes $\varepsilon_{ij} = -pS_{ijkk}$. Using eqn (2.6), the strain in direction **a** is $\varepsilon'_{11} = r_{1i}r_{1j}\varepsilon_{ij} = \varepsilon_{ij}a_ia_j$, and the linear compressibility becomes

$$\beta(\theta,\varphi) = \frac{-\varepsilon'_{11}}{p} = S_{ijkk}a_ia_j. \tag{2.13}$$

2.2.1.4 Two-direction Properties: Shear Modulus and Poisson's Ratio

Other properties depend on two directions (if perpendicular, this requires three angles only), which makes them awkward to represent graphically. A convenient possibility is then to consider three representations: minimum, average, and maximum. For each θ and φ, the angle χ is scanned and the minimum, average, and maximum values are recorded for this direction.

The shear modulus is obtained by applying a pure shear stress in the vector form of eqn (2.4) and results in

$$G(\theta,\varphi,\chi) = \frac{1}{4S'_{66}(\theta,\varphi,\chi)} \tag{2.14}$$

The Poisson's ratio can be obtained using a purely normal stress in eqn (2.4) in its vector form and is given by

$$\nu(\theta,\varphi,\chi) = \frac{-S'_{12}(\theta,\varphi,\chi)}{S'_{11}(\theta,\varphi)} = \frac{-a_ia_jb_kb_lS_{ijkl}}{a_ia_ja_ka_lS_{ijkl}}. \tag{2.15}$$

2.2.2 Complete Elastic Tensors of MOFs

Tables 2.1–2.7 collate the complete elastic stiffness tensors for many MOFs. Almost all have been obtained from theoretical atomistic simulations using either semi-empirical models or density functional theory (DFT). Different methods can be used to extract the elastic tensors, and they are discussed in Section 3.4 of Chapter 3. Experimental determination is very rare, and, to the best of our knowledge, the full elastic tensor of a MOF has only been experimentally measured for a single MOF, cubic ZIF-8, using the Brillouin scattering method.[7] Further details are described in Section 1.5.1 of Chapter 1.

Several analysis tools such as ElAM[8] and ELATE[9] can be used to post-process the elastic tensors and explore anisotropic mechanical properties.

2.2.3 Anisotropy Measures

In order to compare the elastic properties of different materials, it is tempting to use a simple measure of their elastic anisotropy (A). The first quantification of this elastic anisotropy as a single value was the Zener index[31] for cubic crystals,

Table 2.1 Elastic stiffnesses C_{ij} (elastic constants) for cubic MOFs (in GPa).[a]

Compound	C_{11}	C_{44}	C_{12}	Method	Reference
HKUST-1	27.72	5.39	25.72	DFT[b]	10
MOF-C6	44.53	1.82	6.79	MD (10 K)[b]	11
MOF-C10	33.45	1.59	5.49	MD (10 K)[b]	11
MOF-C16	31	1.58	5.24	MD (10 K)[b]	11
MOF-C22	20.11	0.69	3.77	MD (10 K)[b]	11
MOF-C30	14.2	1.17	2.14	MD (10 K)[b]	11
MOF-801 (Zr-FUM)	59.71	20.27	21.47	DFT (0 K)[b]	12
UiO-66 (Zr-BDC)	56.40	18.04	27.20	DFT (0 K)[b]	12
UiO-67 (Zr-BPDC)	25.37	10.24	13.24	DFT (0 K)[b]	12
IRMOF-1 (MOF-5)	21.5	7.5	14.8	DFT	13
IRMOF-1	29.2	1.4	13.1	DFT	14
IRMOF-1	29.4	1.2	12.6	DFT	15
IRMOF-1	28.2	2.7	11.4	DFT	16
IRMOF-1	25.44	1.00	10.92	DFT (0 K)[b]	12
IRMOF-10	15.87	0.27	5.87	DFT (0 K)[b]	12
IRMOF-16	10.82	0.07	3.20	DFT (0 K)[b]	12
ZIF-8	9.52	0.97	6.86	Brillouin scattering	7
ZIF-8	11.04	0.94	8.33	DFT	7
ZIF-8	11.3	2.7	7.6	MD	17

[a]Abbreviations: FUM = fumarate; BDC = benzene-1,4-dicarboxylate; BPDC = 4,4′-biphenyl-dicarboxylate; DFT = density functional theory; MD = molecular dynamics.
[b]Elastic constants at other temperatures or with different model parameterisations are available in the original publication.

Table 2.2 Elastic stiffnesses for tetragonal MOFs (in GPa).[a]

Compound	C_{11}	C_{33}	C_{44}	C_{66}	C_{12}	C_{13}	C_{16}	Method	Reference
NOTT-400	36.67	97.61	16.15	15.51	35.16	20.70	0	DFT	18
CPF-1	37.76	85.76	11.73	12.19	34.82	21.99	0	DFT	18
Al(OH) (1,4-ndc)	84.79	54.99	9.35	2.40	14.92	15.66	0	DFT	18
CAU-10-OCH_3	83.69	3.36	10.80	6.61	10.60	4.41	0.87	DFT	18
NOTT-300	49.34	133.60	26.16	20.39	47.71	25.64	0	DFT	19
NOTT-401	62.98	7.00	6.13	1.52	10.12	7.39	0	DFT	18
CAU-10-CH_3	82.08	5.50	7.91	5.01	9.85	4.15	0	DFT	18
CAU-8	12.67	24.48	3.26	0.69	−0.59	1.79	0.41	DFT	18
CAU-10-Br	84.18	12.09	12.46	5.58	9.64	6.99	−0.14	DFT	18
CAU-10-H	78.58	11.09	8.42	2.70	6.43	3.29	−0.09	DFT	18
ZIF-3	3.80	6.20	0.857	1.528	3.58	3.42	0	DFT	20
ZIF-zni	19.01	23.38	1.56	1.76	13.26	13.38	0	DFT	21

[a]Abbreviation: ndc = naphthalene-1,4-dicarboxylate.

Table 2.3 Elastic stiffness values of hexagonal MOFs (in GPa).

Compound	C_{11}	C_{33}	C_{44}	C_{12}	C_{13}	Method	Reference
$Zn[Au(CN)_2]_2$	36.6	126.8	12.1	29.7	60.6	DFT (0 GPa)[a]	22
$Zn[Au(CN)_2]_2$	48.9	94.3	12.0	25.4	56.3	DFT	23

[a]Elastic constants at other pressures are available in the original publication.

Table 2.4 Elastic stiffness values of trigonal MOFs (in GPa).

Compound	C_{11}	C_{33}	C_{44}	C_{12}	C_{13}	C_{14}	Method	Reference
$KMn[Ag(CN)_2]_3$	34.5	134.8	30.0	19.6	50.9	−12.3	DFT (0 GPa)[a]	24

[a]Elastic constants at other pressures are available in the original publication.

Table 2.5 Elastic stiffness values of orthorhombic MOFs (in GPa).[a]

Compound	C_{11}	C_{22}	C_{33}	C_{44}	C_{55}	C_{66}	C_{12}	C_{13}	C_{23}	Method	Reference
DMOF-1 (sq)	35.33	58.20	58.45	0.11	0.44	0.28	7.32	7.55	11.68	DFT	25
DMOF-1 (loz)	57.15	35.59	17.68	0.62	16.39	0.69	9.85	31.43	5.47	DFT	25
MIL-47	40.69	62.60	36.15	50.83	7.76	9.30	12.58	9.28	46.98	DFT	25
MIL-53 (Al)	90.85	65.56	33.33	7.24	39.52	8.27	20.41	54.28	12.36	DFT	25
MIL-53 (Al)	94.44	65.67	27.09	5.66	41.70	10.39	21.57	53.59	11.02	DFT	26
MIL-53 (Al)	60.17	83.23	25.26	32.24	7.31	8.81	17.50	10.05	45.08	DFT	18
MIL-53 (Ga)	112.32	56.66	18.52	5.48	21.71	6.64	22.87	43.35	10.86	DFT	25
MIL-53 (V)	97.58	43.88	18.59	5.27	24.34	9.73	16.57	40.92	7.06	DFT	18
MIL-61 (V)	83.72	61.87	71.65	4.18	33.98	7.25	14.55	47.16	13.09	DFT	18
MIL-68 (Fe)	65.32	50.09	43.66	6.54	7.48	9.01	20.81	8.37	6.33	DFT	18
MIL-68 (V)	65.32	50.09	43.66	6.54	7.48	9.01	20.81	8.37	6.33	DFT	18
MIL-118 (Al)	22.94	113.84	172.23	9.12	42.31	5.93	9.77	37.03	29.48	DFT	18
ZIF-2	7.69	8.22	4.89	1.266	1.540	1.114	5.41	3.42	4.46	DFT	20
ZIF-4	4.27	3.49	5.02	1.03	1.93	2.45	1.22	1.92	1.53	DFT	21
ZIF-4	3.07	3.36	2.95	0.771	0.903	1.532	0.574	0.603	0.77	DFT	20
ZnPurBr-LT	120.1	290.4	138.2	64.1	54.4	11.7	106.6	26.9	124.9	DFT	27
ZnPurBr-HT	122.0	62.5	107.4	104.3	78.5	75.4	50.4	51.0	18.8	DFT	27

[a]Abbreviations: sq = square; loz = lozenge; LT = low temperature; HT = high temperature.

Table 2.6 Elastic stiffness values of monoclinic MOFs (in GPa).[a]

Compound	C_{11} / C_{15}	C_{22} / C_{25}	C_{33} / C_{35}	C_{44} / C_{46}	C_{55}	C_{66}	C_{12}	C_{13}	C_{23}	Method	Reference
MIL-122 (Al)	252.3	32.6	115.7	28.5	20.9	30.6	34.6	38.9	12.2	DFT	18
	−16.9	−1.91	−9.61	−1.3							
MIL-122 (Ga)	220.7	29.5	96.2	24.0	28.6	30.4	30.9	35.9	9.9	DFT	18
	−18.2	−4.0	−6.8	−2.0							
MIL-121 (Al)	68.6	47.8	62.0	16.0	15.0	41.2	39.5	11.5	17.2	DFT	18
	−1.9	6.4	4.1	1.71							
MIL-116 (Al)	42.9	118.7	229.4	18.3	72.3	7.8	23.9	57.7	44.6	DFT	18
	−1.3	−0.8	−2.2	2.2							
MIL-122 (In)	197.9	54.2	96.3	20.9	22.2	24.4	49.6	38.6	21.0	DFT	28
	−18.2	6.5	−2.4	−2.9							
MIL-140A	49.6	90.5	48.0	0.7	4.0	16.7	20.6	23.8	14.46	DFT	28
	−0.9	2.0	−0.1	0.7							
MIL-53 (Al np)	121.3	4.59	96.05	3.96	24.85	4.85	16.38	3.74	−1.66	DFT	29
	0.79	−0.08	−7.86	0.42							
MIL-53 (Ga)	49.2	23.2	33.3	5.6	8.3	32.0	33.0	5.6	3.7	DFT	18
	0.9	1.7	0.2	0.7							
MIL-53 (Fe)	71.8	39.2	53.7	8.4	11.6	45.3	51.4	14.3	11.5	DFT	18
	1.1	−0.7	1.7	−1.3							
ZIF-1	8.3	4.1	4.6	1.7	1.6	1.2	1.9	2.3	1.25	DFT	20
	−0.9	−0.5	−0.9	0.04							
MIL-140A	94.0	163.0	42.7	3.2	9.1	27.4	42.7	29.6	17.4	DFT	30
	−5.0	0.3	−10.4	0.1							
MIL-140B	80.5	143.1	47.2	5.0	6.6	209	36.4	29.3	12.3	DFT	30
	11.7	4.6	2.1	6.4							
MIL-140C	64.0	129.2	32.6	2.4	4.3	18.4	20.2	17.7	12.6	DFT	30
	−10.5	−4.1	−2.7	−4.2							
MIL-140D	62.2	109.4	29.3	1.6	3.0	17.0	25.2	17.6	8.4	DFT	30
	7.2	2.8	0.1	3.6							

[a]Abbreviation: np = narrow pore.

Table 2.7 Elastic stiffness values of triclinic MOFs (in GPa).

Compound	C_{11} / C_{25} / C_{55}	C_{12} / C_{26} / C_{56}	C_{13} / C_{33} / C_{66}	C_{14} / C_{34}	C_{15} / C_{35}	C_{16} / C_{36}	C_{22} / C_{44}	C_{23} / C_{45}	C_{24} / C_{46}	Method	Reference
MIL-60	57.5	23.0	51.3	−12.4	7.7	20.2	74.2	29.9	1.34	DFT	18
	−8.2	16.8	105.5	−21.6	18.1	11.8	28.8	9.6	−7.6		
	32.9	1.3	25.0								
CAU-13	103.3	13.9	23.5	4.2	1.5	0.8	24.9	38.6	17.5	DFT	19
	0.5	7.31	114.9	34.8	−1.0	−2.4	32.8	3.2	0.5		
	13.2	1.61	10.53								

$$A = \frac{C_{44}}{C_{11} - C_{12}}. \tag{2.16}$$

This was refined and generalised for any crystal class by Ledbetter and Migliori[32] as

$$A^* = \frac{\nu_{\text{max}}^2}{\nu_{\text{min}}^2}, \tag{2.17}$$

where ν_{max} and ν_{min} are the maximum (fast) and minimum (slow) shear wave velocities, respectively. These velocities can be readily calculated from the elastic tensors by diagonalising Christoffel's equation.

Another formulation, the so-called universal anisotropy index, was proposed by Ranganathan and Ostoja-Starzewski[33] as a combination of Voigt and Reuss averages

$$A^{\text{U}} = 5\frac{G_V}{G_R} + \frac{K_V}{K_R} - 6. \tag{2.18}$$

In general, A^* and A^{U} give very similar results. A^* seems more fundamental, but A^{U} is slightly easier to calculate.

Finally, it is also possible to generate *ad hoc* anisotropy measures for each elastic property, for instance for the Young's modulus,

$$A^{\text{E}} = \frac{E_{\text{max}}}{E_{\text{min}}}. \tag{2.19}$$

2.2.4 Thermo-elasticity

The dimensions of a solid are dependent on its temperature. The primary properties describing this phenomenon are the coefficients of thermal expansion α_{ij} that linearly relate the strain tensor ε_{ij} to a temperature change ΔT as[1]

$$\varepsilon_{ij} = \alpha_{ij}\Delta T. \tag{2.20}$$

Like ε_{ij}, α_{ij} is a symmetric tensor, with at most six independent components, but for the higher symmetry isotropic and cubic systems, only one coefficient α_L is required and the volumetric coefficient of thermal expansion α_V is often preferred,

$$\alpha_V = \frac{1}{V}\left(\frac{\partial V}{\partial T}\right)_P = 3\alpha_L. \tag{2.21}$$

Most materials exhibit positive thermal expansion. This is explained by the anharmonicity of the bonds in the solid. Thermal excitations tend to increase the *average* length of the anharmonic bonds, as illustrated in Figure 2.4.

Thermal expansion is a property that originates from the dynamics of the crystal, and can be linked to the natural vibrations or phonons of the crystal through the mode Grüneisen parameters γ_i and their contributions c_i to the volumetric specific heat C_V, $C_V = \sum_i c_i$

$$\alpha_V(T) = \frac{1}{KV}\sum_i \gamma_i \cdot c_i = \frac{1}{KV}\gamma \cdot C_V. \tag{2.22}$$

γ is the mean Grüneisen parameter,

$$\gamma = \frac{\sum_i \gamma_i \cdot c_i}{\sum_i c_i}. \tag{2.23}$$

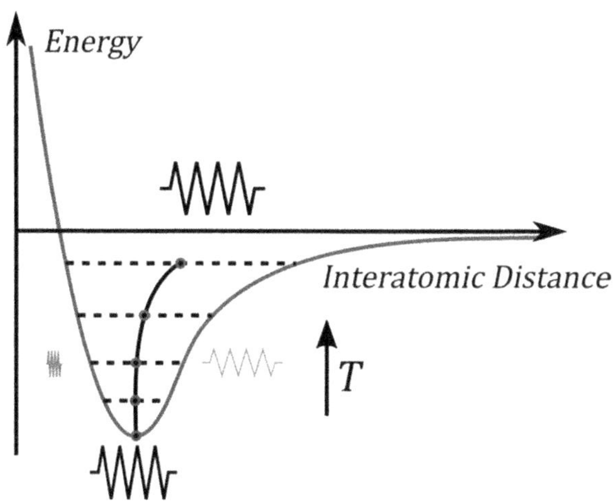

Figure 2.4 Atomistic origin of positive linear expansion. Anharmonicity leads to an increase in average bond length.

The mode Grüneisen parameters are dimensionless and represent the change in phonon frequency with volume change. They are defined by the negative of a logarithmic derivative:

$$\gamma_i = \frac{-V}{\omega_i}\frac{\partial \omega_i}{\partial V} = \frac{-\partial \ln \omega_i}{\partial \ln V}, \tag{2.24}$$

where ω_i is the phonon frequency.

The above exposition is very simplified, not as detailed or rigorous as for elasticity. Thermo-elasticity is complex and subtle, even in the isotropic case, and rigorous treatment of the Grüneisen theory of thermal expansion is beyond the scope of this chapter. For a more detailed but approachable primer, the interested reader can consult the review on the lattice dynamics theoretical foundation of negative thermal expansion (NTE) by Dove and Fang.[34]

The Grüneisen parameters are especially important in the context of NTE: if negative γ_i dominates the summation in eqn (2.22), then α_V must be negative, as all other terms in the equation are positive.

2.3 Anomalous Mechanical Behaviour

Conventional materials under load react in reliable and expected ways: a stretched rubber band becomes thinner, railway tracks expand under the summer sun, and a submarine's hull shrinks under hydrostatic pressure when diving deep. However, some materials behave in more unexpected ways: materials with a negative Poisson's ratio (NPR) become thicker when stretched (often in specific directions), materials with NTE shrink upon an increase in temperature and materials with negative linear or area compressibility (NLC or NAC) expand in a certain direction under hydrostatic pressure. These properties are unusual, but not exactly rare. NPR seems to be quite commonplace, and it has been shown that around a third of single crystals display this property, albeit in often very specific directions,[35] and it often tends to be washed out in macroscopic samples composed of grains oriented randomly. NLC and NPR are probably scarcer in general.

Several MOFs have been shown to exhibit negative elastic or thermo-elastic properties, often with large values. In general, MOFs are discussed within more generic reviews focussing on specific negative properties, NLC,[36] NTE and NPR. Moreover, Coudert and Evans[37] have focussed on MOFs and discussed all negative mechanical properties, among other anomalous behaviour.

In fact, so-called 'anomalous' behaviour is not really surprising as the main mechanisms responsible for the negative properties often rely on an imbalance between the high axial stiffnesses of the ligands and the low rotational stiffnesses of the secondary building unit (SBU)–ligand joints. The following sub-sections examine the MOFs in which the specific negative properties occur as well as the underpinning mechanisms.

2.3.1 Negative Linear Compressibility (NLC)

2.3.1.1 Definitions, Properties and Mechanisms of NLC

Compressibility describes the behaviour of a material or structure under pressure, which is either positive (compressive) or negative (tensile). Volumetric compressibility defines the volume change in a material under applied hydrostatic pressure, whilst linear compressibility describes a change along a specified axis. While most materials will shrink in all directions under hydrostatic pressure (isotropic compressive stress), in NLC materials the dimensions increase in certain directions. Only some dimensions can increase, certainly not all, as negative volume compressibility is thermodynamically not allowed. NLC is often accompanied by a corresponding large positive linear compressibility along other directions, as shown in Figure 2.5.

NLC was first identified in tellurium in 1922[38] but had been the subject of little scientific or industrial interest until 1998 when Baughman reviewed the available existing data for single crystals and identified 13 materials with this property.[39] A few other materials have been identified since, with additional crystals the subject of a review of NLC materials by Cairns and Goodwin[36] and common engineering materials in the meta-analysis of Miller *et al.*,[40] however, this property still appears to be quite rare.

Volumetric compressibility β_V is defined as the relative change in volume V due to hydrostatic pressure p as $\beta_V = -\frac{1}{V}\frac{dV}{dp}$, at constant temperature. It is an invariant of the compliance tensor S_{ij} that can be expressed in Voigt's notation as

$$\beta_V = (S_{11} + S_{22} + S_{33}) + 2(S_{12} + S_{23} + S_{31}). \tag{2.25}$$

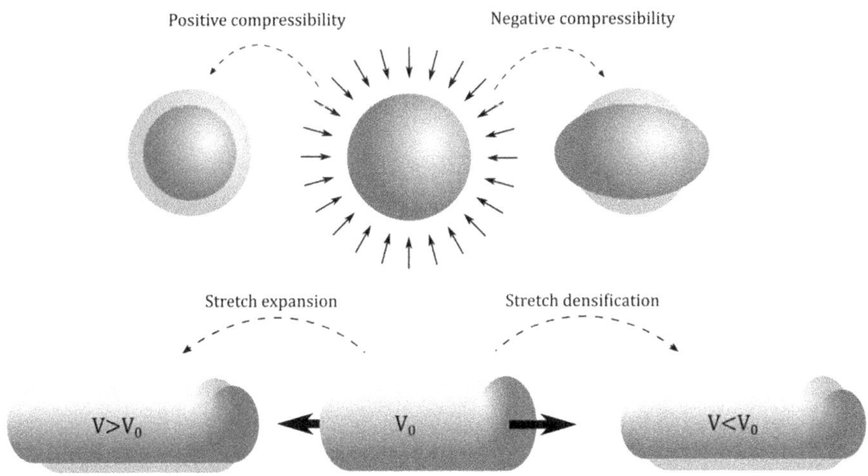

Figure 2.5 Pictorial representation of positive and negative linear compressibility, and of the corresponding stretch expansion and densification.

For isotropic materials, it can be related to Young's modulus E and Poisson's ratio ν by $\beta_V = \frac{3(1-2\nu)}{E}$. Its value varies from a few TPa^{-1} for a hard material such as diamond (2.1 TPa^{-1}) to a few hundred TPa^{-1} for a soft polymer such as polystyrene (330 TPa^{-1}), and up to a thousand or above for a very compliant foam.

The linear compressibility β_l is similarly defined, by $-\frac{1}{l}\frac{dl}{dp}$ or $-\frac{\varepsilon}{p}$ where l and dl are the part's dimension and elementary displacement in a given direction, and ε the corresponding strain (dl/l). Eqn (2.13) gives the general formula in tensor notation, but it can be illuminating to expand it in Voigt's notation for a few crystal symmetries.

For isotropic and cubic materials, the linear compressibility is a constant, independent of direction,

$$\beta_l = S_{11} + S_{12} + S_{13} = \beta_V/3. \tag{2.26}$$

For orthorhombic (orthotropic) materials, the linear compressibility for a given direction (unit vector **a**) can be derived from eqn (2.13) as

$$\begin{aligned} \beta_l &= \beta(\theta,\phi) = S_{ijkk}a_i a_j \\ \beta_l &= (S_{11}+S_{12}+S_{13})a_1^2 + (S_{21}+S_{22}+S_{23})a_2^2 + (S_{31}+S_{32}+S_{33})a_3^2. \end{aligned} \tag{2.27}$$

In direction 1 for instance, this can be simplified as $\beta_{l1} = (1 - \nu_{21} - \nu_{31})/E_1$, which suggests that NLC is likely associated with a large *positive* Poisson's ratio.

The combination of compliance tensor components that describe β_l can also be obtained from a different loading scenario/response combinations: instead of the linear response to a volumetric load, the volumetric response to a linear load. When conventional materials are stretched, their cross-section area shrinks but their volume increases, as they expand longitudinally more than they contract transversely, thus the relative volume change (also known as dilation, dV/V) is positive. This is not the case, however, if the Poisson's ratios are large enough that the material contracts transversely more than it expands longitudinally and thus loses volume, becoming *stretch-densified*, as illustrated in Figure 2.5.

In other words, the *coefficient of stretch-expansion* has the same spatial dependence as the linear compressibility. This is not just a curiosity: it might conceivably be simpler in some circumstances to apply an axial load and measure a volume change than apply a pressure and measure a linear deformation, especially for macroscopic samples. It might also be easier to visualise stretch-expansion to elucidate certain mechanisms of NLC and discover candidate materials and applications.

At present, two main procedures are used to identify NLC materials. The first one involves calculating the linear compressibility (LC) in any direction from a full set of elastic constants (Section 2.2.2) obtained from experiments or calculations. The second approach is more direct and involves measuring the changes in dimensions under pressure by X-ray or neutron diffraction, either using single-crystal or powder samples. Usually, only lattice dimensions are measured, and off-axis LC, which may be more extreme, might be missed. The first method, which can be referred to as elastic, is by essence limited to the elastic limit and to small strain. In addition, the direct method can reach very large strain, for instance up to 5% for $Ag_3[Co(CN)_6]$-II.[41] This also allows the identification of NLC at high pressure (see Section 4.6.2 of Chapter 4), occasionally for materials with conventional LC at standard pressure. Where the two methods have been applied (in $LaNbO_4$ and Se), the elastic and direct values for NLC tend to be comparable (see ref. 40 for a more detailed analysis).

It is not immediately apparent how to rank materials with NLC or what a good or even exceptional value for NLC actually is. For a while, there was a seemingly never-ending escalation in journal articles titles, with immodest adjectives such as 'giant' or even 'colossal'. However, a large absolute value for NLC primarily indicates that the material is very compliant in that direction, which is not especially surprising if the material is also very compliant in general, with a high β_V. Therefore, it is often desirable to also report the volumetric compressibility with values of NLC, or even to calculate a relative NLC, where the third of the volumetric compressibility is the natural reference ($\beta_V/3$ from the isotropic case).

Another subtlety arises from the fact that compressibilities can depend on the magnitude of applied pressure and that NLC often occurs within a limited pressure range. With this in mind, Cairns and Goodwin[36] introduced the concept of *compressibility capacity* χ_K, defined by the integral

$$\chi_K = -\int_0^{p_{\max}} \beta_{\mathrm{NLC}}(p)\,\mathrm{d}p. \tag{2.28}$$

This quantity is dimensionless and represents the fraction by which the material expands in the pressure range. In general, it is <1%, but can reach ~10% in particularly 'soft' materials with a high compliance.

Figure 2.6 relates, in the Ashby style, the (negative) linear compressibility of materials with the property as a function of the pressure range within which it exists. The compressibility capacity is also shown in diagonal bands. The bubble for molecular frameworks (which includes several MOFs) occupies the top left corner, *i.e.*, high absolute NLC and narrow pressure range. It also spans compressibility capacities between 0.1% and 10%.

Several mechanisms have been identified that explain NLC.[36,39] The most intuitive and direct form is topological in nature and based around some form of hinging mechanism. It is often referred to as the 'wine-rack' mechanism by analogy with popular forms of storage systems for bottled alcoholic beverages, as shown in Figure 2.7. The principle is simple: as the applied

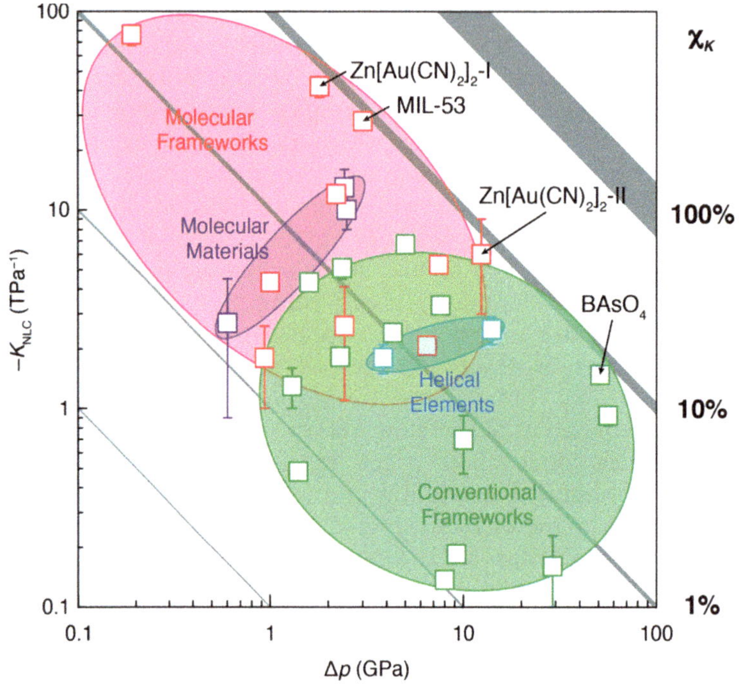

Figure 2.6 Ashby-type plot of NLC materials, linear compressibility (here denoted as K, which is equivalent to β) as a function of pressure range Δp. Adapted from ref. 36 with permission from the Royal Society of Chemistry.

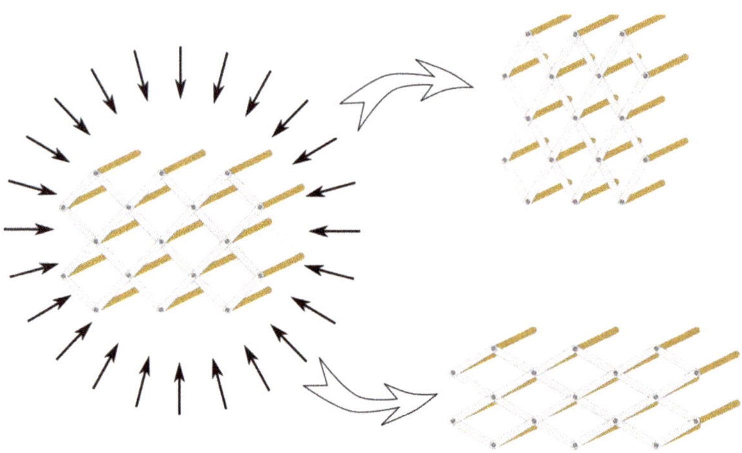

Figure 2.7 The wine rack mechanism for NLC subject to a hydrostatic pressure.

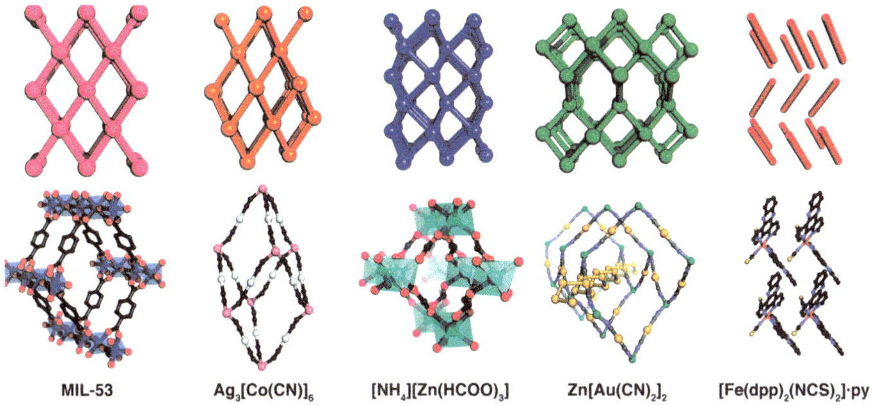

Figure 2.8 Wine-rack-like topologies responsible for NLC in four MOFs and one molecular framework. Reproduced from ref. 36 with permission from the Royal Society of Chemistry.

pressure reduces volume, squares become sharper diamonds with one diagonal expanding and the other diagonal shrinking. This principle does not require a square symmetry, nor a pseudo-2D structure, nor even continuous 'chains', just changes in direction. It is this mechanism that seems to be prominent in MOFs. Figure 2.8 displays five framework structures in which NLC occurs *via* a wine-rack-like mechanism.

Helical structures provide the other topological mechanisms for NLC: under pressure, the radius of the helix contracts, but as the curved length remains constant, the length of the helix must increase. Elemental selenium and tellurium have a helical structure and both display NLC. Some MOFs have a helical structure, and could thus be candidates for NLC.

Another kind of mechanism accounts for most of the non-wine-rack type NLC: phase transition. A pressure-induced phase transition can break the symmetry of a system, naturally decreasing two lattice parameters but increasing the remaining one. Most examples where this occurs are due to a ferroelastic phase transition, but a few more exotic phase transitions can also lead to NLC. A detailed discussion supported by several examples is given in the review of NLC materials by Cairns and Goodwin.[36]

2.3.1.2 NLC in MOFs

Several MOFs exhibit NLC, following the wine-rack mechanism. Table 2.8 lists examples of MOFs with NLC, and includes absolute and relative linear compressibilities, pressure range (when available) and compressibility capacity. In some cases, the value for the experimental linear compressibility is actually an average value over the pressure range and not the most extreme, which usually occurs at the lower pressure boundary. Comparisons can therefore be misleading.

Table 2.8 NLC properties of MOFs.[a]

Compound	β_{NLC}/ TPa^{-1}	$\beta_V/3$/ TPa^{-1}	Pressure range, Δp /GPa	$\frac{3\beta_{NLC}}{\beta_V}$ /%	χ_K/%	Method	Reference
MIL-53 (lp)	−28	—	0–3	—	8.4	PXRD	42
NH_2-MIL-53 (lp)	−28	—	0–2	—	8.4	PXRD	42
$Zn[Au(CN)_2]_2$-I	−42	20.0	0–1.8	−210	7.6		43
$Zn[Au(CN)_2]_2$-I	−52.1	24.5	0–2	−212	10.4	DFT	22
$Zn[Au(CN)_2]_2$-II	−6	12.3	1.8–14.2	−49	7.0		43
$KMn[Ag(CN)_2]_3$	−14.3		0.2–3.5	—	4.72	DFT	24
$Ag_3[Co(CN)_6]$-II	−5.3	28.2	0.19–7.65	−19	4.0		41
$KMn[Ag(CN)_2]_3$	−12	26.2	0–2.2	−46	2.64		44
MCF-34	−47.2	42.1	0–0.53	−112	0.17		45
$Ag_3[Co(CN)_6]$-I	−76	51.3	0–0.19	−148	1.44		41
Ag(mIm)	−4.32	55.6	0–1	−7.8	0.432		46
MIL-53(Al) lp	−257	—	—	—	N/A	DFT	25
MIL-53(Ga) lp	−1441	—	—	—	N/A	DFT	25
MIL-47	−201	—	—	—	N/A	DFT	25
DMOF-1 (loz)	−623	—	—	—	N/A	DFT	25
$KMn[Ag(CN)_2]_3$	−22.6	—	—	—	N/A	DFT (0 GPa)	24
ZAG-4	−2.6	24.9	1.65–5.69	−11	1.1		47
Cu(bpy)·SiF (no PTM)	−1.08	—	0–3.5		0.378	ADXRD	48
Cu(bpy)·SiF (in Ne)	−2.83	—	0–5		1.415	ADXRD	48
Cu(bpy)·SiF (in MEW)	−3.56	—	0–5		1.78	ADXRD	48
Cu(bpy)·SiF (in Ne)	−1.48	—	0–10		1.48	ADXRD	48
Cu(bpy)·SiF (in MEW)	−2.08	—	0–10		2.08	ADXRD	48
CuPyr-II	−5.3	N/A	2.23–4.90	N/A	1.06	SCXRD	49
$[Ag(en)]NO_3$-I	−28.4	9.2	0–0.97	−309	2.75	SCXRD	50
$[Ag(en)]NO_3$-II	−9.3	7.6	0.97–1.5	−123	0.49	SCXRD	50
MFM-133 (Hf)	−7.9	14.9	0–4.9	−53	3.87	SCXRD	51
MFM-133 (Zr)	−5.1	29.6	0–4.4	−17	2.24	SCXRD	51
$[NH_4][Zn(HCOO)_3]$	−1.8	10.2	0–0.93	−18	0.17		52

[a]Abbreviations: PXRD = powder X-ray diffraction; SXRD = single-crystal X-ray diffraction; ADXRD = angle dispersive X-ray diffraction; DFT = density functional theory; PTM = pressure transmitting media; MEW = methanol–ethanol–water.

Table 2.8 shows that NLC is not rare in MOFs, but that several aspects deserve careful consideration. The MFM-133 (Hf/Zr)[51] study illustrates the fact that NLC in MOFs can be tuned by varying the inorganic SBU. It is also very likely that tuneability could derive from changes in the ligand. A peculiarity of experiments at high pressure is that they require a hydrostatic medium, and further details of this are given in Chapter 4. Subsequently, the extent and range of NLC can depend on the medium used, as illustrated by a study on Cu(bpy)·SiF.[48]

2.3.2 Negative Thermal Expansion (NTE)

2.3.2.1 Definitions, Properties and Mechanisms for NTE

Like NLC, NTE was identified at the beginning of the 20th century in quartz and vitreous silica, and like NLC, it was also essentially ignored until the end of the 20th century. It has been shown to be quite a common property in many materials classes, including oxides and cyanides frameworks, zeolites and MOFs. Several reviews have discussed the occurrence and mechanisms of NTE.[34,53]

Like linear compressibility, thermal expansion is isotropic in cubic systems, but a very important difference is that it can be isotropically negative. This combination of high symmetry and negative properties is useful, as it would simplify design and likely be convenient for potential applications.

Unlike NLC where a fully negative volumetric compressibility β_V is impossible, a fully negative volumetric coefficient of thermal expansion α_V is not only possible but common. In fact, most of the research focus has been on negative *volumetric* thermal expansion, and the effect of anisotropy on NTE remains comparatively little studied.

For NTE to appear, eqn (2.22) must be dominated by phonons with negative Grüneisen parameter. Practically, this means the existence of low energy phonons (at THz frequency, see Section 1.7) that tend to close gaps between 'bonds'. The high-level representation of this phenomenon is often presented as variations of Figure 2.9(a); as the corner-sharing polyhedra vibrate, they close the gaps.

Different terminologies have been adopted to describe this effect. Some authors refer to it as a 'tension effect', where increased angular vibration in a relatively straight molecular fragment direction necessarily leads to 'length' reduction, especially if the linear vibrations are comparatively small, with the bonds acting as tensile ties. This is illustrated in Figure 2.9(b).

Another similar interpretation is based on the concept of rigid unit modes (RUMs) as shown in Figure 2.9(c) for $Zn(CN)_2$. Some molecular fragments are comparatively rigid, with stiff linear and angular bonds, but the rigid units can move relatively easily with respect to each other.

What all these somewhat phenomenological descriptions have in common are stiff atom–atom bonds, *i.e.*, with a steep energy function (see Figure 2.1 or Figure 2.4), but compliant bond for dihedral bending or rotations. This suggests a link between NTE and the somewhat nebulous concept of flexibility, which is discussed in Section 2.4.

Another important mechanism, based on phase transition, can account for NTE in some oxides and fluoride framework materials. The phase transition can be attributed to either a coupling between the magnetic degrees of freedom (dofs) and the lattice constants, or to a redistribution of valence electrons between metal centres.

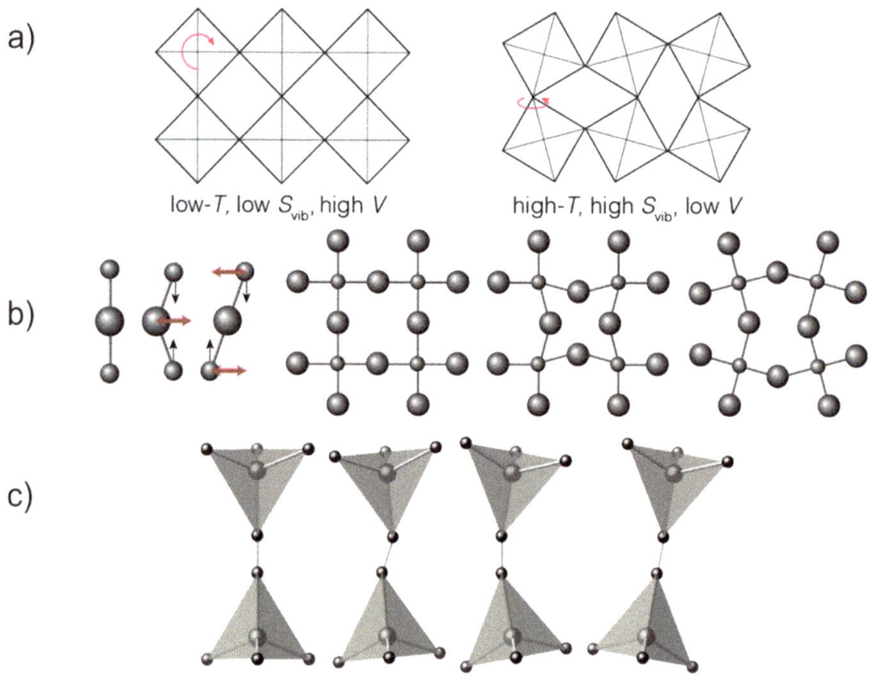

Figure 2.9 (a) Canonical mechanisms for NTE based on concerted rotation of corner-sharing polyhedra. Adapted from ref. 53 with permission from the Royal Society of Chemistry. (b) Tension effect, and (c) rigid unit mode of two adjacent tetrahedra. Adapted from ref. 34 with permission from the Institute of Physics, Copyright 2016.

At first, it would seem obvious as to how to rank materials with NTE, and to decide what a good or even exceptional value for NTE actually is, especially as isotropic NTE is possible. And yes, 'giant' and even 'colossal' have been applied to describe materials with large negative values of the coefficient of thermal expansion.

But another quantity matters; the larger the temperature range within which NTE occurs, the easier it would be to exploit it. With this in mind, and by analogy with the concept of compressibility capacity, Coates and Goodwin[53] introduced the concept of *negative thermal expansion capacity* χ_α, defined by the opposite of the product of the coefficient of thermal expansion by the temperature range,

$$\chi_\alpha = -\alpha_V \Delta T. \tag{2.29}$$

This quantity is dimensionless and represents the fraction by which the material expands in the temperature range. In general, it is <1%, but can reach ~2% in some materials. Theoretical calculations can predict larger values of χ_α, but the predicted range can be highly suspect if phase transitions

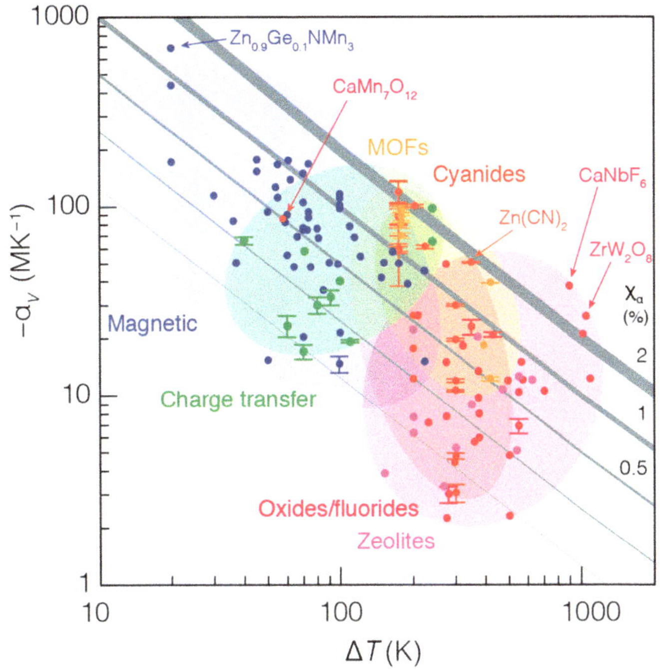

Figure 2.10 Ashby-type plot for NTE. It should be noted how the MOFs (in yellow) are all clustered with an identical value for ΔT. This limitation is due to the setup of the experimental apparatus. Reproduced from ref. 53 with permission from the Royal Society of Chemistry.

are not considered. Experimental ranges can also be unrealistic and limited by experimental setup.

The Ashby-style plot in Figure 2.10 demonstrates the interplay between the NTE properties.

2.3.2.2 NTE in MOFs

MOFs can be thought of as arrays of comparatively rigid SBUs (RUMs) linked by semi-rigid linkers through oftentimes flexible joints. Therefore, it is not surprising that either rotating polyhedra, tension effect or RUMs can very clearly explain NTE in MOFs. Figure 2.11 shows an interpretation of the SBUs of MOF-C22 as rigid units and displays one of the RUMs responsible for NTE. Concerted rotation of polyhedra, tension effect and RUMs can all be observed. It is very likely that this mechanism is responsible for the vast majority of occurrences of NTE in MOFs. However, it is unclear whether a mechanism based on phase transitions contributes to NTE in MOFs.

Table 2.9 collates the volumetric thermal expansion, the temperature range in which NTE occurs, and the NTE capacity for MOFs with NTE. As

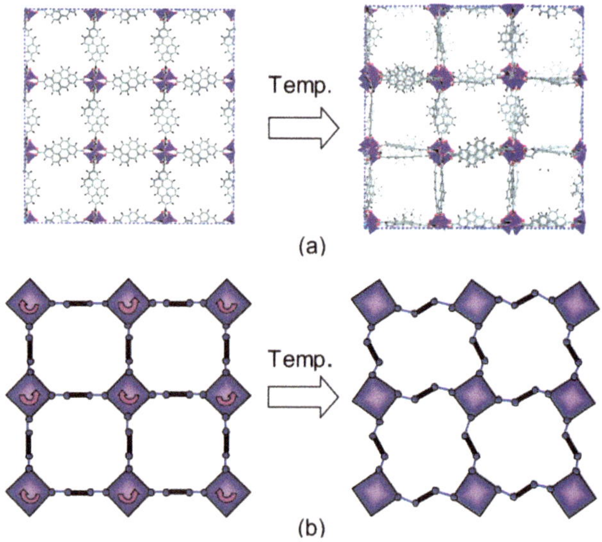

Figure 2.11 NTE mechanism in MOF-C22: (a) atomistic representation and (b) rigid unit representation. Reproduced from ref. 11 with permission from American Chemical Society, Copyright 2007.

Table 2.9 Volumetric NTE values of MOFs.[a]

Compound	$\alpha_V/10^{-6}$ K^{-1}	ΔT/K	χ_α/%	Method	Reference
IRMOF-16	−68.8	1000	6.88	MD	54
$Zn[Au(CN)_2]_2$	−6.5	1000	6.5	DFT	23
IRMOF-10	−59.2	1000	5.92	MD	54
IRMOF-1	−53.3	1000	5.33	MD	54
MOF with 2-butynediodate linkers	−42.2	590	2.49	MD	11
MOF-C22	−33.1	590	1.95	MD	11
MOF-C10	−32.5	590	1.92	MD	11
d-UiO-66(Hf)	−97	180	1.75	PXRD	55
MOF-C30	−29.2	590	1.72	MD	11
d-UiO-66(Hf)	−95	180	1.71	PXRD	55
d-UiO-66(Hf)	−94	180	1.68	PXRD	55
MOF-5	−39.3	420	1.65	PXRD	56
d-UiO-66(Hf)	−88	180	1.58	PXRD	55
d-UiO-66(Hf)	−81	180	1.45	PXRD	55
MOF-C16	−24.2	590	1.43	MD	11
MOF-C6	−23.9	590	1.41	MD	11
d-UiO-66(Hf)	−70	180	1.26	PXRD	55
MOF-14	−18.5	392	0.72	PXRD/NPD	57
HKUST-1 (Cu_3BTC_2)	−12.3	420	0.52	SCXRD/PXRD	58
HKUST-1 (Cu_3BTC_2)	−15.3	210	0.32	PXRD	59
COF-102	−4.53	550	0.25	MD	60
TCNQ@HKUST-1	−8.4	210	0.18	PXRD	59

[a]Abbreviations: NPD = neutron powder diffraction; TCNQ = 7,7,8,8-tetracyanoquinodimethane.

for NLC, in some cases, the value for the experimental coefficient of thermal expansion is actually an average value over the temperature range and not the most extreme, which usually occurs at the lower temperature boundary. Comparisons can therefore be misleading.

Although not strictly speaking part of the MOF family, cyanide-based framework materials share many similarities with MOFs and are some of the more extreme NTE materials. Coates and Goodwin listed 23 cyanide-based framework materials with NTE.[53] Table 2.9 includes the most interesting, and a few instances not covered in ref. 53. COF-102, a covalent organic framework, although not strictly a MOF, is included for comparison.

As with NLC, it is possible to fine-tune NTE in MOFs. Schneider *et al.*[59] used TCNQ to bridge the ligands in HKUST-1 and reduce its coefficient of thermal expansion from $-15.3 \times 10^{-6}\ \mathrm{K}^{-1}$ to $-8.4 \times 10^{-6}\ \mathrm{K}^{-1}$.

2.3.3 Negative Poisson's Ratio (NPR)

2.3.3.1 Definitions, Properties and Mechanisms for NPR

The Poisson's ratio ν is a dimensionless elastic property that relates transverse strain to longitudinal strain according to

$$\varepsilon_{\text{trans}} = -\nu\varepsilon_{\text{long}}. \tag{2.30}$$

For many isotropic materials, its value is around 0.3; this simply means that a material subjected to elongation in a given direction will contract by 30% of that elongation in all directions normal to the direction of loading. This is not surprising and expected in daily life; it is easy to notice that for instance an exercising band expands by a lot, but shrinks proportionally by not as much.

For anisotropic materials, the situation becomes much more complex and Poisson's ratio is now dependent not only on the load direction but also on the transverse direction, as illustrated in Figure 2.12.

Eqn (2.15) shows how the spatial dependence of Poisson's ratio can be obtained from the compliance matrix. The direct experimental determination of Poisson's ratio *for crystals* is not trivial as it depends on axially applied loads. In practice, this requires single crystals large enough to be clamped and put under tension or compression. Bending type experiments where the radii of curvatures are measured have been used for large scale macroscopic systems, but are difficult for MOFs for the same reasons as direct axial loading experiments: sizeable single crystals with a suitable morphology are required.

Indirect experimental measurements are more practical, but still difficult. In theory, Brillouin light scattering experiments can be used to measure the full elastic tensor of a crystal,[61] from which the spatial dependence of the Poisson's ratio can be determined. Unfortunately, the equipment needed is relatively rare these days, and additionally the sample preparation steps are also nontrivial. It has already been pointed out in Section 2.2 that, so far,

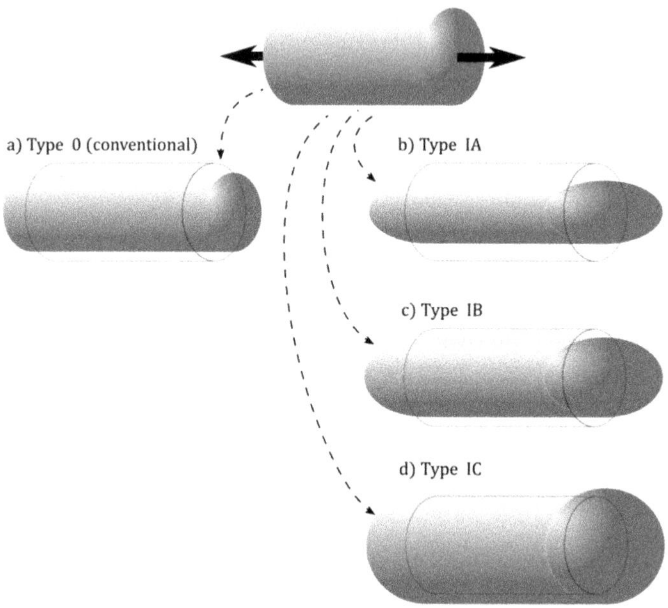

Figure 2.12 Various auxetic materials subjected to a tensile load. (a) Conventional behaviour with only positive Poisson's ratio (PPR). (b) The material meets the minimum requirements to be class 1A as this longitudinal direction exhibits at least some NPR. (c) The material meets the minimum requirements to be class 1B as the mean of the Poisson's ratios around this longitudinal direction is negative. (d) The material meets the minimum requirements to be class 1C as v is negative for all transverse directions.

the full elastic tensor has only been measured for a single MOF, ZIF-8.[7] All the other elastic tensors for MOFs have been calculated, from force field or first-principles methods such as DFT (see Chapter 3).

Materials with NPR are called ‘auxetic’, and the property is referred to as ‘auxeticity’. Auxeticity has been observed in many crystals, mostly through the re-examination of the Brillouin scattering data.[35,62] In fact around a third[35] of single crystals seem to have a negative Poisson’s ratio. Most commonly, an auxetic crystal exhibits NPR in a very narrow range of transverse directions, for a narrow range of longitudinal directions. Rare completely auxetic crystals have NPR for all transverse directions, for every longitudinal deformation. Other intermediate behaviour is possible.

Siddorn *et al.*[63] proposed a fine-grained typology to differentiate between materials with partial, average, or complete auxeticity. This scheme is summarised in Table 2.10.

When applied to 471 single crystals, where the elastic tensors have been obtained from a meta-analysis of experimental studies, the typology reveals that although 37.2% are of type 1A (partial auxeticity), only α-cristobalite displays extended auxeticity, but even then, only up to type 3B. No crystal shows

Table 2.10 Classifications of auxetic materials.

Code	Directions with negative Poisson's ratio (NPR)	Value
0	None	Minimum of the minimum v (positive)
1A	At least one transverse direction around at least one longitudinal direction	Minimum of the minimum v
1B	Negative average (over all transverse directions) around at least one longitudinal direction	Minimum of the average v
1C	All transverse direction around at least one longitudinal direction	Minimum of the maximum v
2iA	At least one transverse direction for more than half of longitudinal directions	Percentage of axes with at least one NPR transverse direction
2iB	Negative average (over all transverse directions) for more than half of longitudinal directions	Percentage of axes with negative average PR over all transverse directions
2iC	For all transverse directions for more than half of longitudinal directions	Percentage of axes with all NPR transverse direction
2iiA	At least one transverse direction for an average of longitudinal direction	Average of the minimum v
2iiB	Negative average (over all transverse directions) for an average of longitudinal direction	Average of the transverse average v
2iiC	Negative average (over all transverse directions) for an average of longitudinal direction	Average of the maximum v
3A	At least one transverse direction for all longitudinal directions	Maximum of the minimum v
3B	Negative average (over all transverse directions) for all longitudinal directions	Maximum of the transverse average v
3C	All transverse direction for all longitudinal directions	Maximum of the maximum v

complete auxeticity (type 3C). The occurrence of auxeticity, partial or average, does not appear to correlate with the crystal system, space group or even atomic structure; for instance, 57% of the BCC (body-centred cubic) metals are auxetic, but 43% are not.

When applied to a sample of all-silica zeolites, where the elastic tensors have been obtained from force fields (5 distinct force fields, 189 distinct zeolites) or DFT (121 distinct zeolites), the typology reveals that around 25% (from DFT, for force fields between 17% and 49%) of zeolites display partial auxeticity (type 1A). There is not much evidence of average auxeticity either, especially from DFT-calculated samples. It appears that as a class of materials, zeolites are not especially prone to auxeticity. There is an exception however, zeolite JST displays complete auxeticity (type 3C), and for all 6 models, suggesting a topological origin based on its underlying net.

More recently, Chibani and Coudert[64] applied the auxetic typology approach to the elastic tensors of 13 621 inorganic materials from the Materials Project Database. They showed that around 30% of these materials exhibit partial auxeticity, which is consistent with the proportions from ref. 35 and 63; confirming that partial auxeticity is really not that special.

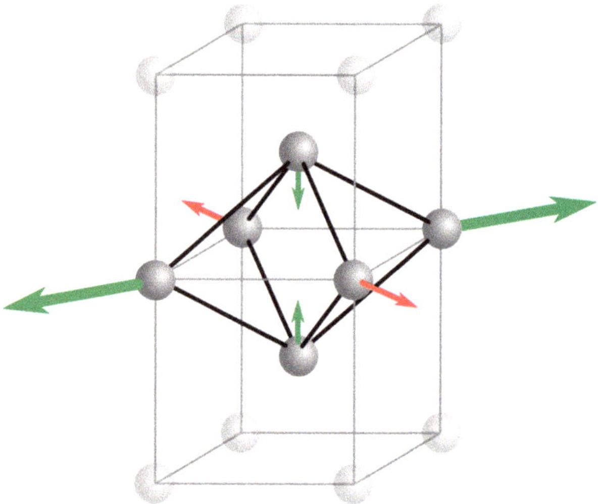

Figure 2.13 The octahedral 'mechanism' for NPR of a BCC unit cell. The pair of thick green arrows oriented along ⟨110⟩ denote the applied forces.

They also found 30 materials with complete auxeticity, but were unable to identify common structural motifs or mechanisms that could be related to their auxeticity.

A small number of thermodynamic, topological or dynamic mechanisms is sufficient to explain the occurrence of NLC and NTE: phase transitions, wine-rack-type topologies or soft phonons (THz modes in Section 1.7) with negative Grüneisen parameters. In contrast, no definitive general mechanisms have been identified for NPR yet. Most analyses rely on *ad hoc* geometrical interpretations. Consider the simple octahedral truss from Figure 2.13. Such an octahedron describes a BCC crystal system with atoms as nodes and bonds as edges. It is immediately apparent that a diagonal load (in the [110] direction) generates PPR (positive Poisson's ratio) in the [001] direction but NPR in the [$\bar{1}$10] direction. Yet, of all the BCC metals investigated, only 57% are auxetic: the detail of the interaction matters. It is therefore likely that some topological nets are capable of harbouring NLC, but that it might be washed out if for instance the balance between bond stretching and bond bending is unfavourable.[65]

2.3.3.2 NPR in MOFs

At the time of writing, there was no *experimental* observation of auxeticity in MOFs. ZIF-8, the only MOF where the elastic tensor has been determined experimentally,[7] is not auxetic, with a minimum Poisson's ratio of 0.33. ZIF-8 is an analogue of the zeolite sodalite, which is also non-auxetic. This non-auxeticity has been confirmed by force-fields[17] and DFT[7] calculations. ZIF-8 and sodalite are cubic, with a low elastic anisotropy (A = 0.73). Considering the link between elastic anisotropy and auxeticity,[35] it is not too surprising that ZIF-8 and sodalite are not auxetic. In addition, ZIF-8 and sodalite have a pseudo-FCC structure, and FCC frameworks do not generate NPR directly.[65,66]

Thus, the only examples of NPR in MOFs derive from calculated elastic tensors, either from force-fields or DFT models. Huang and Chen's review[67] of NPR in modern functional materials reproduces a table from ref. 25.

Table 2.11 lists MOFs with NPR. Most of the instances have been calculated using the ElAM code,[8] utilising the 'raw' values of elastic constants C_{ij}

Table 2.11 NPR of MOFs.

Compound	A*	ν_{min}	ν_{max}	Symmetry	Reference
HKUST-1	5.39	−0.31	1.21	Cubic	10
$Zn[Au(CN)_2]_2$	2.13	−0.54	0.99	Hexagonal	23
NOTT-400	31.93	−1.50	2.26	Tetragonal	18
CPF-1	14.02	−0.52	1.32	Tetragonal	18
Al(OH)	14.56	−0.08	0.91	Tetragonal	18
CAU-10-OCH_3	14.39	−0.98	1.23	Tetragonal	18
NOTT-300	41.33	−1.98	2.71	Tetragonal	19
NOTT-401	17.39	−0.32	1.01	Tetragonal	18
CAU-10-CH_3	8.89	−0.33	0.79	Tetragonal	18
CAU-8	11.42	−0.24	0.91	Tetragonal	18
CAU-10-Br	6.68	−0.23	0.77	Tetragonal	18
CAU-10-H	13.36	−0.21	0.88	Tetragonal	18
ZIF-3	13.89	−0.44	1.34	Tetragonal	20
$KMn[Ag(CN)_2]_3$	20.24	−0.62	1.71	Trigonal	24
DMOF-1(loz)	108.00	−1.73	3.14	Orthorhombic	25
MIL-47	216.00	−2.18	2.89	Orthorhombic	25
MIL-53(Al)	141.90	−2.40	2.98	Orthorhombic	25
MIL-53(Al)	118.00	−2.42	3.11	Orthorhombic	18
MIL-53(Ga)	32.22	−1.19	2.80	Orthorhombic	25
MIL-53(V)	48.24	−1.31	2.79	Orthorhombic	18
MIL-61(V)	8.13	−0.01	0.81	Orthorhombic	18
MIL-118(Al)	10.77	−0.41	1.56	Orthorhombic	18
ZIF-4	2.38	−0.11	0.46	Orthorhombic	21
ZnPurBr-LT	5.48	−0.36	0.95	Orthorhombic	27
ZNPurBr-HT	6.84	−0.62	0.80	Orthorhombic	27
MIL-122(Al)	3.84	−0.16	0.98	Monoclinic	18
MIL-122(Ga)	3.18	−0.11	0.98	Monoclinic	18
MIL-121(Al)	6.19	−0.16	0.91	Monoclinic	18
MIL-116(Al)	13.85	−0.37	1.28	Monoclinic	18
MIL-122 (In)	3.56	−0.16	0.99	Monoclinic	28
MIL-140A	36.96	−0.92	1.70	Monoclinic	28
MIL-53(Al) np	26.99	−0.94	3.61	Monoclinic	29
MIL-53(Ga)	115.20	−1.76	2.68	Monoclinic	18
MIL-53(Fe)	69.97	−1.08	2.08	Monoclinic	18
MIL-140A	12.54	−0.11	1.19	Monoclinic	30
MIL-140B	57.93	−1.08	1.21	Monoclinic	30
MIL-140C	26.01	−0.26	1.15	Monoclinic	30
MIL-140D	36.02	−0.58	1.34	Monoclinic	30
MIL-60	19.76	−0.36	1.27	Triclinic	18
CAU-13	77.73	−1.49	3.09	Triclinic	19

collated in Tables 2.1–2.7. It includes a column for elastic anisotropy, here chosen as A^* based on eqn (2.17),[32] which is readily calculated using the ElAM code.

Out of 19 datasets for 13 cubic MOFs (Table 2.1), only HKUST-1 is auxetic with $\nu_{min} = -0.31$. It appears quite anisotropic with $A^* = 5.39$, but 7 cubic MOFs with higher isotropy (up to 54) do not display auxeticity. The only two sets of elastic tensors in the database covering the hexagonal symmetry are both for the $Zn[Au(CN)_2]_2$ cyanide framework. At first, they look quite similar (see Table 2.3), but only one[23] of them has NPR at −0.56, while the other[22] has a very small positive minimum value of 0.01.

Almost all tetragonal MOFs from Table 2.2 have NPR, 11 out of the 12 of which the elastic tensor is available. The only trigonal crystal, $KMn[Ag(CN)_2]_3$, also appears auxetic with a minimum value of −0.62. Most orthorhombic MOFs have NPR, 11 out of 17, with the minimum values varying between −0.01 and −2.42. Almost all monoclinic MOFs have NPR, 13 out of 14, with values ranging from −0.11 to −1.76. Finally, both triclinic crystals from Table 2.7 are auxetic, with NPR values of −0.36 and −1.49. Although the sample set is still limited, this might suggest that NPR is prevalent in MOFs with low symmetry.

2.4 Framework Flexibility of MOFs

2.4.1 Which Kind of 'Flexibility' Is Relevant for MOFs? Material Flexibility *vs.* Framework Flexibility

The varied and often loosely defined concept of 'flexibility' permeates the previous sections, as well as much of the literature on the mechanical properties of MOFs addressed in this book, where it is alluded to that 'flexible' MOFs behave differently from 'rigid' MOFs and where it is suggested that flexibility is a necessary condition for the unusual mechanical properties discussed in Section 2.3.

In a general chemical context, flexibility is used to refer to the 'ability to move', or 'ability to deform'. This is made clear in the case of computational models where molecules interact with a surface or a porous material, and 'host flexibility' simply describes the fact that the components of the molecule are allowed to change positions while those of the surface or porous material (the host) are not. For a molecule, or molecular linker, conformational flexibility customarily describes the fact that the molecule can most easily change shape, with very low energy barriers, thanks to rotatable dihedral angles.

However, in a more specific 'materials' context, what precisely is flexibility? This is really far from clear. The interest in the flexibility of crystals seem to start with Pauling, who noted in his 1930 study of sodalite[68] that '*The framework, while strong, is not rigid, for there are no strong forces tending to hold it tautly expanded*'. In fact, much of the materials thinking on the subject of structural rigidity follows from the study of silicates, zeolites and glasses, and the concept seems most applicable to, or at least make most sense for, framework-type structures, including glasses: it would be rather meaningless, for instance, to assert that "metal A is more flexible than metal B".

It is perhaps useful to also consider what this material flexibility is not. Flexibility is certainly not a recognised materials property, in the way that stiffness, strength, hardness and toughness are. These properties are well defined and can be measured. Sure, authors without a materials science background often confuse them, and use, for instance, 'soft' when they actually mean 'compliant', or 'strong' when they mean actually 'tough', but misunderstandings often get clarified once the actual materials properties are mentioned (Young's modulus and fracture toughness in these examples).

Could we simply define material flexibility as the ability of a material to deform, following the chemical tradition? That would probably be unwise, as we already have the three uniquely defined elastic moduli for that, Young's, shear and bulk. However, the elastic moduli measure the ability of a material to resist *specific* loads/deformations: uniaxial for the Young modulus, shear for the shear modulus, and isotropic for the bulk modulus. Could flexibility be a measure of the ability of a material to deform, irrespective of the loading type? Again, probably not, as for a given material the elastic moduli tend to be in the same range anyway. A polymeric foam is not more flexible that a ceramic, it is just more compliant.

The obvious difficulty in defining material flexibility, coupled with the fact that it is only really used in relation to framework materials, strongly suggests that it is not possible to use material flexibility as a general, robust and uniquely defined materials property. However, this does not mean that the concept is not useful, and that *ad hoc* combinations of elastic properties cannot be used to estimate flexibility and to meaningfully compare materials. This is similar to measures of elastic anisotropy, which can be defined in many ways (see Section 2.2.3), with most giving comparable results.

The concept of *framework flexibility*, as opposed to *material flexibility*, is much more limited, but also perhaps more fundamental and more predictive. The context here is essentially mechanical in nature, with the underlying net providing the geometry on which a framework of beams, solid bodies and joints can be superimposed. The joints are very important; they can have differing dofs (hinges seem to be frequent for carboxylate linkers). Most of the 'chemical' aspects of bonding in MOFs can thus be ignored (most, because the nature of the joints derives in part from their chemistry). Figure 2.14 illustrates this idea for MOF-5. Following this mechanistic approach, framework flexibility simply describes the existence of deformation modes. Some mechanical frameworks can readily deform and are deemed flexible (they tend to be called a 'mechanism' in mechanical parlance); others cannot deform and are labelled rigid (often called trusses in the study of statics in structural engineering). This network flexibility is a purely binary property, and not a numerical one.

On the one hand, while *material flexibility* is ill defined, it is not too difficult to obtain numerical estimates from conventional modelling tools.

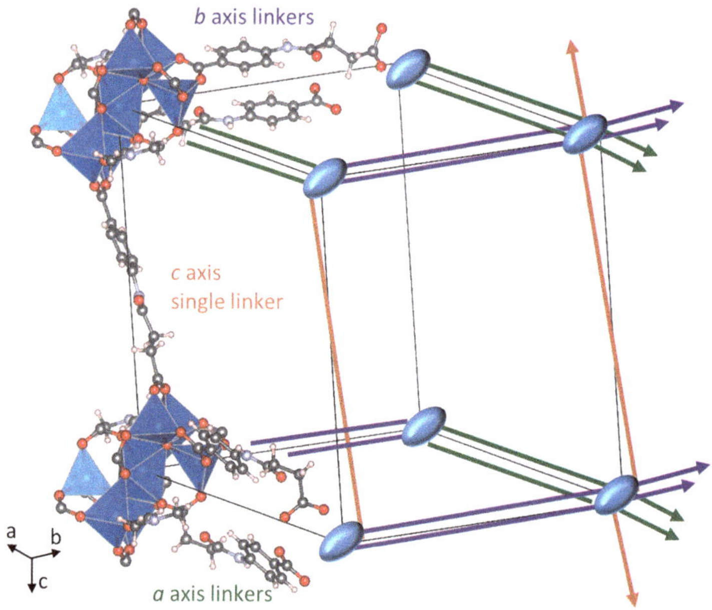

Figure 2.14 Chemical flexibility of MOF-5. Adapted from ref. 69, https://doi.org/10.1021/acs.cgd.9b00558, under the terms of the CC BY 4.0 license https://creativecommons.org/licenses/by/4.0/.

On the other hand, *framework flexibility* is well defined as a binary property, but its determination requires a mechanical type of analysis, one that is not implemented in chemical simulation tools. Fortunately, estimates of material flexibility correlate well with underlying framework flexibility.

2.4.2 Estimates for Material Flexibility

The first attempt to quantify MOF material flexibility appears in the work by Ortiz *et al.*[25] on the elastic anisotropy of the MIL family of MOFs. They suggested that high elastic anisotropy of the Young's modulus is a signature of what they call the structural flexibility of the crystal (framework flexibility in this chapter). They justify this by analogy with molecules, where low-frequency vibrational modes indicate conformational flexibility. The analogy is somewhat forced and indirect, but it makes sense that directions of ultra-low (or at least low compared to other directions) Young's modulus corresponds to the directions of loads that activate modes of framework flexibility, with low resistance (equivalent to a value of near-zero for the Young's modulus). Young's modulus anisotropy, $A^{\mathrm{E}} = E_{\mathrm{max}}/E_{\mathrm{min}}$, is simply calculated as the ratio of maximum value over minimum value. It is also shown that the similarly calculated shear modulus anisotropy, $A^{\mathrm{G}} = G_{\mathrm{max}}/G_{\mathrm{min}}$, can provide a very similar signature of structural (framework) flexibility or more precisely its rigidity. No cut-off values of these anisotropy ratios are suggested, but subsequent works have proposed that crystals with A^E above 100 should be considered flexible.[70] Whilst this approach is easy to implement, strictly speaking, it is just an *ad hoc* measure of elastic anisotropy and valid only for deformations with a small strain (*i.e.*, not exceeding the elastic limit).

Another approach, similar in principle as relying on elastic data, but closer in spirit to the low-frequency molecular vibrational analogy, is based on calculating the eigenvalues of the stiffness matrix: low values are also a signature of framework flexibility.[71] This approach is closely related to the Born stability criteria (further details in Section 3.4.4.1 of Chapter 3), where a necessary condition for a material to be stable is that its stiffness matrix is positive definite. The eigenvectors corresponding to the low stiffness eigenvalue are also indicative of the deformation modes, containing the normal and shear components.

2.4.3 Mathematical Methods for Framework Flexibility

The study of the rigidity/flexibility (also mobility in the mechanical/machine literature) of frameworks/structures is of interest in several disciplines: mathematics,[72–74] engineering,[75,76] chemistry[77–79] and biology.[80,81] Early engineers and architects evidently had some intuitive and practical grasp of the rigidity of structures or just relied on excessive redundancy. The origins of a more formal approach to the rigidity of structures can be traced back to very practical needs in civil engineering, and to the simple constraint counting rules that Maxwell[82] developed to predict whether trusses composed of two force members were flexible or rigid. Maxwell's original counting rule for bar-and-joint framework was later clarified by Calladine,[75] to include the concepts of mechanisms and states of self-stress into the equation

$$m - s = 3j - b - 6, \tag{2.31}$$

where m is the number of mechanisms, s the number of states of self-stress, b the number of bars, and j the number of joints.

The concept of counting rules has also been extended to body-and-bar systems in what is often referred to as the Chebychev, Grübler and Kutzbach (CGK) criterion,

$$m - s = 6(n-1) - 6g + \sum_{i=1}^{g} f_i \tag{2.32}$$

where m is the number of mechanisms, s the number of states of self-stress, n the number of bodies, g the number of joints, and the f_i are the degrees of freedom of each joint. Further details are given in Gogu's review[76] that contains an interesting historical section and provides alternative formulations and references.

2.4.3.1 Group Theoretical Approaches

Improved mathematical mechanics approaches to flexibility have initially been based on setting up the kinematic (compatibility) matrix and solving the corresponding eigenvalue problem to obtain the null-space (see for instance ref. 83–86 for applications to zeolites). These approaches are more rigorous than those based on energy optimisation of systems of 'springs' presented in Section 2.4.4, but algorithms and implementations have been less accessible, for historical reasons.

From these, Guest and Fowler have developed elegant procedures based on group theory that have been applied with success to mechanical[87,88] and chemical[89,90] problems. The main advantages of these schemes are that they are simple enough to be performed 'by hand', and general enough to include all sort of joints and be extended to periodic[90] systems.

Guest and Fowler's method replaces the scalar counting rules with a group theoretical equivalent. In the non-periodic case, body-and-joint case,

$$\Gamma(m) - \Gamma(s) = (\Gamma(v,C) - \Gamma_{\parallel}(e,C) - \Gamma_0) \times (\Gamma_T + \Gamma_R) + \Gamma_f. \tag{2.33}$$

The difference between the representations of mechanisms $\Gamma(m)$ and states of self-stress $\Gamma(s)$ can be obtained from the representations of the vertices $\Gamma(v,C)$, of the edges $\Gamma_{\parallel}(e,C)$ and of the dofs of the joints Γ_f as well as the identity Γ_0, translation Γ_T and rotation Γ_R representations. In practice, the framework is mapped into its contact polyhedron C, the point group of which is identified, and the characters for each symmetry and representation are determined and entered into a table of characters.

When eqn (2.33) is written as a sum of irreducible representations (Irreps), the positive Irreps correspond to mechanisms, and the negative Irreps correspond to states of self-stress. It is of course still possible that states of self-stress cancel out mechanisms, like for the scalar counting rules, but it is a bit less likely. The corresponding scalar counting rule can be read directly from the identity operation column.

The procedure is not especially onerous, but as often, examples are the best way to illustrate it; some can be found in Guest and Fowler's articles[87–90] and specifically for MOFs in the supplementary information of Marmier and Evans' study.[91] The determination of the characters for Γ_f is the most delicate task.

Marmier and Evans have applied the non-periodic version of this group theoretical method to three MOF cells displayed in Figure 2.15: a hypothetical **pcb** framework, and the open and close cells of MOF-5. The resulting group theoretical calculations for the **pcb** framework in the D_{4h} point group are given in Table 2.12.

By inspection of Figure 2.15, the **pcb** framework appears flexible, almost obviously so. And indeed, the irreducible representation of $\Gamma(m) - \Gamma(s)$ has one positive term that is not obscured by a state of self-stress. It is important to note that the corresponding scalar equation gives $m - s = -6$, suggesting rigidity. This is not surprising when considering that the scalar counting

Figure 2.15 Unit cells for **pcb** net (top), MOF-5 large-pore cell (bottom left) and MOF-5 small-pore cell (bottom right). The arrows in the MOF-5 models display the flexible mechanism. Adapted from ref. 91 with permission from the Royal Society of Chemistry.

Table 2.12 Character table of eqn (2.33) for a **pcb** framework in the D_{4h} group.

D_{4h}	E	$2C_4$	C_2	$2C_2'$	$2C_2''$	I	$2S_4$	σ_h	$2\sigma_v$	$2\sigma_d$
$\Gamma(v,C)$	10	2	2	0	0	—	—	—	—	—
$\Gamma_{\parallel}(\bar{e},C)$	12	0	0	−2	0	—	—	—	—	—
Γ_0	1	1	1	1	1	—	—	—	—	—
=	−3	1	1	1	−1	—	—	—	—	—
×										
$\Gamma_T + \Gamma_R$	6	2	−2	−2	−2	0	0	0	0	0
=	−18	2	−2	−2	2	0	0	0	0	0
+										
Γ_f	12	0	0	2	0	0	0	4	6	0
$= \Gamma(m) - \Gamma(s)$	−6	2	−2	0	2	0	0	4	6	0
	$\Gamma(m) - \Gamma(s) = [A_{1g}] - A_{2g} - B_{2g} - E_g - A_{1u} - 2B_{1u}$									

rules ignore the framework geometry: in the **pcb** case the fact that the bending arms are all of the same length. A similar framework, but with bending arms of differing lengths, would indeed be locked.

For MOF-5, $\Gamma(m) - \Gamma(s)$ equals $-T_{1g} - A_{1u} + [A_{2u}] - T_{2u}$ for the large pore framework (Figure 2.15, bottom-left) and $-A_{1g} + [A_{2g}] - T_{2g} - T_{1u}$ for the small pore framework (Figure 2.15, bottom-right). Both frameworks have the O_h point group, and each exhibits a deformation mechanism, obvious by inspection in the large pore case (rhombohedral deformation, along the cube diagonal), less so in the small pore case (contra-rotation of opposing sides, see Figures 2.15 and 2.19). But when the full unit cell of MOF-5 is considered (4 small pores and 4 large pores arranged alternately), it has the T_d point group and its $\Gamma(m) - \Gamma(s)$ is $-3A_1 - 3A_2 - 6E - 7T_1 - 7T_2$, without positive Irreps, which suggests the absence of mechanisms. The main limitation of this approach is that it can only offer proof of flexibility. Rigidity remains ambiguous, as it remains possible that a mechanism is obscured (subtracted) by a state of self-stress of the same symmetry.

We are not aware of published studies using the full periodic method to investigate the framework flexibility of MOFs, but the principles are the same as for non-periodic methods.

2.4.3.2 *Rigidity Theory*

Since the 1970s, mathematicians and materials scientists interested in glasses and amorphous materials have developed a formal, rigorous and theorem-based rigidity theory (also structural rigidity) as a branch of combinatorial graph theory that can be used to predict which parts of an ensemble formed of bars or rigid bodies connected by flexible hinges or linkages can bend or flex. Graver's short volume, 'Counting on Frameworks'[92] provides an accessible introduction to the topic albeit centred on tensegrity structures. 'Rigidity Theory and Applications',[93] edited by Thorpe and Duxbury, is also accessible, with chapters dedicated to molecular aspects of rigidity theory. 'Combinatorial Rigidity'[94] by Graver, Servatius and Servatius goes into more details and is often more abstract. But, all considered, the review article on rigidity theory applied to biomolecules by Hermans *et al.*[80] is lighter on abstract mathematics and relevant for MOFs.

Rigidity theory has evolved way beyond the simple Maxwell-style scalar counting rules and has developed into a generic theory with many contemporary research avenues and with complex applications into biology (protein folding) and engineering (swarm robotics[95]) to name a few. It concerns itself with frameworks, which in this context comprise both a graph and a realisation (a physical model, or a set of coordinates) and accommodate both bar-and-joint and body-and-joint models. The theory has also been extended to periodic graphs,[96] which is very promising for MOFs or other framework materials.

The theory can consider different types of rigidity: global, local, minimal, redundant, finite, infinitesimal and static. But properly defining these requires the kind of mathematics that goes way beyond the scope of this chapter.

Several algorithms have been developed to assess whether a graph or subgraph is (infinitesimally) rigid or flexible. The most advanced seem to be of the pebble game[97] variety. In the 2D bar-and-joint implementation, each node is given 2 pebbles (representing its dofs) and is considered free if it retains both pebbles. The algorithm then considers sequentially each edge: if it links two free nodes (with 2 pebbles), it is then allocated one (in general it does not matter which) of the 4 'free' pebbles. Pebbles can be shuffled through the graphs following certain rules. Redundant edges, those that link nodes for which it is impossible to find 2 pebbles are not allocated any pebbles. The remaining untethered pebbles define the number of dofs of the framework.

In general, pebble game algorithms do not work in 3D for bar-and-joint frameworks, but they do work for body-and-joint frameworks.

Software is available to carry out rigidity analysis of biomolecules, and there are no reasons why they could not be adapted for MOFs. The review by Hermans *et al.*[80] has a comprehensive list with descriptions, which we provide links for here:

- ProFlex/FIRST:[98] https://kuhnlab.natsci.msu.edu/software/proflex/
- CNAnalysis:[99] https://cpclab.uni-duesseldorf.de/cna/
- KINARI:[100] http://kinari.cs.umass.edu

2.4.4 Atomistic Methods for Framework Flexibility

The following methods follow a much more chemical tradition and eschew graphs and matroids for atomistic sites and harmonic potentials. They all highlight a certain tendency in computational chemistry to formulate problems that can be solved by tools such as energy optimisation and lattice dynamics for which algorithms are widely known and often already implemented in libraries and packages.

2.4.4.1 Rigid Unit Modes

The rigid unit modes (RUMs) method is another method with a strong chemical flavour, based on frameworks flexibility. It was designed in the 1990s to explore phonon spectra and displacive phase transitions in framework materials such as silicates and perovskites.[101–104] More recently, it has been applied to MOF-5[105] and several members of the ZIF family.[106] Figure 2.16 displays plausible rigid units for MOF-5, two possible descriptions for the Zn_4O_{13} secondary building unit (SBU) and three for the BDC (1,4-benzenedicarboxylate) ligands.

The implementation of the RUMs method is strikingly elegant and involves defining groups of atoms, the so-called rigid units, and splitting the atoms that link them (often oxygen atoms, red spheres in Figure 2.16) into two particles, which are joined by a stiff linear spring of zero equilibrium length, thus mimicking a revolute joint. The total energy can be expressed as a sum of harmonic potentials, easily accounting for periodicity, and all the tools available to more precise atomistic models as discussed in Chapter 3, such as lattice dynamics and molecular dynamics, are in principle also applicable.

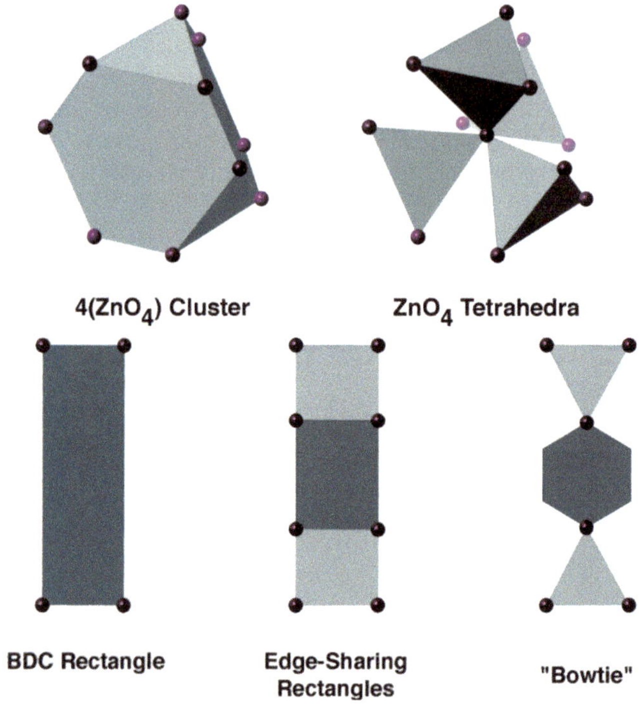

Figure 2.16 Flexibility models for the components of MOF-5 (top: secondary building units (SBUs), bottom: ligands). Adapted from ref. 105 with permission from the Royal Society of Chemistry.

The total energy is clearly meaningless, but relative energy changes can indicate which deformations are more or less favourable. The dynamic matrix is easily calculated and can be diagonalised for any wave-vector to obtain phonon dispersion curves.

It might be possible to implement this very simple but unusual RUM potential model into standard atomistic simulation codes, but the generation of the rigid units might be quite challenging. To the best of our knowledge, the only reported publications explicitly following the RUMs model have mostly used a dedicated software named CRUSH.[103,104]

CRUSH has mostly been used to determine the RUM phonon modes of zeolites Zr_2O_8 and $Zn(CN)_2$, but also of a few MOFs, the ubiquitous MOF-5[105] and four ZIFs,[106] specifically to understand which modes have a negative Grüneisen parameter and contribution to NTE. More specifically, the phonon spectra for the RUM models of Figure 2.16 (for MOF-5) are compared with a more realistic spectrum obtained from DFT; how well the NTE modes match reveals the type of motion that control NTE. Interestingly, while most of the NTE in MOF-5 is due to RUM modes, a non-negligible contribution is due to a mode that cannot be reproduced by any of the RUM models shown in Figure 2.16.

We are not aware of studies using the RUMs idea to specifically determine framework flexibility. On the one hand this is surprising, as software is available and automation does not seem overly complex. On the other hand, it might actually be even simpler to just use more realistic potential models (even at the DFT level) and materials databases with a well-established atomistic modelling code, obtain a whole raft of properties, and determine the framework flexibility using an estimate as described in Section 2.4.2. Such a study has been carried out for zeolites[107] using elastic anisotropy as a proxy. No such analysis has been carried for MOFs thus far.

CRUSH is available from the CCP14 website at

- CRUSH: http://ccp14.cryst.bbk.ac.uk/ccp/web-mirrors/crush/mineral_sciences/crush/

A novel group theoretical/algebraic approach to RUMs has been recently developed by Campbell *et al.*[108] It uses conditions of connectedness to construct a linear system of equations in the rotational symmetry-mode amplitudes. The procedure relies on some of the group-theoretical tools of the ISOTROPY software suite within which it has been implemented as ISOTILT.

- ISOTILT: https://iso.byu.edu/iso/

2.4.4.2 Template-based Geometric Simulation

Template-based geometric simulation[109] is a development and generalisation of the RUM ideas. It is based on ideal templates (regular polyhedral, or molecular fragments) superimposed onto an actual atomic configuration, or clusters. The algorithm first allocates the templates onto the relevant atomic clusters, locating (rotation being the difficult part) them in a way that minimises mismatches. In the likely situation where the initial configuration contains non-ideal clusters, some mismatch remains and the ideal templates are not joined. An iterative relaxation of the atomic position is then carried out, to reduce the mismatches. These ideas are implemented in the GASP software.[110]

This approach has mostly been used for zeolites, specifically to study the so-called flexibility window,[77] the range of density within which zeolites are realisable. Experimental zeolites exist toward the low density edge of this range, the pore being inflated by Coulombic repulsion, in contrast to most materials which tend to be as dense as local chemical and geometrical considerations allow.

Recent developments have made possible the study of MOFs by geometric simulations. In the pilot study,[110] the lattice parameter variation/optimisation facilities of GASP successfully predicted stress-free motion and therefore framework flexibility for MIL-47 and MIL-53. GASP was also used to study the flexibility and probe possible interconversion of three MOFs based on Zn and a tripodal tricarboxylic ligand (H_3**cbt**).[111] The GASP software is available at: GASP: https://github.com/EssayWells/GASP_6

2.4.4.3 Molecular Truss Models

Another numerical approach for checking the flexibility of carboxylate-based MOF[78] has been proposed by Sarkisov *et al.*[78] (note that all the interesting details are in the supporting information material). A specialised algorithm simplifies the molecular structure and creates rigid bonds (in practice stiff harmonic springs) between metallic, carbon and oxygen atoms, subject to various cut-off distances. In some cases, for instance, to reproduce the trimesic acid linkers in HKUST-1, 'ghost atoms' are added (Figure 2.17, bottom right). This procedure leads to truss structures, as seen in Figure 2.17, containing quasi-rigid SBUs and ligands, but also free hinges. Thirteen unit-cell perturbations (1 bulk, 6 uniaxial, 3 shear, 3 biaxial) are quantified by an order parameter whose value is 1.0 at equilibrium. These perturbations are applied with a range of order parameters spanning the equilibrium (from 0.9 to 1.1), and the system is relaxed using potential energy minimisation techniques, leading to increases in energy.

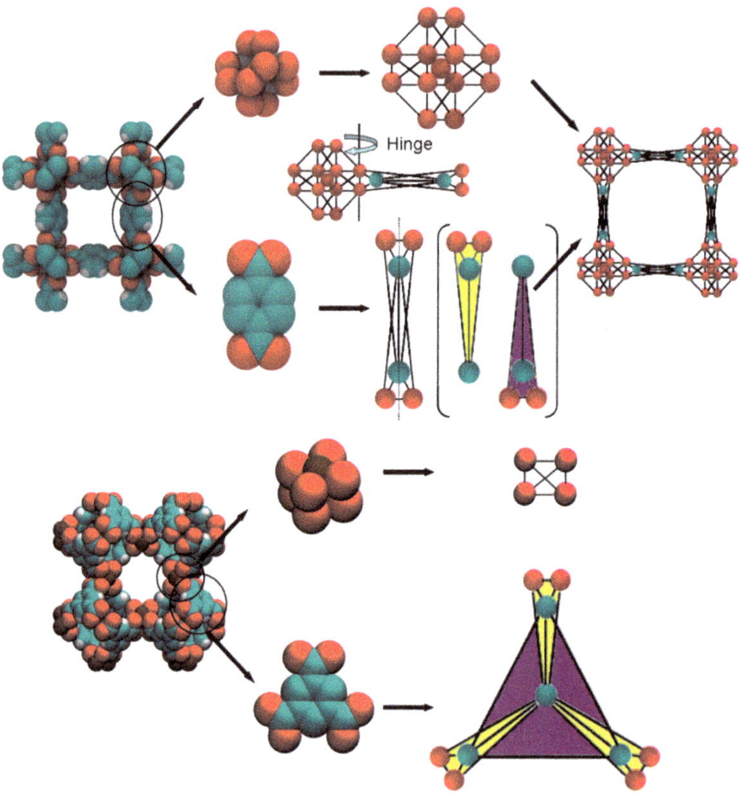

Figure 2.17 Molecular truss models of MOF-5 (top) and HKUST-1 (bottom). Adapted from ref. 78 with permission from American Chemistry Society, Copyright 2014.

A rigid framework is defined as one which experiences high energy variation ΔE (termed a 'penalty' in the article) for all perturbations, while a flexible framework has a very flat energy curve for at least one deformation mode. The authors provide a criterion of $\Delta E < 0.01$ for this, but it is based on an arbitrary 'stiffness constant' and is not generalisable.

Using this approach, 4 out of 10 experimental MOF structures display flexibility (MIL-47, MIL-53, MIL-88 and DYB), although for one perturbation only, and 7 out of 13 hypothetical MOFs display flexibility, often for several perturbations.

2.4.5 Macro-mechanical Models

In the same article,[78] Sarkisov *et al.* also built several very interesting physical mechanical models using wooden components for the SBUs and ligands and metallic hinges for the hinges. For obvious reasons, periodic structures are not possible, but this very tactile method can demonstrate very nicely the flexibility of unit cells, as shown in Figure 2.18.

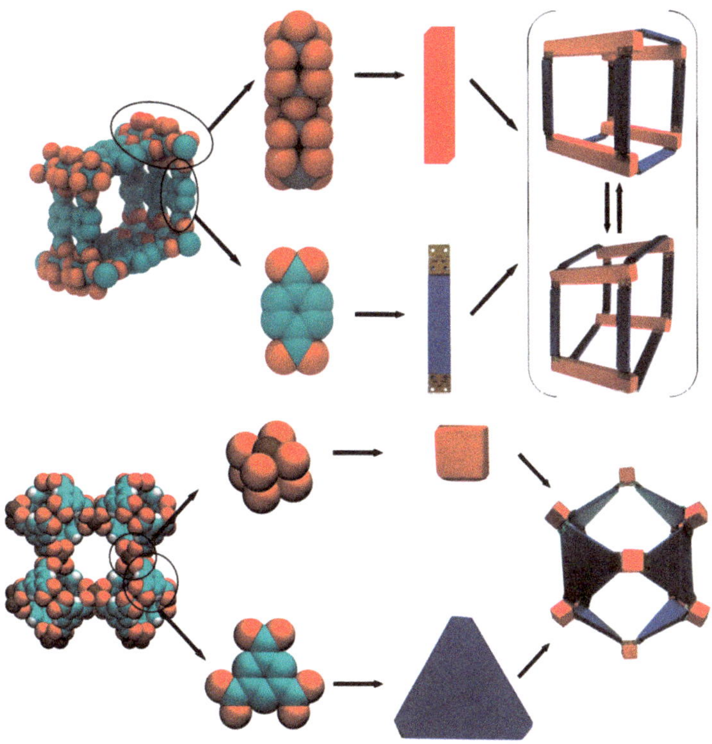

Figure 2.18 Macro-mechanical models for MOF-5 (top) and HKUST-1 (bottom). Adapted from ref. 78 with permission from American Chemistry Society, Copyright 2014.

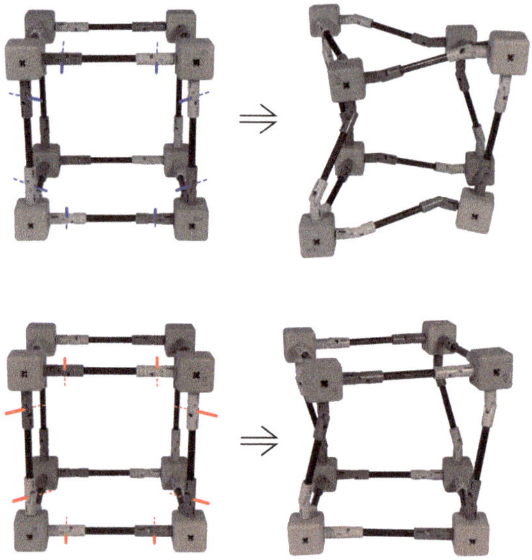

Figure 2.19 Macro-mechanical models for MOF-5, large pore cell (top) and small-pore cell (bottom). Notice how the orientation of the hinge differs between the small-pore and large-pore cell configurations.

This approach is being partially replicated using standard LEGO® Technics components for the ligands and crucially the joints, coupled to custom 3D printed 'SBUs'. Figure 2.19 showcases the LEGO® models and their deformation modes for the small-cell and large-cell models of MOF-5.

Unfortunately, the available hinge components all have drawbacks. The initial models used universal joints (LEGO® ID 61903) locked-in into hinge motion in a single plane with epoxy adhesive, but they proved too compliant with no neutral position and too fragile. The second iteration used ratcheted 'click' hinges (ID 30552/30553) which were very good for the neutral/straight positions, but with only nine angular positions available made deformations stiff and over constrained, as can be seen in Figure 2.19. For the third iteration, we are developing 3D printed flexible hinge joints for a future study.

2.5 Conclusions and Outlook

Many MOFs seem to display anomalous elastic and thermo-elastic properties.

The mechanisms for NLC are essentially well understood and are based on the interplay between rigid, stiff building blocks and compliant joints flexing in 'wine-rack' motions or analogues thereof. The mechanisms for NTE are also well understood, and are also based on similar, but dynamic, motions of rigid building blocks.

NPR still seems different. The meta-analysis of Section 2.3.3.2 implies that NPR is very common in MOFs with symmetry lower than cubic. It will be interesting to expand this preliminary study to ascertain the type of auxeticity, investigate the links between auxeticity, anisotropy and flexibility, and identify the underlying mechanisms. Based on past experience with zeolites, it is unlikely that a unifying mechanism will emerge. Moreover, it is possible that new mechanisms based on new geometries will be discovered.

The relationship between framework flexibility and unusual properties is not entirely clear yet, beyond the obvious that the existence of flexible mechanisms allows larger framework deformation (whether due to pressure, temperature or directional load). It is entirely possible that rigid frameworks can still generate unusual, negative elastic properties, especially NPR. After all, many dense cubic metals are auxetic.

Tables 2.1–2.7 list the elastic constants for more than 67 MOF samples (from 56 distinct MOFs), but there is little doubt that these tables are already, if not obsolete, at least incomplete. New MOFs are synthesised at a fast rate, and improved computational tools, chiefly DFT packages, currently make the determination of their elastic properties relatively straightforward. The availability of elastic tensor exploration tools has also made the identification of unusual elastic properties painless. Experimental validations of the computed elastic constants, however, are not so straightforward.

Besides these serendipitous bottom-up approaches, the emergence of more systematic methods, either based on brute force atomistic modelling[63,64,107] or on machine learning,[112,113] promises to speed up the discovery of more MOFs with unusual elastic properties. Another challenge lies in the translation of these theoretically determined properties to yield functional materials with real-world applications.

Acknowledgements

AM would like to thank Siti Raudah Binti Nasruddin who designed and printed the LEGO® compatible cubic SBUs of Figure 2.19 as part of her Masters Thesis.

References

1. J. F. Nye, *Physical Properties of Crystals*, Clarendon Press, Oxford, 1985.
2. T. C. T. Ting, *Anisotropic Elasticity*, Oxford University Press, New York, 1996.
3. W. Voigt, *Lehrbuch der Kristallphysik*, Teuber, Leipzig, 1928.
4. J. Mandel, *Int. J. Solids Struct.*, 1965, **1**, 273.
5. A. Reuss, *Z. Angew. Math. Mech.*, 1929, **9**, 49.
6. R. Hill, *Proc. Phys. Soc., London*, 1952, **65**, 349.
7. J.-C. Tan, B. Civalleri, C.-C. Lin, L. Valenzano, R. Galvelis, P.-F. Chen, T. D. Bennett, C. Mellot-Draznieks, C. M. Zicovich-Wilson and A. K. Cheetham, *Phys. Rev. Lett.*, 2012, **108**, 095502.

8. A. Marmier, Z. A. D. Lethbridge, R. I. Walton, C. W. Smith, S. C. Parker and K. E. Evans, *Comput. Phys. Commun.*, 2010, **181**, 2102.
9. R. Gaillac, P. Pullumbi and F.-X. Coudert, *J. Phys.: Condens. Matter*, 2016, **28**, 275201.
10. M. R. Ryder, B. Civalleri, G. Cinque and J.-C. Tan, *CrystEngComm*, 2016, **18**, 4303.
11. S. S. Han and W. A. Goddard, *J. Phys. Chem. C*, 2007, **111**, 15185.
12. K. Banlusan and A. Strachan, *J. Chem. Phys.*, 2017, **146**, 184705.
13. M. Mattesini, J. Soler and F. Yndurain, *Phys. Rev. B*, 2006, **73**, 094111.
14. A. Samanta, T. Furuta and J. Li, *J. Chem. Phys.*, 2006, **125**, 084714.
15. W. Zhou and T. Yildirim, *Phys. Rev. B*, 2006, **74**, 180301.
16. D. F. Bahr, J. A. Reid, W. M. Mook, C. A. Bauer, R. Stumpf, A. J. Skulan, N. R. Moody, B. A. Simmons, M. M. Shindel and M. D. Allendorf, *Phys. Rev. B*, 2007, **76**, 184106.
17. A. U. Ortiz, A. Boutin, A. H. Fuchs and F.-X. Coudert, *J. Phys. Chem. Lett.*, 2013, **4**, 1861.
18. E. V. Alexandrov, A. V. Goltsev, R. A. Eremin and V. A. Blatov, *J. Phys. Chem. C*, 2019, **123**, 24651.
19. A. U. Ortiz, A. Boutin and F. X. Coudert, *Chem. Commun.*, 2014, **50**, 5867.
20. M. R. Ryder and J.-C. Tan, *Dalton Trans.*, 2016, **45**, 4154.
21. J.-C. Tan, B. Civalleri, A. Erba and E. Albanese, *CrystEngComm*, 2015, **17**, 375.
22. M. K. Gupta, B. Singh, R. Mittal, M. Zbiri, A. B. Cairns, A. L. Goodwin, H. Schober and S. L. Chaplot, *Phys. Rev. B*, 2017, **96**, 214303.
23. L. Wang, H. Luo, S. Deng, Y. Sun and C. Wang, *Inorg. Chem.*, 2017, **56**, 15101.
24. K. Kamali, C. Ravi, T. Ravindran, R. Sarguna, T. Sairam and G. Kaur, *J. Phys. Chem. C*, 2013, **117**, 25704.
25. A. U. Ortiz, A. Boutin, A. H. Fuchs and F.-X. Coudert, *Phys. Rev. Lett.*, 2012, **109**, 195502.
26. V. Saleev and A. Shipilova, *Russ. J. Phys. Chem. A*, 2018, **92**, 1940–1946.
27. N. Li, K. Blankson, Y. Yang, P. Zhang and X. Zhao, *APL Mater.*, 2020, **8**, 111101.
28. A. Ortiz, PhD Thesis, Université Pierre et Marie Curie-Paris VI, 2014.
29. M. Wang, X. Zhang, Y. Chen and D. Li, *J. Phys. Chem. C*, 2016, **120**, 5059.
30. M. R. Ryder, B. Civalleri and J.-C. Tan, *Phys. Chem. Chem. Phys.*, 2016, **18**, 9079.
31. C. Zener, *Elasticity and Anelasticity of Metals*, University of Chicago Press Chicago, 1948.
32. H. Ledbetter and A. Migliori, *J. Appl. Phys.*, 2006, **100**, 063516.
33. S. I. Ranganathan and M. Ostoja-Starzewski, *Phys. Rev. Lett.*, 2008, **101**, 055504.
34. M. T. Dove and H. Fang, *Rep. Prog. Phys.*, 2016, **79**, 066503.
35. Z. A. D. Lethbridge, R. I. Walton, A. S. H. Marmier, C. W. Smith and K. E. Evans, *Acta Mater.*, 2016, **58**, 6444.

36. A. B. Cairns and A. L. Goodwin, *Phys. Chem. Chem. Phys.*, 2015, **17**, 20449.
37. F.-X. Coudert and J. D. Evans, *Coord. Chem. Rev.*, 2019, **388**, 48.
38. P. W. Bridgman, *Proc. Natl. Acad. Sci. U. S. A.*, 1922, **8**, 361.
39. R. H. Baughman, S. Stafstrom, C. X. Cui and S. O. Dantas, *Science*, 1998, **279**, 1522.
40. W. Miller, K. E. Evans and A. Marmier, *Appl. Phys. Lett.*, 2015, **106**, 231903.
41. A. L. Goodwin, D. A. Keen and M. G. Tucker, *Proc. Natl. Acad. Sci. U. S. A.*, 2008, **105**, 18708.
42. P. Serra-Crespo, A. Dikhtiarenko, E. Stavitski, J. Juan-Alcañiz, F. Kapteijn, F.-X. Coudert and J. Gascon, *CrystEngComm*, 2015, **17**, 276.
43. A. B. Cairns, J. Catafesta, C. Levelut, J. Rouquette, A. van der Lee, L. Peters, A. L. Thompson, V. Dmitriev, J. Haines and A. L. Goodwin, *Nat. Mater.*, 2013, **12**, 212.
44. A. B. Cairns, A. L. Thompson, M. G. Tucker, J. Haines and A. L. Goodwin, *J. Am. Chem. Soc.*, 2012, **134**, 4454.
45. Q. Zeng, K. Wang and B. Zou, *ACS Mater. Lett.*, 2020, **2**, 291.
46. J. M. Ogborn, I. E. Collings, S. A. Moggach, A. L. Thompson and A. L. Goodwin, *Chem. Sci.*, 2012, **3**, 3011.
47. K. J. Gagnon, C. M. Beavers and A. Clearfield, *J. Am. Chem. Soc.*, 2013, **135**, 1252.
48. Z. Chen, B. Xu, Q. Li, Y. Meng, Z. Quan and B. Zou, *Inorg. Chem.*, 2020, **59**, 1715.
49. C. J. McMonagle, P. Comar, G. S. Nichol, D. R. Allan, J. González, J. A. Barreda-Argüeso, F. Rodríguez, R. Valiente, G. F. Turner and E. K. Brechin, *Chem. Sci.*, 2020, **11**, 8793.
50. W. Cai and A. Katrusiak, *Nat. Commun.*, 2014, **5**, 1.
51. Y. Yan, A. E. O'Connor, G. Kanthasamy, G. Atkinson, D. R. Allan, A. J. Blake and M. Schröder, *J. Am. Chem. Soc.*, 2018, **140**, 3952.
52. W. Li, M. R. Probert, M. Kosa, T. D. Bennett, A. Thirumurugan, R. P. Burwood, M. Parinello, J. A. Howard and A. K. Cheetham, *J. Am. Chem. Soc.*, 2012, **134**, 11940.
53. C. S. Coates and A. L. Goodwin, *Mater. Horiz.*, 2019, **6**, 211.
54. D. Dubbeldam, K. S. Walton, D. E. Ellis and R. Q. Snurr, *Angew. Chem.*, 2007, **119**, 4580.
55. M. J. Cliffe, J. A. Hill, C. A. Murray, F.-X. Coudert and A. L. Goodwin, *Phys. Chem. Chem. Phys.*, 2015, **17**, 11586.
56. N. Lock, Y. Wu, M. Christensen, L. J. Cameron, V. K. Peterson, A. J. Bridgeman, C. J. Kepert and B. B. Iversen, *J. Phys. Chem. C*, 2010, **114**, 16181.
57. Y. Wu, V. K. Peterson, E. Luks, T. A. Darwish and C. J. Kepert, *Angew. Chem.*, 2014, **126**, 5275.
58. Y. Wu, A. Kobayashi, G. J. Halder, V. K. Peterson, K. W. Chapman, N. Lock, P. D. Southon and C. J. Kepert, *Angew. Chem., Int. Ed.*, 2008, **47**, 8929.

59. C. Schneider, D. Bodesheim, M. G. Ehrenreich, V. Crocellà, J. Mink, R. A. Fischer, K. T. Butler and G. Kieslich, *J. Am. Chem. Soc.*, 2019, **141**, 10504.
60. L. Zhao and C. Zhong, *J. Phys. Chem. C*, 2009, **113**, 16860.
61. S. Speziale, H. Marquardt and T. S. Duffy, *Rev. Mineral. Geochem.*, 2014, **78**, 543.
62. R. H. Baughman, J. M. Shacklette, A. A. Zakhidov and S. Stafstrom, *Nature*, 1998, **392**, 362.
63. M. Siddorn, F.-X. Coudert, K. E. Evans and A. Marmier, *Phys. Chem. Chem. Phys.*, 2015, **17**, 17927.
64. S. Chibani and F.-X. Coudert, *Chem. Sci.*, 2019, **10**, 8589.
65. T. P. Hughes, A. Marmier and K. E. Evans, *Int. J. Solids Struct.*, 2010, **47**, 1469.
66. T. P. Hughes, PhD thesis, University of Exeter, 2012.
67. C. Huang and L. Chen, *Adv. Mater.*, 2016, **28**, 8079.
68. L. Pauling, *Z. Kristallogr.*, 1930, **74**, 213.
69. E. J. Carrington, R. Pétuya, R. K. Hylton, Y. Yan, D. Antypov, G. R. Darling, M. S. Dyer, N. G. Berry, A. P. Katsoulidis and M. J. Rosseinsky, *Cryst. Growth Des.*, 2019, **19**, 5604.
70. S. M. Rogge, M. Waroquier and V. Van Speybroeck, *Acc. Chem. Res.*, 2018, **51**, 138.
71. A. U. Ortiz, A. Boutin, A. H. Fuchs and F. X. Coudert, *J. Chem. Phys.*, 2013, **138**, 174703.
72. C. S. Borcea and I. Streinu, *Proc. R. Soc. A*, 2010, **466**, 2633.
73. C. Borcea, I. Streinu and S. Tanigawa, *SIAM J. Discrete Math.*, 2015, **29**, 93.
74. S. C. Power, *N. Y. J. Math.*, 2014, **20**, 665.
75. C. Calladine, *Int. J. Solids Struct.*, 1978, **14**, 161.
76. G. Gogu, *Mech. Mach. Theory*, 2005, **40**, 1068.
77. A. Sartbaeva, S. A. Wells, M. M. J. Treacy and M. F. Thorpe, *Nat. Mater.*, 2006, **5**, 962.
78. L. Sarkisov, R. L. Martin, M. Haranczyk and B. Smit, *J. Am. Chem. Soc.*, 2014, **136**, 2228.
79. K. D. Hammonds, V. Heine and M. T. Dove, *J. Phys. Chem. B*, 1998, **102**, 1759.
80. S. M. Hermans, C. Pfleger, C. Nutschel, C. A. Hanke and H. Gohlke, *Wiley Interdiscip. Rev.: Comput. Mol. Sci.*, 2017, 7, e1311.
81. M. E. Gáspár and P. Csermely, *Briefings Funct. Genomics*, 2012, **11**, 443.
82. J. C. Maxwell, *Lond. Edinb. Dublin Philos. Mag. J. Sci.*, 1864, **27**, 294–299.
83. S. Pellegrino and C. R. Calladine, *Int. J. Solids Struct.*, 1986, **22**, 409.
84. S. D. Guest and J. W. Hutchinson, *J. Mech. Phys. Solids*, 2003, **51**, 383–391.

85. V. Kapko, C. Dawson, M. M. J. Treacy and M. F. Thorpe, *Phys. Chem. Chem. Phys.*, 2010, **12**, 8531.
86. V. Kapko, C. Dawson, I. Rivin and M. M. J. Treacy, *Phys. Rev. Lett.*, 2011, **107**, 164304.
87. S. D. Guest and P. W. Fowler, *Mech. Mach. Theory*, 2005, **40**, 1002.
88. P. W. Fowler, S. D. Guest and T. Tarnai, *Symmetry*, 2014, **6**, 368.
89. P. W. Fowler and S. D. Guest, *Proc. R. Soc. A*, 2005, **461**, 1829.
90. S. D. Guest and P. W. Fowler, *Philos. Trans. R. Soc., A*, 2014, **372**, 20130358.
91. A. Marmier and K. E. Evans, *Dalton Trans.*, 2016, **45**, 4360.
92. J. E. Graver, *Counting on Frameworks: Mathematics to Aid the Design of Rigid Structures*, The Mathematical Association of America, Washington D.C, 2001.
93. *Rigidity Theory and Applications*, ed. M. F. Thorpe and P. M. Duxbury, Kluwer Academics Publisher, New York, 2002.
94. J. E. Graver, B. Servatius and H. Servatius, *Combinatorial Rigidity*, American Mathematical Society, 1993.
95. J. M. Hendrickx, B. D. Anderson, J. C. Delvenne and V. D. Blondel, *Int. J. Robust Nonlinear Control*, 2007, **17**, 960.
96. J. C. Owen and S. C. Power, *N. Y. J. Math.*, 2011, **17**, 445.
97. D. J. Jacobs and B. Hendrickson, *J. Comput. Phys.*, 1997, **137**, 346.
98. D. J. Jacobs, A. J. Rader, L. A. Kuhn and M. F. Thorpe, *Proteins: Struct., Funct., Bioinf.*, 2001, **44**, 150.
99. D. M. Krüger, P. C. Rathi, C. Pfleger and H. Gohlke, *Nucleic Acids Res.*, 2013, **41**, W340.
100. N. Fox, F. Jagodzinski, Y. Li and I. Streinu, *Nucleic Acids Res.*, 2011, **39**, W177.
101. F. S. Tautz, V. Heine, M. T. Dove and X. Chen, *Phys. Chem. Miner.*, 1991, **18**, 326.
102. A. P. Giddy, M. T. Dove, G. S. Pawley and V. Heine, *Acta Crystallogr., Sect. A*, 1993, **49**, 697.
103. K. D. Hammonds, M. T. Dove, A. P. Giddy and V. Heine, *Am. Mineral.*, 1994, **79**, 1207.
104. K. D. Hammonds, M. T. Dove, A. P. Giddy, V. Heine and B. Winkler, *Am. Mineral.*, 1996, **81**, 1057.
105. L. H. Rimmer, M. T. Dove, A. L. Goodwin and D. C. Palmer, *Phys. Chem. Chem. Phys.*, 2014, **16**, 21144.
106. M. Gao, A. J. Misquitta, L. H. Rimmer and M. T. Dove, *Dalton Trans.*, 2016, **45**, 4289.
107. F.-X. Coudert, *Phys. Chem. Chem. Phys.*, 2013, **15**, 16012.
108. B. Campbell, C. J. Howard, T. B. Averett, T. A. Whittle, S. Schmid, S. Machlus, C. Yost and H. T. Stokes, *Acta Crystallogr., Sect. A*, 2018, **74**, 408.
109. S. A. Wells and A. Sartbaeva, *Mater.*, 2012, **5**, 415.

110. S. A. Wells and A. Sartbaeva, *Mol. Simul.*, 2015, 1.
111. W. J. Gee, S. A. Wells, S. J. Teat, P. R. Raithby and A. D. Burrows, *New J. Chem.*, 2021, **45**, 8728.
112. R. Gaillac, S. Chibani and F.-X. Coudert, *Chem. Mater.*, 2020, **32**, 2653.
113. S. Chibani and F.-X. Coudert, *APL Mater.*, 2020, **8**, 080701.

CHAPTER 3

Computational Modelling of MOF Mechanics: From Elastic Behaviour to Phase Transformations

SVEN M. J. ROGGE*[a]

[a]Center for Molecular Modeling, Ghent University, Technologiepark-Zwijnaarde 46, 9052 Zwijnaarde, Belgium
*E-mail: Sven.Rogge@UGent.be

3.1 Introduction

The structure of metal–organic frameworks (MOFs) is governed by an intricate interplay of disparate interactions, from weak dispersion forces to strong covalent bonds, and from isotropic to strongly directional interactions. This gives rise to several attractive but often poorly understood mechanical phenomena in these 'soft' solid-state materials, including negative linear compressibility (NLC),[1] negative thermal expansion (NTE),[2] extreme anisotropy that may even lead to a negative Poisson's ratio or auxetic behaviour (Chapter 2), and a wide variety of both displacive and reconstructive phase transitions induced by mechanical stress or other external stimuli including gas sorption and temperature, among others.[3,4] Furthermore, the experimentally observed macroscopic response of MOFs to external stimuli is not only defined by this intrinsic interplay of interactions at the atomic, microscopic level, but is also impacted by effects occurring on larger length scales that

Inorganic Materials Series No. 12
Mechanical Behaviour of Metal–Organic Framework Materials
Edited by Jin-Chong Tan

Published by the Royal Society of Chemistry, www.rsc.org

may supersede this intrinsic response. Examples of such long-range effects include the interplay between different types of structural disorder,[4,5] the impact of crystallite size and morphology,[6] interactions occurring between different MOF crystals present in a macroscopic sample, and interactions between the MOF and the experimental probe interrogating its mechanical behaviour.[7] Therefore, isolating and understanding the salient features that define the macroscopic mechanical behaviour of a MOF is not only inspired by a fundamental scientific curiosity, but is also vital to enable the design of MOFs with predefined macroscopic functionalities not encountered in other materials.

Computational MOF research plays an important role in this respect, especially when adopted in close feedback with experimental research. Given the relative compliance of MOFs to external stimuli, the aforementioned complexity in different interactions at the microscopic level, and the hierarchical and multiscale nature of interactions from the unit cell level to the macroscopic MOF material, it is clear that the applicability of any given modelling tool strongly depends on the specific question at hand. Furthermore, since anomalous mechanical behaviour is encountered much more frequently in MOFs than in other solid-state materials, modelling techniques that are deemed firmly established in solid-state computational research may be too inaccurate to be used in MOF research. As a result, the advent of MOFs also necessitated – and continues to necessitate – the development of fundamentally new *in silico* techniques, as well as the adaptation of established techniques, to account for their soft behaviour. In this respect, computational MOF research is located at the exciting intersection where newly developed modelling tools are put to a stringent test to reproduce the mechanical behaviour of experimentally characterised MOFs on the one hand, and where newly observed anomalous behaviour in MOFs (Chapter 2) forms an impetus to develop new computational techniques on the other.

In Chapter 3, we shall critically review the computational tools that are being developed and used to model MOF mechanics. An extensive list of literature examples is provided to discuss each of these methods, focussing on the computational parameters affecting their accuracy, their scope of applicability and limitations, and how they compare to alternative techniques. This discussion is divided into techniques that probe the equilibrium or elastic behaviour of MOFs in Section 3.4 and techniques that model phase transitions or the plastic behaviour of MOFs, discussed in Sections 3.5 and 3.6 for single-crystal-to-single-crystal transitions and single-crystal-to-amorphous transitions, respectively. While these sections also briefly summarise the theoretical cornerstones on which these techniques are built, a more elaborate theoretical underpinning of how to model the mechanical behaviour of MOFs, which transcends the specific techniques discussed in the second half of this chapter, is provided in Sections 3.2 and 3.3. Section 3.2 focuses on how to identify mechanical properties, which are mostly taught from a macroscopic continuum point of view, with an atomistic view of MOFs. Section 3.3 discusses the ingredients needed to extract mechanical properties from computational MOF simulations as accurately as possible. The chapter closes in Section 3.7 with a personal outlook on the critical challenges that computational MOF research needs to overcome to further advance the research field, and provide a holistic and multiscale approach to predict the mechanical performance of MOFs.

3.2 From a Continuum to an Atomic Description of Stress and Strain

Variables such as the Cauchy stress or constitutive relations such as Hooke's law are ubiquitously used in computational MOF research, although they find their origin in continuum theory.[8] In continuum mechanics, a macroscopic body is regarded as an assembly of continuum particles or voxels such that the state variables that define the mechanics of this body – such as stress and strain – vary smoothly over adjacent particles and are governed by macroscopic continuum field equations.[8] While continuum particles need to be infinitesimally small from a macroscopic point of view, the state variables associated with such a continuum particle are still derived from a finite microscopic region surrounding the location of the particle to ensure this smoothness.[8] This contrasts with the typical atomistic viewpoint of materials adopted in computational MOF research, in which for instance the mass density fluctuates wildly between nonzero values for locations where nuclei are present and zero values elsewhere. To reconcile these two points of view, the length scale of continuum particles – and hence the length scale on which state variables such as Cauchy stress are defined – should be much smaller than the length scale on which variations in the continuum fields occur and, simultaneously, much larger than typical atomic length scales to ensure the smooth behaviour over adjacent continuum particles.[8]

While continuum mechanics does not make any assumptions on the underlying atomistic framework of the material, one may ask how these phenomenological continuum equations are governed by the particles constituting the material and their interactions. This would allow one not only to define these phenomenological parameters, such as the stiffness tensor or the point at which a material deforms irreversibly, from a quantum mechanical point of view, but also to draw casual relations between the atomic structure of a MOF and its mechanical properties and hence accelerate the discovery of exciting anomalous behaviour in tuneable functional materials such as MOFs.

Such a statistical mechanics picture of continuum theory was pioneered by Irving and Kirkwood in 1950, when they proposed atomic expressions for continuum properties including the stress tensor and the heat current density.[9] In order for these properties to be true atomistic descriptions of the material consistent with the continuum variables defined earlier, these so-defined point functions should satisfy two requirements:

1. When properly averaged over space (over the continuum particle length scale defined earlier on) and time, the atomistic definitions should coincide with the continuum definitions;

2. The atomistic definitions should satisfy the same hydrodynamical equations – continuity equation, equation of motion, and equation of energy transport – as the continuum variables. This also implies that, for instance, an elasticity tensor defined at the atomic scale can be interpreted as a local equivalent of an elasticity tensor defined for a macroscopic body, forming a bridge between atomic and continuum theories.

In this section, this consistent approach will be adopted to derive atomic-level definitions of strain and stress. While one could, alternatively, simply postulate these definitions, this explicit derivation will help the further discussion in two ways. First, it will unveil the different approximations when defining stress and strain at the atomic level and hence help to understand under which inherent limitations the mechanical properties of MOFs can be simulated. Second, this discussion provides the ingredients necessary to computationally simulate MOFs under constant pressure or stress in Section 3.3.3.

Let us begin by considering a conservative atomic system of N particles and a right-handed Cartesian frame of reference with basis vectors $(\boldsymbol{e}_1,\boldsymbol{e}_2,\boldsymbol{e}_3)$, as depicted in Figure 3.1. With respect to this reference frame, the system is defined by the positions of its N particles $\boldsymbol{r}_i = x_i\boldsymbol{e}_1 + y_i\boldsymbol{e}_2 + z_i\boldsymbol{e}_3$, $i \in [1,N]$ and their momenta $\boldsymbol{p}_i$. To simplify the notation, define the 3N-component vectors $\boldsymbol{r}^N = (x_1,y_1,z_1,x_2,y_2,z_2,\ldots,x_N,y_N,z_N)$ and $\boldsymbol{p}^N = (p_{1x},p_{1y},p_{1z},p_{2x},p_{2y},p_{2z},\ldots,p_{Nx},p_{Ny},p_{Nz})$. An alternative way to describe the positions of these atoms is by defining them with respect to a right-handed set of vectors $(\boldsymbol{a}_1,\boldsymbol{a}_2,\boldsymbol{a}_3)$ that forms a parallelepiped $\left(\boldsymbol{a}_i = \sum a_{ij}\boldsymbol{e}_j\right)$. Note that, in contrast to the basis vectors defined before, the vectors $\boldsymbol{a}_i$ do not need to be orthogonal nor normalised, although

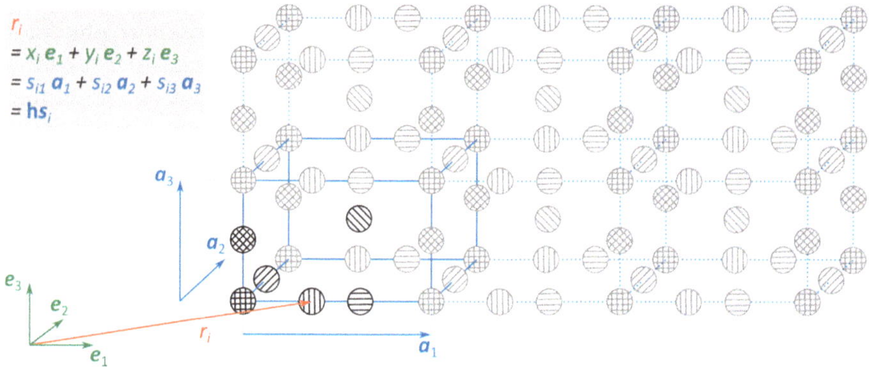

Figure 3.1 Schematic illustration of a crystalline material for which a unit cell $\mathbf{h} = [\boldsymbol{a}_1\ \boldsymbol{a}_2\ \boldsymbol{a}_3]$ (blue parallelepiped) containing six atoms can be defined. The material can then be described by a periodic repetition of this unit cell (five additional repetitions are shown in grey). Each atom can be defined either with respect to a Cartesian basis (in green) or in fractional coordinates, with respect to the unit cell vectors (in blue).

they do need to form a linearly independent set. These vectors can be collected in the 3 × 3 matrix **h** in which these vectors appear as the columns (observe that **h** is *not* a tensor):

$$\mathbf{h} = \begin{bmatrix} \boldsymbol{a}_1 & \boldsymbol{a}_2 & \boldsymbol{a}_3 \end{bmatrix} = \begin{bmatrix} a_{11} & a_{21} & a_{31} \\ a_{12} & a_{22} & a_{32} \\ a_{13} & a_{23} & a_{33} \end{bmatrix}. \tag{3.1}$$

Note that both matrices and tensors of rank two or higher will be denoted by bold-faced upright symbols. For (nearly) crystalline materials such as MOFs, the vectors $\boldsymbol{a}_i$ are often defined such that **h** defines the simulation cell matrix which is periodically repeated in space to mimic bulk conditions, as shown in Figure 3.1 and as discussed more extensively in Section 3.3.1. However, one could also define them based on the external surfaces of finite MOF crystallites in nonperiodic simulations. With respect to **h**, the atomic positions can be defined in fractional coordinates $\boldsymbol{s}_i$, such that $\boldsymbol{r}_i = s_{i1}\boldsymbol{a}_1 + s_{i2}\boldsymbol{a}_2 + s_{i3}\boldsymbol{a}_3$, $i \in [1,N]$ or, in matrix notation, $\boldsymbol{r}_i = \mathbf{h}\boldsymbol{s}_i$. Furthermore, we will assume that the atoms interact with one another through a potential energy function that only depends on the positions of these atoms: $\mathcal{V} = \mathcal{V}(\boldsymbol{r}^N)$. This potential energy forms a surface in the 3*N*-dimensional configuration space, the construction of which will be further discussed in Section 3.3.2. Equipped with these definitions, it is now possible to proceed to atomistic definitions of strain and stress.

3.2.1 The Atomic Definition of Deformation Gradient and Strain

As in the continuum definition, atomic-level strain requires a reference or undeformed state against which a deformation is defined. Herein, properties of this reference state will be denoted by a subscript '0' symbol. Hence, the positions, fractional coordinates, and cell matrix of the reference state will be indicated by $\boldsymbol{r}_0^N$, $\boldsymbol{s}_0^N$, and $\mathbf{h}_0$, respectively, as illustrated in Figure 3.2(a).

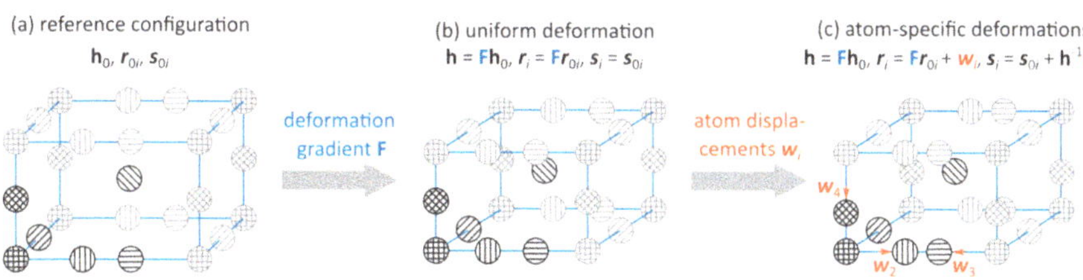

Figure 3.2 A general deformation of a crystal with respect to a reference configuration (a) can be divided into a uniform deformation (b), defined by the deformation gradient (in blue), and atom-specific deformations (c, in orange).

Following Cauchy and Born,[10–12] suppose that the material under study deforms to the state defined by $\boldsymbol{r}^N$, $\boldsymbol{s}^N$, and $\mathbf{h}$, shown in Figure 3.2(c). If the atoms would be deformed uniformly, as shown in Figure 3.2(b), the fractional coordinates in the deformed configuration would remain unchanged with respect to their values in the reference state, $\boldsymbol{s}_{0i} = \mathbf{h}_0^{-1}\boldsymbol{r}_{0i}$, where $\mathbf{h}^{-1}$ is the matrix inverse of $\mathbf{h}$. Hence, under a uniform deformation, the positions of the particle are given by $\boldsymbol{r}_i = \mathbf{h}\mathbf{h}_0^{-1}\boldsymbol{r}_{0i}$. This motivates the identification of $\mathbf{h}\mathbf{h}_0^{-1}$ as the atomic definition of the continuum deformation gradient $\mathbf{F}$:

$$\mathbf{F} = \mathbf{h}\mathbf{h}_0^{-1}. \tag{3.2}$$

Note that this definition ensures that $\mathbf{F}$ is a second-rank tensor, even though $\mathbf{h}$ is not.

A general deformation will not be completely uniform over the cell matrix. For quasi-uniform deformations, this departure from a uniform deformation is defined by the atomic displacement vectors $\boldsymbol{w}_i$, such that $\boldsymbol{r}_i = \mathbf{F}\boldsymbol{r}_{0i} + \boldsymbol{w}_i = \mathbf{h}\mathbf{h}_0^{-1}\boldsymbol{r}_{0i} + \boldsymbol{w}_i$, as shown in Figure 3.2(c).[8] When following the system through time, for instance in a molecular dynamics simulation, this property is time dependent:

$$\boldsymbol{r}_i(t) = \mathbf{F}(t)\boldsymbol{r}_{0i} + \boldsymbol{w}_i(t) = \mathbf{h}(t)\mathbf{h}_0^{-1}\boldsymbol{r}_{0i} + \boldsymbol{w}_i(t). \tag{3.3}$$

This so-called Cauchy–Born rule was originally derived to relate the mean positions of a deformed lattice to its undeformed reference state under a homogeneous deformation; but it has since been extended to also describe noncrystalline structures. Importantly, in the formulation given above, a quasiuniform deformation of $\mathbf{h}_0$ is assumed, which limits the field of applicability of eqn (3.3) to a region around the undeformed state, the extent of which is, inconveniently, *a priori* unknown.[8] As a result, the Cauchy–Born rule has been shown to no longer hold under phase transformations, either displacive or reconstructive, among other limitations.[13] When studying large-amplitude responses in MOFs, care should therefore be exerted to establish a proper reference state to define the deformation gradient of eqn (3.3). Taking MIL-53(Al) as a textbook example,[14] it cannot be expected that both the large-pore (lp) and closed-pore (cp) states of this material can be described by the same reference cell. When discussing a simulation protocol to probe MOFs under a constant external stress in Section 3.3.3, this statement will be further refined through the introduction of an adaptive modelling scheme. For a more in-depth discussion on the limitations and extensions of this approach, the interested reader is referred to Section 11.2 in ref. 8 and references therein.

From the atomic definition of the deformation gradient in eqn (3.1), one can define the Lagrangian strain $\boldsymbol{\eta}$, in the same fashion as in continuum mechanics:

$$\boldsymbol{\eta} = \frac{1}{2}\left(\mathbf{F}^T\mathbf{F} - \mathbf{I}\right) = \frac{1}{2}\left(\mathbf{h}_0^{-T}\mathbf{h}^T\mathbf{h}\mathbf{h}_0^{-1} - \mathbf{I}\right), \tag{3.4}$$

which coincides with the definition by Ray and Rahman.[15] Herein, $\mathbf{I}$ is the second-rank unit tensor, and $\mathbf{a}^{-T} = (\mathbf{a}^{-1})^T = (\mathbf{a}^T)^{-1}$, with T denoting a matrix transposition. Similarly, the Eulerian strain,† $\varepsilon^{\mathrm{eul}}$, is given by

$$\varepsilon^{\mathrm{eul}} = \frac{1}{2}\left(\mathbf{I} - \mathbf{F}^{-T}\mathbf{F}^{-1}\right) = \frac{1}{2}\left(\mathbf{I} - \mathbf{h}^{-T}\mathbf{h}_0^T\mathbf{h}_0\mathbf{h}^{-1}\right). \tag{3.5}$$

3.2.2 The Atomic Definition of Stress

Deriving an atomic equivalent to the continuum stress tensors is slightly more intricate. An intuitive derivation, based on ref. 8, starts from the thermodynamic observation that an N-atom system strained by a deformation gradient $\mathbf{F}$ and under temperature and stress control is described by the Gibbs free energy (or free enthalpy) G:[16]

$$G(N, \mathbf{P}, T; \mathbf{F}) = F(N, T; \mathbf{F}) - V_0\mathbf{P}{:}\mathbf{F}. \tag{3.6}$$

In this expression, T is the absolute temperature, $V_0 = \det(\mathbf{h}_0)$ is the volume of the cell matrix $\mathbf{h}_0$ describing the undeformed system, $\mathbf{P}$ is the first Piola–Kirchhoff stress tensor, and F is the Helmholtz free energy that describes the system under constant deformation and temperature. Furthermore, $\mathbf{P}{:}\mathbf{F}$ denotes the tensor contraction of the two second-rank tensors $\mathbf{P}$ and $\mathbf{F}$, such that $\mathbf{P}:\mathbf{F} = \sum_{i=1}^{3}\sum_{j=1}^{3} P_{ij}F_{ij}$.‡ The Gibbs free energy reveals the (meta)stable states of the system under temperature and stress control as those states for which the derivative of the Gibbs free energy with respect to the deformation gradient vanishes:

$$\frac{\partial G(N,\mathbf{P},T;\mathbf{F})}{\partial \mathbf{F}} = \mathbf{0} \Leftrightarrow \frac{\partial F(N,T;\mathbf{F})}{\partial \mathbf{F}} - \mathbf{P}V_0 = \mathbf{0} \Leftrightarrow \mathbf{P}(T;\mathbf{F}) = \frac{1}{V_0}\frac{\partial F(N,T;\mathbf{F})}{\partial \mathbf{F}}. \tag{3.7}$$

Note that eqn (3.7) defines a tensor equality, which needs to hold for all nine components of $\mathbf{F}$. Obtaining an atomic definition of the stress tensor

†These two different definitions for the same 'strain' property find their origin in the different ways one can look at continuum deformations: the material or the spatial perspective. In the materials perspective, properties are assigned to a material point, which moves through space upon deforming the material. In the spatial perspective, properties are assigned to a fixed point in space, irrespective of the material deformation. These two points of view give rise to the Lagrangian and the Eulerian strains as defined in eqn (3.4) and (3.5) respectively. Herein, we will adopt the material point of view when discussing materials properties and hence use the Lagrangian strain as the correct strain descriptor.

‡A similar description for the Gibbs free energy can be obtained by considering the finite Lagrangian strain $\boldsymbol{\eta}$, defined in eqn (3.4), as independent variable in eqn (3.6) instead of the deformation gradient $\mathbf{F}$. In that case, the second term in the right-hand side of eqn (3.6) equals $-V_0\mathbf{S}{:}\boldsymbol{\eta}$ instead of $-V_0\mathbf{P}{:}\mathbf{F}$, with $\mathbf{S}$ being the second Piola–Kirchhoff stress tensor, also denoted (the negative of) the thermodynamic tension.[15] In addition, it should be noted that the existence of a Gibbs free energy as a thermodynamic potential under a general deformation has been questioned, in contrast to the universally accepted definition of free energy under a hydrostatic pressure.[253]

P near equilibrium thus boils down to finding an atomic definition of the Helmholtz free energy F, which is possible through the partition function Z of this system:

$$F(N,T;\mathbf{F}) = -k_{\mathrm{B}}T\ln Z_{NFT} = -k_{\mathrm{B}}T\ln\left[\frac{1}{N!h^{3N}}\int_{\Gamma(\mathbf{F})}\exp\left(\frac{-\mathcal{H}(\boldsymbol{r}^N,\boldsymbol{p}^N)}{k_{\mathrm{B}}T}\right)\mathrm{d}\boldsymbol{r}^N\mathrm{d}\boldsymbol{p}^N\right]. \tag{3.8}$$

Herein, k_{B} is Boltzmann's constant, h is Planck's constant,

$$\mathcal{H}(\boldsymbol{r}^N,\boldsymbol{p}^N) = \sum_{i=1}^{N}\frac{\boldsymbol{p}_i\cdot\boldsymbol{p}_i}{2m_i} + \mathcal{V}(\boldsymbol{r}^N) \tag{3.9}$$

is the Hamiltonian describing the system, $\Gamma(\mathbf{F})$ describes the subregion of the 6N-dimensional phase space spanned by $\boldsymbol{r}^N$ and $\boldsymbol{p}^N$ that is accessible under the given deformation gradient, and m_i is the mass of particle i. In this equation, $\boldsymbol{p}_i{\cdot}\boldsymbol{p}_i$ denotes the scalar product of two vectors: $\boldsymbol{a}\cdot\boldsymbol{b} = \sum_{i=1}^{3}a_i b_i$. The **F**-dependence in the integration boundaries of eqn (3.8) arises as the deformation gradient is defined based on the matrix **h** and hence restricts the positions where the atoms can be located. However, its presence complicates taking the derivative of eqn (3.8) with respect to **F**. To move the **F**-dependence of eqn (3.8) to the integrand of the partition function, Ray and Rahman proposed a canonical transformation of the Hamiltonian using the generating function[15]

$$G(\boldsymbol{p}^N,\boldsymbol{r}_0^N) = -\sum_{i=1}^{N}\boldsymbol{p}_i\cdot(\mathbf{F}\boldsymbol{r}_{i,0}), \tag{3.10}$$

in which the subscript '0', as usual, indicates properties of the system in its undeformed configuration. Eqn (3.10) generates the canonical transformation

$$\boldsymbol{r}_i = \mathbf{F}\boldsymbol{r}_{i,0}\,\boldsymbol{p}_{i,0} = \mathbf{F}^T\boldsymbol{p}_i \tag{3.11}$$

and leaves the form of the Hamiltonian invariant, such that

$$\mathcal{H}_0(\boldsymbol{r}_0^N,\boldsymbol{p}_0^N,\mathbf{F}) = \mathcal{H}(\boldsymbol{r}^N(\boldsymbol{r}_0^N,\mathbf{F}),\boldsymbol{p}^N(\boldsymbol{p}_0^N,\mathbf{F})) = \sum_{i=1}^{N}\frac{(\mathbf{F}^{-T}\boldsymbol{p}_{i,0})\cdot(\mathbf{F}^{-T}\boldsymbol{p}_{i,0})}{2m_i} + \mathcal{V}(\mathbf{F}\boldsymbol{r}_0^N). \tag{3.12}$$

Importantly, this transforms the Helmholtz free energy of eqn (3.8) in such a way that the deformation gradient no longer appears in the integration limits but in the integrand instead:

$$F(N,T;\mathbf{F}) = -k_{\mathrm{B}}T\ln\left[\frac{1}{N!h^{3N}}\int_{\Gamma_0}\exp\left(\frac{-\mathcal{H}_0(\boldsymbol{r}_0^N,\boldsymbol{p}_0^N,\mathbf{F})}{k_{\mathrm{B}}T}\right)\mathrm{d}\boldsymbol{r}_0^N\mathrm{d}\boldsymbol{p}_0^N\right]. \tag{3.13}$$

As a result, the first Piola–Kirchhoff tensor from eqn (3.6) can be evaluated as follows:

$$P(T;\mathbf{F}) = \frac{1}{V_0}\frac{\partial F(N,T;\mathbf{F})}{\partial \mathbf{F}} = -\frac{k_B T}{V_0}\frac{1}{Z_{NFT}}\frac{\partial Z_{NFT}}{\partial \mathbf{F}} = \frac{1}{V_0}\left\langle \frac{\partial \mathcal{H}_0\left(\boldsymbol{r}_0^N, \boldsymbol{p}_0^N, \mathbf{F}\right)}{\partial \mathbf{F}} \right\rangle, \tag{3.14}$$

or, after some lengthy calculus:

$$\mathbf{P}(T;\mathbf{F}) = -\frac{1}{V_0}\sum_{i=1}^{N}\left\langle \frac{\boldsymbol{p}_i \otimes \left(\mathbf{F}^{-1}\boldsymbol{p}_i\right)}{m_i} - \frac{\partial \mathcal{V}}{\partial \boldsymbol{r}_i} \otimes \boldsymbol{r}_{i,0} \right\rangle, \tag{3.15}$$

where ⊗ denotes the dyadic product of two vectors, $(\boldsymbol{a}\otimes\boldsymbol{b})_{ij} = a_i b_j$, yielding a second-rank tensor. Both in eqn (3.14) and (3.15), $\langle\cdot\rangle$ indicates an ensemble average, although an instantaneous stress $\mathbf{P}^{\text{inst}}$ can be defined by omitting this average in eqn (3.15), such that $\langle\mathbf{P}^{\text{inst}}\rangle = \mathbf{P}$. Similar to the definition of an instantaneous temperature, however, it is important to recognise that only the ensemble-averaged definition of the Piola–Kirchhoff tensor enters thermodynamic relations.

From this definition of the first Piola–Kirchhoff tensor **P** and following the continuum relations between this tensor and the second Piola–Kirchhoff stress tensor **S** and Cauchy stress tensor $\boldsymbol{\sigma}$,[8] one obtains:

$$\mathbf{S}(T;\mathbf{F}) = \mathbf{F}^{-1}\mathbf{P}(T;\mathbf{F}) = -\frac{1}{V_0}\sum_{i=1}^{N}\left\langle \frac{\left(\mathbf{F}^{-1}\boldsymbol{p}_i\right)\otimes\left(\mathbf{F}^{-1}\boldsymbol{p}_i\right)}{m_i} - \left(\mathbf{F}^{-1}\frac{\partial \mathcal{V}}{\partial \boldsymbol{r}_i}\right)\otimes \boldsymbol{r}_{i,0} \right\rangle; \tag{3.16}$$

$$\boldsymbol{\sigma}(T;\mathbf{F}) = \frac{1}{\det \mathbf{F}}\mathbf{P}(T;\mathbf{F})\mathbf{F}^{T} = -\frac{1}{V}\sum_{i=1}^{N}\left\langle \frac{\boldsymbol{p}_i \otimes \boldsymbol{p}_i}{m_i} - \frac{\partial \mathcal{V}}{\partial \boldsymbol{r}_i}\otimes \boldsymbol{r}_i \right\rangle. \tag{3.17}$$

The Cauchy stress tensor of eqn (3.17) is the stress tensor most often calculated during each integration step of a molecular simulation to obtain a measure for the internal stress. From the above definition, one can also determine the pressure of the system as (the negative of) the trace (Tr) of the above expression:

$$P(T;\mathbf{F}) = -\frac{1}{3}\text{Tr}\left\{\boldsymbol{\sigma}(T;\mathbf{F})\right\} = \frac{1}{3V}\sum_{i=1}^{N}\left\langle \frac{p_i^2}{m_i} - \frac{\partial \mathcal{V}}{\partial \boldsymbol{r}_i}\cdot \boldsymbol{r}_i \right\rangle, \tag{3.18}$$

although it is more straightforward to obtain an atomic definition of the pressure by realising that the appropriate Gibbs free energy expression for a system under pressure and temperature control is given by

$$G(N, P, T; V) = F(N, T; V) + PV, \tag{3.19}$$

which should be minimised with respect to the volume V to obtain an atomic expression for the pressure. This alternative procedure yields the same results as eqn (3.18).

Eqn (3.4) and (3.5) for the strain and eqn (3.15)–(3.18) for the stress define the sought-after connection between the continuum stress and strain and the atomic properties of a system: the masses m_i, positions $\boldsymbol{r}_i$, and momenta $\boldsymbol{p}_i$ of each of the atoms, as well as interactions $\mathcal{V}$ acting between the atoms and the deformation gradient $\mathbf{F}$ defining the current cell configuration $\mathbf{h}$ compared to the reference cell $\mathbf{h}_0$.

3.3 Ingredients Necessary to Atomically Model MOF Mechanics

Computational modelling is a bottom-up approach in which the macroscopic behaviour of a material is determined starting from knowledge about the positions of its constituents – electrons and nuclei – and the interactions between them. The common goal of any computational modelling endeavour is to understand how interactions at the atomic scale give rise to macroscopically observed properties and to adopt this information to design materials with predesigned functionalities (*e.g.*, susceptible to phase transitions, NLC, NTE). In order to achieve this ambitious target, the following three main questions need to be considered, irrespective of the details of the problem at hand:

1. How can the experimental, macroscopic material consisting of an immense number of atoms be represented by a much smaller number of particles, varying from a few thousands to a few million atoms depending on the available computational power and the choice of model to approximate the interatomic interactions and external conditions (see points 2 and 3)?
2. How can the interatomic interactions, which are inherently many-body and quantum mechanical in nature, be approximated to an extent that they can be used to computationally model the phenomena under study while allowing both for sufficiently large structural models (see point 1) and for a sufficient similarity to the real materials behaviour under the given external conditions (see point 3)?
3. How can, for a given structural model (see point 1) and approximation of the interatomic interactions (see point 2), the material be simulated in such a way that properly averaging the atomic properties over the computer simulation allows for a direct comparison with experimental results?

In this section, these three fundamental questions will be discussed separately. The importance of each of these key aspects in order to obtain reliable results that can further our understanding and design capabilities of the mechanical behaviour of MOFs will be provided, although much of this discussion can be readily generalised to other materials and other properties.

3.3.1 Generating an Accurate Structural Model of the Material

The size of the largest atomic MOF models to date remains limited to a few tens of thousands of atoms and a few tens of nanometres in diameter,[7,17–23] save for a very recent study of a 1 million atom model of MIL-53(Al).[24] Despite this recent increase in computationally accessible length scales, atomic models remain substantially smaller than experimental MOF crystallites at the moment. Therefore, the majority of computational MOF research nowadays starts from the assumption that the properties of a macroscopic MOF material can be approximated well by considering a smaller simulation cell that is repeated *ad infinitum* along the three cell vectors $\boldsymbol{a}_1$, $\boldsymbol{a}_2$, and $\boldsymbol{a}_3$, which define the cell matrix **h** of eqn (3.1) (see Figure 3.1). By construction, periodic boundary conditions cannot be adopted to model surface effects in finite MOF crystals. Nevertheless, this periodic ansatz is being used extensively to approximate bulk properties.

The increasing realisation that MOFs come in various degrees of crystallinity[4,25,26] – from almost perfectly crystalline MOFs, over MOFs with local defects or mesopores locally deviating from crystallinity,[5,27–34] to amorphous or glassy MOFs that lack any long-range spatial order[35–41] – has made it increasingly difficult to define accurate structural MOF models. First, the lack of (perfect) order is straining the ansatz of periodic boundary conditions as a *passe-partout* in computational MOF research. Several recent studies have sought to limit the effect of these spurious long-range correlations in MOF simulation cells by considering increasingly larger cells, giving rise to exciting new phenomena not accessible in smaller simulation cells (*vide infra*). However, for glassy or amorphous MOFs, without any long-range order, or when interested in surface effects, even these larger simulation cells cannot be expected to yield accurate predictions, and new and finite MOF crystallite models need to be constructed. Second, even when making abstraction of the validity of periodic boundary conditions to mimic MOF properties, it is not straightforward to come up with appropriate atomic simulation cells for disordered MOFs in the first place. As argued recently,[17,42] this requires a concerted effort from experimentalists and theoreticians to interrogate the MOF structure from the atomic to the macroscopic scale with a varied toolbox of high-resolution techniques and high-throughput automated computational screenings of potential (disordered) MOF structures.[17,42–44] Until such a toolbox has been developed and becomes mainstream accessible, disorder in MOFs can only be treated in an *ad hoc* fashion (*vide infra*), in which the comparison between theoretically predicted and experimentally measured properties quantifies the validity of those approximations – thereby limiting the predictive power of computational research. In what follows, three common pitfalls associated with defining an accurate structural model will be discussed, demonstrating the impact of (i) including disorder at the atomic scale in the periodic simulation cell, (ii) choosing sufficiently large simulation cells, and (iii) going beyond the periodic simulation cell for surface-dominated effects.

A first potential pitfall when modelling MOFs is neglecting the impact of local defects – with a spatial extent of only a few nanometres – on the macroscopic MOF properties. The most prominent example in this category is UiO-66(Zr).[45] Perceived as an exceptionally stable material, a result of its relatively strong zirconium–oxygen coordination bonds and the up to twelvefold coordination of its inorganic building blocks,[45–47] it has been the protagonist in many experimental and computational studies. Thanks to the rather straightforward synthesis of large UiO-66 crystals, it was also one of the first MOFs for which spatial disorder could be characterised extensively, revealing a variety of deviations from periodicity: from local inorganic node and linker vacancies to the formation of correlated nanodomains in which these local vacancies tend to cluster together.[5,29,32] Various works have tried to incorporate, to a different extent, these deviations from perfect crystallinity while still adopting periodic boundary conditions. In 2016, Thornton *et al.* calculated the *ab initio* elastic constants and elastic moduli for a variety of defect-containing UiO-66 simulation cells with a unit cell length of around 2.1 nm, which was sufficiently large to incorporate isolated linker and node vacancies.[48] As visualised in Figure 3.3(a), an appreciable increase in mechanical anisotropy and a decrease in the Young's and shear moduli were observed upon increasing linker vacancy concentration. In a force-field based study on a similar-sized simulation cell, Rogge *et al.* demonstrated that the bulk modulus and the mechanical stability of UiO-66 are also strongly impacted by linker vacancies.[49] Importantly, the extent to which the mechanical behaviour is impacted is dictated not only by the concentration of defects, but also by their distribution throughout the framework.[49] They illustrated that incorporating a realistic amount of defects is necessary to correctly predict the amorphisation pressure of ~1.4 GPa obtained from high-pressure mercury intrusion experiments.[50] Failing to take these defects

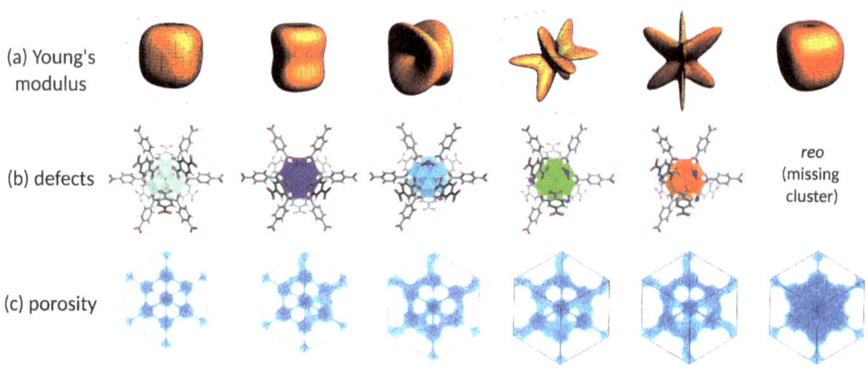

Figure 3.3 (a) Spatially-dependent Young's modulus, (b) visualisation of defects as observed from the perspective of a single cluster and (c) porosity available for adsorption of a 3 Å-diameter probe. Each cluster is coloured according to its coordination number. Adapted from ref. 48 with permission from the Royal Society of Chemistry.

into account overestimates this experimental value by ~0.4 GPa.[49] A similar pitfall also manifests itself for defect-free MOFs in which high-symmetric lattice points can be decorated by building blocks that have a lower symmetry. For instance, Ehrling *et al.* demonstrated that the response of DUT-8(Ni) on gas adsorption depends on the configuration of the nonlinear linkers present in this framework, the configuration of which can be guided during synthesis.[44] This opens up the possibility to design these MOFs,[51] by deriving relations between the local symmetry and structure of the MOF on the one hand, and the macroscopic material behaviour on the other. However, a *conditio sine qua non* for such structure–property relationships is the proper inclusion of these local deviations from crystallinity in the atomistic model.

A second pitfall is underestimating the required size of the atomic model, which leads to spurious effects in the simulation. This pitfall is very easily recognised when studying the stimuli-induced switching behaviour of flexible MOFs or soft porous crystals (SPCs),[52] such as MIL-53(Al)[14] and DMOF-1(Cu).[53,54] Both materials have been experimentally demonstrated to 'breathe' between a large-pore (lp) and a closed-pore (cp) form, and the ability of atomic simulations to reproduce such phase transformations was one of the early successes of MOF modelling.[55,56] However, when adopting periodic boundary conditions, the mechanism through which the phase transition takes place is forced to be a cooperative one, in which all periodic images of the simulation cell undergo the same transition at the same time. By systematically increasing the simulation size in these materials and by removing the periodic boundary conditions,[18,19] it was recently suggested that this cooperative mechanism is artificially imposed by the periodic ansatz, as phase transformations in larger crystals occur gradually instead. In this way, the transition from one phase to another can nucleate locally in the MOF, then propagate and grow through the material *via* a layer-by-layer mechanism.[18,19] The extra degrees of freedom present in the larger simulation cells lead to the proposition of a gradual transition mechanism that is energetically favoured over the cooperative mechanism assumed before, leading to stimuli-induced spatial disorder under the form of phase coexistence.[19] The latter was recently also confirmed experimentally for DUT-8(Ni),[57] a different SPC, through Raman spectroscopy.[58]

Finally, a third pitfall occurs when neglecting the surface of the MOF crystal, which becomes more prominent as the MOF crystallite becomes smaller, given its larger surface-to-volume ratio. For surface-dominated properties, accurate atomistic models need to explicitly account for the MOF surface by removing the periodic boundary conditions along at least one of the directions (or, equivalently, introducing a MOF slab model with a large vacuum region) and properly terminating the interface. Simulating finite MOF crystallites remains to date a largely unexplored area, as the much smaller scale of computational MOF crystallites compared to experimental crystals implies a much larger surface-to-volume ratio and an overestimation of surface effects. However, such finite crystal simulations do provide qualitatively interesting observations, such as the mechanism through which phase transformations

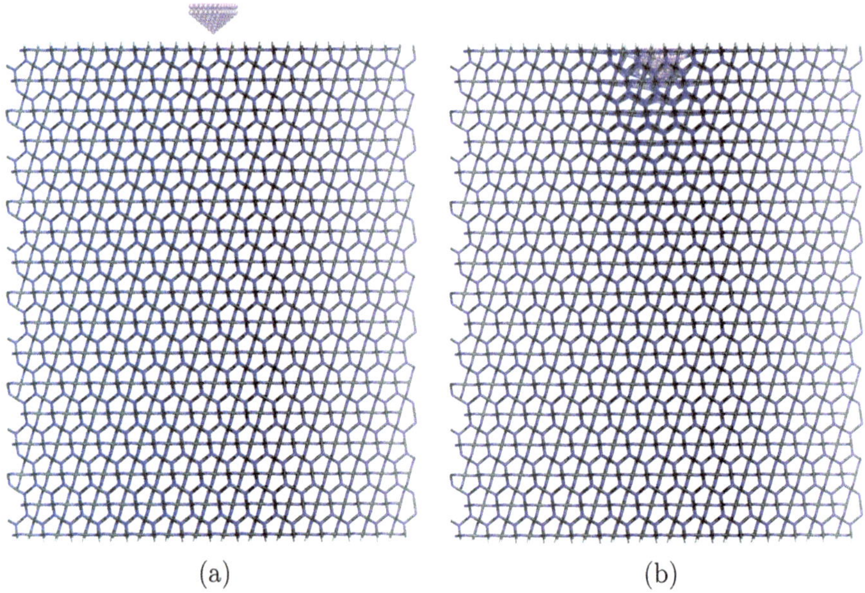

Figure 3.4 Penetration of an atomistic indenter 'tip' into the [111] surface of a coarse grained slab model of HKUST-1 (5 × 5 × 6 supercell), where (a) shows the initial and (b) the final state (colour scheme: black: inorganic moiety; blue: linker; white: terminal CH_3 bead; brown: tungsten atoms). Reproduced from ref. 7 with permission from the Royal Society of Chemistry.

develop in finite DMOF-1 particles.[18] Similarly, mechanical properties that are probed at the surface of the material – such as nanoindentation studies that probe the hardness of the material (see Sections 1.4 and 1.8.1 of Chapter 1) – require a proper termination of the MOF surface. A first step in this direction was set in 2016, when Dürholt *et al.* computationally determined the energy needed to penetrate the [111] surface of HKUST-1 with a tungsten tip, schematically depicted in Figure 3.4.[7] They demonstrated that converged results for the penetration energy could only be reached when both properly terminating the surface and when ensuring that the lateral surface dimensions, which are still treated periodically, are sufficiently large to avoid spurious interactions, thereby connecting to the second pitfall mentioned above.

3.3.2 Approximating the Interactions in a Material: the Level of Theory

As a bottom-up technique, molecular modelling requires an accurate description of the interactions between the nuclei and electrons that make up MOFs to predict macroscopic properties. Many different approaches exist to model these interactions, starting from a fundamental quantum mechanical or *ab initio* treatment, over classical potentials used in force fields, to coarse-grained (CG) and continuum descriptions. In this order, these models become computationally less expensive – often at the cost of an appreciable loss in accuracy – which opens up the possibility to treat much larger atomic models (see Section 3.3.1) on much longer time scales. Although studies are emerging in which the mechanical behaviour at longer time and length scales becomes accessible *via* CG techniques[7,59,60] or through finite-elements modelling (FEM),[61,62] the mechanical behaviour of MOFs is nowadays mostly characterised using either *ab initio* or atomistic force field approaches.[63] Therefore, these methods and their application to MOF research will be briefly outlined below, with appropriate references to more in-depth reviews.

As the interactions between electrons and nuclei in MOFs are quantum-mechanical in nature, completely characterising the (micro)state of a MOF requires knowledge of the positions (and possibly intrinsic spin coordinates) of all its nuclei and the positions and intrinsic spin coordinates of all its electrons, collected in the many-body wavefunction $|\Psi\rangle$, and this for every time instance of interest.§ In the absence of time-dependent external fields, the stationary many-body wavefunction $|\psi_n\rangle$ can be obtained by solving the time-independent Schrödinger equation:[64]

$$\hat{H}|\psi_n\rangle = E_n|\psi_n\rangle, \tag{3.20}$$

in which E_n and $|\psi_n\rangle$ are the eigenvalues and eigenstates, respectively, of the Hamiltonian $\hat{H}$ that describes the interactions in the system. The quantum label n is used to differentiate between the different states. Under the assumption that the $N^{(e)}$ electrons and N nuclei in the system only interact through Coulomb interactions, this Hamiltonian can be written as

$$\hat{H} = \hat{T}_N + \hat{T}_e + \hat{\mathcal{V}}_{NN} + \hat{\mathcal{V}}_{ee} + \hat{\mathcal{V}}_{Ne}. \tag{3.21}$$

Herein,

$$\hat{T}_N = \sum_{i=1}^{N} \frac{\boldsymbol{p}_i \cdot \boldsymbol{p}_i}{2m_i} \quad \text{and} \quad \hat{T}_e = \sum_{i=1}^{N^{(e)}} \frac{\boldsymbol{p}_i^{(e)} \cdot \boldsymbol{p}_i^{(e)}}{2m_i^{(e)}} \tag{3.22}$$

are the kinetic energy operators associated with the nuclei and the electrons, respectively,

$$\hat{\mathcal{V}}_{NN} = \frac{1}{4\pi\varepsilon_0} \sum_{i=1}^{N} \sum_{j=i+1}^{N} \frac{Z_i Z_j}{|\boldsymbol{r}_i - \boldsymbol{r}_j|} \quad \text{and} \quad \hat{\mathcal{V}}_{ee} = \frac{1}{4\pi\varepsilon_0} \sum_{i=1}^{N^{(e)}} \sum_{j=i+1}^{N^{(e)}} \frac{1}{|\boldsymbol{r}_i^{(e)} - \boldsymbol{r}_j^{(e)}|} \tag{3.23}$$

are the potential energy operators associated with the repulsive Coulomb interactions between each pair of nuclei and each pair of electrons, respectively, and

$$\hat{\mathcal{V}}_{Ne} = -\frac{1}{4\pi\varepsilon_0} \sum_{i=1}^{N} \sum_{j=1}^{N^{(e)}} \frac{Z_i}{|\boldsymbol{r}_i - \boldsymbol{r}_j^{(e)}|} \tag{3.24}$$

§Of course within the limitations set out by the Heisenberg uncertainty principle.

is the potential energy operator associated with the attractive Coulomb interaction between each nucleus–electron pair. In these expressions, $m_i^{(e)}$, $\boldsymbol{r}_i^{(e)}$, and $\boldsymbol{p}_i^{(e)}$ are the mass, position, and momentum of the ith electron, Z_i is the nuclear charge of the ith nucleus, and ε_0 is the permittivity of the vacuum.

If our computational model (see Section 3.3.1) consists of N nuclei and $N^{(e)}$ electrons and if we can neglect the spin of the nuclei, the wavefunctions $|\psi_n\rangle$ depend on $3N + 4N^{(e)}$ degrees of freedom. For typical MOF unit cells, which contain about 100 to 10 000 nuclei and about one order of magnitude more electrons, it would be an extraordinary task to fully determine this wavefunction. Therefore, different approximations exist to determine the eigenenergies E_n, and especially the ground-state energy E_0, of eqn (3.20).

A first approximation, the Born–Oppenheimer (BO) approximation, starts from the realisation that electrons are much lighter than nuclei, and will therefore equilibrate on a much shorter timescale under typical conditions.[65] From the perspective of the electrons, the nuclei can be assumed as 'clamped'. If the excited electronic states are furthermore much higher in energy than the ground state so that nuclear excitations cannot induce electronic excitations (the so-called adiabatic approximation), the stationary Schrödinger equation (eqn (3.20)) can be separated into an electronic and a nuclear eigenvalue problem. The former reads

$$\hat{H}_{\mathrm{e}}\left|\psi_n^{(e)}\right\rangle = \left(\hat{T}_e + \hat{\mathcal{V}}_{NN} + \hat{\mathcal{V}}_{ee} + \hat{\mathcal{V}}_{Ne}\right)\left|\psi_n^{(e)}\right\rangle = \mathcal{V}_n^{\mathrm{BO}}\left|\psi_n^{(e)}\right\rangle, \tag{3.25}$$

in which the stationary electronic wavefunctions $|\psi_n^{(e)}\rangle$ (with $4N^{(e)}$ degrees of freedom) and associated eigenenergies $\mathcal{V}_n^{\mathrm{BO}}$ are determined from the electronic Hamiltonian $\hat{H}_{\mathrm{e}}$. Importantly, although the nuclear coordinates still enter the electronic Hamiltonian through the terms $\hat{\mathcal{V}}_{NN}$ and $\hat{\mathcal{V}}_{Ne}$, they do so not as degrees of freedom but as fixed parameters because of the clamped nuclei approximation. By varying the nuclear coordinates and again solving eqn (3.25), the eigenenergies $\mathcal{V}_n^{\mathrm{BO}}$ can be determined as a function of the $3N$ nuclear coordinates. These hypersurfaces are known as the Born–Oppenheimer surfaces or potential energy surfaces (PES) of the system on which the nuclei move. Knowledge of these surfaces then allows one to solve the nuclear eigenvalue problem, the eigenstates of which describe the vibrations and rotations of the nuclei.

While the Born–Oppenheimer approximation decouples the electronic and the nuclear eigenvalue problem, the number of degrees of freedom remains formidable, especially for the electronic eigenvalue problem in eqn (3.25). As a result, approaches aiming to immediately determine the electronic wavefunctions – so-called wavefunction-based techniques such as Hartree–Fock – are computationally very expensive and are only seldomly used in MOF research. When they are used, first a small finite cluster is cut out of the MOF material to limit the number of electrons in the simulation model, as outlined in more detail in ref. 66.

A much more popular quantum-mechanical method to solve the electronic eigenvalue problem for solid-state materials (such as MOFs) was pioneered by

Hohenberg and Kohn, whose theorems form the basis of density functional theory (DFT).[67] These theorems state that the ground state energy $\mathcal{V}_0^{\text{BO}}$ of eqn (3.25) is a unique functional of the ground state electron density, which can be found variationally as the density that minimises this (unknown) functional.[68] Hence, instead of solving eqn (3.25) for the highly multidimensional wavefunction, the Hohenberg–Kohn theorems allow one to solve for the three-dimensional ground state electron density instead, and to obtain all relevant ground state properties from this density. In the traditionally used Kohn–Sham formalism, this problem is hugely simplified by considering a fictitious system of non-interacting electrons moving in an effective potential, which is chosen such that the system has the same ground-state electron density as the actual interacting system.[69] In this way, the variational principle gives rise to single-particle equations, the so-called Kohn–Sham equations, the solutions of which yield the ground-state electron density. While this method is formally exact up to this point, it features unknown contributions to the effective potential, arising from the unknown exchange-correlation functional. Hence, to actually find the ground-state electron density, one needs to postulate an appropriate form of this exchange-correlation functional. A vast number of different functionals have been proposed to this end, which strongly vary in computational cost and accuracy. However, as DFT is in general much faster than wavefunction-based approaches, it has been used extensively in MOF research. For a thorough discussion of the applications and limitations of DFT in modelling MOFs, the recent review by Mancuso *et al.* forms an excellent starting point.[70]

The atomic interactions in MOFs can also be modelled classically. The aim of classical interaction potentials, also called force fields, is to approximate the ground-state Born–Oppenheimer surface $\mathcal{V}_0^{\text{BO}}$ of eqn (3.25), which depends on all nuclear coordinates, by a sum of analytical functions that each depend on a limited number of nuclear coordinates (typically two, three, or four) and on a limited number of fitting parameters. Force fields can typically be divided into bonded and nonbonded interaction terms. The former are often simple functions of the internal coordinates of covalently bonded nuclei, such as bonds, bends, and torsion terms, while the latter describe interactions also between nonbonded nuclei, such as Coulomb and van der Waals interactions. Thanks to their analytical form, these classical potentials can be used to efficiently calculate the forces on each of the nuclei without the need to resort to expensive electronic structure methods. However, most force fields preclude the description of bond reorganisation, which limits their applicability. Different force fields differ not only in the form of the analytical functions they contain – some do inherently account for anharmonic interaction potentials, for instance – but also in the way the fitting parameters are determined. MOF force fields are often parametrised based on data obtained from electronic structure methods. For instance, in the first flexible MOF force field, derived by Greathouse *et al.*, an appropriate parametrisation of terms associated with the inorganic node in MOF-5 was obtained from preceding DFT calculations.[71] Recently, more systematic parametrisation approaches have been developed, in which the force field (FF) parameters are determined based on the *ab initio* optimised structure and the local shape of the PES around this optimised structure. Examples include MOF-FF,[72,73] BTW-FF,[74] QuickFF,[75,76] and VMOF.[77] In the context of this work, also the approach introduced by Heinen *et al.*, in which a force field was derived by fitting to the *ab initio* elastic constants,[78] is noteworthy to mention. For a more in-depth discussion of force fields for MOFs, the reader is referred to ref. 76 and 79.

3.3.3 Exploring the Potential Energy Surface for a System under Stress Control

When experimentally extracting the mechanical behaviour of a MOF material, the MOF sample is often held under temperature control, while simultaneously controlling either the pressure (or, more generally, stress) or the strain. To obtain comparable properties from a molecular simulation, it is necessary to impose the same control conditions while the system samples the PES constructed in Section 3.3.2, *i.e.*, to ensure that the system samples every region of the PES with a probability dictated by the experimental conditions. One way to trace such a representative trajectory on the PES is by performing a molecular dynamics (MD) simulation, in which the nuclear structure of the MOF is followed through time.[80,¶] In an MD simulation, temperature or stress control means that the Hamiltonian equations of motion following from eqn (3.9), which are simply the Newton equations for the nonrelativistic nuclei, need to be adapted. While temperature and pressure control is well developed, general stress control in MD simulations is substantially less straightforward as different subtleties arise for soft anisotropic materials such as MOFs, as shown below.

Central in the discussion of temperature and stress control of systems *in thermodynamic equilibrium* is the ensemble concept. An ensemble denotes the collection of microstates (here the positions and momenta of each of the N nuclei of the MOF and, if present, its simulation cell matrix $\mathbf{h}$) that are accessible under the given macroscopic conditions (here typically the temperature T and stress $\boldsymbol{\sigma}$ under which the experiment takes place). Ensembles are labelled based on the macroscopic variables that are controlled. These control variables come in conjugate pairs, including temperature T *vs.* entropy S (which also determines the energy E), pressure P *vs.* volume V, and chemical potential μ *vs.* number of particles N. An ensemble is fully defined by choosing one out of each pair of conjugate control variables, in which one needs to ensure that at least one of the variables scales with the size of the sample (either the entropy/energy, volume, or particle number). For gases

¶A very popular and versatile alternative to MD are Monte Carlo (MC) simulations,[309] which are very often used to study adsorption in MOFs. However, MC routines are typically less suited to describe collective structural framework deformations discussed here, as the associated MC trial moves are only seldomly accepted.

and liquids, the most often used ensembles are the microcanonical or *NVE* ensemble, the canonical or *NVT* ensemble, the isothermal–isobaric or *NPT* ensemble, and the grand canonical or μVT ensemble.[80]

For typical (Newtonian) fluids, the pressure P is the only stress state the material can withstand, and the volume forms an appropriate variable to describe the response of such a system to pressure control. For solids, however, this situation differs. Given that a solid is typically anisotropic, the volume alone does not suffice to describe the simulation cell, and instead the six independent components of the simulation cell matrix $\mathbf{h}$ introduced in eqn (3.1) need to be specified.[81,82,||] In addition, for anisotropic solids, it also makes sense to probe their response to anisotropic stresses $\boldsymbol{\sigma}$, as defined by the deviatoric Cauchy stress tensor $\boldsymbol{\sigma}_a$ that remains after subtracting the hydrostatic pressure from the Cauchy stress:**

$$\boldsymbol{\sigma}_a = \boldsymbol{\sigma} + P\mathbf{I}. \tag{3.26}$$

As a result, for anisotropic solids, instead of simply choosing between the scalar variables P and V as conjugate variables, one needs to choose between the matrix variables $\boldsymbol{\sigma}$ and $\mathbf{h}$, the latter of which defines the Lagrangian strain through eqn (3.4). However, $\boldsymbol{\sigma}$ and $\mathbf{h}$ do not form a pair of conjugate variables, which makes the definition of constant-stress ensembles substantially less straightforward (*vide infra*). While this distinction between ensembles for fluids and anisotropic solids was already recognised in 1984,[15] the nomenclature used in computational research can often obscure the exact ensemble – and hence the exact control variables – used during the simulation. For instance, the label '$N\boldsymbol{\sigma}T$ ensemble' has been used both to denote an ensemble in which the full stress $\boldsymbol{\sigma}$ is controlled (in addition to the number of particles and the temperature) and to denote an ensemble in which only the pressure P is controlled but in which the cell matrix $\mathbf{h}$ could change anisotropically (which is a more restrictive ensemble than the previous one).[83] Therefore, a more appropriate ensemble classification to study MOFs and other anisotropic solids has been proposed in ref. 83, which can be found in Table 3.1 and which will be used in this chapter.

Given the above discussion, it may come as no surprise that stress control or barostat algorithms were initially derived to control only the pressure, while allowing only for isotropic volume deformations. The first such barostat algorithm was proposed by Andersen.[84] Andersen recognised that pressure control could be achieved by describing the system as a function of the volume and the fractional coordinates defined in Section 3.2 and by

---

||Although the cell matrix contains nine components, three of them define the cell orientation with respect to the frame of reference and hence carry no intrinsic physical meaning. Therefore, the cell matrix is often assumed to be either symmetrised or upper triangular to fix these three degrees of freedom.

**Note that the pressure was defined in eqn (3.18) as *the negative* of the trace of the stress tensor. If this minus sign is not used, the plus sign in eqn (3.26) should also be replaced by a minus sign.

Table 3.1 Overview of the different thermodynamic ensembles considered for MOFs at a constant number of particles N according to the unified notation proposed in ref. 83, where crosses indicate which variables are controlled. Next to the number of particles, the ensemble is defined by controlling the volume V or the pressure P, the cell shape matrix $\mathbf{h}_0$ or the deviatoric stress $\boldsymbol{\sigma}_a$, and the energy/enthalpy (E or H) or temperature T. The last column indicates the notation used in popular software packages such as DL_POLY,[312] LAMMPS,[313] and CP2K/QUICKSTEP,[314] which may lead to confusion in some cases due to making no distinction between constant-pressure and constant-stress ensembles. Adapted from ref. 83 with permission from American Chemical Society, Copyright 2015.

Ensemble	N	V	$\mathbf{h}_0$	P	$\boldsymbol{\sigma}_a$	E or H	T	Other notations
$(N,V,\mathbf{h}_0,E)$	×	×	×			×		NVE
$(N,V,\mathbf{h}_0,T)$	×	×	×				×	NVT
$(N,V,\boldsymbol{\sigma}_a,H)$	×	×			×	×		
$(N,V,\boldsymbol{\sigma}_a,T)$	×	×			×		×	
$(N,P,\mathbf{h}_0,H)$	×		×	×		×		NPH, NPE_I
$(N,P,\mathbf{h}_0,T)$	×		×	×			×	NPT, NPT_I
$(N,P,\boldsymbol{\sigma}_a,H)$	×			×	×	×		NPH
$(N,P,\boldsymbol{\sigma}_a,T)$	×			×	×		×	NPT
$(N,V,\boldsymbol{\sigma}_a=\mathbf{0},H)$	×	×			$\boldsymbol{\sigma}_a=\mathbf{0}$	×		
$(N,V,\boldsymbol{\sigma}_a=\mathbf{0},T)$	×	×			$\boldsymbol{\sigma}_a=\mathbf{0}$		×	
$(N,P,\boldsymbol{\sigma}_a=\mathbf{0},H)$	×			×	$\boldsymbol{\sigma}_a=\mathbf{0}$	×		$N\boldsymbol{\sigma}H$, NPE_F
$(N,P,\boldsymbol{\sigma}_a=\mathbf{0},T)$	×			×	$\boldsymbol{\sigma}_a=\mathbf{0}$		×	$N\boldsymbol{\sigma}T$, NPT_F

extending the system's Hamiltonian of eqn (3.9) with two terms describing the kinetic energy and the (elastic) potential energy of a piston acting on the system's volume. This 'extended Hamiltonian' approach was generalised by Parrinello and Rahman by letting the piston define not only the volume of the simulation cell, but all components of the simulation cell matrix $\mathbf{h}$ instead (while still only accepting isotropic pressure as a control variable).[85] In 1981, Parrinello and Rahman extended this approach to allow imposing the whole stress tensor $\boldsymbol{\sigma}$,[86] which was later further refined to allow for simultaneous temperature control and cast in a true Hamiltonian form by Ray and Rahman.[15,87,88] In this latter formulation, which is implemented in many MD codes as the Parrinello–Rahman (PR) barostat, the system's Hamiltonian of eqn (3.9) is extended to read[††]

$$\mathcal{H}^{\mathrm{PR}}\left(\boldsymbol{s}^N,\boldsymbol{\pi}^N,\mathbf{h},\mathbf{p}_g\right)=\sum_{i=1}^{N}\frac{\left(\mathbf{h}^{-T}\boldsymbol{\pi}_i\right)\cdot\left(\mathbf{h}^{-T}\boldsymbol{\pi}_i\right)}{2m_i}+\mathcal{V}\left(\mathbf{h}\boldsymbol{s}^N\right)+\frac{\mathbf{p}_g:\mathbf{p}_g}{2W}-V_0\mathbf{S}:\boldsymbol{\eta}. \quad (3.27)$$

[††]The elastic term in eqn (3.27) is slightly more cumbersome in the original description, since Parrinello and Rahman made a distinction between applying a purely isotropic stress, in which case the elastic term should simply read $-P(V-V_0)$, and a stress state that is not purely isotropic, in which case eqn (3.27) is valid. As the interest here is specifically for the latter case, eqn (3.27) will be adopted throughout. It is important to note, however, that the PR barostat for isotropic stresses does not suffer from the same approximations as the anisotropic PR barostat.

In this expression, which can be extended to simultaneously allow for temperature control,[88] $\boldsymbol{\pi}_i = \mathbf{h}^T \boldsymbol{p}_i$ is the momentum conjugate to the fractional coordinates $\boldsymbol{s}_i$ of particle i, and $\mathbf{p}_g$ is the cell momentum matrix conjugate to the cell matrix $\mathbf{h}$. The third and fourth terms in eqn (3.27) describe the kinetic and elastic energy associated with cell deformations, the latter being described by the product of the reference volume V_0 and the double contraction of the second Piola–Kirchhoff tensor of eqn (3.16) and the finite Lagrangian strain of eqn (3.4). As a result, this Hamiltonian depends on a reference cell matrix $\mathbf{h}_0$ through both the reference cell volume and the strain, raising the same limitations on its applicability as discussed in Section 3.2.1.

Although being widely used, the PR barostat suffers from two inaccuracies which may become nonnegligible for soft framework materials such as MOFs. First, since the transformation from the original Hamiltonian in eqn (3.9) to that of eqn (3.27) is not canonical, as shown in ref. 8, the PR equations of motion are only satisfied exactly if the time derivative of the cell matrix can be neglected with respect to the time derivatives of the fractional coordinates ($|\dot{\mathbf{h}}\boldsymbol{s}_i| \ll |\mathbf{h}\dot{\boldsymbol{s}}_i|$).[15] Second, the direct dependence of the equations of motion on the cell matrix is unsatisfactory. Consider a modular transformation $\mathbf{T}$ that transforms the cell matrix $\mathbf{h}$ to $\mathbf{h}' = \mathbf{hT}$, where $\mathbf{T}$ is a 3×3 matrix only containing integer elements and chosen such that $\det\mathbf{T} = 1$. This transformation is a simple volume-conserving redefinition of the periodic cell vectors. Describing the system in terms of $\mathbf{h}'$ is completely interchangeable with describing it in terms of $\mathbf{h}$. However, the PR Hamiltonian and equations of motion do explicitly depend on this choice. As a result, the PR equations of state are not modularly invariant (or T-invariant), as recognised by Cleveland and Wentzcovitch, and a system's trajectory through time in a PR-controlled MD simulation will artificially depend on the chosen cell matrix.[89,90]

A solution to both drawbacks is to select the deformation gradient $\mathbf{F}$ of eqn (3.2) instead of $\mathbf{h}$ as the fundamental variable describing the cell deformation. By construction with respect to a reference cell matrix, $\mathbf{F}$ is modularly invariant, and any Hamiltonian that only depends on $\mathbf{F}$ is hence also modularly invariant.[91] This approach was followed by Tadmor *et al.*, who used the deformation gradient to define the generating function of eqn (3.10).[8] Given that, in an MD simulation, the deformation gradient explicitly depends on time, this generating function yields the transformed Hamiltonian

$$\mathcal{H}_0\left(\boldsymbol{r}_0^N, \boldsymbol{p}_0^N, \mathbf{F}\right) = \sum_{i=1}^{N} \frac{\left(\mathbf{F}^{-T} \boldsymbol{p}_{i,0}\right) \cdot \left(\mathbf{F}^{-T} \boldsymbol{p}_{i,0}\right)}{2m_i} + \mathcal{V}\left(\mathbf{F}\boldsymbol{r}_0^N\right) - \sum_{i=1}^{N} \left(\mathbf{F}^{-T} \boldsymbol{p}_{i,0}\right) \cdot \left(\dot{\mathbf{F}}\boldsymbol{r}_{i,0}\right), \quad (3.28)$$

Extending this Hamiltonian with the kinetic and potential energy terms of the barostat yields

$$\begin{aligned} \mathcal{H}\left(\boldsymbol{r}_0^N, \boldsymbol{p}_0^N, \mathbf{F}, \mathbf{p}_g\right) = &\sum_{i=1}^{N} \frac{\left(\mathbf{F}^{-T} \boldsymbol{p}_{i,0}\right) \cdot \left(\mathbf{F}^{-T} \boldsymbol{p}_{i,0}\right)}{2m_i} + \mathcal{V}\left(\mathbf{F}\boldsymbol{r}_0^N\right) + \frac{\mathbf{p}_g : \mathbf{p}_g}{2W} \\ &- V_0 \mathbf{S} : \boldsymbol{\eta} - \sum_{i=1}^{N} \left(\mathbf{F}^{-T} \boldsymbol{p}_{i,0}\right) \cdot \left(\dot{\mathbf{F}}\boldsymbol{r}_{i,0}\right). \end{aligned} \quad (3.29)$$

Compared to the PR Hamiltonian of eqn (3.27), an additional last term appears in eqn (3.29) which ensures that modularly invariant equations of motion are obtained under stress control.[8]

A remaining issue, which was also present in the PR barostat, is that both Hamiltonian formulations – and hence also the equations of motion – feature the second Piola–Kirchhoff tensor $\mathbf{S}$ as the stress control variable. As a result, the second Piola–Kirchhoff tensor is also the property to which the system converges, as demonstrated in Figure 3.5 for an MD simulation of an aluminium crystal, although it is the Cauchy stress $\boldsymbol{\sigma}$ that is controlled experimentally.[92] While this difference between both properties is negligible for small deformations, it becomes important for larger deformations from the cell matrix $\mathbf{h}_0$, as can be observed from eqn (3.16)–(3.17). This difference may hence play a crucial role during phase transitions in MOFs. To overcome this, Miller *et al.* developed an adaptive 'Cauchystat', in which the Cauchy stress is directly controlled through proportional feedback control.[92]

Besides these Hamiltonian approaches to impose stress control, many other barostats have been introduced. Among those still used extensively today are the Berendsen barostat,[93] the Hoover barostat,[94,95] the barostat developed by Martyna, Tobias, and Klein (MTK),[96] and the Langevin barostat.[97] Given that these approaches do not start from a canonical transformation of the Hamiltonian, they allow for more freedom in defining the equations of motion which may also benefit their convergence.[92] Recently, stress control achieved through the Berendsen, MTK, and Langevin barostats was compared to study phase transformations in MOFs, demonstrating that the Berendsen barostat does not sample the correct constant-stress ensemble, whereas the MTK and Langevin barostats lead to interchangeable results, as shown in Figure 3.6.[83]

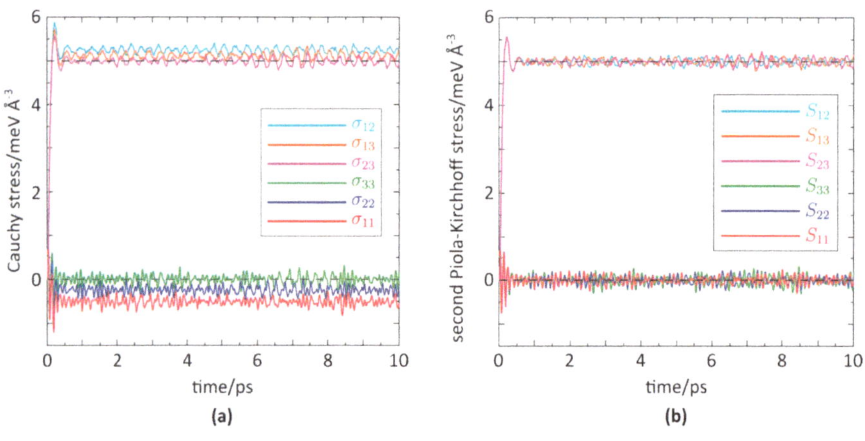

Figure 3.5 Instantaneous values of (a) the Cauchy stress and (b) the second Piola–Kirchhoff stress during a short MD simulation of a perfect aluminium crystal using the PR barostat. The horizontal dashed lines indicate the stress values supplied to the PR barostat (5 meV Å^{-3} for the off-diagonal elements, 0 meV Å^{-3} for the diagonal elements), showing that the components of the second Piola–Kirchhoff tensor rather than those of the Cauchy stress converge to these values. Adapted from ref. 92 with permission from the American Institute of Physics, Copyright 2016.

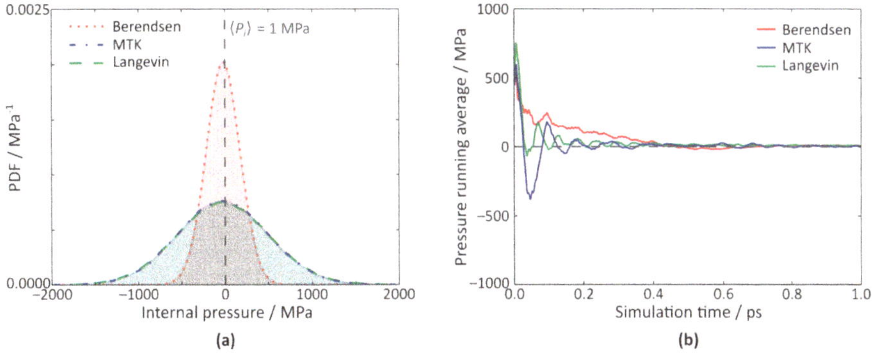

Figure 3.6 Internal pressure exhibited by MIL-53(Al) during an (N, P, $\boldsymbol{\sigma}_a = \mathbf{0}$, T) MD simulation at 1 MPa and 300 K using the Berendsen, MTK, and Langevin barostats and starting from the lp phase of this material at 0 K. (a) Probability density function (PDF) generated over a simulation time of 800 ps. (b) Running average of the internal pressure generated for the first picosecond. Adapted from ref. 83 with permission from American Chemical Society, Copyright 2015.

3.4 The Equilibrium Mechanical Behaviour of MOFs: the Elastic Regime

MOFs exhibit an extremely rich versatility in their response to mechanical stimuli, especially when the magnitude of the stimulus increases (see Sections 3.5 and 3.6). However, near an equilibrium configuration, the mechanical response of a MOF can be completely characterised by the stiffness or elasticity tensor **C**, which relates the (Cauchy) stress $\boldsymbol{\sigma}$ that needs to be exerted on the material to realise an infinitesimal (Cauchy) strain $\boldsymbol{\varepsilon}$:

$$\boldsymbol{\sigma} = \mathbf{C} : \boldsymbol{\varepsilon} \Leftrightarrow \sigma_{\alpha\beta} = \sum_{\gamma=1}^{3}\sum_{\delta=1}^{3} C_{\alpha\beta\gamma\delta}\varepsilon_{\gamma\delta}. \tag{3.30}$$

Eqn (3.30) can be regarded as the 3D generalisation of Hooke's law for a 1D spring.‡‡ The infinitesimal strain $\boldsymbol{\varepsilon}$ entering this expression is obtained from both the finite Lagrangian and Eulerian strains defined in eqn (3.4) and eqn (3.5) in the limit of small (infinitesimal) deformations.[98] Since both the stress and strain are 3 × 3 second-rank tensors, the stiffness tensor **C** as well as its inverse, the compliance tensor $\mathbf{S} = \mathbf{C}^{-1}$, are 3 × 3 × 3 × 3 fourth-rank tensors, the elements of which completely govern the response of a material to 'small' stresses.

‡‡Eqn (3.30) is not the only way to define the stiffness tensor. Alternative definitions are based on the energy density of a deformed material in the harmonic approximation, eqn (3.33), or through the equation of motion of an elastic wave propagating through the material, an approach that is often followed experimentally,[247] but which is seldomly used computationally.[299] For an unstressed material, all three definitions are interchangeable. Only when a material is held under external stress does care need to be taken to distinguish between these three definitions.[247]

To properly define 'small' stresses, it suffices to expand the Helmholtz free energy of eqn (3.8) with respect to the Lagrangian strain $\boldsymbol{\eta}$ around the unstressed equilibrium configuration:

$$F(N,T;\boldsymbol{\eta}) = F(N,T;\boldsymbol{\eta}_0 = \mathbf{0}) + \sum_{\alpha,\beta=1}^{3} \left.\frac{\partial F(N,T;\boldsymbol{\eta})}{\partial \eta_{\alpha\beta}}\right|_{\boldsymbol{\eta}_0=\mathbf{0}} \eta_{\alpha\beta}$$
$$+\frac{1}{2}\sum_{\alpha,\beta,\gamma,\delta=1}^{3} \left.\frac{\partial^2 F(N,T;\boldsymbol{\eta})}{\partial \eta_{\alpha\beta}\partial \eta_{\gamma\delta}}\right|_{\boldsymbol{\eta}_0=\mathbf{0}} \eta_{\alpha\beta}\eta_{\gamma\delta} + \mathcal{O}(\eta^3). \tag{3.31}$$

By combining eqn (3.7) with the definitions of eqn (3.4) and (3.16), one finds that $\partial F/\partial \eta_{\alpha\beta} = V_0 S_{\alpha\beta}$, where V_0 is the volume of the reference configuration and $S_{\alpha\beta}$ are the elements of the second Piola–Kirchhoff tensor. Hence, in an unstressed configuration, the second term in the right-hand side of eqn (3.31) vanishes. Furthermore, $F(N, T; \boldsymbol{\eta}_0 = \mathbf{0}) \equiv F_0$ is a constant that can be absorbed by the left-hand side of eqn (3.31). As a result, eqn (3.31) can be rewritten as:

$$F(N,T;\boldsymbol{\eta}) - F_0 = \frac{V_0}{2}\sum_{\alpha,\beta,\gamma,\delta=1}^{3} \left.\frac{\partial S_{\alpha\beta}(N,T;\boldsymbol{\eta})}{\partial \eta_{\gamma\delta}}\right|_{\boldsymbol{\eta}_0=\mathbf{0}} \eta_{\alpha\beta}\eta_{\gamma\delta} + \mathcal{O}(\eta^3). \tag{3.32}$$

For a configuration sufficiently close to equilibrium, we may replace the Lagrangian strain $\boldsymbol{\eta}$ in eqn (3.32) by the infinitesimal strain $\boldsymbol{\varepsilon}$, and the second Piola–Kirchhoff tensor $\mathbf{S}$ by the Cauchy tensor $\boldsymbol{\sigma}$. Together with the definition of the stiffness tensor in eqn (3.30), this yields:

$$F(N,T;\boldsymbol{\varepsilon}) - F_0 = \frac{V_0}{2}\sum_{\alpha,\beta,\gamma,\delta=1}^{3} C_{\alpha\beta\gamma\delta}\varepsilon_{\alpha\beta}\varepsilon_{\gamma\delta} + \mathcal{O}(\varepsilon^3) = \frac{V_0}{2}\boldsymbol{\varepsilon}:\mathbf{C}:\boldsymbol{\varepsilon} + \mathcal{O}(\varepsilon^3). \tag{3.33}$$

Eqn (3.33) demonstrates the importance of the stiffness tensor: its elements completely define the (change in) free energy of the material upon infinitesimal strain, and hence completely define how the material will respond to a small mechanical stimulus. Because of the truncation of eqn (3.33) up to second order, the 81 elements that enter the stiffness tensor are called second-order elastic constants (higher-order elastic constants exist but have to date not been reported for MOFs). A very similar derivation starting from the internal energy instead of the free energy gives an expression similar to eqn (3.33); the difference being that the internal energy U is the appropriate thermodynamic potential when entropy is controlled instead of temperature. As a result, the elastic constants derived when starting from the internal energy are defined as the *adiabatic* elastic constants $V_0 C_{\alpha\beta\gamma\delta} = \partial^2 U/\partial\varepsilon_{\alpha\beta}\partial\varepsilon_{\gamma\delta}$, whereas eqn (3.33) gives rise to the *isothermal* elastic constants $V_0 C_{\alpha\beta\gamma\delta} = \partial^2 F/\partial\varepsilon_{\alpha\beta}\partial\varepsilon_{\gamma\delta}$.

The 81 elements within the stiffness and compliance tensors are not all independent. As these tensors are the second-order derivatives of the free energy with respect to the strain and the order of differentiation is unimportant, $C_{\gamma\delta\alpha\beta} = C_{\alpha\beta\gamma\delta}$ and only 45 independent elements remain (the major symmetry of **C**). Furthermore, because of the symmetry of the stress and strain tensors, eqn (3.30) implies that $C_{\beta\alpha\gamma\delta} = C_{\alpha\beta\delta\gamma} = C_{\alpha\beta\gamma\delta}$ (the minor symmetries of **C**).[8] As a result, for a general material, 21 independent elastic constants exist, although the symmetry of the unit cell can reduce this number further – *e.g.*, the elastic tensor of a cubic material is defined by only three independent constants.[99] Because of the general symmetries of the stress, strain, and stiffness tensors, one often writes the infinitesimal strain and Cauchy stress tensors as 6 × 1 vectors that inherently take into account these symmetries using the so-called Voigt contraction,[100] as detailed further in Section 2.2.1 of Chapter 2:

$$\boldsymbol{\sigma} = [\sigma_{11}\ \sigma_{22}\ \sigma_{33}\ \sigma_{23}\ \sigma_{13}\ \sigma_{12}]^T \quad \boldsymbol{\varepsilon} = [\varepsilon_{11}\ \varepsilon_{22}\ \varepsilon_{33}\ 2\varepsilon_{23}\ 2\varepsilon_{13}\ 2\varepsilon_{12}]^T. \tag{3.34}$$

Likewise, the fourth-rank compliance and stiffness tensors can be reduced to 6 × 6 matrices that account for the minor symmetries:

$$\mathbf{C} = \begin{bmatrix} C_{1111} & C_{1122} & C_{1133} & C_{1123} & C_{1113} & C_{1112} \\ C_{2211} & C_{2222} & C_{2233} & C_{2223} & C_{2213} & C_{2212} \\ C_{3311} & C_{3322} & C_{3333} & C_{3323} & C_{3313} & C_{3312} \\ C_{2311} & C_{2322} & C_{2333} & C_{2323} & C_{2313} & C_{2312} \\ C_{1311} & C_{1322} & C_{1333} & C_{1323} & C_{1313} & C_{1312} \\ C_{1211} & C_{1222} & C_{1233} & C_{1223} & C_{1213} & C_{1212} \end{bmatrix}$$
$$\mathbf{S} = \begin{bmatrix} S_{1111} & S_{1122} & S_{1133} & 2S_{1123} & 2S_{1113} & 2S_{1112} \\ S_{2211} & S_{2222} & S_{2233} & 2S_{2223} & 2S_{2213} & 2S_{2212} \\ S_{3311} & S_{3322} & S_{3333} & 2S_{3323} & 2S_{3313} & 2S_{3312} \\ 2S_{2311} & 2S_{2322} & 2S_{2333} & 4S_{2323} & 4S_{2313} & 4S_{2312} \\ 2S_{1311} & 2S_{1322} & 2S_{1333} & 4S_{1323} & 4S_{1313} & 4S_{1312} \\ 2S_{1211} & 2S_{1222} & 2S_{1233} & 4S_{1223} & 4S_{1213} & 4S_{1212} \end{bmatrix}. \tag{3.35}$$

Due to the major symmetry of **C**, the matrices in eqn (3.35) are symmetric. The factor of two appearing in the last three elements of the strain vector of eqn (3.34) and the factors of two and four appearing in the compliance matrix of eqn (3.35) arise so that Hooke's law in 3D, eqn (3.30), can be written in Voigt notation as $\boldsymbol{\sigma} = \mathbf{C}\boldsymbol{\varepsilon}$ or $\boldsymbol{\varepsilon} = \mathbf{S}\boldsymbol{\sigma}$. In what follows, the Voigt notation will be systematically adopted.

Although eqn (3.35) fully defines the elastic response of a material, these matrices are often not very practical to interpret, especially for low-symmetry materials. For instance, anomalous mechanical properties may not be directly clear from inspecting these matrices and it may be cumbersome to predict and visualise the response of a material to a specific deformation based on eqn (3.35), although it contains all necessary ingredients. Therefore, studies on the elastic behaviour of MOFs often report the elastic moduli, such as the bulk modulus K (an isotropic property), the Young's modulus

$E(\boldsymbol{u})$ (the value of which depends on the direction $\boldsymbol{u}$ along which the stress is applied, which coincides with the direction along which the deformation is measured), and the shear modulus $G(\boldsymbol{u},\boldsymbol{v})$ (which depends both on the direction of the applied stress and the orthogonal direction of the deformation measurement). Programs such as ElAM[101] and ELATE[102] form an instrumental tool in this regard, as they allow one to easily visualise these directional elastic *moduli* – as well as the directional Poisson's ratio and linear compressibility – using the elastic *constants* as input. Alternatively, directional moduli can be averaged out to obtain a 'mean' elastic modulus, using either the Voigt (based on the stiffness tensor, assuming a uniform strain[100]), Reuss (based on the compliance tensor, assuming a uniform stress[103]), or Hill (the arithmetic average of the Voigt and Reuss values[104]) averaging schemes. For anisotropic materials, such as most MOFs, however, the usefulness of these average values is limited.

Based on eqn (3.33), the elastic regime of a material – or more precisely, of a given phase of a material – can now be rigorously defined as that region in deformation space for which eqn (3.33) is (approximately) valid; this elastic regime will be the focus of this section. First, in Sections 3.4.1 and 3.4.2, the two main methods to extract elastic constants from a simulation will be discussed, illustrated with examples from MOF literature. Subsequently, in Section 3.4.3, we will focus our attention on how the equilibrium elastic constants can be used to predict whether a MOF is prone to single-crystal-to-single-crystal phase transitions under moderate pressures. In Section 3.4.4, the limits of the elastic regime will be explored to identify under which conditions the material becomes unstable, at which point it enters the inelastic regime discussed in Sections 3.5 and 3.6.

3.4.1 Extracting Elastic Constants Through Explicit Deformations

The most popular method to derive elastic constants starts with considering a set of linearly independent deformation modes and calculating either the energy or the stress of the material strained along this deformation mode. In Section 3.4.1.1, the general methodology behind this approach will be outlined. Given the relatively large body of work on the mechanics of isoreticular MOFs (IRMOFs) and ZIF-like materials, these studies will be used in Sections 3.4.1.2 and 3.4.1.3 to illustrate the different flavours of the explicit deformation approach and how these choices affect the extracted elastic moduli. Afterwards, the UiO-66 family and HKUST-1 will be discussed in Sections 3.4.1.4 and 3.4.1.5, focussing on the importance of taking into account the intrinsic disorder in these materials. Section 3.4.1.6 focuses on the MIL-53/MIL-47 family of wine-rack type materials, given that they exhibit pressure-induced flexibility. Finally, Section 3.4.1.7 provides a brief overview of other MOFs for which elastic constants have been extracted using the explicit deformation approach.

3.4.1.1 General Methodology

The adiabatic elastic constants of a material, $C_{\alpha\beta\gamma\delta} = 1/V_0(\partial^2 U/\partial\varepsilon_{\alpha\beta}\partial\varepsilon_{\gamma\delta})$, can be completely defined using six independent deformation modes, which in turn can be constructed by considering deformations in which all but one of the elements in the strain vector of eqn (3.34) are put to zero. For each deformation mode, the nonzero entry is set to $n\Delta\varepsilon$, where $\Delta\varepsilon$ is a small strain increment and n takes on $2n_{\text{max}} + 1$ different discrete values centred around $n = 0$ (corresponding to the undeformed reference state). Appropriate values of $\Delta\varepsilon$ depend on the system and elastic modulus; for smaller elastic moduli and moduli corresponding to volume-preserving deformations, typically larger increments need to be chosen to ensure the obtained energy difference between the strained and unstrained structure exceeds the computational inaccuracy. A comprehensive example of how the strain increment size and the value of n_{max} may impact the stiffness tensor of a MOF can be found in ref. 78. For each of the resulting strains $n\Delta\varepsilon$, either the stress or the energy of the system is calculated through a relaxation of the fractional coordinates while keeping the deformed cell matrix fixed. The stiffness tensor **C** is then found by requiring that the imposed strains and the resulting stresses satisfy eqn (3.30) in a least-squares sense or, in the case the energy was measured, that the adiabatic version of eqn (3.33) is satisfied in a least-squares sense.[105] Any excessive remaining error after this fit may indicate that too large a strain increment $\Delta\varepsilon$ was chosen, so that the maximum deformation $n_{\text{max}}\,\Delta\varepsilon$ no longer falls in the elastic regime, or that too small a strain increment was chosen, so that the obtained data is marred by the inaccuracy of the used level of theory. Typically, this boils down to deformations of a few percent. For more symmetric MOFs, such as the cubic IRMOF or UiO-66 series, the number of deformation modes can be reduced given the lower number of independent elastic constants. The basic algorithm outlined here gives rise to *adiabatic* elastic constants, at 0 MPa and at 0 K, although approximate approaches to obtain isothermal elastic constants are also discussed below. In contrast, Sections 3.4.2 and 3.4.4 will focus on dedicated techniques by which to obtain *isothermal* elastic constants at finite temperatures and/or pressures.

In many cases, one is especially interested in the volumetric contraction of the MOF upon a small hydrostatic pressure, as expressed through the bulk modulus K. While the bulk modulus can be determined from the elastic constants calculated above, it can also be calculated directly by considering only one type of deformation, namely a volumetric deformation. In this case, a number of initial structures are prepared with volumes $V = V_0 + n\Delta V$, where V_0 is the reference volume and ΔV is a small volume increment. As before, n takes on discrete values around zero, corresponding to the reference state. Typically, volumes of up to around 10% above and below the equilibrium volume are sampled, with the main criterion being that the elastic regime should be sampled. At each volume, a full optimisation of the nuclear positions and the cell shape is performed, and the final energy is tabulated. From the resulting energy *versus* volume profile, the bulk modulus can then be obtained by the second-order derivative of the profile around its equilibrium value, as§§

$$K = V_0 \left. \frac{\partial^2 E}{\partial V^2} \right|_{V=V_0} \tag{3.36}$$

As this second-order derivative may lead to large inaccuracies, a more appropriate way to extract the bulk modulus from the $E(V)$ profile, is to fit the simulated $E(V)$ data to a known equation of state (EOS), such as the Murnaghan,[106] Birch–Murnaghan,[107] or Rose–Vinet[108] equation of state. In all these equations of state, the bulk modulus (as well as its pressure derivative) appear as fitting parameters. Appropriately fitting the simulated $E(V)$ data therefore yields an alternative way of calculating the bulk modulus and inherently takes into account the anharmonicities described by the EOS, whereas bulk moduli extracted from the elastic stiffness tensor are necessarily harmonic approximations. Furthermore, this fitting procedure has the added benefit that it allows determination of the equilibrium volume without the Pulay effect that mars volume optimisations in plane-wave DFT codes.[109] While all three equations of state are used interchangeably, the Rose–Vinet equation of state has been observed to better describe the response of a solid material upon contraction, and is hence the most appropriate choice, especially when treating flexible MOFs.[109] To account for temperature, the energy E in eqn (3.36) should be replaced by the free energy F, which can be obtained using dedicated simulation techniques, as discussed in Section 3.5.3.

Finally, a last direct approach to calculate the 0 K elastic stiffness tensor **C** is by splitting the tensor into two components (in Voigt notation):[110,111]

$$\mathbf{C} = \bar{\mathbf{C}} - \frac{1}{V_0} \mathbf{\Lambda}^T \mathbf{H}^- \mathbf{\Lambda}. \tag{3.37}$$

In this expression, $\bar{\mathbf{C}}$ is the Born term (also called the clamped-ion or frozen-ion elastic stiffness tensor): this is the stiffness tensor that is obtained when straining the material *without allowing the ions to equilibrate in this strained unit cell* (*i.e.*, at fixed fractional coordinates, without performing a nuclear optimisation in the above procedure). As this would overestimate the true stiffness of the material, the second term in eqn (3.37), the 'nuclear relaxation term', corrects for this *via* the $3N \times 6$ force-response internal strain tensor $\mathbf{\Lambda}$ and the $3N \times 3N$ Hessian matrix **H**, both evaluated in the equilibrium configuration and defined component-wise as

$$\Lambda_{ij} = -\frac{\partial^2 U}{\partial x_i \partial \varepsilon_j} = -\frac{\partial \sigma_j}{\partial x_i} \quad \text{and} \quad H_{ij} = \frac{\partial^2 U}{\partial x_i \partial x_j}, \tag{3.38}$$

§§The symbol E instead of U will be used here to denote the 0 K (internal) energy of the system, in line with the general nomenclature when discussing equations of state. The reader must not confuse this with the Young's modulus denoted also by symbol E.

where x_i is the ith element of the vector $\boldsymbol{r}^N$ defined in Section 3.2. The eigenmodes and eigenvalues of the Hessian matrix $\mathbf{H}$ define the $3N-3$ vibrational modes of the undeformed periodic system and the associated harmonic frequencies (the three 'missing' eigenvectors correspond to global translations of the periodic system, which have a zero frequency). The force-response internal strain tensor $\boldsymbol{\Lambda}$ contains derivatives of the energy with respect to both the nuclear displacement x_i and the strain component ε_j, and can be regarded as the off-diagonal elements of an 'extended' or 'generalised' Hessian matrix:[110]

$$\mathbf{H}^{\text{ext}} = \begin{bmatrix} \mathbf{H} & -\boldsymbol{\Lambda} \\ -\boldsymbol{\Lambda}^T & V_0\bar{\mathbf{C}} \end{bmatrix}. \tag{3.39}$$

With these definitions, it is clear that the second term in eqn (3.37) can be interpreted as the product of the transpose of the displacement-response internal strain tensor $\boldsymbol{\Gamma}^T = \boldsymbol{\Lambda}^T\mathbf{H}^-$, which describes the first-order displacements resulting from a first-order strain, and the force-response internal strain tensor $\boldsymbol{\Lambda}$.[110] Both here and in eqn (3.37), $\mathbf{H}^-$ denotes the pseudo-inverse of the Hessian matrix $\mathbf{H}$, as the aforementioned three zero frequency modes precludes one from taking the proper inverse. This 'extended Hessian' approach is only seldomly used for MOFs, given that these materials typically exhibit many low-frequency modes (Section 1.7 of Chapter 1) that are difficult to calculate accurately yet dominate its pseudo-inverse. As a result, the direct calculation of the elastic stiffness tensor is more straightforward in most cases, although the extended Hessian approach does allow partitioning the nuclear part of the elastic constants into different vibrational contributions, as pursued by Maul *et al.* to understand the nuclear origin for shear deformation in ZIF-8.[112]

3.4.1.2 *The IRMOF Family*

As a first case study, we draw our attention to MOF-5 (IRMOF-1), which is composed of Zn_4O inorganic nodes that are sixfold connected through benzene-1,4-dicarboxylate (BDC) organic ligands (see Figure 3.7(a)).[113] MOF-5 was originally introduced as a very promising material for hydrogen storage, with a hydrogen uptake of up to 4.5 wt% at 78 K.[114] For this purpose, the material should also be stable under (relatively low) mechanical pressures, explaining why MOF-5 and its sister materials in which the BDC ligand is replaced by other dicarboxylate ligands, the IRMOF series, were the first MOFs whose mechanical behaviour was extensively characterised both experimentally and theoretically. As MOF-5 exhibits a cubic conventional unit cell in equilibrium ($Fm\bar{3}m$ space group), its stiffness tensor only features three independent elastic constants: C_{11}, C_{12}, and C_{44}. As a result, its elastic constants can be determined using three independent deformation modes, typically an isotropic contraction (yielding $K = (C_{11} + 2C_{12})/3$), a volume-conserving orthorhombic strain (yielding C_{11}–C_{12}), and a volume-conserving monoclinic shear (yielding C_{44}).[115]

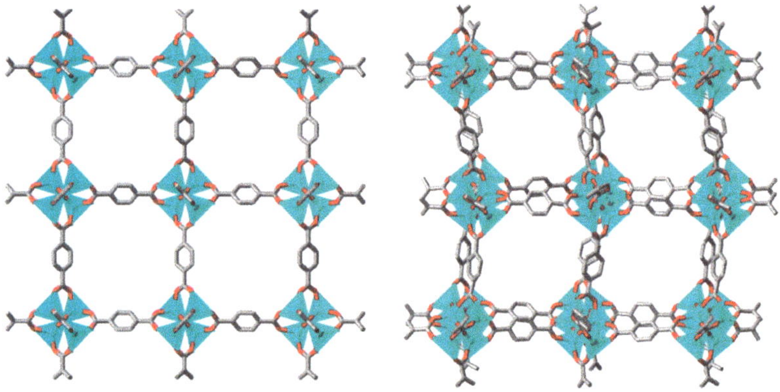

Figure 3.7 Structure of (a) MOF-5 or IRMOF-1 and (b) the Δ enantiomer of chiral MOF-5 (CMOF-5). Hydrogen atoms have been omitted for clarity, ZnO_4 tetrahedra, oxygen atoms, and carbon atoms are coloured blue, red, and grey, respectively. Adapted from ref. 144 with permission from American Chemical Society, Copyright 2016.

The first DFT studies on MOF-5 mechanics consistently revealed a very low C_{44} shear elastic constant. As explored in more detail in Section 3.4.4, this indicates that the material is prone to shear instability. However, the exact value of this constant – and hence the exact point at which the material becomes unstable – depends substantially on the technical aspects of the simulation. While Mattesini *et al.* reported a value of 7.54 GPa using a local density approximation (LDA) of the exchange-correlation functional and a double-zeta atomic-centred basis set,[116] a plane-wave basis set with an LDA functional resulted in a much lower value of 1.4 GPa or 1.16 GPa,[117,118] depending on the implementation code, the size of the basis set, and the size of the sampling grid in reciprocal space. At a higher level of theory, using a generalised gradient approximation (GGA), Bahr *et al.* reported an elastic constant C_{44} of 3.6 GPa.[119] The same variations in magnitude arise for the other two elastic constants, with C_{11} varying between 21.5 GPa and 29.4 GPa and C_{12} varying between 10.6 GPa and 14.8 GPa,[116–119] although the relative importance of these variations is evidently smaller than for the stability-defining C_{44} elastic constant. This first observation immediately emphasises that the impact of technical aspects in these simulations, including the choice of functional, the type of basis (plane wave basis *vs.* atomic-centred basis), the basis set size, the reciprocal sampling grid, and the unit cell size, should be carefully investigated to ensure that convergence has been reached for each of these aspects before calculating the stiffness tensor.[120] To give an idea of the magnitude of such possible deviations, Nazarian *et al.* performed a benchmark study including twelve MOFs and six DFT functionals and observed deviations of up to 20 GPa in elastic moduli depending on the choice of functional.[121]

To obtain a more accurate prediction of the MOF-5 stiffness tensor, Banlusan *et al.*[120] used the Perdew–Burke–Ernzerhof (PBE) functional[122] including Grimme D3 dispersion corrections,[123] which has become a very popular choice to account for long-range dispersion interactions in periodic calculations. In addition, they also used larger triple-zeta valence basis sets including polarisation functions for all elements except for zinc, obtaining results – C_{11} = 25.44 GPa, C_{12} = 10.92, and C_{44} = 1.00 GPa – that are largely in agreement with the aforementioned LDA values. Based on these elastic constants, they also calculated the MOF-5 directional Young's modulus. As illustrated in Figure 3.8(a), the highest Young's modulus in MOF-5, obtained when uniaxially deforming the material along the ⟨100⟩ directions along which the ligands are located, is about 6.5 times larger than its value along the ⟨111⟩ directions, which points along the body diagonal into the MOF-5 cavity.[120]

In their 2007 study, Bahr *et al.* initially aimed to follow the general procedure sketched above to determine the elastic constants.[119] However, the energy *versus* deformation curve they obtained was not sufficiently smooth to extract the elastic constants. They hypothesised that this could be caused by local dynamic effects – *in casu* the rotation of the linker about its symmetry axis – that are not accurately captured by DFT optimisations.[119] While they circumvented this issue by determining the stiffness tensor *via* only a single deformation and calculating the resulting strain, the question remained as to how the increased importance of entropy at higher temperatures would affect the elastic behaviour of MOF-5. Traditional materials, which expand upon heating as covalent bonds tend to expand due to anharmonicities, typically become less rigid at higher temperatures. For MOF-5, which instead shows NTE,[124] this reasoning would predict a hardening of the material at higher temperatures.

To verify this hypothesis, the MOF-5 elastic constants should be calculated at finite temperatures, for instance by performing MD simulations for a series of deformed structures in the $(N,V,\mathbf{h}_0,T)$ ensemble at the target temperature and calculating the time-averaged stress. Given the computational cost associated with these dynamic simulations compared to optimisations,

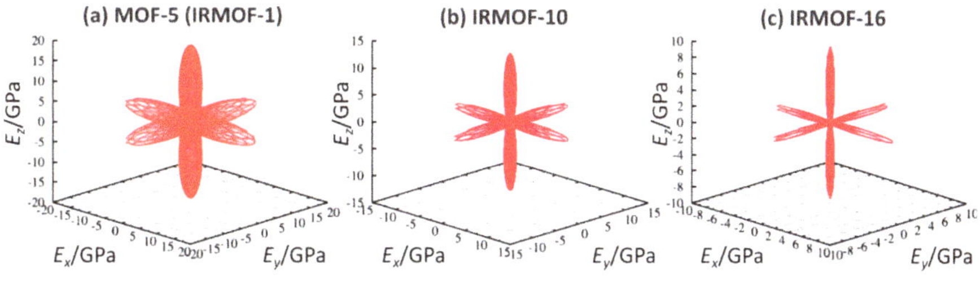

Figure 3.8 Directional 0 K Young's modulus of (a) MOF-5, (b) IRMOF-10, and (c) IRMOF-16 as calculated using the PBE+D3 level of theory. Adapted from ref. 120 with permission from the American Institute of Physics, Copyright 2017.

the first isothermal elastic constants at room temperature were determined using a variety of force fields instead of *ab initio* techniques.[125–129] These simulations confirmed the experimental NTE behaviour of MOF-5, but also predicted the material to become *softer* at higher temperatures despite this NTE. For instance, the generic DREIDING force field[130] predicted a decrease in both C_{11} and C_{44}, from 44.53 GPa to 26.27 GPa and from 1.82 GPa to 1.04 GPa, respectively, when increasing the temperature from 10 K to 600 K, while C_{12} was predicted to increase from 6.79 GPa to 9.83 GPa under the same conditions.[125] Similar decreases in elastic moduli were also predicted using the CVFF force field[131] by Greathouse *et al.*[126] However, the force field values of the elastic constants at 0 K differ appreciably from the ones obtained with DFT mentioned earlier, which was partially explained by Dubbeldam *et al.* as they observed that some of the aforementioned 0 K elastic constants were incorrectly calculated at a stationary point on the PES that does not correspond to a minimum.[129] In addition, it is important to mention that different force fields can give substantially different elastic moduli, even surpassing the deviations between different functionals.[132] When comparing both generic force fields, such as UFF,[133] UFF4MOF,[134] and DREIDING,[130] as well as force fields that were specifically derived to describe MOFs, such as BTW-FF[74] and the one constructed in ref. 135, Boyd *et al.* observed variations in the largest MOF-5 elastic constant, C_{11}, of up to several tens of GPa between these different force fields.[132] This discrepancy can be traced back to the generic character of most of these force fields, or to the fact that they were fitted to experimental data, which makes it difficult to assess their accuracy for different materials.

Therefore, Tafipolsky *et al.* presented a force field specifically derived for MOF-5, starting from a higher-level *ab initio* description of the equilibrium structure and using a genetic algorithm approach that would later be used to generate the MOF-FF library.[127] Using this *ab initio*-derived force field, they obtained 0 K elastic constants amounting to 25.3 GPa, 8.9 GPa, and 2.3 GPa for C_{11}, C_{12}, and C_{44}, respectively. These results are in much closer agreement to the DFT results of Banlusan *et al.*,[120] again demonstrating the importance of using an appropriate and sufficiently accurate level of theory. Furthermore, at 300 K, Tafipolsky *et al.* observed a decrease of 15% and 40% in C_{11} and C_{12}, confirming the reduced stiffness of the material.[127] In 2017, Banlusan *et al.* also calculated the 300 K elastic constants directly *via ab initio* MD, obtaining values of 21.6 GPa, 7.6 GPa, and 0.8 GPa for C_{11}, C_{12}, and C_{44},[120] which coincide well with the 300 K results obtained by an *ab initio*-derived QuickFF force field (22.3 GPa, 8.9 GPa, and 4.0 GPa) except for the shear coefficient C_{44}.[136] Also at 300 K, Banlusan *et al.* observed very large anisotropy in the Young's modulus, which drops significantly upon small deviations from the stiff ⟨100⟩ directions (see Figure 3.9). This observation motivated them to hypothesise that the remaining mismatch between experimentally and theoretically determined elastic moduli in MOF-5 could, at least partially, be explained by small tilts of the crystal during the nanoindentation experiments.[120] In this regard, however, also the potential degradation of the

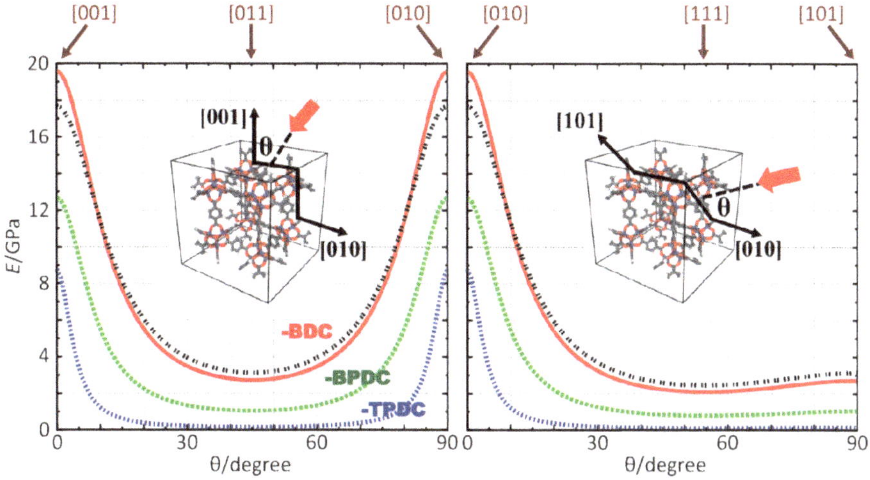

Figure 3.9 Directional 300 K Young's modulus for MOF-5 (red), IRMOF-10 (green), and IRMOF-16 (blue) as a function of the deviation angle (left) from the [001] towards the [010] direction and (right) from the [010] towards the [101] direction, as calculated using the PBE+D3 level of theory. The dashed black lines represent MOF-5 results obtained at a longer simulation time of 5 ps. Adapted from ref. 120 with permission from the American Institute of Physics, Copyright 2017.

moisture-dependent MOF-5 sample during preparation or the nanoindentation experiment should not be neglected,[137,138] which would reduce the stiffness of the material.[139]

Besides MOF-5, a whole series of IRMOF materials have been synthesised that share the MOF-5 architecture but feature a different ligand. Several of these materials have also been the subject of simulation studies to interrogate their equilibrium mechanical behaviour, using either DFT,[120] density functional based tight binding (DFTB),[140] generic force fields,[125,128,141] or system-specific force fields.[78] Adding extra phenyl rings in the BDC ligand leads to the extended biphenyl-4,4′-dicarboxylate (BPDC, IRMOF-10) and terphenyl-4,4″-dicarboxylate (TPDC, IRMOF-16) ligands. As shown in Figure 3.8(b and c), DFT calculations indicate that such linker expansions decrease the stiffness of the material and enhance the anisotropy in elastic moduli.[120] Similar weakening for larger ligands was observed also when using the DFTB framework and with various force fields;[78,125,128,140,141] the only exceptions being hypothetical cage-containing ligands for which the DFTB framework predicts slightly stiffer materials.[140]

In addition to determining the full stiffness tensor, the bulk modulus of MOF-5 and other IRMOF materials have also been widely studied. Given the relative softness of MOFs and the relative ease with which the bulk modulus can be determined, these studies are also widely used to validate new force field and DFTB models that approximate the PES,[74,76,132,142] to compare different equations of state,[143] to study the impact of symmetry,[144] and, at finite temperatures, to quantify the impact of thermostat and barostat schemes.[83] In 2005, Fuentes-Cabrera *et al.* used an LDA functional to calculate the bulk modulus of different MOF-5 analogues, in which the zinc ions were replaced by either cadmium, beryllium, magnesium, or calcium.[145] They found that the bulk modulus of MOF-5, amounting to around 18.0 GPa, drops when going down the periods in the periodic table; a feature they attributed to the simultaneous increase in porosity and equilibrium volume when incorporating larger cations.[145] This was further investigated in a series of studies by Yang *et al.*[143,146–154] Regarding the effect on the bulk modulus, their work can be summarized in five main points. First, the bulk modulus of IRMOF-type materials is insensitive to the exact form of the equation of state, be it the one proposed by Birch,[107] Birch–Murnaghan,[106] or Rose–Vinet.[108] Second, the aforementioned observation that the bulk modulus decreases when replacing zinc with heavier cations also holds for all alkaline earth metals and for different IRMOF materials, including IRMOF-10,[147] IRMOF-14 (based on pyrene-2,7-dicarboxylate),[148] IRMOF-993 (based on anthracene-9,10-dicarboxylate),[149] IRMOF-M0 (in which the inorganic nodes are directly connected through formate units),[152] and the IRMOF constructed with fumarate ligands.[150] Third, upon increasing the length of the organic ligand and hence the porosity of the material, the bulk modulus decreases. Fourth, when replacing the oxygens in the Zn_4O node with heavier chalcogens, the bulk modulus decreases.[153] Fifth, when fully halogenating the BDC ligand in MOF-5, the bulk modulus decreases for fluorine, but increases systematically when replacing fluorine with chlorine, bromine, or iodine.[154]

In contrast to the full elastic tensor, the MOF-5 bulk modulus is quite insensitive to the level of theory, with 0 K values between 15.5 and 18.2 GPa for different LDA and GGA functionals,[118,145,155] even when including dispersion interactions,[156,157] and between 13.6 and 22.0 GPa for different force fields.[132] Even within the DFTB framework, a similar bulk modulus was found for MOF-5, which decreased when incorporating longer ligands, yielding MOF-177 and MOF-205 (also known as DUT-6).[142] The only somewhat deviating DFT value was reported by Evans and Coudert, who used the PBESOL0 functional[158] and an atom-centred triple-zeta valence and polarisation basis set as implemented in CRYSTAL14[159] to obtain a MOF-5 bulk modulus of 13.5 GPa.[144] Interestingly, they also investigated CMOF-5, a MOF-5 analogue in which chirality is induced by guest adsorption (see Figure 3.7(b)). This lowers the symmetry of the cubic unit cell and also lowers the bulk modulus from 13.5 GPa to 11.1 GPa, showing the importance of choosing a sufficiently large unit cell to investigate MOFs (see Section 3.3.1).

Similar to the elastic constants, Banlusan *et al.* showed through *ab initio* MD (AIMD) simulations that the MOF-5 bulk modulus decreases with increasing temperature, from 15.8 GPa at 0 K to 12.3 GPa at 300 K.[120] Using the CVFF force field description, Greathouse *et al.* overestimated this temperature-induced softening, obtaining a decrease in bulk modulus from 20 GPa to 4 GPa when increasing the temperature from 0 K to 300 K. The latter value was obtained by fitting a Birch–Murnaghan equation of state to the average energy of an isotropically scaled MOF-5 system that was simulated in the $(N,V,\mathbf{h}_0,T)$ ensemble, which constrained the cubic symmetry.[126] By allowing temporary deviations from this cubic cell shape while constraining the cell volume, Rogge *et al.* obtained a higher bulk modulus of 6.6 to 7.3 GPa at 300 K, depending on the precise barostat implementation.[83] This approach to determine the bulk modulus and, more generally, the full pressure and free energy equation of state of MOFs at finite temperatures is further detailed in Section 3.5.3. A decrease in bulk modulus from *ca.* 15 GPa at 0 K to *ca.* 13 GPa at 350 K – which approaches the AIMD results – was obtained using the quasi-harmonic approximation (QHA) to the free energy.[157] In the QHA framework, anharmonicities in interatomic vibrations or phonon modes, which become more prominent at higher temperatures and are especially present in low-frequency vibrations, are partially taken into account by assuming that the temperature dependence of the vibrational frequencies is completely caused by the temperature-induced change in volume.[160] However, this QHA framework was found to be very sensitive, as an earlier QHA study observed an increase in bulk modulus upon increasing temperature instead.[161]

3.4.1.3 Zeolitic Imidazolate Frameworks (ZIFs) and Associated Materials

Zeolitic imidazolate frameworks (ZIFs) are a subclass of MOFs that, since they are composed of tetrahedrally coordinated cations – typically zinc or cobalt – connected through imidazolate-based ligands, can adopt the same topologies as zeolites.[162,163] Compared to the IRMOF series discussed above, they are chemically and thermally more stable,[163] and are more easily synthesised in the form of single crystals, making them ideal subjects for nanoindentation studies. From a theoretical point of view, ZIF-zni, composed of zinc cations surrounded by imidazolate ligands in a rather dense framework, was the first ZIF for which the bulk modulus was obtained through DFT optimisations, yielding a 0 K bulk modulus of 13.3 GPa.[164] By replacing the zinc cations in ZIF-zni alternatingly with lithium and boron ions, BIF-1(Li) is obtained, for which a slightly higher 0 K bulk modulus of 16.5 GPa was calculated. Both moduli are in excellent agreement with the experimental values of ~14 GPa and ~16.6 GPa collected under ambient conditions.[164]

Tan *et al.* carried out the first full tensorial DFT analysis of a ZIF material by investigating ZIF-8, a cubic material consisting of zinc cations and 2-methylimidazolate ligands that synthesises into the sodalite (**sod**) topology.[165] By adopting the hybrid B3LYP functional[166] using a double-zeta atom-centred basis set, as implemented in CRYSTAL09,[167] they obtained C_{11} = 11.04 GPa, C_{12} = 8.33 GPa, and C_{44} = 0.94 GPa, resulting in a moderately anisotropic material that is most compliant to shear deformations, to an even greater extent than MOF-5.[165] The simulated elastic constants correspond very well with the experimental values of 9.522 GPa, 6.865 GPa, and 0.967 GPa determined *via* Brillouin scattering,[165] in contrast to the larger discrepancies between experiment and simulation for MOF-5 discussed above.

The remaining mismatch between the theoretically and experimentally obtained values was attributed to the absence of thermal effects in the simulations.[165] Very similar elastic constants were obtained using the ZIF-FF force field derived by Weng and Schmidt[168] and the MOF-FF force field derived by Dürholt *et al.*[73] The bulk modulus of 9.23 GPa predicted by Tan *et al.*[165] is also in good agreement with the value of 8.9 GPa obtained by Zhang *et al.* using the hybrid M06-2X functional.[169,170] MAF-7, a ZIF-8 analogue containing 3-methyl-1,2,4-triazolate instead of 2-methylimidazolate ligands (MAF = metal-azolate framework), was found to exhibit higher mechanical resilience than ZIF-8 ($K = 10.29$ GPa), an effect that was attributed to the electron-donating effect of the additional nitrogen atom in the triazolate ligand.[171] Maul *et al.* established that the elastic constants of ZIF-67, the cobalt analogue of ZIF-8, were very similar to those obtained for ZIF-8.[112] Interestingly, they adopted the extended Hessian approach to determine that only a selected amount of vibrational modes substantially contribute to (the nuclear part of) the ZIF-8 elastic moduli within the harmonic approximation. All identified modes exhibit low frequencies, in the THz regime (see Section 1.7.2 of Chapter 1), being associated with the symmetric gate-opening phenomenon for C_{11} and C_{12}, and with the antisymmetric gate opening phenomenon and breathing deformations of the four-membered ring (4MR) windows for the shear coefficient C_{44}.[112]

Following a similar approach but using the less expensive PBE functional,[122] Tan *et al.* also calculated the elastic tensors of ZIF-4 and ZIF-zni, both composed of zinc cations surrounded by unsubstituted imidazolate ligands, but exhibiting different topologies (**cag** for ZIF-4 *versus* **zni** for ZIF-zni).[172] Similar to ZIF-8, a very good correspondence with the experimental bulk and Young's moduli was obtained.[172] As shown in Figure 3.10, ZIF-zni is more anisotropic than ZIF-4 and ZIF-8, whereas both ZIF-zni and ZIF-4 exhibit greater resistance towards shear deformation than the prototypical ZIF-8 material.[172] Even so, the lowest shear modulus of ZIF-zni, the densest ZIF structure, only lies at ~1.6 GPa, indicating that even dense ZIF materials are prone to shear deformation.[172] Just as other ZIFs that feature 4MR windows, this low shear modulus can be understood by the compliance of the four ZnN_4 tetrahedra surrounding the 4MR windows.[172] In addition, this full tensorial analysis revealed certain directions along which ZIF-4 exhibits (limited) auxetic behaviour, which is uncommon in rigid materials.[172] Using the hybrid B3LYP functional, three more materials with the same composition as ZIF-4 and ZIF-zni were studied in a follow-up study: ZIF-1, ZIF-2, and ZIF-3, exhibiting the **crb** (ZIF-1 and ZIF-2) or **dft** (ZIF-3) topology.[173] Using the hybrid B3LYP functional, Ryder *et al.* observed an almost zero but non-negative Poisson ratio for ZIF-1, ZIF-2, and ZIF-4, whereas ZIF-3 showed both significant auxetic behaviour and an exceptionally low minimum shear modulus of 0.11 GPa, lower than any other ZIF to date.[173] In addition, limited NLC was observed along certain directions in ZIF-1 and ZIF-2.[173] Comparing both studies, it is clear that the choice of exchange-correlation functional impacts the emergence of auxetic behaviour in ZIF-4, and that care should be taken

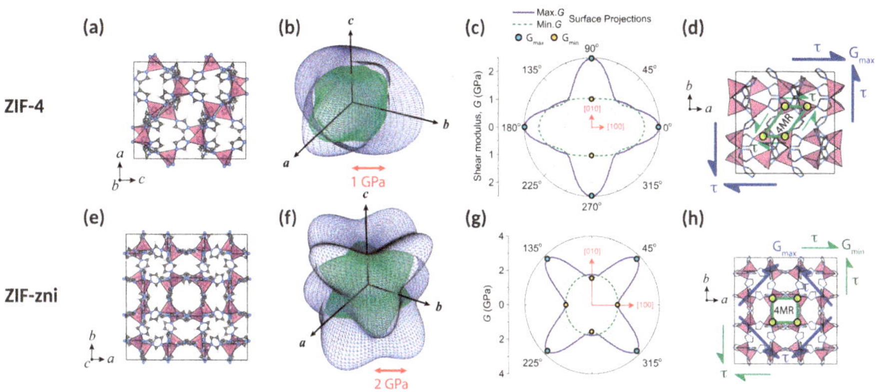

Figure 3.10 Comparison of the directional 0 K shear moduli of ZIF-4 and ZIF-zni calculated at the PBE level of theory. (a and e) Orthorhombic and tetragonal unit cells of ZIF-4 and ZIF-zni, respectively, with ZnN_4 tetrahedra, nitrogen atoms, carbon atoms, and hydrogen atoms indicated in purple, blue, grey, and white, respectively. (b and f) Shear modulus representation surface, where blue and green colour coding denotes the maximum and minimum moduli. (c and g) Polar plots obtained *via* projections through the origin and down the *c* axis, showing the positions of maximum and minimum shear moduli. (d and h) Structure–property relationships illustrating the source of the maximum shear moduli in relation to the position of the 4MR, which has low rigidity against shear deformations. Shear stresses in blue generate G_{max}, while the opposite pairs in green result in G_{min}. Adapted from ref. 172 with permission from the Royal Society of Chemistry.

to ensure that the used level of theory is sufficiently accurate to describe the phenomena under study, as discussed in Section 3.3.2.

The above statement can be further exemplified by the study of Zheng *et al.*, who investigated the elastic constants of five sodalite ZIFs containing zinc cations and either unsubstituted imidazolate (SALEM-2), 2-chloroimidazolate (ZIF-Cl), 2-bromoimidazolate (ZIF-Br), 2-iodoimidazolate (ZIF-I) or imidazolate-2-carboxaldehyde (ZIF-90) as a ligand.[174] Using the PBE functional with dispersion corrections,[122] they found that the mechanical resilience of ZIF-Cl outperformed that of ZIF-Br, which in turn outperformed that of both SALEM-2 and ZIF-90, despite the larger functional group on the ligand of ZIF-90.[174] The authors were able to negatively correlate the magnitude of the elastic constants with the (positive) charge on the zinc cations, which in turn is a function of the electron withdrawing character of the functional group. They hypothesised that the smaller the electron density surrounding the zinc cations, the more flexible the ZnN_4 tetrahedra, and the easier the tetrahedra may deform under stress.[174] Interestingly from a theoretical standpoint, the authors also investigated the effect of the choice of the exchange-correlation functional and the inclusion of dispersion corrections on the reported elastic moduli. While the aforementioned trends were observed for all theoretical models, their absolute magnitude varied by more than 100%, highlighting the importance of validating and benchmarking these choices.[174] A similar trend was observed by Dürholt *et al.*, *via* both DFT calculations and calculations performed with newly derived MOF-FF force fields.[73] This observed correlation between functionalisation and mechanical stability was generalised by Moosavi *et al.*, who investigated the effect of ligand functionalisation in 50 different ZIF topologies, considering unsubstituted imidazolate, 2-methylimidazolate, 4,5-dichloroimidazolate, and 2-nitroimidazolate as ligands, using a DREIDING-based force field.[130,175] They observed that the nonbonding interactions, which are strengthened by these functional groups, give rise to a secondary network that can impart the material with additional mechanical stability beyond the primary network, defined by the bonding interactions and hence the topology.[175] These functional groups therefore act as 'chemical caryatids' that may strengthen the ZIF structure.

Finally, Zhang *et al.* derived a system-specific force field to determine the bulk modulus of a mixed-matrix membrane (MMM) composed of polybenzimidazole (PBI) and various amounts of ZIF-7 as a filler material.[176] To speed up their calculations, they assumed rigid ZnN_4 tetrahedra, formed from the coordination of the benzimidazolate ligands around the zinc ions in ZIF-7 and only allowed for flexibility in the ligands themselves. They observed that the bulk modulus of neat PBI, amounting to 4.1 GPa, could be substantially increased with increasing ZIF-7 filler content, up to 5.9 GPa for a MMM with 45 wt% of ZIF-7. However, given the exceptionally large value of the pure ZIF-7 bulk modulus they obtained (29.5 GPa) and the observed flexibility of the ZnN_4 tetrahedra in the aforementioned studies, the mechanical properties of this MMM are most likely overestimated by the here imposed constraint that the ZnN_4 tetrahedra are rigid.[176]

3.4.1.4 The UiO-66 Series

The UiO-66 series has been lauded for its exceptional stability compared to other MOFs, a result of its $Zr_6(\mu_3\text{-}OH)_4(\mu_3\text{-}O)_4$ inorganic nodes that are, in its ideal structure, twelvefold connected by either BDC (UiO-66), BPDC (UiO-67), or TPDC (UiO-68) through relatively strong zirconium–oxygen bonds.[45] However, UiO-66 is also one of the first MOFs for which it became apparent that it exhibits tuneable disorder, in which different types and concentrations of defects impact the macroscopic properties of the material, including its mechanical behaviour.[5,29,32] In this respect, the interest in modelling the mechanical properties of UiO-66 lies both in understanding *why* it has such a high mechanical resistance and, especially in more recent years, *how* this is impacted by disorder. As for MOF-5, the conventional cells of the UiO-66 series in equilibrium exhibit cubic symmetry, so that the same three independent deformations can be used to obtain the elastic constants.

As expected given its highly coordinated structure, UiO-66 exhibits high elastic constants, with 0 K elastic constants amounting to 59.4 GPa, 31.9 GPa, and 17.6 GPa for C_{11}, C_{12}, and C_{44}, and a low anisotropy.[177] Wu *et al.*

furthermore observed that replacing the zirconium cations in UiO-66 with either hafnium or titanium cations does not change its 0 K elastic stiffness tensor appreciably.[177] In contrast, increasing the linker length to obtain UiO-67 or UiO-68 does have a profound effect, with the elastic constants and moduli systematically decreasing by about 52–58% for UiO-67 and by about 58–66% for UiO-68.[177] Exchanging the BPDC ligands in UiO-67 with azobenzene-4,4′-dicarboxylate (ABDC) ligands results in a further small decrease in elastic moduli of ~12%.[178] In contrast, exchanging the BDC ligands in UiO-66 by shorter fumarate ligands increases the elastic moduli by ~10%, except for the C_{12} elastic modulus, which decreases by ~20%, presumably due to exchanging an aromatic for an aliphatic ligand.[120] Using system-specific force fields, Moghadam *et al.* generalised these observations for 14 different ligands, obtaining a strong negative correlation between the equilibrium cell length of the UiO-66-type material on the one hand and its elastic constants and elastic moduli on the other hand.[141] Increasing the temperature to 300 K results only in a small decrease in elastic moduli and retains the weak anisotropy in the material.[120]

In the aforementioned studies, UiO-66 and its analogues were treated as defect-free materials, which led to an overestimation of the mechanical stiffness compared to experimental samples that contained ligand and/or node vacancies. To remedy this, Cliffe *et al.* calculated in their landmark study on node defects in UiO-66 the impact of such node vacancies on the elastic constants of UiO-66(Zr) and UiO-66(Hf).[5] While little difference was found between the zirconium and hafnium analogues, the node-defective **reo** materials exhibit *ab initio* elastic moduli that are halved compared to the defect-free **fcu** materials discussed above,[5] clearly demonstrating the importance of correctly accounting for structural defects when modelling UiO-66, as elucidated in Section 3.3.1. Thornton *et al.* further investigated this issue, considering defect-free **fcu** and node-defective **reo** materials, as well as four materials that were obtained from the defect-free structure by systematically removing one, two, three, or four ligands per node, without creating node vacancies.[48] Given that these linker vacancies alter the cubic symmetry of the material (in contrast to the **fcu** and **reo** materials, which are both cubic), six instead of three distinct deformation modes had to be considered in order to characterise all different elastic constants. As illustrated in Figure 3.3, the elastic constants decreased and the anisotropy in elastic moduli increased with an increasing number of linker vacancies, even leading to auxetic behaviour for the most defective structures.[48] Furthermore, they also showed that larger substituents used to cap the vacancy locations, such as acetate and trifluoroacetate, yield elastic moduli that are an order of magnitude larger than smaller substituents such as formate, hydroxyl, or chloride.[48]

As for the elastic constants, the bulk modulus drops significantly upon linker expansion, from K = 37–42 GPa for UiO-66, over 17–22 GPa for UiO-67 and 15–17 GPa for UiO-ABDC, to 14–15 GPa for UiO-68, depending on the DFT level of theory.[177–180] This bulk modulus is quite insensitive to the metal cation, with similar bulk moduli reported for the zirconium, hafnium, and titanium analogues of these materials.[177,179] The sole exception appears to be UiO-66(Ce), for which a small drop in bulk modulus was observed compared to UiO-66(Zr), from 42 GPa to 37 GPa.[180] From a force-field perspective, Bristow *et al.* reported substantially smaller bulk moduli at 1 K of 27.2 GPa and 19.2 GPa for UiO-66 and UiO-67, respectively.[74] At elevated temperatures, a force-field investigation of the bulk modulus of 14 UiO materials with different ligands demonstrated a decrease in bulk modulus upon increasing linker length, from ~35 GPa for a dicarboxylate linker with a C≡C moiety to ~6 GPa for a dicarboxylate linker with two phenyl rings and two C≡C moieties.[141]

When including node vacancies in UiO-66, the 0 K bulk modulus was observed to decrease from ~39.5 GPa for the defect-free **fcu** material to ~18.3 GPa for the node-defective **reo** material.[5] A very similar decrease was observed when including four ligand vacancies per conventional cubic unit cell instead, although the 0 K bulk modulus is impacted not only by the concentration of ligand vacancies but also by their distribution throughout the framework.[181] Additionally, the dehydrated form of UiO-66 was observed to have a slightly smaller bulk modulus than the hydrated form discussed up to this point.[181] Force-field studies confirmed and generalised these observations to 300 K.[49,182] By considering all inequivalent ways in which one or two ligand vacancies could be distributed throughout the conventional UiO-66 unit cell, Rogge *et al.* observed a decrease in bulk modulus from 22.2 GPa for the defect-free UiO-66 material to 19.9 GPa for UiO-66 in which one out of every 24 ligand positions was vacant.[49] Removing an additional ligand results in a bulk modulus of 15.5–18.9 GPa, depending on the distribution of the two linker vacancies,[49] which coincides with the experimental results obtained by Yot *et al.* with a similar defect concentration.[50] In a combined experimental/theoretical study, the effect of systematically replacing one or multiple zirconium cations in the inorganic node with either hafnium or cerium cations was investigated. Figure 3.11 reveals a negligible effect for hafnium incorporation but a slight decrease in bulk modulus upon cerium incorporation.[182] Finally, node vacancies lead to a distinct drop in 300 K bulk modulus, from 25.5 GPa to 13.9 GPa for UiO-66(Zr),[182] which is somewhat lower than the 0 K results obtained by Cliffe *et al.*[5]

3.4.1.5 HKUST-1

From a mechanical point of view, the cubic HKUST-1 framework, composed of copper paddlewheel units connected through benzene-1,3,5-tricarboxylate (BTC) ligands,[183] shows two distinct changes with respect to the aforementioned MOFs. First, the inorganic nodes are connected through tritopic instead of ditopic ligands, which should impart additional stability to the material. Second, the undercoordinated square-planar copper paddlewheel units in HKUST-1 feature two axial positions to which molecules can adsorb.

Despite being first synthesised in 1999, the first study of the HKUST-1 elastic stiffness tensor only took place in 2010, when Watanabe *et al.* simulated

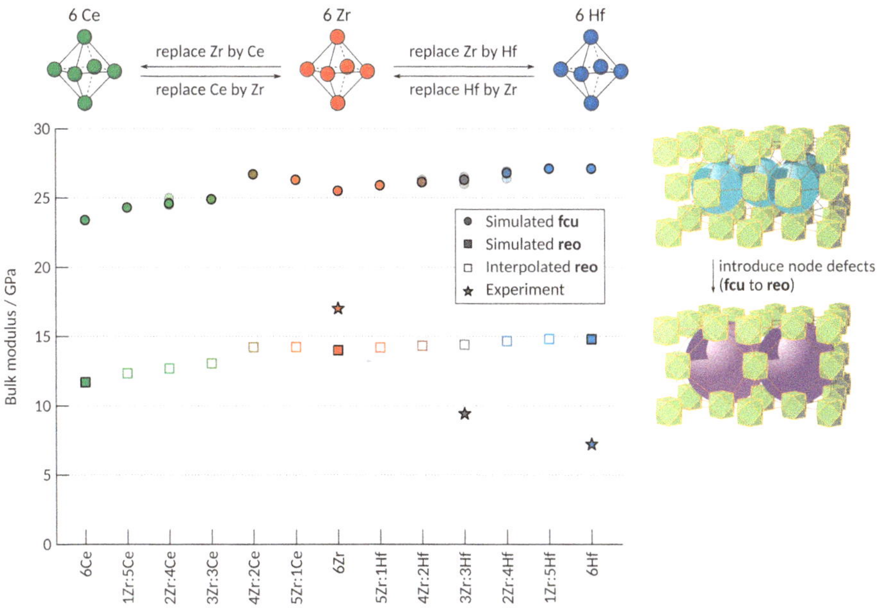

Figure 3.11 Equilibrium bulk moduli at 300 K for a series of bimetallic UiO-66 materials as determined from the pressure equations of state (see Section 3.5.3), both in the pristine **fcu** (circles) and the defective **reo** (squares) topologies, as obtained through force field calculations. For mixed-metal bricks for which two inequivalent bricks can be obtained, the weighted average is shown, whereas the two independent results are included as semi-transparent data points. Experimental results are indicated with stars. Reproduced from ref. 182 with permission from American Chemical Society, Copyright 2020.

its elastic constants within the DFT framework using the PW91 GGA functional.[184] They obtained elastic constants C_{11} = 31.8 GPa, C_{12} = 21.2 GPa, and C_{44} = 12.7 GPa, and a bulk modulus of 24.4 GPa determined *via* a Birch–Murnaghan equation of state.[185] Very similar values were reported by Tafipolsky *et al.*, who derived a MOF-FF force field that could also predict the NTE of the material.[186] While the 0 K bulk modulus was also well reproduced by BTW-FF and more generic force fields,[74,132] DFTB calculations obtained a much larger value of 34.66 GPa,[142] close to the value Zhao *et al.* obtained at 300 K using a newly derived system-specific force field.[187]

The high elastic constants, indicative of a mechanically resistant material, were also calculated using the PBE functional,[122] retrieving good agreement for C_{11} and C_{12} with deviations of up to 20% compared to the aforementioned studies, although a much lower value of C_{44} amounting to 4.41 GPa was found.[177] These latter values were also confirmed independently by Ryder *et al.* using the hybrid B3LYP functional,[166] showing that the inclusion of dispersion corrections has only a minor influence on the elastic stiffness tensor in this case.[188] In addition, they obtained an average difference of less than 1% between the elastic constants of the ferromagnetic and antiferromagnetic form of HKUST-1.[188] Interestingly, Ryder *et al.* also showed that HKUST-1 exhibits auxetic behaviour, which is most pronounced in the ⟨110⟩ directions upon application of a uniaxial strain in the ⟨1$\bar{1}$0⟩ direction, which could be explained by the theoretically observed low-frequency twisting of the copper paddlewheel clusters shown schematically in Figure 3.12.[188]

In order to probe the structural and mechanical properties of HKUST-1 at longer length scales – and hence to be able to describe the experimentally observed mesopores in the material[34] – Dürholt *et al.* developed the first coarse-grained MOF force field that builds upon their earlier MOF-FF framework.[7] In their coarsest HKUST-1 model, the ⟨100⟩ Young's modulus and bulk modulus were found to be 20.3 GPa and 17.6 GPa, respectively, comparable with the values of 23.8 GPa and 19.6 GPa obtained using an atomistic MOF-FF force field.[7] In addition, they found that the inclusion of nonbonded interactions, *i.e.*, van der Waals and Coulomb interactions, did not alter these values appreciably, and that a hypothetical HKUST-1 material that adopts the **pto** topology would be about 20–40% more rigid than the experimentally observed **tbo** topology, although the latter is energetically more favourable.[7] This coarse-grained force field was then adopted in a multiscale investigation to determine the 0 K bulk modulus of defective HKUST-1 samples, in which spherical mesopores of different sizes and with different concentrations were created.[23] As expected, increasing the mesopore volume

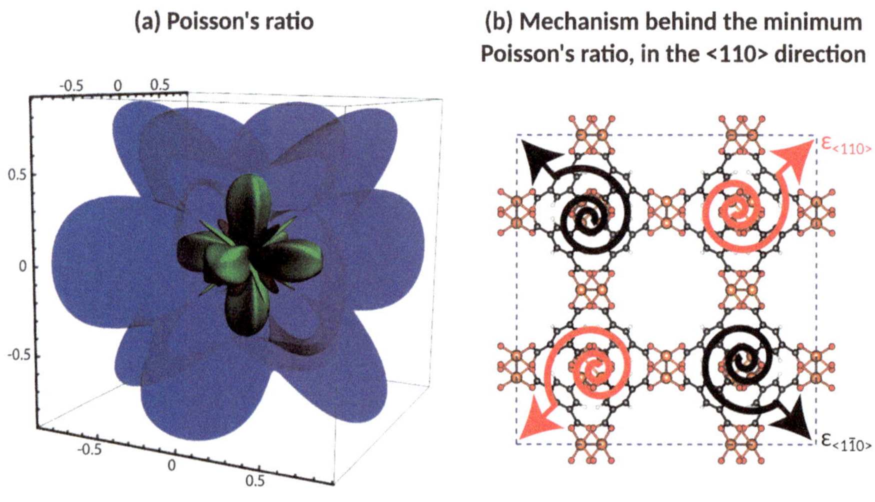

Figure 3.12 (a) 0 K Poisson's ratio representation surface obtained at the B3LYP-D level of theory, where blue signifies the maximum Poisson's ratio and green and red denote the positive and negative (= auxetic behaviour) minimum Poisson's ratios, respectively. (b) Demonstration of the plausible mechanism associated with cluster rotational dynamics responsible for the auxetic response. Adapted from ref. 188 with permission from the Royal Society of Chemistry.

leads to a continuous decrease in bulk modulus, although the decrease in bulk modulus remains modest.[23] Interestingly, Dürholt *et al.* observed that many smaller mesopores in the material impact the bulk modulus more strongly than one larger mesopore with the same total mesopore volume,[23] an important guideline when designing defect-engineered MOFs.

3.4.1.6 Wine-Rack Type MOFs: MIL-47 and MIL-53-type Materials

Wine-rack type MOFs are composed of extended one-dimensional inorganic chains that are connected through dicarboxylate ligands, in such a way that their structure viewed along the 1D chain resembles that of a wine rack (Figure 1.1(e)). The most prominent examples of this family are the MIL-47 and the MIL-53 series,[189,190] which can be synthesised with different metal cations. Because of their wine rack structure, these materials can undergo large-amplitude phase transitions between a large-pore (lp) and closed-pore (cp) structure, so-called 'breathing', induced by mechanical stress,[191,192] among other external stimuli,[193] and feature NLC, NTE, and auxeticity (Chapter 2).[1,194,195] Since these MOFs have a lower symmetry than the MOFs mentioned before, they exhibit more independent elastic constants and hence more independent deformations are needed to fully characterise their elastic tensor.

A full tensorial analysis of the elastic tensors of the lp structures of MIL-53(Al), MIL-53(Ga), and MIL-47 was first reported by Ortiz *et al.*[196] using the B3LYP functional.[166] The nine elastic constants for these orthorhombic structures are reproduced in Table 3.2, while Figure 3.13 visualises the resulting directional Young's modulus of the lp structures of MIL-53(Al) and MIL-47.[196,197] This visualisation shows very strong anisotropy, with stiffer directions along the inorganic chains and the organic ligands, and very compliant directions along the diagonals of the lozenge-shaped pore.[196] As discussed further in Section 3.4.3, this strong anisotropy in Young's modulus and other elastic moduli was posited as a revealing signature of structural flexibility in MOFs.[196] In addition, Ortiz *et al.* showed that all investigated MIL-53 and MIL-47 materials exhibit NLC along a direction pointing into the channel, with a magnitude of up to $-257\ \mathrm{TPa}^{-1}$ for MIL-53(Al) lp.[196] Finally,

Table 3.2 0 K stiffness constants for five wine-rack type materials in Voigt notation, obtained at the B3LYP level of theory. Reproduced from ref. 196 with permission from the American Physical Society, Copyright 2012.

C_{ij}/GPa	C_{11}	C_{22}	C_{33}	C_{44}	C_{55}	C_{66}	C_{12}	C_{13}	C_{23}
MIL-53(Al) lp	90.85	65.56	33.33	7.24	39.52	8.27	20.41	54.28	12.36
MIL-53(Ga) lp	112.32	56.66	18.52	5.45	21.71	6.64	22.87	45.35	10.86
MIL-47 lp	40.68	62.60	36.15	50.83	7.76	9.30	12.58	9.28	46.98
DMOF-1 loz	57.15	35.59	17.68	0.62	16.39	0.69	9.85	31.43	5.47
DMOF-1 sq	35.33	58.20	58.45	0.11	0.44	0.28	7.32	7.55	11.68

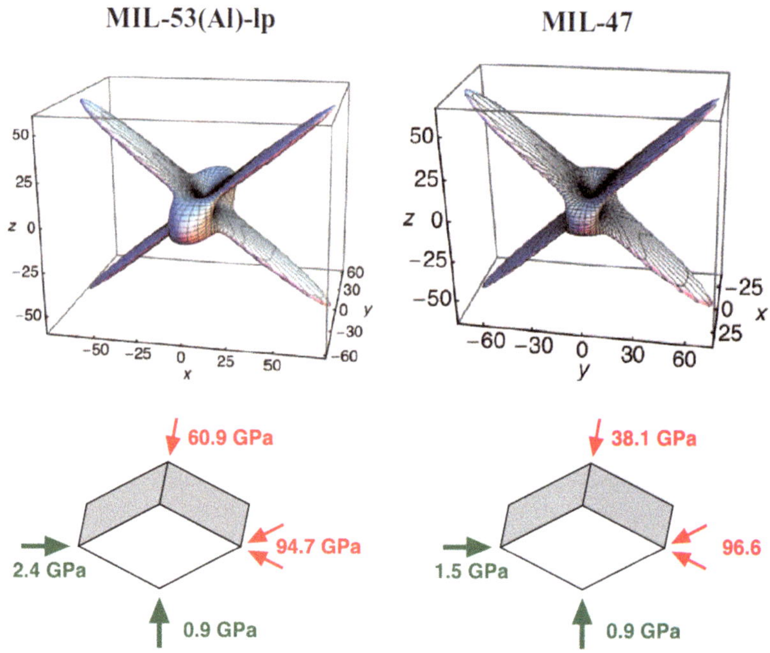

Figure 3.13 Top: Directional 0 K Young's modulus for the lp phases of MIL-53(Al) (left) and MIL-47 (right) represented as 3D surfaces, with axes tick labels in GPa. Bottom: Scheme of the Young's modulus values for the stiffest and most compliant directions, with Young's modulus values indicated. Reproduced from ref. 197 with permission from the American Institute of Physics, Copyright 2013.

they reported that the bulk modulus calculated from these elastic constants depends on the averaging scheme, in contrast to the less anisotropic MOFs discussed before,[196] with values between 1 and 20 GPa for the MOFs considered here. In line with earlier results on rigid MOFs, DFTB strongly overestimates the true bulk modulus also for MIL-53.[142] For MIL-47(V), similar elastic constants as those reported in Table 3.2 were obtained by Vanpoucke *et al.* and Heinen *et al.* using the generalised Hessian approach.[78,109] As noted earlier for MOF-5, Vanpoucke *et al.* also observed that a proper energy minimisation is essential to obtain the true unstrained equilibrium structure and its elastic constants.[109] The MIL-47 *ab initio* elastic constants from the three aforementioned studies were used by Heinen *et al.* to propose an alternative force field fitting procedure based on the elastic stiffness tensor.[78] The elastic constants predicted using this MIL-47 force field correspond remarkably well with those obtained from DFT calculations,[78] in contrast to more *ad hoc* based force fields.[191] Heinen *et al.* furthermore observed that amino functionalising the BDC ligands in MIL-47 leads, on average, to a decrease in elastic constants, although the exact magnitude of the elastic constants depends on the location of the amino group, with the most stable amino-functionalised structures being those with the highest elastic constants.[198] Finally, also the fumarate-based analogues of MIL-53(Al) and MIL-53(Ga) were studied at the DFT level using the PBE functional,[122] showing that the gallium analogue is softer than the aluminium variant.[199]

Regarding the cp phases, DFT calculations using the PBE functional[122] reveal that the MIL-53(Al) cp structure exhibits a lower anisotropy in Young's and shear moduli compared to its lp structure, although it still features directions of NLC and auxetic response.[200] A similar result was obtained for fumarate-based MIL-53(Al) and MIL-53(Ga).[199] Furthermore, it was observed that low water loadings in the MIL-53(Al) lp structure soften some of the deformation modes of the material,[200] a counterintuitive response upon adsorption that was observed earlier by Mouhat *et al.*[201] A similar effect was observed at 300 K for MIL-53(Cr), where a force-field study showed a decrease in bulk modulus from ~1.8 GPa to ~0.75 GPa at low CO_2 loading.[202] For MIL-47, the cp structure was found to exhibit both directions of NLC and of auxetic behaviour.[109] In addition, the MIL-47 cp structure shows higher anisotropy in Young's modulus and a lower anisotropy in shear modulus than its lp structure.[109] While DFT predicts its 0 K bulk modulus to decrease from 6.1 GPa to 2.8 GPa when transitioning from the lp to the cp structure,[109] a force field study predicts that MIL-53(Al) does the opposite, with an increase in 300 K bulk modulus from ~1.6 GPa to ~3.7 GPa when transitioning to the denser structure.[83] A similar increase in bulk modulus for the denser phase was observed for COMOC-2,[203] an isoreticularly expanded version of MIL-47 in which the BDC ligands are replaced by BPDC ligands.[204] Additionally, Wieme *et al.* observed a decrease in bulk modulus for the COMOC-2 cp state and an increase for the COMOC-2 lp state with increasing temperature, whereas the bulk moduli of MIL-47 and COMOC-3 (obtained by replacing the BDC ligand in MIL-47 by a naphthalene-2,6-dicarboxylate (NDC) ligand) are largely temperature-independent.[204] The MIL-47 bulk modulus furthermore depends on the magnetic ordering of the vanadium ions in the inorganic chain, with antiferromagnetic chains leading to bulk moduli that are ~2 GPa larger than for ferromagnetic chains.[205] Finally, Hoffman *et al.* observed only a modest decrease in bulk modulus with temperature for the MIL-53(Al) cp phase within the QHA framework, while a much larger increase in bulk modulus was observed for the lp phase, from 2.1 GPa at 0 K to 8.4 GPa at 500 K.[206] The latter effect, however, was assigned to an overestimation of the thermal expansion behaviour of the lp phase in this study.[206] For both phases, they were able to pinpoint the magnitude of the bulk modulus to a limited set of low-frequency vibrational modes (see Figure 1.19).[206]

In a follow-up study,[197] Ortiz *et al.* compared the behaviour observed for MIL-53 and MIL-47 to that of two other wine-rack type MOFs, MIL-122[207] and MIL-140.[208] MIL-122 is structurally similar to MIL-53, with the exception that the ditopic BDC ligands in MIL-53 are replaced by tetratopic naphthalene-1,4,5,8-tetracarboxylate ligands, which precludes wine-rack type hinging. While MIL-122 was observed to still exhibit auxetic behaviour and directions of NLC, a feature of wine-rack type MOFs, the anisotropy in its Young's modulus is about two orders of magnitude smaller than for the flexible MIL-53 and MIL-47 materials.[197] The zirconium-based MIL-140 material, in turn, can be regarded as a 'reinforced' or 'retrofitted' wine-rack structure, which features additional linkers that span the pores, hence again precluding flexibility. Although MIL-140 does exhibit strong anisotropy in its Young's and shear moduli, the most compliant directions are not located along the pores and hence predict that the material is not flexible.[197] While Ortiz *et al.* did not observe any directions with NLC,[197] Ryder *et al.* observed NLC directions both in MIL-140, albeit rather of limited magnitude, and an isoreticular framework with a BPDC ligand.[209] By systematically increasing the length of the ligand, they observed a continuous increase in anisotropy for the elastic moduli of these materials as well as a continuous increase in auxetic behaviour and a continuous decrease in bulk modulus.[209] However, no such correlation was observed regarding the occurrence of NLC and the shape of the linear compressibility representation surface in general.[209] Furthermore, as for the MIL-53 materials, it was noted that the bulk modulus calculated from the elastic constants is dependent on the chosen averaging scheme.[209]

3.4.1.7 Other Systematic Observations in MOF Mechanics

Alongside the prototypical MOFs given above, several other studies have appeared in recent years in which the elastic constants or moduli of MOFs have been determined. Noteworthy examples include 2D MOFs such as MOF-901 (the most compliant directions of which are located in the direction perpendicular to the 2D planes),[210] MOFs with differing degrees of interpenetration, such as in the MOF-14/DUT-34 pair of structures (where interpenetration increases the mechanical stability),[211] thin film SURMOFs,[212] flexible 3D MOFs such as ZnPurBr,[213] DUT-49,[214] DMOF-1,[215] ZAG-4,[216] ZAG-6,[216] CoBDP,[217] and others.[218] While it is impossible to discuss them exhaustively here, two systematic sets of 'high-throughput' studies have appeared in recent years that warrant a more in-depth discussion.

In 2019, Alexandrov *et al.* performed a comparative DFT study of 22 MOFs assembled from chain-like inorganic building blocks and calculated their full stiffness tensor.[219] They observed that many of these MOFs exhibit both NLC and auxetic behaviour, even though not all of them were flexible.[219] This work generalises earlier observations of such anomalous behaviour in rod-like MOFs,[220] such as the MIL-47 and MIL-53 series discussed in Section 3.4.1.6, ZAG-4 (which shows NLC but no auxetic behaviour),[216] ZAG-6,[216] SION-8,[221] CoBDP,[217] the lozenge-pore phase of DMOF-1 (although a similar effect is missing in its square-pore structure),[196] MOF-74(Zn),[222] and its analogue with a shorter 2,3-dihydroxyfumarate (which shows auxetic behaviour but no NLC),[223] CAU-13,[224] and NOTT-300 (which shows auxetic behaviour but no NLC).[224] Based on this anomalous behaviour, CAU-13 and NOTT-300 were predicted to be flexible,[224] which was later confirmed experimentally for CAU-13.[225]

On a larger scale, Anderson *et al.* constructed a library of 122 copper-paddlewheel MOFs, focusing on how their mechanical properties are altered as a function of their organic ligands and the topology in which the material assembles.[226] 19 different topologies were included in this study, as it was observed that previous high-throughput studies lacked diversity in this aspect. This path was further explored by Moghadam *et al.*, who constructed a database of 3385 MOFs consisting of 41 different topologies and various types of organic and inorganic building blocks, making it the largest MOF database up to this point for which the mechanical properties have been fully calculated.[141] Given the size of these databases, they were explored using the UFF4MOF force field with the aim of understanding what makes certain MOFs more robust than others.[134] Both studies revealed the topology of the material to be the most important predictor of its mechanical properties,[141,226] to the extent that a neural network could not satisfactorily predict the mechanical behaviour of MOFs without considering topology as a feature.[141] Additionally, these studies revealed that MOFs tend to become more compliant when they exhibit low-connected inorganic nodes or when the size of the organic ligand increases,[141,226] in line with the observation that the densest MOFs are also the most mechanically rigid.[139,227] Efforts to discover mechanically stable MOFs with exceptional internal surface areas therefore need to find building blocks and topologies that balance these two aspects, such as found in the recently synthesised DUT-60 framework that adopts the less common **ith-d** topology.[228]

3.4.2 Extracting Elastic Constants Through Fluctuation Formulae to Predict Temperature Effects

Most mechanical studies on MOFs in the elastic regime use the explicit deformation methodology outlined in Section 3.4.1. However, the elastic constants derived using the relation $V_0C_{ij} = \partial^2U/\partial\varepsilon_i\partial\varepsilon_j$ are obtained at 0 K, whereas the mechanical properties of MOFs are strongly temperature dependent. An extension of the explicit deformation methodology to finite temperature is, in principle, possible by replacing the internal energy U by the Helmholtz free energy F, but obtaining the free energy from a simulation requires dedicated simulation protocols, as explained in Section 3.5.3, and is therefore not straightforward. The same limitation arises for the energy equation of state: to obtain the finite-temperature bulk modulus, the *free* energy equation of state should be constructed. In Section 3.4.2.1, an alternative approach based on fluctuation formulae will be outlined from which the finite-temperature elastic stiffness tensor can be obtained. Its application to MOF research will be discussed in Section 3.4.2.2. Given that this methodology requires taking appropriate ensemble averages, it is computationally more expensive than the explicit deformation approach, explaining the rather limited MOF literature on the topic and the focus on force field rather than DFT studies.

3.4.2.1 General Methodology

Two main types of fluctuation formula arise, which differ in whether the stress is imposed and variations in strain are measured, or whether the strain is imposed and variations in stress are measured. Also mixed forms exist, but are not discussed here as they have not yet been used in MOF research. For the first category, Parrinello and Rahman showed that the elements of the fourth-rank compliance tensor **S** of a material around an equilibrium structure with volume V_0 and at a temperature T can be obtained as:[229]

$$S_{\alpha\beta\mu\nu} = \frac{V_0}{k_B T}\langle \Delta\varepsilon_{\alpha\beta}\Delta\varepsilon_{\mu\nu}\rangle = \frac{V_0}{k_B T}\left(\langle \varepsilon_{\alpha\beta}\varepsilon_{\mu\nu}\rangle - \langle \varepsilon_{\alpha\beta}\rangle\langle \varepsilon_{\mu\nu}\rangle\right). \tag{3.40}$$

In this expression, $\langle\cdot\rangle$ denotes an ensemble average. When this average is taken in the (N, P, $\boldsymbol{\sigma}_a$, H) ensemble, the adiabatic compliance tensor is obtained, whereas taking the average in the (N, P, $\boldsymbol{\sigma}_a$, T) ensemble leads to the isothermal compliance tensor.[229] The corresponding stiffness tensor then follows through matrix inversion. This formula can be extended to also calculate higher-order stiffness tensors.[230]

For the second category, Squire *et al.* and Ray *et al.* showed that a material's fourth-rank stiffness tensor **C** around an equilibrium structure with volume V_0 and at a temperature T can be obtained as:[15,231,232]

$$C_{\alpha\beta\mu\nu} = \langle \bar{C}_{\alpha\beta\mu\nu}\rangle - \frac{V_0}{k_B T}\left(\langle \sigma_{\alpha\beta}\sigma_{\mu\nu}\rangle - \langle \sigma_{\alpha\beta}\rangle\langle \sigma_{\mu\nu}\rangle\right) + \frac{2Nk_B T}{V_0}\left(\delta_{\alpha\nu}\delta_{\beta\mu} + \delta_{\alpha\mu}\delta_{\beta\nu}\right). \tag{3.41}$$

In this expression, where the notation of ref. 111 is used, δ_{ij} is the Kronecker delta, being one if $i = j$ and zero otherwise. $\bar{C}_{\alpha\beta\mu\nu}$ is the Born term, which was first encountered in eqn (3.37). The second and third terms are the fluctuation term, which converges to the relaxation term in eqn (3.37) in the zero temperature (0 K) limit, and the ideal gas term, which vanishes in the same limit.[233] The ensemble over which the average is taken again determines which elastic tensor is calculated: the (N, V, $\mathbf{h}_0$, E) ensemble yields the adiabatic stiffness tensor, whereas the (N, V, $\mathbf{h}_0$, T) ensemble yields the isothermal stiffness tensor.[15,231,232]

While both eqn (3.40) and (3.41) can be used interchangeably to determine the elastic constants, they differ from a practical perspective. First, eqn (3.40) requires pressure control during the simulation, and as a result the barostat simulation parameters may impact the observed fluctuations, as discussed in Sections 3.3.3 and 3.5.1. This was illustrated in ref. 83, in which the 300 K bulk modulus of MIL-53(Al) was determined based on fluctuations in the unit cell volume, similar to eqn (3.40). For the Langevin and MTK barostats, bulk moduli of 8.5–12.1 GPa for the cp phase and 3.2–10.1 GPa for the lp phase were predicted.[83] Those variations were caused only by a variation in coupling strength between the barostat and the system, as the barostat relaxation time was varied between 1 ps and 10 ps.[83] This effect was two orders of magnitude worse for the Berendsen barostat, where the same procedure led to bulk moduli of 275–1150 GPa (cp phase) and 93–215 GPa (lp phase) due to the artificial suppression of volume fluctuations in this barostat coupling scheme.[83] Second, eqn (3.40) was observed to converge more slowly than eqn (3.41),[82] although this argument is less important for force-field based simulations. The main disadvantage of eqn (3.41) lies in the Born term, which requires calculating the second-order derivatives of the potential energy with respect to the strain along the simulation to calculate the first term. While first-order derivatives of the potential energy are readily available in molecular software as they determine the forces on the system, this is often not the case for second-order derivatives which quickly become very intricate (see ref. 129, 233, and 234 for examples).

3.4.2.2 *Application to MOFs*

A first study of the temperature-dependent elastic properties of MOFs was performed for the cubic ZIF-8 material by Ortiz *et al.*, showing an increase in C_{11} when increasing the temperature from 77 K to 500 K, while the C_{12} and C_{44} elastic constants, the last of which determines the most compliant deformation mode in ZIF-8, were largely temperature-independent.[235] Despite the ZIF-FF force field showing an overall good agreement with *ab initio* data, it could not reproduce this temperature-induced increase in C_{11}.[168] Ying *et al.* used eqn (3.40) to predict the effect of linker functionalisation in ZIF-8 on its 300 K shear elastic constant C_{44}, showing that C_{44} is smaller for ZIF-Cl and ZIF-Br than for ZIF-8, while it is larger in SALEM-2.[236] In comparison, the bulk modulus was found to increase from SALEM-2 over ZIF-8 and ZIF-Cl to ZIF-Br,[236,237] in contrast to earlier observations obtained at 0 K.[73,174] In contrast to the temperature-induced stiffening in ZIF-8, the three elastic constants of HKUST-1 were predicted to decrease when increasing the temperature from 0 K to 400 K, as shown in the inset of Figure 3.14, thereby lowering the structural stability of the material.[238] These theoretical predictions of temperature-induced softening in HKUST-1 were confirmed by the nanoindentation experiments illustrated in Figure 3.14.[238]

A comparative study of the elastic moduli in UiO-66, MOF-5, MIL-47, and the two phases of MIL-53 demonstrated that the impact of temperature on the elastic moduli is strongly material dependent.[136] For UiO-66, the maximum Young's and shear moduli were predicted to increase with temperature, while their minimum values decreased slightly, thereby increasing the anisotropy with increasing temperature.[136] For MOF-5, only the minimum shear modulus was found to decrease appreciably upon increasing temperature, making the material even less stable, whereas other elastic moduli were found to remain more or less constant, thereby reducing the anisotropy with increasing temperature.[136] For the lp phases of MIL-47 and MIL-53, an increase in maximum values of the Young's and shear moduli were obtained with increasing temperature, whereas the minimum values either remained more or less constant (for MIL-47) or decreased (for MIL-53), in both cases leading to an increase in anisotropy.[136] Finally, for the MIL-53 cp phase, a mixed picture was obtained where the minimum Young's modulus,

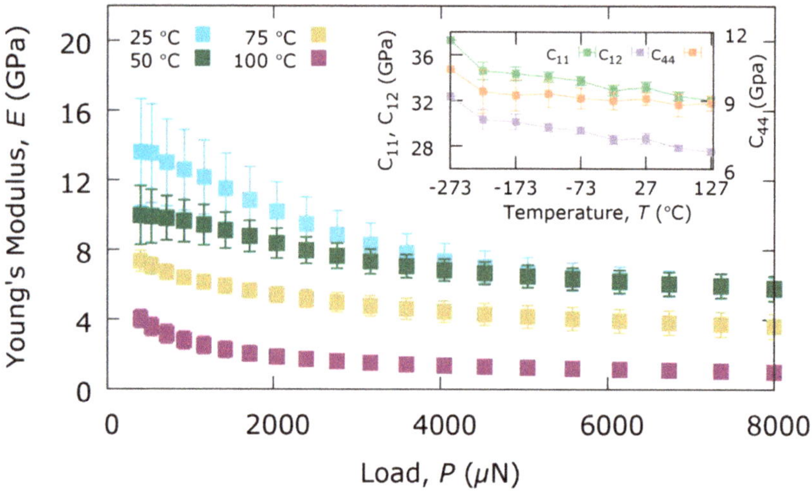

Figure 3.14 Young's modulus as a function of load at various temperatures obtained from nanoindentation experiments. Inset: Variable-temperature elastic constants obtained from classical molecular dynamics simulations. Lines provide guidance for the eye. Adapted from ref. 238 with permission from American Chemical Society, Copyright 2018.

the maximum Young's modulus and the minimum shear modulus increased with temperature, whereas the other values remained constant, leading to a net decrease in anisotropy for these two elastic moduli upon an increase in temperature.[136]

In two force field studies, Ying *et al.* used the strain fluctuation formula of eqn (3.40) to calculate the 300 K elastic moduli of DUT-49 and four isoreticular analogues obtained by changing the organic ligand.[237,239] In agreement with earlier 0 K observations discussed in Section 3.4.1, they found that longer ligands decrease the magnitude of all 300 K moduli, both in the open-pore (op) and cp phase, as shown in Figure 3.15.[237,239] In addition, Figure 3.15 reveals that the cp phase exhibits lower elastic moduli than the op phase for all investigated DUT materials.[239] Finally, a full elastic tensorial analysis of eleven $Zn(CN)_2$ polymorphs revealed auxetic behaviour in all these structures, although only one also showed NLC.[240]

3.4.3 Predicting Flexibility from Equilibrium Elastic Properties

Up to this point, we have been interested in characterising the elastic regime of MOFs, determined through the second-order elastic stiffness tensor. Based on the derivation in eqn (3.33), this limits the region for which the properties and observations in Sections 3.4.1 and 3.4.2 hold to small strains around an

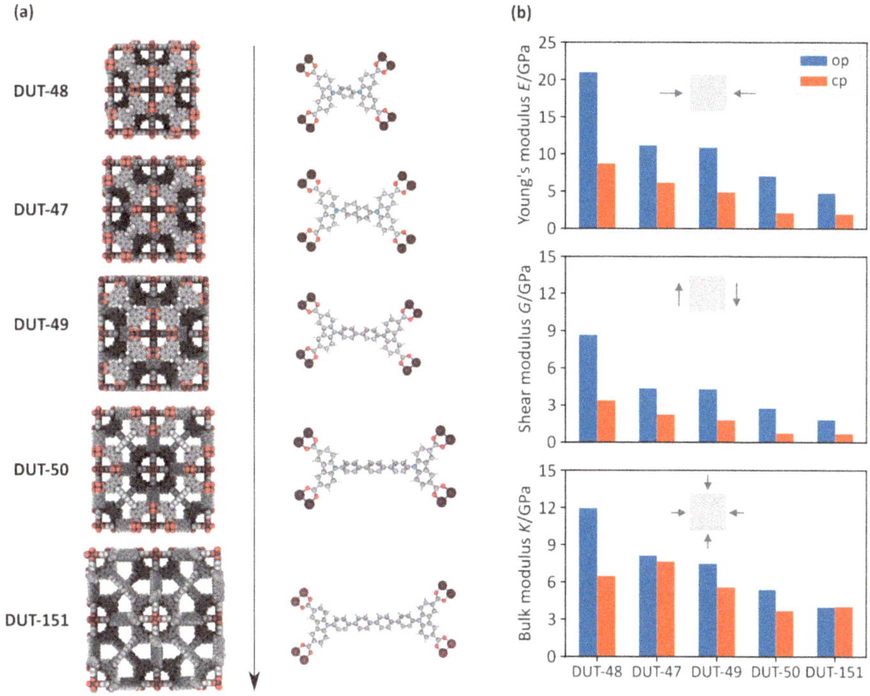

Figure 3.15 (a) Atomic structures of five isoreticular DUT materials with increasing ligand length. Copper, oxygen, nitrogen, carbon, and hydrogen atoms are shown in brown, red, cyan, grey, and white, respectively. (b) Averaged Young's, shear, and bulk moduli for these five materials in both the op (blue) and cp (orange) phase. The insets show the deformation mode corresponding to each modulus. Adapted from ref. 239 with permission from American Chemical Society, Copyright 2021.

equilibrium structure of the MOF, either at 0 K or at elevated temperatures. It is only natural, however, to ponder to which extent these elastic properties can be correlated with 'plastic' mechanical properties (exceeding the elastic limit) – for instance, in how far can elastic constants distinguish between flexible materials, exhibiting a single-crystal-to-single-crystal phase transition under plastic deformation, and rigid materials, exhibiting a single-crystal-to-amorphous phase transition instead? Finding such correlations would be very powerful, both from an experimental and a theoretical perspective. Experimentally, such correlations would circumvent the need to perform destructive tests to establish the plastic behaviour of a material. Theoretically, it is often far easier to describe the equilibrium structure of a material rather than its structure during plastic deformation, as discussed in Sections 3.5 and 3.6.

Empirically, Zener found that the anisotropy index $A = 2C_{44}/(C_{11} - C_{12})$, which is the ratio of the resistance of a cubic material against monoclinic shear to its resistance to an orthorhombic shear, is an important quantity by which to predict mechanical processes in metals beyond the elastic regime.[241] This Zener anisotropy index forms a signature for phase transformations in alloys and for plastic anisotropy in sheet metals: the higher the anisotropy, the less energetically costly the phase transformation becomes.[242,243] The question remains, however, if a similar measure can also be introduced for MOFs, which often exhibit noncubic structures.

In this regard, Ortiz *et al.* systematically calculated the 0 K elastic moduli of the flexible MIL-53, MIL-47, and DMOF-1 materials, and observed that the ratio of the maximum and minimum Young's modulus, as well as the ratio of the maximum and minimum shear modulus, is about two orders of magnitude higher for these flexible MOFs than for relatively rigid MOFs such as MOF-5 and ZIF-8.[196] This motivated them to conclude that high anisotropy in these moduli, and hence the presence of directions with rather low Young's and shear moduli, is an indicator for structural flexibility in MOFs – hence predicting their plastic behaviour.[196] Later, they noted that the occurrence of such 'soft' deformations modes along which the material can deform with only a small energy penalty can also be directly extracted by determining the eigenvalues and eigenmodes of the stiffness tensor **C**, as the eigenmode with the lowest eigenvalue describes the softest deformation mode of the material.[197] Their elastic signature for flexibility in MOFs can therefore be more generally expressed as the simultaneous occurrence of soft deformation modes and large anisotropy in eigenvalues of the elastic stiffness tensor.[197] Although they recognised that this signature should be used with caution – counterexamples were found in the rigid MIL-140,[197] which shows unexpectedly high anisotropy, and in a series of flexible DUT-49 analogues, which show unexpectedly low anisotropy[237] – this criterion remains to date a convenient criterion to check for flexibility in MOFs based on their computed equilibrium properties, both at 0 K and at finite temperatures.[136,216] When a material is brought closer to instability, its anisotropy was furthermore observed to increase substantially.[136] In 2014, Sarkisov *et al.* demonstrated that soft deformation modes in a material, and hence large-amplitude flexibility, could also be predicted by representing the MOF as a mechanical model constructed from rigid trusses that are connected through flexible hinges (Section 2.4.5).[244] The flexibility of this mechanical model and hence the associated MOF, can then be predicted by considering the energy needed to deform the model under various deformation modes.[244]

3.4.4 Determining the Stability Range of MOFs: Elasticity Under Pressure

Predicting the thermodynamic conditions – temperature, stress, adsorption – under which a certain phase of a MOF is stable is essential in order to know how these MOFs can be shaped and handled, as well as for which applications they can be adopted. This can be directly achieved by modelling the material outside of its elastic equilibrium, which is the topic of Sections 3.5 and 3.6. In contrast, herein two main methods will be outlined to determine the stability of MOFs – focussing on their 'mechanical stability' – based on properties derived in the *elastic* regime: either the elastic stiffness tensor (Section 3.4.4.1) or the vibrational modes (Section 3.4.4.2).

To do so, however, it is first necessary to generalise the definition of the stiffness tensor of eqn (3.33) to account for finite pressures. Recall that finite temperatures are already taken care of in eqn (3.33), given that the elastic tensor is therein defined based on the Helmholtz free energy F. At a finite stress $\boldsymbol{\sigma}$, the Helmholtz free energy of eqn (3.33) should be replaced by the Gibbs free energy G, defined in eqn (3.6) for a general stress and in eqn (3.19) for a hydrostatic pressure.[245] As a result, the first-order expansion coefficients in eqn (3.31) will no longer disappear at finite stress, and the unloaded elastic stiffness tensor **C** should be replaced by the elastic stiffness tensor **B** under a load $\boldsymbol{\sigma}$, defined component-wise as:[246,247,¶¶]

$$B_{ijkl} = C_{ijkl} + \frac{1}{2}\left(\sigma_{ik}\delta_{jl} + \sigma_{il}\delta_{jk} + \sigma_{jk}\delta_{il} + \sigma_{jl}\delta_{ik} - 2\sigma_{ij}\delta_{kl}\right). \tag{3.42}$$

In general, $B_{klij} \neq B_{ijkl}$ due to the last term between brackets in eqn (3.42). Hence, the *loaded* elastic stiffness tensor **B** has, in general, a lower symmetry than the *unloaded* elastic stiffness tensor **C**. As a result, also the Voigt notation cannot be used in this case. Under a hydrostatic pressure $\sigma_{ij} = -P\delta_{ij}$, eqn (3.42) simplifies to

$$B_{ijkl} = C_{ijkl} - P(\delta_{ik}\delta_{jl} + \delta_{il}\delta_{jk} - \delta_{ij}\delta_{kl}), \tag{3.43}$$

and the Voigt symmetry of **C** is preserved.[246–248]

From a practical perspective, one can hence determine the loaded elastic stiffness tensor **B** at a given temperature and stress by first determining the unloaded elastic stiffness tensor **C** under the same conditions and then applying eqn (3.42) or eqn (3.43). Alternatively, the elements of the loaded elastic stiffness tensor can be determined directly through fluctuation formulae similar to those presented in Section 3.4.2 but including a correction term, and by setting the reference cell matrix necessary to define the strain in eqn (3.4) equal to the average cell matrix at the applied stress and temperature, $\mathbf{h}_0 = \langle \mathbf{h} \rangle$, as derived in ref. 249 and 250.

3.4.4.1 Determining Elastic Stability through the Born Stability Criteria

Intuitively, a MOF structure is said to be elastically stable at a given temperature T if any arbitrary infinitesimally small strain $\boldsymbol{\varepsilon}$ increases its free energy F.[251] For such small strains, the harmonic approximation of eqn (3.33) is valid

¶¶As mentioned before, the different definitions of the elastic stiffness tensor (based on the stress–strain, stress–energy density, and elastic wave propagation) no longer coincide when the material is subjected to an external load. Herein, we will apply the definition necessary to check the Born stability criteria (based on strain–energy density) and refer the interested reader to further discussions on this topic.[245,248,310,311] and the aforementioned statement requires **ε**:**C**:**ε** to be positive for all small deformations **ε**, the so-called Born stability criterion. As a result, a necessary and sufficient condition for a given MOF structure to be elastically stable at a given temperature T is that its elastic stiffness tensor **C** is *positive definite* or, in other words, that the eigenvalues of **C** are all positive or that all its leading principal minors are positive (Sylvester's criterion).[251,252] Using the terminology introduced in Section 3.4, this implies that energy needs to be supplied to deform the system, even along the softest or most compliant deformation mode, which is the eigenmode corresponding to the lowest eigenvalue of **C**. From the point of view of the fluctuation formula of eqn (3.40), Parrinello and Rahman noted that elastic instability of a crystal would lead to a divergence in some of the correlations $\langle \varepsilon_{\alpha\beta}\varepsilon_{\mu\nu}\rangle - \langle \varepsilon_{\alpha\beta}\rangle\langle \varepsilon_{\mu\nu}\rangle$.[229]

For a cubic crystal, calculating the eigenvalues of the elastic stiffness tensor and requiring them to be positive leads to the well-known elastic stability criteria[251]

$$C_{11} + 2C_{12} > 0,\ C_{11} - C_{12} > 0,\ C_{44} > 0. \tag{3.44}$$

In turn, these inequalities can be interpreted as requiring the material to be stable against a volumetric compression or expansion, against orthorhombic deformation, and against monoclinic shear.[253] Important, however, is to realise that eqn (3.44) is a specific form of the Born stability criteria valid *only when a cubic crystal system* is considered. For other crystal systems, with other nonzero elastic constants, the requirement that the elastic stiffness tensor is positive definite leads to different Born stability criteria, as summarised by Mouhat and Coudert.[252] The lower the symmetry of the system, the more elastic constants appear in the stiffness tensor and the more intricate these expressions become.[252]

For materials subjected to an external stress **σ**, the Born stability criteria should be expressed in terms of the loaded elastic stiffness tensor **B** (see eqn (3.42)) or its symmetric variant $2\mathbf{A} = \mathbf{B} + \mathbf{B}^T$: a given material at a certain temperature and under a certain external stress is stable with respect to small elastic deformations if, and only if, **B** or **A** is positive definite.[253–255] As an example, for a cubic system under a hydrostatic pressure P (which retains the cubic symmetry), the elastic stability criteria of eqn (3.44) transform to

$$C_{11} + 2C_{12} + P > 0,\ C_{11} - C_{12} - 2P > 0,\ C_{44} - P > 0. \tag{3.45}$$

One has to take care, however, that a general stress may lower the symmetry of the crystal. The requirement that the loaded elastic stiffness tensor is positive definite, may hence involve a more cumbersome expression of its elastic constants than requiring the unloaded elastic stiffness tensor to be positive definite.

Using these definitions, the elastic stability region of a MOF can be defined as that region in parameter space (including, *e.g.*, temperature, stress, and gas loading) for which the Born stability criteria are satisfied. A first study exploring this stability region for a MOF was performed in 2012, by calculating the Born stability criteria for the cubic ZIF-8 material as a function of the pressure for different temperatures and methane loadings.[235] As shown in Figure 3.16, which illustrates the two smallest out of the three Born stability criteria for this cubic material, $C_{44} - P > 0$ is the first criterion to be violated for ZIF-8, irrespective of the methane loading or temperature.[235] As could be predicted qualitatively by the extremely low 0 K shear modulus of ZIF-8 discussed in Section 3.4.1, the material first becomes mechanically unstable under shear deformation.[235] However, while the 0 K shear constant C_{44} of ZIF-8 amounts to about 1 GPa, the guest-free material becomes unstable already at a lower pressure of ~0.4 GPa,[235] a result which was later also verified *via* DFT calculations.[112] This illustrates the importance of determining the elastic constants and verifying the positive definiteness of **B** at or close to the thermodynamic conditions of interest, at finite temperatures and pressures (using the technique in Section 3.4.2), rather than obtaining the elastic constants from simulations without external stress (using the technique in Section 3.4.1) and applying eqn (3.44) or its analogues to determine the stability region.[245] While Figure 3.16 demonstrates that the pressure at which this instability occurs is rather insensitive to temperature, the inclusion of methane in this material does expand its region of stability appreciably.[235]

A similar shear-induced elastic instability was also observed for ZIF-4,[235] several ZIF-8 analogues,[236] and the denser $Zn(CN)_2$ materials,[240] whereas

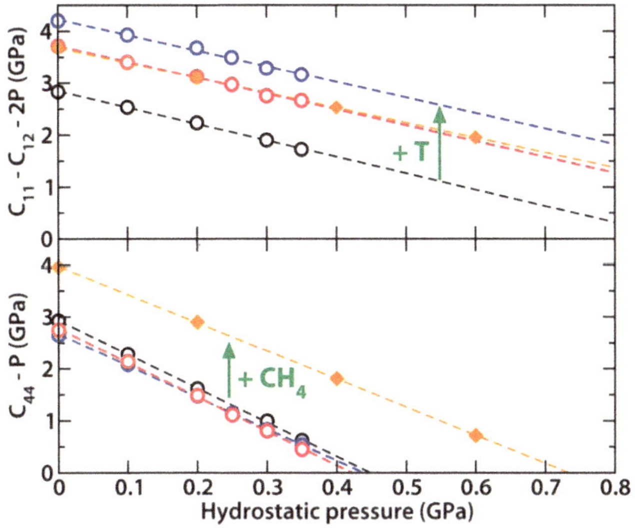

Figure 3.16 Evolution of the Born stability criteria for compression (upper panel) and shear (lower panel) of ZIF-8 as a function of hydrostatic pressure. Black: empty ZIF-8 at 77 K; red: empty ZIF-8 at 300 K; blue: empty ZIF-8 at 500 K; orange: ZIF-8 with 18 methane molecules per unit cell at 300 K. Reproduced from ref. 235 with permission from American Chemical Society, Copyright 2013.

the softest deformation mode for the defect-free cubic UiO-66 MOF was observed to be an axial deformation, which violates the Born stability conditions starting from a pressure of ~1.85 GPa onwards.[49] In contrast, in the UiO-66 analogue obtained by replacing the phenyl ring in the BDC ligand by a C≡C moiety, the first Born stability criterion to be violated is the isotropic compression mode.[141] This difference with respect to UiO-66 was attributed to the weaker steric hindrance when compressing the MOF containing this aliphatic linker compared to the aromatic linker in UiO-66.[141] A similar variation in the weakest deformation mode leading to mechanical instability was observed in DUT-49 and reticular analogues, although the underlying mechanism remains uncertain at this point.[237]

In 2018, this procedure to determine the mechanical stability of MOFs was contrasted with the anisotropy signature described in Section 3.4.3 and the construction of pressure-*versus*-volume equations of state, which will be discussed in more depth in Sections 3.5 and 3.6. To this end, next to the cubic and rigid UiO-66 and MOF-5 materials, MIL-47 and MIL-53 were also considered, for which the lp phases belong to the orthorhombic crystal system.[136] For MOF-5, this analysis showed that the first Born stability criterion to be violated is not due to shear deformation, as one may expect based on the low C_{44} value and based on predictions within the QHA framework,[157] but is rather induced by isotropic compression, given that Figure 3.17 reveals that the associated Born stability criterion shows a

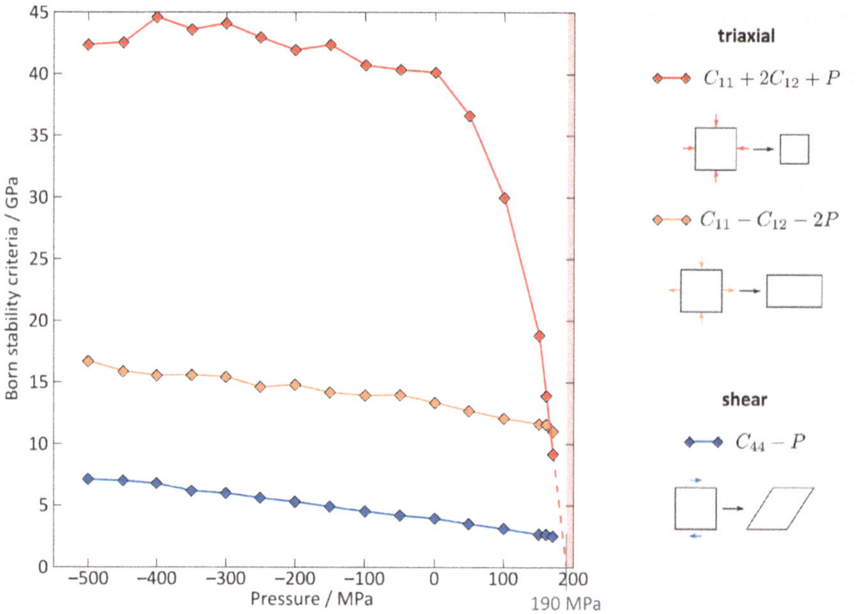

Figure 3.17 Born stability criteria for MOF-5 as a function of the hydrostatic pressure, at 300 K. Adapted from ref. 136 with permission from American Chemical Society, Copyright 2018.

substantial pressure-induced softening.[136] Despite the similarities between the lp phases of MIL-47(V) and MIL-53(Al), their mechanical instability was observed to be induced by two different deformation modes: a triaxial deformation mode for MIL-47 and a shear deformation mode for MIL-53, presumably due to the weaker inorganic chain present in the latter material.[136] In contrast to the cubic UiO-66 and MOF-5 materials, however, the weakest axial deformation mode in MIL-47 is not easily interpretable, due to the lower symmetry of this orthorhombic structure.[136] This was further exemplified by the even less-symmetric monoclinic cp phase of MIL-53(Al). In this case, the Born stability criteria need to be evaluated by calculating the eigenvalues of the stiffness tensor and requiring them to be positive, giving only limited and less-interpretable information about the deformation mode leading to mechanical instability compared to more symmetric phases.[136]

3.4.4.2 Determining Dynamic Stability through the Vibrational Spectrum

While the elastic stability criteria in Section 3.4.4.1 give insight into the macroscopic deformation mode of the crystal that first induces mechanical instability, complementary microscopic insight can in principle be obtained through the vibrational spectrum. The Hessian matrix, introduced in eqn (3.38), describes the energy associated with small displacements of the different nuclei from their equilibrium structure, in the harmonic approximation. The Hessian matrix therefore takes a very similar role to that of the stiffness tensor: while the elastic stiffness tensor should be positive definite to ensure elastic stability, the Hessian matrix should be positive definite to ensure dynamic stability. If the Hessian matrix contains negative eigenvalues, corresponding to imaginary vibrational frequencies, displacing the nuclei along the associated vibrational modes would lower the energy of the system, indicating its mechanical instability. However, theoretically obtaining these low-frequency or terahertz vibrations is highly challenging, as is their interpretation, given that these collective vibrational modes contain contributions from many atoms in the unit cell.[188,256]

As an instructive example, Ryder *et al.* investigated the terahertz vibrations in ZIF-4, ZIF-7, and ZIF-8,[256] as well as in HKUST-1,[188] which were discussed earlier in Section 3.4.1.5. For ZIF-4, two soft phonon modes were identified: one associated with a distortion of the 4MR aperture, discussed earlier in Section 3.4.1.3, and one associated with a gate-opening phenomenon.[256] For ZIF-7 and ZIF-8, several soft modes were found that could be associated with gate-opening and breathing mechanisms, as well as a soft mode that could explain the experimentally observed phase transition in these materials.[256–258] Also the phase transition in the related $Zn(CN)_2$ material could be associated with the instability of several low-frequency vibrational modes.[259] While promising, extending this approach to finite temperatures is not straightforward due to the difficulties in accurately assigning vibrational modes in those cases. Recently, Maul *et al.* presented such an approach that combines this vibrational analysis with the Born stability criteria to obtain both macroscopic insight into the weakest deformation mode and atomic insight into which nuclei play a major role in determining mechanical instability.[112] The Born stability criteria revealed that the shear deformation, defined by C_{44}, is the weakest deformation mode in ZIF-8.[112] Subsequently, they determined through the relaxation term in eqn (3.37) that the antisymmetric breathing of the 4MR aperture and the antisymmetric gate opening contribute most to the relaxation term associated with the C_{44} elastic constant and hence to instability.[112]

3.5 The Response of Flexible MOFs to Large Pressures: the Inelastic Regime

Thanks to their porous structure, many MOFs undergo large-amplitude structural phase transitions under external stimuli.[3] Studying the mechanical response of these flexible MOFs or SPCs beyond the equilibrium regime requires dedicated techniques differing from those outlined in Section 3.4. While the approaches discussed in Sections 3.4.3 and 3.4.4 give insight into whether a MOF is flexible or not and, in the case of the Born stability criteria, at which pressure it may undergo a phase transition, they do not model the phase transition explicitly. As a result, none of these aforementioned techniques provide information about the structure of the MOF during and after the phase transition or the energy needed to induce these transitions.

In this section, three general techniques will be discussed to access that information. The most straightforward technique, outlined in Section 3.5.1, consists in probing the response of the MOF structure during an MD simulation in the $(N,P,\boldsymbol{\sigma}_a,T)$ ensemble at various pressures and following the phase transition that is observed once the external pressure P is sufficiently high. Due to the inability of this technique to determine the energy required to induce the phase transition or the difference in energy between the initial and final phases, as well as due to inaccuracies in predicting the transition pressure, two other approaches will be outlined in Sections 3.5.2 and 3.5.3. In Section 3.5.2, the energy-*versus*-volume equation of state, introduced in Section 3.4.1 to extract the equilibrium bulk modulus, will be extended so to cover the different phases the material can attain. In Section 3.5.3, a technique will be outlined that generalises this approach to finite temperatures, giving rise to pressure and free energy equations of state that govern the dynamics of MOFs under experimental conditions. While flexibility in MOFs can be induced by many stimuli, we will focus on mechanical stress and pressure as stimuli herein to keep the discussion succinct and refer the reader interested in a broader range of stimuli-induced transitions to ref. 3 and 260.

3.5.1 Directly Modelling Phase Transformations in Flexible MOFs and Its Limitations

Initial force field studies on MOFs focussed on their exceptional adsorption properties, which were described by keeping the atomic structure of the MOF rigid and only accounted for noncovalent interactions between the MOF and the adsorbed gases. Given that these so-called 'rigid' force fields are unable to describe phase transitions, and given that *ab initio* MD simulations lasting several picoseconds – necessary to observe a phase transition – were computationally too costly in the early years of MOF research, the first direct $(N,P,\boldsymbol{\sigma}_a,T)$ simulations that demonstrate pressure-induced phase transitions in MOFs were only reported in the early 2010s. For MIL-47(V), Yot *et al.* combined high-pressure mercury intrusion experiments with direct force field simulations to discover that this material, which was known to remain rigid under gas adsorption, can undergo phase transitions under high mechanical pressures.[191] The simulated lp-to-cp transition pressure of 137 MPa and the simulated cp-to-lp transition pressure of 66 MPa were both in good agreement with the experimental range of pressures, amounting to 85–125 MPa and 55–75 MPa, respectively, and are rather high compared to other flexible MOFs.[191] For isostructural MIL-53(Cr), lower transition pressures were observed, amounting to 53.5 MPa and 2.4 MPa for the lp-to-cp and cp-to-lp transitions, respectively, which decrease further when low amounts of guest molecules are present inside the pores.[202,261,262] This latter observation is in line with the equivalence between internal adsorption stress and external mechanical stress postulated earlier.[263] These computational studies also identified the rotation of the phenyl ring with respect to the carboxylate group in the BDC ligand as one of the main rearrangements during breathing in MIL-47(V) and MIL-53(Cr),[191,202,261] next to the experimentally observed 'knee cap' motion in which the carboxylate O–O axis acts as a hinge.[264] For MIL-53(Al), an even lower lp-to-cp transition pressure of 19 MPa was obtained.[265]

In addition to the aforementioned force field studies, also several *ab initio* MD simulations have been performed to study pressure-induced phase transformations in MOFs. In cases where the studied phase transformation is reconstructive in nature rather than displacive, such an *ab initio* approach is vital to describe the bond rearrangement. However, as with force fields, extreme care needs to be exerted regarding technical details, given that the occurrence of a phase transformation may depend on, among other things, the basis set size, the size of the unit cell, and the functional and dispersion scheme used, and this to an even larger extent than for the static simulations described in Section 3.4.1.[266] Using the B3LYP functional,[166] Ortiz *et al.* followed this approach to predict the flexibility of NOTT-300 and CAU-13.[224] For NOTT-300, an lp-to-cp transformation was observed from pressures of 700 MPa onwards.[224] Although the authors noted that large volume fluctuations were already visible starting from 500 MPa, no transformation was observed in the 6 ps simulation time at this pressure, an important point discussed later in this section.[224] Also CAU-13 was predicted to be flexible, although the transformation to its lp phase was observed for *negative* mechanical pressures.[224] However, as mentioned before, such negative pressures can be achieved by means of adsorption stress, when a low amount of suitable guest molecules adsorb in the MOF.[263]

Similarly, ZAG-4 and ZAG-6 were predicted to undergo proton transfer at high pressures (3.7 GPa and 3.0 GPa, respectively), which in the case of ZAG-6 is accompanied by a coiling of the linker.[216] While these results were obtained through minimisation of the enthalpy rather than through a direct $(N,P,\boldsymbol{\sigma}_a,T)$ simulation, it is important to mention that the proton transfer describing the phase transformation cannot be modelled *via* force-field techniques due to the bond rearrangement, necessitating the more costly *ab initio* technique used in this study.[216] A similar minimisation procedure was employed to theoretically predict a displacive phase transition in ZIF-8 at around 3.8 GPa, under the condition that the well-known reconstructive phase transition at lower pressures is suppressed, for instance by theoretically constraining the symmetry of the material.[169] For CoBDP, an lp-to-cp phase transition was predicted at an extremely low pressure of 0.5 MPa.[217]

While these direct simulation approaches are attractive to gain insight into the possible occurrence of phase transitions in a material, there are a few pitfalls associated with them. Direct simulations explore only one possible trajectory through phase space at the given temperature and pressure, while discarding all other potential trajectories. This also implies that repeating the same simulation with slightly different initial conditions – atomic positions, velocities, or slight changes in the initial cell matrix – may fundamentally alter the simulation results, making interpretations based on a single simulation speculative. Instead, a proper analysis of phase transitions requires taking an ensemble average over all possible trajectories, weighted with their proper statistical importance. This is key in the approaches outlined in Sections 3.5.2 and 3.5.3.

However, even when repeating the direct $(N,P,\boldsymbol{\sigma}_a,T)$ simulations of this section several times with slightly different initial conditions to stimulate ergodicity, taking a statistical average may still underestimate the transition pressure. To understand this, Figure 3.18 shows the simulation time needed to observe an lp-to-cp transition in a $1 \times 2 \times 1$ unit cell of MIL-53(Al), containing 152 atoms, at 300 K and at various pressures between 0.1 MPa and 1000 MPa.[83] For each pressure, 100 independent simulations were performed starting in the lp phase.[83] In addition, each simulation was conducted for two different barostat implementations, the MTK barostat and the Langevin barostat,[96,97] and three different barostat relaxation times of 1 ps, 5 ps, and 10 ps.[83] For simulations performed at pressures higher than the experimental lp-to-cp transition pressure, an lp-to-cp transition was observed in all simulations, as expected.[83] Moreover, these transitions occurred on a subpicosecond timescale, as shown in Figure 3.18(b), which are also accessible in AIMD simulations.[83] However, an lp-to-cp transition was also unexpectedly observed for all simulations at pressures *below* the experimental transition pressure and *below* the transition pressure extracted *via* the thermodynamic

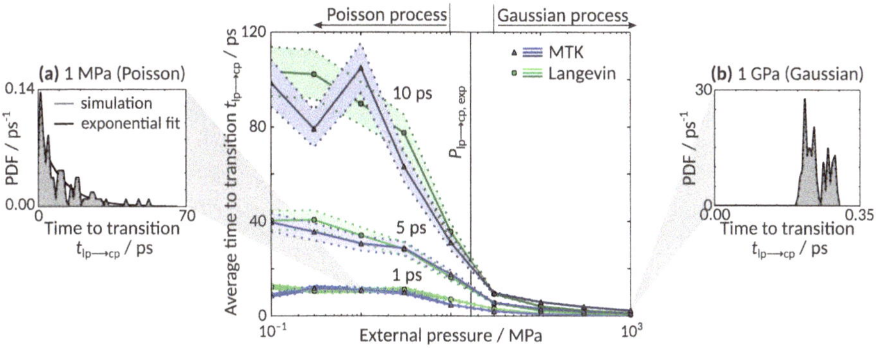

Figure 3.18 Average simulation time needed to observe the MIL-53(Al) lp-to-cp transition in a $(N,P,\boldsymbol{\sigma}_a = 0,T)$ ensemble at 300 K as a function of the applied pressure and carried out for different barostats and barostat relaxation times of 1, 5, and 10 ps. The shaded regions indicate the 1σ confidence interval for this average simulation time. The vertical line indicates the experimental transition pressure. The side panes (a) and (b) display the probability density function (PDF) of the time to transition for two selected pressures, each constructed based on 100 independent simulations carried out with the MTK barostat and a relaxation time of 1 ps. Adapted from ref. 83 with permission from American Chemical Society, Copyright 2015.

approach outlined in Section 3.5.3.[83] Compared to transitions at higher pressures, the simulation time needed to induce transitions at these lower pressures is substantially higher – up to several tens of picoseconds, which are not reached in the aforementioned AIMD simulations – and also more strongly impacted by the barostat relaxation time.[83] Furthermore, as shown in Figure 3.18(a), the time to observe the transition at a certain pressure below the experimental transition pressure was found to be exponentially distributed, indicating that the mechanism inducing this transition is stochastic in nature.[83]

These premature phase transformations were found to be induced by fluctuations in the internal pressure of the material.[83] Even though the average internal pressure converges to the value supplied to the barostat scheme, simulations with pressure control show substantial relative fluctuations in the instantaneous pressure, as demonstrated in Figure 3.6(a). These fluctuations, which scale inversely with the square root of the number of atoms, are necessary to properly sample the isobaric ensemble; a similar effect is also present in temperature control.[267] In the case of the 152-atom unit cell of MIL-53(Al), these pressure fluctuations were found to amount to several gigapascals, thereby exceeding the experimental lp-to-cp transition pressure by several orders of magnitude.[83] When a sufficiently large sequence of positive pressure fluctuations are observed during a simulation, they may hence trigger a premature phase transition.[83] A similar underestimation of the transition pressure *via* direct $(N,P,\boldsymbol{\sigma}_a,T)$ simulations was also observed for a series of IRMOFs by Ying *et al.*[237] They also noted that this effect was much less pronounced in DUT-49-like materials, most likely because of their larger conventional unit cells, as the conventional DUT-49 unit cell contains 1728 atoms.[237] Besides providing an underestimation of the transition pressure, these premature phase transitions also limit how close to mechanical instability one can simulate the vibrational spectra or Born stability criteria of Section 3.4.4. As a result, to use these techniques, one needs to rely on extrapolations to obtain the Born stability criteria or vibrational modes at the point of instability.[136]

3.5.2 Modelling Flexible MOFs at 0 K: Energy Equations of State

A first approach to overcome the limitations of direct $(N,P,\boldsymbol{\sigma}_a,T)$ simulations outlined in Section 3.5.1 – as well as the Pulay stress for plane-wave codes mentioned earlier in Section 3.4.1 – is to model the energy-*versus*-volume equation of state of the material that connects the different (meta)stable states. Such an energy equation of state can be obtained by optimising the cell shape and nuclear coordinates of the MOF while keeping its volume fixed for different structures dispersed along a volume grid, and collecting the internal energy as a function of the volume. This gives access both to the energy difference between the different phases and to the transition pressure necessary to induce transformations between them, given that the 0 K pressure equation of state can be obtained as the negative of the first derivative of the $E(V)$ profile with respect to the volume. The approach outlined here assumes that the different phases can be distinguished based on the volume and is in its basic form limited to 0 K results. Although extensions to both limitations will be discussed in Section 3.5.3, these 0 K profiles are computationally less expensive than those discussed in Section 3.5.3 because they only require optimisations. As a result, they are more easily accessible through *ab initio* techniques, thereby providing a first set of guidelines on how to model the flexibility of MOFs at 0 K, which are directly transferable to finite temperatures.

Although Vanpoucke *et al.* only reported partial energy equations of state for the cp and lp phases of MIL-47, they extracted an *ab initio* lp-to-cp transition pressure between 82 and 125 MPa, depending on the magnetic ordering of the inorganic chains, and an lp phase that is around 1.3 kJ mol^{-1} more stable than the cp phase.[109,205,‖‖] Through the harmonic approximation, they predicted that entropic factors would increase the relative stability of the lp phase to around 13.7 kJ mol^{-1} at 300 K.[109] These small energy differences emphasise the need to use very accurate models to estimate the energy, given that changes in the choice of functional or dispersion scheme can give rise to errors that are substantially larger than these energy differences.[55,268,269] This can be exemplified by MIL-53(Cr), shown in Figure 3.19, where both the

‖‖Energy differences are defined per unit cell of the material, which contain four metal atoms for MIL-47 and MIL-53.

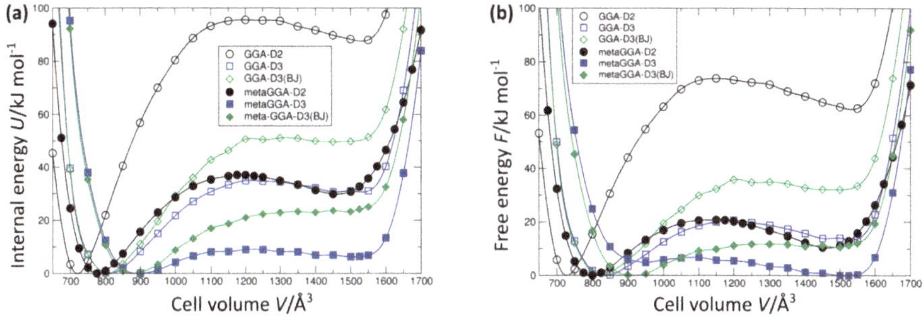

Figure 3.19 Equation of state of (a) the internal energy, neglecting zero-point motion, and (b) the Helmholtz free energy at 293 K for MIL-53(Cr) as a function of the volume, for different density functionals. Each curve is scaled so that its minimum is zero. Adapted from ref. 268 with permission from American Chemical Society, Copyright 2017.

internal energy and the Helmholtz free energy at 293 K obtained through QHA strongly depend on the functional and dispersion scheme, which also affect the prediction of the most stable phase in this temperature range.[268]

The aforementioned observations were the motivation behind a more extensive study by Wieme *et al.*, in which the random phase approximation (RPA) to the correlation energy, including single-excitation (SE) effects, was adopted to model the 0 K energy equation of state for MIL-53(Al).[269] The RPA + SE equation of state showed that the MIL-53(Al) cp phase was only around 7.4 kJ mol^{-1} more stable than the lp phase, separated by an lp-to-cp barrier amounting to around 8 kJ mol^{-1}.[269] This bistability was found to occur because of an interplay between entropic effects (favouring the lp phase) and dispersion interactions (favouring the cp phase).[269] They used this knowledge to show that lowering the extent of dispersion interactions, such as by replacing the BDC ligand in MIL-53(Al) with a fumarate ligand, gives rise to an energy equation of state with only an lp minimum, while replacing the ligand with either BPDC or NDC ligands, with larger aromaticity and hence higher dispersion interactions, leads to energy profiles with only a cp minimum.[269] In contrast, altering the metal atom only has a minor effect on the energy profile, with MIL-53(Ga) showing the same bistability as MIL-53(Al).[269] The absence of a 0 K cp minimum in the fumarate-based MIL-53(Al) material as well as the bistability of MIL-53(Al) were also reproduced using a QuickFF force field,[265,270] although the stability of the cp phase in MIL-53(Al) was overestimated at both 0 K and 300 K.[265]

A similar MIL-53(Al) cp overstabilisation was also observed when applying the quasi-harmonic approximation[206] or through umbrella sampling simulations.[271] For the fumarate-based MIL-53, it was found that the gallium analogue stabilises the cp phase compared to the aluminium analogue, forming a metastable phase at 0 K.[199] A similar relative stabilisation was also observed by replacing the copper cation in DMOF-1(Cu) with zinc, although the lp phase remained the most stable phase at 0 K.[215]

These observations led to the theoretical prediction that the cp phase in DMOF-1 could be accessed by applying appropriate mechanical pressure, which was subsequently confirmed experimentally.[215] Finally, for a different wine-rack type material, CUK-1(Mg), the slope of the 0 K energy equation of state as a function of the volume is almost constant between the cp and lp phase.[272] This indicates that the cp-to-lp and lp-to-cp transition pressures coincide for CUK-1(Mg), hence forming the first pressure-driven molecular MOF spring.[272]

3.5.3 Modelling Flexible MOFs at Finite Temperature: Free Energy Equations of State

3.5.3.1 General Methodology

To incorporate the full impact of temperature on the mechanical properties of a MOF, the energy-*versus*-volume equation of state in Section 3.5.2 should be replaced by a Helmholtz free energy equation of state as a function of the volume.[273,274] Free energy profiles are ubiquitously adopted to describe transformations in biomolecular complexes, using a variety of so-called enhanced sampling techniques such as metadynamics,[275,276] umbrella sampling,[277] multistate Bennett acceptance ratio estimation,[278] thermodynamic integration,[279] and free energy perturbation.[280] When the free energy profile is obtained as a function of the volume, the pressure profile can be retrieved as the negative of its first derivative:

$$P(N,T;V) = -\frac{\partial F(N,T;V)}{\partial V}. \tag{3.46}$$

This pressure profile directly reveals the mechanical response of the material upon pressure.

An alternative way that circumvents taking this derivative – and hence avoids the numerical noise that may be introduced when predicting transition pressures this way – is to directly simulate the pressure-*versus*-volume or $P(V)$ equation of state, as first derived in ref. 83. In Section 3.5.2, the 0 K pressure profile was obtained by optimising the cell shape and nuclear coordinates at fixed volume for a variety of structures along a predefined volume grid. Analogously, the pressure profile at finite temperatures can be obtained by determining the ensemble-averaged instantaneous pressure of a material at a range of fixed volumes, while simultaneously allowing the nuclear positions and cell shape to dynamically fluctuate in line with the applied temperature and so that the average deviatoric stress is zero (only yielding a hydrostatic pressure).[83] Hence, the appropriate ensemble to directly simulate these $P_i(V)$ or $P(V)$ profiles at finite temperatures is the $(N,V,\boldsymbol{\sigma}_\mathrm{a} = \mathbf{0},T)$ ensemble, in which the temperature T and the deviatoric stress $\boldsymbol{\sigma}_\mathrm{a}$ are controlled and both the number of particles N and the cell volume V are kept fixed.[83] As was demonstrated in ref. 271, this thermodynamic integration procedure parallels the accuracy of alternative enhanced sampling methods, while being, together with umbrella sampling, among the most efficient methods for this purpose.

As shown in Figure 3.20, any pressure equation of state can be divided into regions with a positive slope, which are mechanically unstable regions as they show a thermodynamically forbidden negative volume compressibility, that separate mechanically stable regions (with a negative slope), which can be stabilised under pressure. More specifically, the different (meta)stable states and transition states at any given pressure P_{ext} can be obtained by determining the intersections of the $P(V)$ profile with a horizontal line at $P = P_{ext}$ and classifying these intersections into mechanically (meta)stable states and mechanically unstable transition states. This is demonstrated in Figure 3.20, which shows three pressures that lead only to a stable cp state (at a pressure P_1), only to a stable lp state (at a pressure P_3), or to the presence of both a (meta)stable cp and lp state separated by a transition state (at a pressure P_2). Furthermore, the transition pressures between the different phases follow as the minima and maxima in the pressure profile (see Figure 3.20).

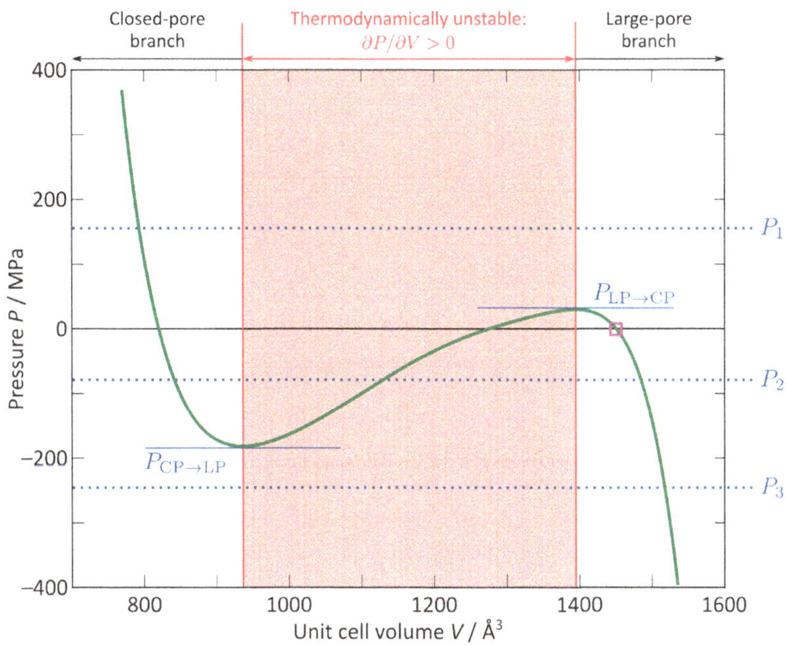

Figure 3.20 Generic pressure-*versus*-volume equation of state (green) for a bistable material, exhibiting a closed-pore (cp) and a large-pore (lp) phase, separated by a thermodynamically unstable region (red). The two transition pressures, lp-to-cp and cp-to-lp, are indicated by solid blue lines, whereas the magenta square indicates the (met)stable lp state at atmospheric pressures. The intersections between the equation of state and a blue dotted line indicate the different (meta)stable and, if present, transition state at that pressure. Adapted from ref. 136 with permission from American Chemical Society, Copyright 2018.

From the pressure profile, the free energy equation of state can be constructed *via* thermodynamic integration:[83]

$$F(N,T;V)-F(N,T;V_0)=-\int_{V_0}^{V}P(N,T;V')\mathrm{d}V'=-\int_{V_0}^{V}\langle P_i(N,T;V')\rangle\mathrm{d}V'. \qquad (3.47)$$

Finally, from the above Helmholtz free energy, also the Gibbs free energy or free enthalpy profile at a pressure P can be determined as

$$G(N,P,T;V) = F(N,T;V) + PV \qquad (3.48)$$

thereby revealing the relative stability of the different (meta)stable phases at any given pressure.[83] By repeating this procedure at different temperatures, also temperature-induced phase transitions can be predicted.[274] It is important to emphasise here that the cell shape $\mathbf{h}_0$ should be able to fluctuate when constructing $P(V)$ profiles.[281] If, instead of the appropriate $(N,V,\boldsymbol{\sigma}_a = 0,T)$ ensemble, the more popular but shape-restricted $(N,V,\mathbf{h}_0,T)$ ensemble is used, cell shape contributions to the free energy are neglected, which introduces a free energy error comparable to the free energy difference between the lp and cp phases in MIL-53(Al).[281] Finally, one should realise that the volume may not always be the most optimal parameter to describe pressure-induced phase transitions in MOFs.[282,283] In that case, one has to employ one of the other enhanced sampling techniques mentioned above to construct the free energy profile,[282,283] from which the pressure profile can then be obtained by taking its derivative with respect to the volume.

3.5.3.2 *Application to MOFs*

The general procedure described above was first adopted to determine the transition pressures for MIL-53(Al), revealing lp-to-cp and cp-to-lp transition pressures of *ca.* 30 MPa and −180 MPa, respectively, yielding much better agreement with the experimental transition pressures than those obtained through the direct method of Section 3.5.1.[83,136] For MIL-47, both transition pressures were found to be higher and the corresponding free energy equation of state only revealed a stable lp phase at atmospheric pressure and for temperatures between 100 K and 400 K.[136,204] Increasing the aromaticity of the ligand increases the relative stability of the cp phase, resulting in a bistable BPDC-containing COMOC-2 material, similar to MIL-53(Al), whereas the NDC-containing COMOC-3 material exhibits only a stable cp phase between 100 K and 400 K.[204] For CUK-1(Mg), a similar procedure demonstrated that it behaves as a spring-like material in which the lp-to-cp and cp-to-lp transition pressures coincide.[272] Similar to their 0 K behaviour discussed in Section 3.5.2, MIL-53(Ga) and the fumarate-based MIL-53(Al) show only a single stable phase at 300 K and atmospheric pressure.[274] For the MIL-53(Al) analogue in which the hydroxide anions in the inorganic chain are replaced by fluoride anions, a similar bistability as in MIL-53(Al) is found at room temperature, although only a stable lp phase remains at temperatures of around 450 K and higher.[274]

For the unfunctionalised DMOF-1(Zn), mercury intrusion experiments indicate an lp-to-cp transition at around 51 MPa without exhibiting a cp-to-lp transition upon lowering the pressure again, indicating that the material is bistable at 300 K in the absence of an external pressure.[215] Although this bistability was not retrieved using a QuickFF force field fitted to the *ab initio* Hessian matrix and geometry,[215] Keupp *et al.* did correctly obtain this bistability using a MOF-FF force field that was derived in a similar fashion,[18] demonstrating how small differences in the theoretical description of the interatomic interactions may greatly affect the predicted macroscopic behaviour. This is true to an even larger extent when considering fu-MOFs, which are obtained from DMOF-1 by the side chain functionalisation of its BDC linkers.[284] Keupp *et al.* noticed that the volume of the fu-MOF cp phase as well as its relative stability with respect to the lp phase strongly depend on the alkoxy group present on the BDC linker.[285] Even more importantly, for the longest functional groups considered in this study, the $P(V)$ profile depends on the conformation of the functionalised BDC linker, where some conformations exhibit a metastable cp phase, while others only retrieve a monostable lp material, which indicates that the volume alone may not be appropriate to describe the phase transition.[285] By combining experimental and computational tools, they further explored this observation for the copper-containing DMOF-1 and fu-DMOF-1 analogues.[286] This study revealed that the stability of the cp and lp phases of these materials, as probed through pressure equations of state, is governed both by the configurational entropy associated with the different conformations of the alkoxy groups as well as by the dispersion interactions acting between them.[286]

To explore the extreme volume contraction during the op-to-cp transition in DUT-49, Evans *et al.* highlighted the similarities between the elastic buckling of the ligand in this material and the (plastic) buckling of a column upon application of a critical load in macroscopic mechanics.[20] In addition, they constructed pressure-*versus*-volume equations of state at fixed guest loadings to understand the thermodynamics underpinning negative gas adsorption (NGA) in DUT-49.[20] Although these simulations were performed in the $(N,V,\mathbf{h}_0,T)$ ensemble instead of the $(N,V,\boldsymbol{\sigma}_a = \mathbf{0},T)$ ensemble and hence neglect cell shape sampling (*vide supra*), this has no substantial impact on the predicted free energy profiles given the large number of nuclear degrees of freedom in DUT-49.[281] These observations were later generalised to other members of the DUT family, revealing that the linker should both be stiff enough to stabilise the op phase of the structure under guest-free and overloaded conditions, while simultaneously being sufficiently soft to allow for the guest-induced op-to-cp transformation.[237,287,288]

Pressure equations of state have also been adopted to investigate the phase transition mechanism in flexible MOFs. In 2019, Rogge *et al.* investigated how increasing the simulation cell size affects the thermodynamics of MIL-53(Al), demonstrating that cp and lp phases may coexist as metastable configurations in between the pure-phase cp and lp phases, but only for crystals that are sufficiently large, as visualised in Figure 3.21.[19] For the studied MIL-53(Al)

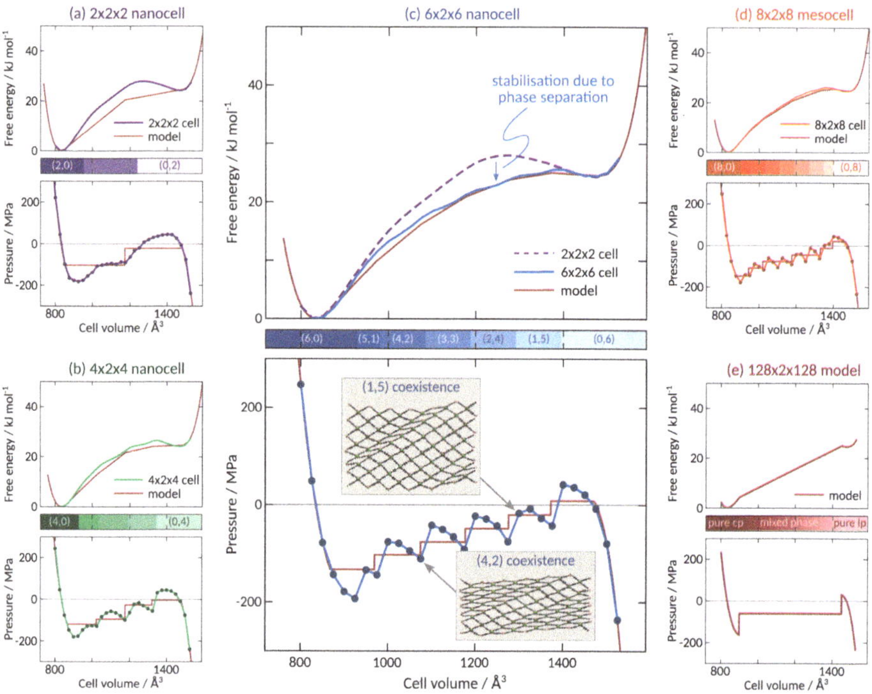

Figure 3.21 Stabilisation of the metastable cp/lp coexistence regions with increasing cell size. Pressure (bottom) and free energy (top) equations of state, with indication of the metastable (n_{cp},n_{lp}) phase coexistence regions, in which n_{cp} cp and n_{lp} lp layers coexist, as a function of the volume (middle) for four different cell sizes: (a) a $2 \times 2 \times 2$ supercell, (b) a $4 \times 2 \times 4$ supercell, (c) a $6 \times 2 \times 6$ supercell, and (d) an $8 \times 2 \times 8$ supercell, all at 300 K. The red lines indicate the fitted thermodynamic model, which is extrapolated in (e) for a $128 \times 2 \times 128$ supercell ($208.0 \times 1.4 \times 171.2$ nm^3), similar in size to experimental MIL-53(Al)-NH_2 crystals. Reproduced from ref. 19, https://doi.org/10.1038/s41467-019-12754-w, under the terms of the CC BY 4.0 license https://creativecommons.org/licenses/by/4.0/.

mesocells, with a critical dimension larger than 10 nm, the possibility of phase coexistence also impacts the phase transition mechanism. While periodic boundary conditions on small unit cells imply a transition mechanism in which all unit cells transform collectively, phase coexistence in mesocells allow for a layer-by-layer transition mechanism[289] that is energetically more favourable than the concerted transition mechanism, as shown in Figure 3.21, and which becomes more likely the larger the crystal.[19] These phase coexistence regions, shown in the insets of Figure 3.21 for MIL-53(Al), can also be stabilised by dedicated temperature, pressure, or adsorption quenching experiments, as shown for DMOF-1(Zn), MIL-53(Al)-F, and CoBDP.[19] In a follow-up study on a one million atom simulation cell of MIL-53(Al), it was observed that the phase transition could also proceed *via* discrete nucleation points instead of this layer-by-layer transition mechanism, but only for sufficiently high excess pressures.[24]

For DMOF-1(Zn), a similar layer-by-layer transition mechanism was also observed by Keupp *et al.* when discarding periodic boundary conditions altogether and modelling the free energy of various finite DMOF-1 crystallites with different sizes.[18] They found that this transition mechanism becomes more complex for fu-MOFs with larger alkoxy functionalised ligands.[285] However, by going beyond periodic boundary conditions, and hence also beyond the definition of a simulation cell, it also becomes less straightforward to define pressure in a finite crystallite simulation, in line with the discussion in Section 3.2.[18] To circumvent this problem, they calculated the free energy equations of state *via* umbrella sampling instead.[18] More recently, they proposed an alternative approach in which the pressure medium was modelled explicitly and distinguished between four different transition mechanisms in finite crystallites of DMOF-1, DUT-8, and DUT-128.[290]

While pressure equations of state derived in the $(N,V,\boldsymbol{\sigma}_a = \mathbf{0},T)$ ensemble can directly model pressure- and temperature-induced phase transitions, guest-induced transitions require a different approach. To model the appropriate thermodynamic potential in this case – the osmotic potential – the chemical potential μ of the guest molecules needs to be controlled next to the number of atoms of the host material N_{host}, giving rise to the osmotic ensemble.[291] However, directly simulating in this $(N_{host},\mu,P,\boldsymbol{\sigma}_a = \mathbf{0},T)$ ensemble may give rise both to unsurmountable free energy barriers between the different metastable states and to large volume fluctuations in the MOF unit cell, limiting the accuracy of this method similar to the earlier discussion in Section 3.5.1.[292] Therefore, Rogge *et al.* proposed in ref. 292 a hybrid MC/MD scheme in which MD simulations in the $(N_{host},N_{guest},V,\boldsymbol{\sigma}_a = \mathbf{0},T)$ ensemble are considered as an extra type of trial move during grand canonical Monte Carlo simulations in the $(N_{host},\mu,V,\mathbf{h}_0,T)$ ensemble. The flexible-host osmotic potential can then be directly calculated from the obtained flexible adsorption isotherms.[292] By comparing this method with three existing approaches, it was shown that it features a strongly improved accuracy thanks to isolating the volume – the coordinate along which the largest free energy barriers are typically encountered in flexible MOFs – from the other dynamic variables.[292] This approach was also successfully used to construct the osmotic potential in DUT-49 and explain how the occurrence of different (meta)stable states as a function of (gas) pressure and volume give rise to the attractive NGA phenomenon.[293]

3.6 The Response of 'Rigid' MOFs to Large Pressures: the Inelastic Regime

The distinction between rigid and flexible MOFs is not an absolute one (see Section 2.4 of Chapter 2) as many 'rigid' MOFs still contain some flexible modes, albeit not as extreme in amplitude as flexible MOFs. However, in contrast to the flexible MOFs discussed in Section 3.5, the rigid MOFs discussed in this section all undergo a transition to a noncrystalline state when subjected to high pressures. Noncrystalline states, which are for instance encountered in amorphous MOFs, MOF liquids, or MOF glasses, show no long-range order.[26,37,294] This makes obtaining an accurate structural model of the material, as discussed in Section 3.3.1, substantially more difficult and limits the use of periodic boundary conditions, as they assume perfect translational order at a length scale corresponding to the simulation cell. Furthermore, classical force fields cannot describe these phase transformations if they are reconstructive in nature. As a result, many of the techniques mentioned in Section 3.5 can be adopted only to a limited extent for these noncrystalline states, and pressure-induced transitions of rigid MOFs are studied in a more *ad hoc* way compared to transitions in flexible MOFs. Herein, three important aspects will be covered: how plastic deformation nucleates and propagates through MOFs (Section 3.6.1, using MOF-5 as a case study), which deformation modes are responsible for MOF amorphisation and melting (Section 3.6.2, using ZIFs as a case study), and how different types of disorder and defects alter this plastic deformation behaviour (Section 3.6.3, using UiO-66 as a case study). Other important results concerning the inelastic behaviour of rigid MOFs will be discussed in Section 3.6.4.

3.6.1 Nucleation and Propagation of Mechanical Instability in the IRMOF Series

As the prototypical rigid MOF that undergoes amorphisation under relatively low pressures, different complementary approaches have been adopted to study amorphisation in MOF-5. In 2011, Graham *et al.* performed a series of *ab initio* optimisations under various external pressures, observing that mainly the zinc–oxygen bond lengths that connect the MOF-5 inorganic brick with the carboxylate moieties decrease upon increasing pressure.[155] In a classical force-field-based study in which the phase transition was modelled directly, Biswas *et al.* observed that MOF-5 undergoes a pressure-induced transition towards a ~60% denser structure at around 250 MPa,[128] which is below the amorphisation pressure determined experimentally.[36] This transition was found to be irreversible for the empty MOF-5, since releasing the pressure did not allow them to retrieve the original unstressed structure, although reversibility was obtained upon low hydrogen gas loadings.[36] In contrast to ref. 155, they did not find any substantial effect on the bond lengths under pressure, but rather attributed the observed phase transition to a cooperative rotation of the ligands and a hinge-like motion of the metal oxide clusters around the carboxylate joints.[36] A similar phase transition was also obtained by Rogge *et al.*, Ryder *et al.*, and Erkartal *et al.*,[136,157,295] although this latter study only observed amorphisation at pressures between 1 and 2 GPa.[295] Even though no bond reorganisation was observed, strong local distortions of the framework led to disorder within the unit cell, as evidenced from radial distribution functions.[136,295] In all aforementioned studies, however, periodic boundary conditions were assumed and relatively small unit cells were used, thereby preventing the direct simulation of the loss of long-range order that accompanies amorphisation and limiting the range of transition mechanisms that could be observed, as discussed in Section 3.5.3.2.

To allow for disorder to take place on longer length scales, Banlusan *et al.* generated a 6 × 6 × 12 supercell of MOF-5, containing around 180 000 atoms.[296] They directly followed the structural transformation of the framework upon different uniaxial deformations *via* the ReaxFF force field,[296] which overcomes some of the limitations of classical force fields as it can be adopted to describe reactive events such as reconstructive phase transitions.[297] As an example, Figure 3.22 illustrates various snapshots of the plastic deformation in MOF-5 upon increasing compression along the [001] direction, using a colour bar to localise the positions in the material that show the largest volume compression and the largest plastic deformation.[296] For all deformation directions, Banlusan *et al.* observed that plastic deformation nucleates through local slip-collapse events, in which the ligand hinges with respect to the inorganic cluster resulting in a local slip of adjacent planes and an associated collapse of the unit cell,[296] while a later study showed that the exact deformation mechanism depends on the defects present in the material.[298] The propagation of local slip-collapse events releases

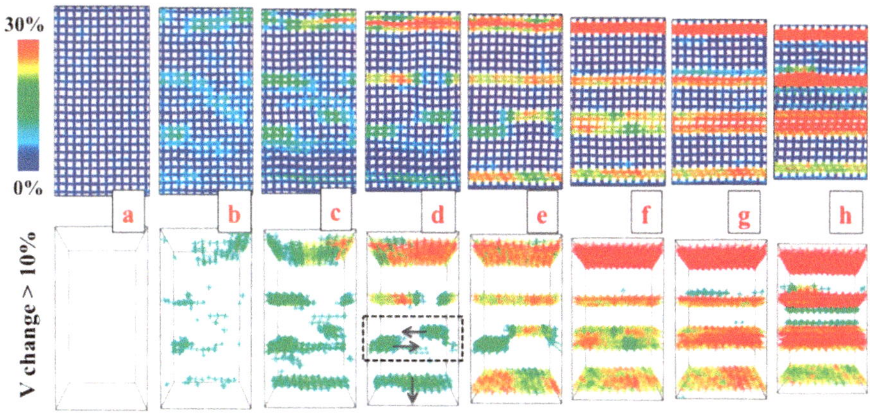

Figure 3.22 Atomistic snapshots showing plastic deformation of MOF-5 during the compression along the [001] direction at various strains, from 0% in panel a to 20% in panel h. Colours indicate the percentage of local volume decrease compared to the initial value in as-equilibrated systems. The bottom panels show only groups of atoms with a volume decrease of more than 10% to highlight the nucleation and propagation of plasticity. The arrows in the bottom panel (d) illustrate the propagation in ⟨100⟩ and ⟨010⟩ directions on the (001) plane of volume collapse. Reproduced from ref. 296 with permission from American Chemical Society, Copyright 2015.

the stress in neighbouring planes, which may even lead to the recovery from plastic deformation, as shown in Figure 3.22. Furthermore, while the initial nucleation of such slip-collapse events requires significant stress, additional slip-collapse events are facilitated by the heterogeneous nature of the now partially collapsed MOF-5, requiring lower stress.[296] Banlusan *et al.* also predicted that plastic deformation makes the material more compliant along certain directions.[296] Using a similar model, they later generalised these quasistatic loading conditions to simulate how a shock wave propagates through the MOF-5 framework.[299] Such a shock propagates through the material in two waves: a leading elastic wave followed by a pore-collapse wave that leads to plastic deformation and significant local heating of the material.[299] For sufficiently fast shocks, the pore-collapse wave also induced chemical decomposition of the material which partially dissipated the energy of the shockwave.[299]

Finally, Pallach *et al.* investigated how functionalising the MOF-5 ligand with different alkoxy sidechains affects its amorphisation behaviour using a variety of experimental and computational tools.[300] Besides calculating the pressure equation of state for these alkoxy-functionalised materials, they also performed pressure ramp force field simulations on much larger simulation cells to directly observe loss of spatial order upon increasing pressure, as evidenced by the simulated X-ray diffraction patterns and pair distribution functions.[300] These simulations confirmed the experimental observations that large volumetric contractions of these frameworks, induced by the alkoxy functionalisation, require random distortions of the framework which conflict with the rigidity of the material, leading to the concept of frustrated flexibility.[300]

3.6.2 Failure Modes in ZIFs and their Melting Behaviour

Given the variety of crystalline, and especially amorphous ZIF states, that can be obtained by controlling the temperature and pressure, these materials form a rich but largely untapped playground to computationally investigate how these different (dis)ordered states can be obtained. To study the mechanism behind the mechanical failure of ZIF-8, Hegde *et al.* investigated whether tensile or shear failure is most likely in this material.[301] Along the first mode, the material is cleaved upon failure, requiring twice the surface energy associated with the cleavage plane. Along the second mode, the material shears upon failure, forming stacking faults, and the free energy barrier to get from the material in equilibrium to the material with stacking faults needs to be overcome.[301] The authors determined computationally that the energy to form intrinsic stacking faults in ZIF-8 is rather low, comparable to that of metals like copper and gold. However, as the lowest-energy barrier to form these stacking faults from a fault-free material was found to require more energy than the energy needed to cleave the material, the authors concluded that ZIF-8 is nonetheless a brittle rather than a ductile material.

Furthermore, they suggested that ZIF-8 might energetically prefer to form a high-density amorphous state under shear stress rather than forming stacking faults.[301]

In 2014, Bouëssel du Bourg *et al.* directly modelled the mechanical stability of a set of ten ZIFs, each composed of zinc cations and unsubstituted imidazolate (Im) ligands but exhibiting a different topology using a force-field approach.[302] At room temperature, nine of these materials mechanically failed under compression at pressures of 0.4 GPa or lower, the sole exception being the material in the **coi** topology, confirming the weak mechanical stability of ZIFs also discussed in Section 3.4.4.1.[302] Despite the clear impact topology has on the mechanical stability of the material, no straightforward relation between the density and stability of the framework was observed.[302] For ZIF-4, one of the materials investigated in this study, Gaillac *et al.* later performed AIMD simulations of the crystalline unit cell at various high temperatures – up to 2250 K – to understand the mechanism behind its melting behaviour.[38] Radial distribution functions revealed that ZIF-4 glasses retain the chemical configuration, porosity, and short-range order of the crystalline parent material, while simultaneously being disordered on longer length scales (while still satisfying periodic boundary conditions).[38] A follow-up study revealed that this mechanism behind ZIF melting is largely unaffected by applying a pressure in the GPa regime during the heating process.[303] However, they did observe a decrease in energy to rupture a zinc–imidazolate bond at higher pressures due to the softening of the Im-Zn-Im angle upon pressurisation, leading to more frequent cleavage events.[303] As a result, AIMD simulations showed that increasing the pressure lowers the temperature necessary to melt ZIF-4; similar observations also hold for ZIF-62.[303]

3.6.3 The Impact of Defects on the Amorphisation of the UiO-66 Series

While localised defects and other types of spatial disorder are expected to profoundly impact the (mechanical) stability and amorphisation of all MOFs, this effect has until now only been extensively studied for MOF-5 (see Section 3.6.1) and the UiO-66 series of materials. In a force-field based study, the pressure-*versus*-volume equations of state of UiO-66, UiO-67, and UiO-68, as well as of a series of low-defective UiO-66 materials were investigated.[49] For the defect-free UiO-66, this equation of state reveals that mechanical instability is induced at 1.83 GPa.[49] In addition, although the force-field description precludes a direct simulation of the amorphisation process, the radial distribution functions and space group of the deformed materials show a distinct loss of crystalline order upon increasing the pressure above this instability, something that is not observed when applying this protocol to the flexible MOFs discussed in Section 3.5.3.[49] By extending the length of the ligand in UiO-66 to achieve the more porous UiO-67 and UiO-68 MOFs, a sharp decrease in loss-of-crystallinity pressure was observed, from 1.83 GPa over 0.45 GPa to 0.2 GPa.[49] A similar observation holds when increasing the number of defects, although the pressure at which the material becomes unstable is not only determined by the concentration of defects, but also by their distribution throughout the framework (a similar conclusion was also drawn concerning the bulk modulus in Section 3.4.1.4).[49] Both for UiO-66 and UiO-67, the obtained results coincide well with those obtained through mercury intrusion experiments[50] and through computationally optimising the structure at increasing pressures and at 0 K.[178]

When accounting for node defects, the mechanical stability of UiO-66 is compromised even further, yielding a loss-of-crystallinity pressure of only 0.8 GPa for the **reo**-defective material, similar to the decrease in bulk modulus upon increasing defectivity.[182] However, incorporating cerium in the zirconium nodes of UiO-66 affects the loss-of-crystallinity pressure differently than it affects the bulk modulus: while it barely impacts the bulk modulus of the bimetallic UiO-66 materials, as discussed earlier in Figure 3.11, their loss-of-crystallinity pressure strongly decreases upon increasing cerium content, as shown in Figure 3.23.[182] In contrast, no substantial changes were

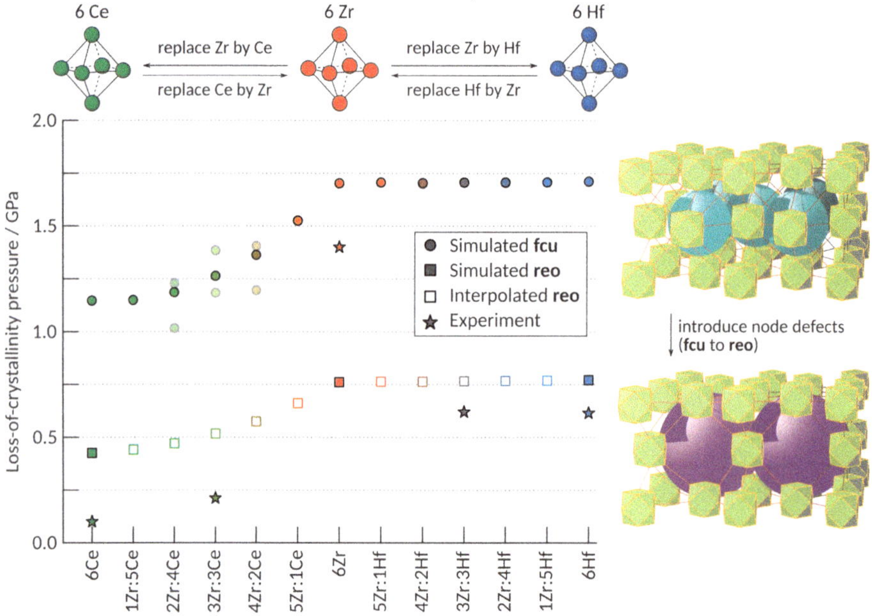

Figure 3.23 Loss-of-crystallinity pressures at 300 K for a series of bimetallic UiO-66 materials as determined from the pressure equations of state, both in the pristine **fcu** (circles) and node-defective **reo** (squares) topologies. For mixed-metal bricks for which two inequivalent bricks can be obtained, the weighted average is shown, whereas the two independent results are included as semi-transparent data points. Experimental results are indicated with stars. Reproduced from ref. 182 with permission from American Chemical Society, Copyright 2020.

observed upon hafnium incorporation instead, showing that the lower mechanical stability of UiO-66(Hf) that was observed experimentally in this study is solely a consequence of the higher defect concentration in this material and not an intrinsic materials property.[182] The observation that the mechanical properties in equilibrium, such as the bulk modulus, are not perfect predictors of the mechanical properties near instability was further underscored by the findings of Moghadam *et al.*[141] They noted that replacing the BDC ligand in UiO-66 with a shorter ligand leads to an increase in bulk modulus, as expected, but to a decrease in loss-of-crystallinity pressure, despite the lower porosity of the material.[141] This counterintuitive result was explained based on the two materials featuring different softest modes of deformation.[141]

3.6.4 Other Computational Studies on the Plastic Deformation of MOFs

In 2013, Fang *et al.* derived a force field for $Zn(CN)_2$ and adopted it to study the high-pressure behaviour of this NTE material that features the same tetrahedrally coordinated zinc cations as ZIFs, but with smaller ligands.[259] Direct $(N,P,\boldsymbol{\sigma}_a=\mathbf{0},T)$ simulations revealed that, at room temperature, $Zn(CN)_2$ undergoes a phase transition at pressures between 1.2 and 2.1 GPa, which could be explained by the substantially pressure-induced softening of several vibrational modes.[259] However, since increasing the temperature tends to stiffen these low-frequency modes, the phase transition could be postponed to higher pressures upon increasing the temperature, as confirmed by MD simulations.[259] This study was later extended by Trousselet *et al.*, considering ten additional topologies in which $Zn(CN)_2$ could potentially be synthesised. They clearly observed a correlation between porosity on the one hand, and the bulk modulus and the critical pressure needed to induce a structural transition on the other hand, with the densest topologies showing the highest mechanical stability, while the less dense topologies already failed in the sub-gigapascal regime.

As the absence of crystallinity in amorphous states pushes classical simulation methods to its limits, it forms an impetus to develop new approaches to model the plastic deformation in MOFs without resorting to the concept of periodic boundary conditions altogether. Two approaches can be distinguished at this point. First, similar to the finite DMOF-1 crystallites discussed earlier,[18] the Schmid group also cleaved a HKUST-1 surface to investigate how a tungsten tip penetrates into the material, mimicking nanoindentation experiments.[7] Although the classical force field description used in that work precludes modelling plastic deformation, their setup holds promise to model other surface-related mechanical properties. Second, FEM approaches have not only been used to model the elastic behaviour of mixed-matrix membranes,[61] but also to understand how stresses exerted on a macroscopic MOF crystal distribute within the particle, both in the elastic and plastic regimes.[62] Although both approaches are still in their infancy at this point, the development of these and other methods that bridge the gap towards macroscopic MOF crystals is essential to fully comprehend how the intricate interplay of interactions at the atomic level gives rise to the attractive but complex mechanical behaviour of macroscopic MOF materials.

3.7 Conclusions and Outlook

Although MOFs exhibit a variety of attractive responses to mechanical stimuli, computational research on MOF mechanics remains largely in its infancy. The available literature is concentrated overwhelmingly on pressure-induced responses near equilibrium, such as the NLC and auxetic behaviour discussed in Section 3.4, with a strong focus on their 0 K behaviour. However, the inelastic responses of MOFs to large mechanical pressures, which give rise to flexibility and amorphisation as discussed in Sections 3.5 and 3.6, form an even more attractive field of computational research that remains largely untapped to date. In these concluding remarks, it is my intention to provide a personal reflection on the challenges that need to be overcome to mature computational research on MOF mechanics and the opportunities that the availability of such established and commonly available computational tools would engender.

A first challenge that was formulated throughout this chapter is how the occurrence of spatial disorder on various length scales can be designed to engineer MOF mechanics. While disorder and defects are often regarded as detrimental to the mechanical stability of a MOF, the large degree to which this disorder can be tuned as well as its strong effect on the macroscopic performance of MOFs open up enormous opportunities to design defect-engineered materials.[25] A first example of this was provided in Section 3.3.1, when discussing how the different orientation of low-symmetric ligands on the high-symmetric DUT-8(Ni) topology gives rise to distinct adsorption behaviour,[44] a functional property coined as 'adaptive flexibility' by Reynolds *et al.*[304] Two additional case studies, focusing on the inelastic response of MOFs, were presented in Sections 3.6.1 and 3.6.3, when discussing frustrated flexibility induced by the sidechain functionalisation of ligands in otherwise rigid MOFs and when discussing the impact of intrinsic defects in UiO-66 materials on their elastic and inelastic behaviour.[49,182,300] While forming an important guidance, these studies only form the tip of the iceberg, as (correlated) disorder is expected to be the norm rather than the exception in MOFs.[4] Seizing this opportunity requires an enormous effort from computational MOF researchers, as modelling such correlated phenomena requires both sufficiently large simulation cells and a proper inclusion of defects in these models. As with the discussion of phase coexistence in Section 3.5.3.2, where distinct phase transition mechanisms could be identified based on the size of the unit cell,[18,19,24,290] different realisations of correlated disorder inside a material could in general amplify different sets of mechanical responses, from among a library of available responses.[304] This phenomenon, termed combinatorial mechanics,[304] emphasises the unparalleled design opportunities offered by MOFs; computational research will therefore be vital to isolate and identify the salient features that drive the macroscopic MOF response and to predict how these features interact with one another in macroscopic materials.

A second challenge that closely relates with this first one is how to realistically model disordered MOFs, from the atomic to the macroscopic level and from local defects over amorphous structures to beyond-particle disorder found in mixed-matrix membranes, MOF monoliths, and other shaped MOFs.[17,305] The wildly varying computational cost of different techniques such as DFT, atomistic force fields, CG force fields, and FEM approaches motivates the development of a hierarchical toolbox of computational techniques in order to cover MOF mechanics across these disparate length scales. Crucial in this regard is the proper theoretical foundation and validation of such a hierarchical bottom-up approach. This is determined by the information that is transferred between more accurate but also more expensive methods on the one hand and less accurate methods that can be adopted for larger systems on the other. Examples of such hierarchical approaches are dispersed throughout this chapter and include FEM techniques that take as input the elastic moduli calculated at the *ab initio* or force field level,[61,62,306] the CG force field of HKUST-1 that was fitted based on an atomistic force field,[7] the HKUST-1 atomistic force field derived based on the *ab initio* stiffness tensor,[78] and the micromechanical model, in which the force-field based elastic stiffness tensor is used as input to derive the mechanics of MOFs at much larger length scales.[60] As these examples show, mechanical properties such as strain fields and the elastic stiffness tensor play a central role in bridging these different length scales, evidencing the need for more in-depth and hierarchical computational studies on MOF mechanics.

A third challenge for computational MOF mechanics is to overcome the focus of current literature on the response of these materials to hydrostatic pressures and to account for their response to a more general stress state. Since stress is a tensorial quantity, compared to pressure as a scalar quantity, subjecting a MOF to a general stress state would most likely pave the way to discover an even larger versatility of attractive stress-induced MOF phenomena that are not expressed under pure hydrostatic pressure. However, general stress control in molecular simulations remains the exception today due to the confusion that arises from the different definitions of stress during a molecular simulation, as mentioned in Section 3.2.2, and the apparent lack of a proper definition of a Gibbs free energy under general stresses and for general deformations, as discussed in Section 3.3.3. As a result, current barostat coupling schemes have only been tested for hydrostatic pressures, save for some very limited non-MOF related examples whose applicability to MOFs is uncertain, given the mechanical softness of these materials that may reveal inconsistencies in general stress coupling schemes that remain hidden for more rigid materials. Therefore, there is an urgent need to develop and validate more general stress coupling schemes, applicable to both periodic simulations and isolated MOF crystallites, both from a fundamental theoretical curiosity and in order to fully explore the entire library of structural responses MOFs can exhibit when subjected to such a general stress.

While this book focuses on MOF mechanics, the mechanical behaviour of MOFs cannot be fully decoupled from their adsorption or thermal behaviour. This is clear when considering for instance the coupling between NTE and purely mechanical phenomena such as NLC and auxetic behaviour, the temperature-induced melting of ZIFs which is affected by the pressure exerted on the material,[303] as discussed in Section 3.6.2, and the impact of temperature on the limit of mechanical instability, which was examined more closely in Sections 3.4.2 and 3.4.4. As a result, a fourth challenge arises to rationalise how different external stimuli – temperature, sorption, mechanical stress, among others – can amplify each other to access MOF phenomena inaccessible when applying only a single stimulus. This idea has been explored through the equivalency between mechanical stress and adsorption-induced stress[263,307] and through thermal stresses,[308] although these examples remain scarce at the moment.

A final challenge for the computational research on MOF mechanics is the general lack of high-throughput studies. Compared to, *e.g.*, adsorption phenomena, such high-throughput studies on the mechanical behaviour of MOFs are much scarcer and the library of MOF structures investigated in these studies is typically much smaller and less diverse. While this connects to the computational cost associated with accurately calculating finite-temperature mechanical properties in MOFs, which is higher than for purely structural properties, such high-throughput studies are vital to identify overarching trends that may be obscured when only focussing on specific MOFs. While the accuracy of these high-throughput projects is typically lower than for studies focussing on one material or one class of materials, the qualitative trends obtained in this way are pivotal to go beyond phenomenological observations of attractive mechanical behaviour in MOFs. In addition, this accuracy limitation may be partially overcome by adopting stepwise screening, as in ref. 141, in which less expensive levels of theory are used first to filter the most promising materials, which are subsequently considered at a higher level of theory.

While this chapter demonstrates that the challenges that need to be overcome to mature computational MOF mechanics to a fully developed field of research are numerous, so are the variety of new mechanical phenomena hiding in these materials. In this respect, our overarching aim should be to develop computational MOF mechanics into a holistic and multiscale research field that is able (i) to provide an in-depth understanding of what drives these phenomena – both in the elastic and the inelastic regime, and (ii) to predict and identify MOF materials with exceptional mechanical behaviour that can be adopted for practical applications. It is my hope that the critical discussion herein, on both the technical aspects of the computational simulations of MOFs as well as on the mechanical phenomena that can be extracted from these simulations, not only summarises the current state-of-the-art in the research field but also forms an impetus to further its development into uncharted but undoubtedly exciting territories.

Acknowledgements

This work was supported by the Fund for Scientific Research Flanders (FWO) through a senior postdoctoral fellowship (grant no. 12T3522N) and the Research Board of Ghent University (BOF).

References

1. A. B. Cairns and A. L. Goodwin, *Phys. Chem. Chem. Phys.*, 2015, **17**, 20449.
2. N. C. Burtch, S. J. Baxter, J. Heinen, A. Bird, A. Schneemann, D. Dubbeldam and A. P. Wilkinson, *Adv. Funct. Mater.*, 2019, **29**, 1904669.
3. A. Schneemann, V. Bon, I. Schwedler, I. Senkovska, S. Kaskel and R. A. Fischer, *Chem. Soc. Rev.*, 2014, **43**, 6062.
4. T. D. Bennett, A. K. Cheetham, A. H. Fuchs and F.-X. Coudert, *Nat. Chem.*, 2017, **9**, 11.
5. M. J. Cliffe, W. Wan, X. Zou, P. A. Chater, A. K. Kleppe, M. G. Tucker, H. Wilhelm, N. P. Funnell, F.-X. Coudert and A. L. Goodwin, *Nat. Commun.*, 2014, **5**, 4176.
6. Y. Sakata, S. Furukawa, M. Kondo, K. Hirai, N. Horike, Y. Takashami, H. Uehara, N. Louvin, M. Meilikhov, T. Tsuruoka, S. Isoda, W. Kosaka, O. Sakata and S. Kitagawa, *Science*, 2013, **339**, 193.
7. J. P. Dürholt, R. Galvelis and R. Schmid, *Dalton Trans.*, 2016, **45**, 4370.
8. E. B. Tadmor and R. E. Miller, *Modeling Materials: Continuum, Atomistic and Multiscale Techniques*, Cambridge University Press, New York, 2011.
9. J. H. Irving and J. G. Kirkwood, *J. Chem. Phys.*, 1950, **18**, 817.
10. A. Cauchy, *Exercises du Mathématique*, Chez de Bure Frères, Paris, 1828, pp. 227–252.
11. A. Cauchy, *Exercises du Mathématique*, Chez de Bure Frères, Paris, 1828, pp. 253–277.
12. M. Born, *Atomtheorie des Festen Zustandes*, B. G. Teubner, Leipzig, 1923.
13. P. Steinmann, A. Elizondo and R. Sunyk, *Modell. Simul. Mater. Sci. Eng.*, 2007, **15**, S271.

14. T. Loiseau, C. Serre, C. Huguenard, G. Fink, F. Taulelle, M. Henry, T. Bataille and G. Férey, *Chem. – Eur. J.*, 2004, **10**, 1373.
15. J. R. Ray and A. Rahman, *J. Chem. Phys.*, 1984, **80**, 4423.
16. E. B. Tadmor, R. E. Miller and R. S. Elliott, *Continuum Mechanics and Thermodynamics: From Fundamental Concepts to Governing Equations*, Cambridge University Press, New York, 2014.
17. V. Van Speybroeck, S. Vandenhaute, A. E. J. Hoffman and S. M. J. Rogge, *Trends Chem.*, 2021, **3**, 605.
18. J. Keupp and R. Schmid, *Adv. Theory Simul.*, 2019, **2**, 1900117.
19. S. M. J. Rogge, M. Waroquier and V. Van Speybroeck, *Nat. Commun.*, 2019, **10**, 4842.
20. J. D. Evans, L. Bocquet and F.-X. Coudert, *Chem*, 2016, **1**, 873.
21. L. Kollias, D. C. Cantu, M. A. Tubbs, R. Rousseau, V.-A. Glezakou and M. Salvalaglio, *J. Am. Chem. Soc.*, 2019, **141**, 6073.
22. A. Ozcan, R. Semino, G. Maurin and A. O. Yazaydin, *Chem. Mater.*, 2020, **32**, 1288.
23. J. P. Dürholt, J. Keupp and R. Schmid, *Eur. J. Inorg. Chem.*, 2016, **2016**, 4517.
24. S. Vandenhaute, S. M. J. Rogge and V. Van Speybroeck, *Front. Chem.*, 2021, **9**, 718920.
25. Z. Fang, B. Bueken, D. E. De Vos and R. A. Fischer, *Angew. Chem., Int. Ed.*, 2015, **54**, 7234.
26. J. Fonseca, T. Gong, L. Jiao and H.-L. Jiang, *J. Mater. Chem. A*, 2021, **9**, 10562.
27. L. Valenzano, B. Civalleri, S. Chavan, S. Bordiga, M. H. Nilsen, S. Jakobsen, K. P. Lillerud and C. Lamberti, *Chem. Mater.*, 2011, **23**, 1700.
28. C. A. Trickett, K. J. Gagnon, S. Lee, F. Gándara, H.-B. Bürgi and O. M. Yaghi, *Angew. Chem., Int. Ed.*, 2015, **54**, 11162.
29. G. C. Shearer, S. Chavan, S. Bordiga, S. Svelle, U. Olsbye and K. P. Lillerud, *Chem. Mater.*, 2016, **28**, 3749.
30. W. Zhang, M. Kauer, P. Guo, S. Kunze, S. Cwik, M. Muhler, Y. Wang, K. Epp, G. Kieslich and R. A. Fischer, *Eur. J. Inorg. Chem.*, 2017, **2017**, 925.
31. O. Kozachuk, M. Meilikhov, K. Yusenko, A. Schneemann, B. Jee, A. V. Kuttatheyll, M. Bertmer, C. Sternemann, A. Pöppl and R. A. Fischer, *Eur. J. Inorg. Chem.*, 2013, **2013**, 4546.
32. D. N. Johnstone, F. C. N. Firth, C. P. Grey, P. A. Midgley, M. J. Cliffe and S. M. Collins, *J. Am. Chem. Soc.*, 2020, **142**, 13081.
33. S. Diring, S. Furukawa, Y. Takashima, T. Tsuruoka and S. Kitagawa, *Chem. Mater.*, 2010, **22**, 4531.
34. Z. Fang, J. P. Dürholt, M. Kauer, W. Zhang, C. Lochenie, B. Jee, B. Albada, N. Metzler-Nolte, A. Pöppl, B. Weber, M. Muhler, Y. Wang, R. Schmid and R. A. Fischer, *J. Am. Chem. Soc.*, 2014, **136**, 9627.
35. T. D. Bennett, T. K. Todorova, E. F. Baxter, D. G. Reid, C. Gervais, B. Bueken, B. Van de Voorde, D. De Vos, D. A. Keen and C. Mellot-Draznieks, *Phys. Chem. Chem. Phys.*, 2016, **18**, 2192.

36. K. W. Chapman, G. J. Halder and P. J. Chupas, *J. Am. Chem. Soc.*, 2009, **131**, 17546.
37. T. D. Bennett and A. K. Cheetham, *Acc. Chem. Res.*, 2014, **47**, 1555.
38. R. Gaillac, P. Pullumbu, K. A. Beyer, K. W. Chapman, D. A. Keen, T. D. Bennett and F.-X. Coudert, *Nat. Mater.*, 2017, **16**, 1149.
39. J. M. Tuffnell, C. A. Ashling, J. Hou, S. Li, L. Longley, M. L. Ríos Gómez and T. D. Bennett, *Chem. Commun.*, 2019, **55**, 8705.
40. L. Frentzel-Beyme, M. Kloß, P. Kolodzeiski, R. Pallach and S. Henke, *J. Am. Chem. Soc.*, 2019, **141**, 12362.
41. S. Horike, S. S. Nagarkar, T. Ogawa and S. Kitagawa, *Angew. Chem., Int. Ed.*, 2020, **59**, 6652.
42. J. D. Evans, V. Bon, I. Senkovska, H.-C. Lee and S. Kaskel, *Nat. Commun.*, 2020, **11**, 2690.
43. S. Borgmans, S. M. J. Rogge, J. S. De Vos, C. V. Stevens, P. Van Der Voort and V. Van Speybroeck, *Angew. Chem., Int. Ed.*, 2021, **60**, 8913.
44. S. Ehrling, E. M. Reynolds, V. Bon, I. Senkosvka, T. E. Gorelik, J. D. Evans, M. Rauche, M. Mendt, M. S. Weiss, A. Pöppl, E. Brunner, U. Kaiser, A. L. Goodwin and S. Kaskel, *Nat. Chem.*, 2021, **13**, 568.
45. J. H. Cavka, S. Jakobsen, U. Olsbye, N. Guillou, C. Lamberti, S. Bordiga and K. P. Lillerud, *J. Am. Chem. Soc.*, 2008, **42**, 13850.
46. K. Leus, T. Bogaerts, J. De Decker, H. Depauw, K. Hendrickx, H. Vrielinck, V. Van Speybroeck and P. Van Der Voort, *Microporous Mesoporous Mater.*, 2016, **226**, 110.
47. M. Kandiah, M. H. Nilsen, S. Usseglio, S. Jakobsen, U. Olsbye, M. Tilset, C. Larabi, E. A. Quadrelli, F. Bonino and K. P. Lillerud, *Chem. Mater.*, 2010, **22**, 6632.
48. A. W. Thornton, R. Babarao, F. Trousselet and F.-X. Coudert, *Dalton Trans.*, 2016, **45**, 4352.
49. S. M. J. Rogge, J. Wieme, L. Vanduyfhuys, S. Vandenbrande, G. Maurin, T. Verstraelen, M. Waroquier and V. Van Speybroeck, *Chem. Mater.*, 2016, **28**, 5721.
50. P. G. Yot, K. Yang, F. Ragon, V. Dmitriev, T. Devic, P. Horcajada, C. Serre and G. Maurin, *Dalton Trans.*, 2016, **45**, 4283.
51. A. Simonov and A. L. Goodwin, *Nat. Rev. Chem.*, 2020, **65**, 657.
52. S. Horike, S. Shimomura and S. Kitagawa, *Nat. Chem.*, 2009, **1**, 695.
53. S. Kenji, T. Satoshi and M. Wasuke, *Chem. Lett.*, 2001, **30**, 332.
54. D. N. Dybtsev, H. Chun and K. Kim, *Angew. Chem., Int. Ed.*, 2004, **43**, 5033.
55. A. M. Walker, B. Civalleri, B. Slater, C. Mellot-Draznieks, F. Corà, C. M. Zicovich-Wilson, G. Román-Pérez, J. M. Soler and J. D. Gale, *Angew. Chem., Int. Ed.*, 2010, **49**, 7501.
56. J. S. Grosch and F. Paesani, *J. Am. Chem. Soc.*, 2012, **134**, 4207.
57. N. Klein, C. Herzog, M. Sabo, I. Senkovska, J. Getzschmann, S. Paasch, M. R. Lohe, E. Brunner and S. Kaskel, *Phys. Chem. Chem. Phys.*, 2010, **12**, 11778.

58. A. Krylov, I. Senkovska, S. Ehrling, M. Maliuta, S. Krylova, E. Slyusareva, A. Vtyurin and S. Kaskel, *Chem. Commun.*, 2020, **56**, 8269.
59. R. Semino, J. P. Dürholt, R. Schmid and G. Maurin, *J. Phys. Chem. C*, 2017, **121**, 21491.
60. S. M. J. Rogge, *Faraday Discuss.*, 2021, **225**, 271.
61. J. D. Evans and F.-X. Coudert, *J. Phys. Chem. Lett.*, 2017, **8**, 1578.
62. Z. Wang, K. M. Schmalbach, R. L. Combs, Y. Chen, R. L. Penn, N. A. Mara and A. Stein, *ACS Appl. Mater. Interfaces*, 2020, **12**, 49971.
63. F.-X. Coudert and A. H. Fuchs, *Coord. Chem. Rev.*, 2016, **307**, 211.
64. E. Schrödinger, *Ann. Phys.*, 1926, **384**, 361.
65. M. Born and R. Oppenheimer, *Ann. Phys.*, 1927, **389**, 457.
66. S. O. Odoh, C. J. Cramer, D. G. Truhlar and L. Gagliardi, *Chem. Rev.*, 2015, **115**, 6051.
67. D. S. Sholl and J. A. Steckel, *Density Functional Theory: A Practical Introduction*, John Wiley & Sons, Inc., Hoboken, 2009.
68. P. Hohenberg and W. Kohn, *Phys. Rev. B*, 1964, **136**, 864.
69. W. Kohn and L. J. Sham, *Phys. Rev. A*, 1965, **140**, 1133.
70. J. L. Mancuso, A. M. Mroz, K. N. Le and C. H. Hendon, *Chem. Rev.*, 2020, **120**, 8641.
71. J. A. Greathouse and M. D. Allendorf, *J. Am. Chem. Soc.*, 2006, **128**, 10678.
72. S. Bureekaew, S. Amirjalayer, M. Tafipolsky, C. Spickermann, T. K. Roy and R. Schmid, *Phys. Status Solidi B*, 2013, **250**, 1128.
73. J. P. Dürholt, G. Fraux, F.-X. Coudert and R. Schmid, *J. Chem. Theory Comput.*, 2019, **15**, 2420.
74. J. K. Bristow, D. Tiana and A. Walsh, *J. Chem. Theory Comput.*, 2014, **10**, 4644.
75. L. Vanduyfhuys, S. Vandenbrande, T. Verstraelen, R. Schmid, M. Waroquier and V. Van Speybroeck, *J. Comput. Chem.*, 2015, **36**, 1015.
76. L. Vanduyfhuys, S. Vandenbrande, J. Wieme, M. Waroquier, T. Verstraelen and V. Van Speybroeck, *J. Comput. Chem.*, 2018, **39**, 999.
77. J. K. Bristow, J. M. Skelton, K. L. Svane, A. Walsh and J. D. Gale, *Phys. Chem. Chem. Phys.*, 2016, **18**, 29316.
78. J. Heinen, N. C. Burtch, K. S. Walton and D. Dubbeldam, *J. Chem. Theory Comput.*, 2017, **13**, 3722.
79. J. Heinen and D. Dubbeldam, *Wiley Interdiscip. Rev. Comput. Mol. Sci.*, 2018, **8**, e1363.
80. D. Frenkel and B. Smit, *Understanding Molecular Simulation: From Algorithms to Applications*, Elsevier, London, 2012.
81. S. Nosé and M. L. Klein, *Mol. Phys.*, 1983, **50**, 1055.
82. J. R. Ray, *Comput. Phys. Rep.*, 1988, **8**, 109.
83. S. M. J. Rogge, L. Vanduyfhuys, A. Ghysels, M. Waroquier, T. Verstraelen, G. Maurin and V. Van Speybroeck, *J. Chem. Theory Comput.*, 2015, **11**, 5583.
84. H. C. Andersen, *J. Chem. Phys.*, 1980, **72**, 2384.

85. M. Parrinello and A. Rahman, *Phys. Rev. Lett.*, 1980, **45**, 1196.
86. M. Parrinello and A. Rahman, *J. Appl. Phys.*, 1981, **52**, 7182.
87. J. R. Ray, *J. Chem. Phys.*, 1983, **79**, 5128.
88. J. R. Ray and A. Rahman, *J. Chem. Phys.*, 1985, **82**, 4243.
89. C. L. Cleveland, *J. Chem. Phys.*, 1988, **89**, 4987.
90. R. M. Wentzcovitch, *Phys. Rev. B*, 1991, **44**, 2358.
91. J. V. Lill and J. Q. Broughton, *Phys. Rev. B*, 1992, **46**, 12068.
92. R. E. Miller, E. B. Tadmor, J. S. Gibson, N. Bernstein and F. Pavia, *J. Chem. Phys.*, 2016, **144**, 184107.
93. H. J. C. Berendsen, J. P. M. Postma, W. F. van Gunsteren, A. DiNola and J. R. Haak, *J. Chem. Phys.*, 1984, **81**, 3684.
94. W. G. Hoover, *Phys. Rev. A*, 1986, **34**, 2499.
95. S. Melchionna, G. Ciccotti and B. L. Holian, *Mol. Phys.*, 1993, **78**, 533.
96. G. J. Martyna, D. J. Tobias and M. L. Klein, *J. Chem. Phys.*, 1994, **101**, 4177.
97. D. Quigley and M. I. J. Probert, *J. Chem. Phys.*, 2004, **120**, 11432.
98. N. C. Burtch, J. Heinen, T. D. Bennett, D. Dubbeldam and M. D. Allendorf, *Adv. Mater.*, 2018, **30**, 1704124.
99. J. F. Nye, *Physical Properties of Crystals: Their Representation by Tensors and Matrices*, University Press, Oxford, 2006.
100. W. Voigt, *Lehrbuch der Kristallphysik*, B. G. Teubner, Leipzig, 1910.
101. A. Marmier, Z. A. D. Lethbridge, R. I. Walton, C. W. Smith, S. C. Parker and K. E. Evans, *Comput. Phys. Commun.*, 2010, **181**, 2102.
102. R. Gaillac, P. Pullumbi and F.-X. Coudert, *J. Phys.: Condens. Matter*, 2016, **28**, 275201.
103. A. Reuss, *Z. Angew. Math. Mech.*, 1929, **9**, 49.
104. R. Hill, *Proc. Phys. Soc., London, Sect. A*, 1952, **65**, 349.
105. Y. Le Page and P. Saxe, *Phys. Rev. B*, 2001, **63**, 174103.
106. F. D. Murnaghan, *Proc. Natl. Acad. Sci. U. S. A.*, 1944, **30**, 244.
107. F. Birch, *Phys. Rev.*, 1947, **71**, 809.
108. P. Vinet, J. H. Rose, J. Ferrante and J. R. Smith, *J. Phys.: Condens. Matter*, 1989, **1**, 1941.
109. D. E. P. Vanpoucke, K. Lejaeghere, V. Van Speybroeck, M. Waroquier and A. Ghysels, *J. Phys. Chem. C*, 2015, **119**, 23752.
110. X. Wu, D. Vanderbilt and D. R. Hamann, *Phys. Rev. B*, 2005, **72**, 035105.
111. J. F. Lutsko, *J. Appl. Phys.*, 1989, **65**, 2991.
112. J. Maul, M. R. Ryder, M. T. Ruggiero and A. Erba, *Phys. Rev. B*, 2019, **99**, 014102.
113. H. Li, M. Eddaoudi, M. O'Keeffe and O. M. Yaghi, *Nature*, 1999, **402**, 276.
114. N. L. Rosi, J. Eckert, M. Eddaoudi, D. T. Vodak, J. Kim, M. O'Keeffe and O. M. Yaghi, *Science*, 2003, **300**, 1127.
115. M. J. Mehl, *Phys. Rev. B*, 1993, **47**, 2493.
116. M. Mattesini, J. M. Soler and F. Ynduráin, *Phys. Rev. B*, 2006, **73**, 094111.
117. A. Samanta, T. Furuta and J. Li, *J. Chem. Phys.*, 2006, **125**, 084714.
118. W. Zhou and T. Yildirim, *Phys. Rev. B*, 2006, **74**, 180301.

119. D. F. Bahr, J. A. Reid, W. M. Mook, C. A. Bauer, R. Stumpf, A. J. Skulan, N. R. Moody, B. A. Simmons, M. M. Shindel and M. D. Allendorf, *Phys. Rev. B*, 2007, **76**, 184106.
120. K. Banlusan and A. Strachan, *J. Chem. Phys.*, 2017, **146**, 184705.
121. D. Nazarian, P. Ganesh and D. S. Sholl, *J. Mater. Chem. A*, 2015, **3**, 22432.
122. J. P. Perdew, K. Burke and M. Ernzerhof, *Phys. Rev. Lett.*, 1996, **77**, 3865.
123. S. Grimme, J. Antony, S. Ehrlich and H. Krieg, *J. Chem. Phys.*, 2010, **132**, 154104.
124. J. L. C. Rowsell, E. C. Spencer, J. Eckert, J. A. K. Howard and O. M. Yaghi, *Science*, 2005, **309**, 1350.
125. S. S. Han and W. A. Goddard III, *J. Phys. Chem. C*, 2007, **111**, 15185.
126. J. A. Greathouse and M. D. Allendorf, *J. Phys. Chem. C*, 2008, **112**, 5795.
127. M. Tafipolsky and R. Schmid, *J. Phys. Chem. B*, 2009, **113**, 1341.
128. M. M. Biswas and T. Cagin, *Mater. Chem. Phys.*, 2011, **131**, 44.
129. D. Dubbeldam, R. Krishna and R. Q. Snurr, *J. Phys. Chem. C*, 2009, **113**, 19317.
130. S. L. Mayo, B. D. Olafson and W. A. Goddard, *J. Phys. Chem.*, 1990, **94**, 8897.
131. P. Dauber-Osguthorpe, V. A. Roberts, D. J. Osguthorpe, J. Wolff, M. Genest and A. T. Hagler, *Proteins*, 1988, **4**, 31.
132. P. G. Boyd, S. M. Moosavi, M. Witman and B. Smit, *J. Phys. Chem. Lett.*, 2017, **8**, 357.
133. A. K. Rappe, C. J. Casewit, K. S. Colwell, W. A. Goddard III and W. M. Skiff, *J. Am. Chem. Soc.*, 1992, **114**, 10024.
134. M. A. Addicoat, N. Vankova, I. F. Akter and T. Heine, *J. Chem. Theory Comput.*, 2014, **10**, 880.
135. D. Dubbeldam, K. S. Walton, D. E. Ellis and R. Q. Snurr, *Angew. Chem., Int. Ed.*, 2007, **46**, 4496.
136. S. M. J. Rogge, M. Waroquier and V. Van Speybroeck, *Acc. Chem. Res.*, 2018, **51**, 138.
137. S. S. Kaye, A. Dailly, O. M. Yaghi and J. R. Long, *J. Am. Chem. Soc.*, 2007, **129**, 14176.
138. Y. Ming, J. Purewal, J. Yang, C. Xu, R. Soltis, J. Warner, M. Veenstra, M. Gaab, U. Müller and D. J. Siegel, *Langmuir*, 2015, **31**, 4988.
139. J. C. Tan and A. K. Cheetham, *Chem. Soc. Rev.*, 2011, **40**, 1059.
140. A. Kuc, A. Enyashin and G. Seifert, *J. Phys. Chem. B*, 2007, **111**, 8179.
141. P. Z. Moghadam, S. M. J. Rogge, A. Li, C.-M. Chow, J. Wieme, N. Moharrami, M. Aragones-Anglada, G. Conduit, D. A. Gomez-Gualdron, V. Van Speybroeck and D. Fairen-Jimenez, *Matter*, 2019, **1**, 219.
142. B. Lukose, B. Supronowicz, S. P. P. J. Frenzel, A. B. Kuc, G. Seifert, G. N. Vayssilov and T. Heine, *Phys. Status Solidi B*, 2012, **249**, 335.
143. L.-M. Yang, P. Vajeeston, P. Ravindran, H. Fjellvåg and M. Tilset, *Inorg. Chem.*, 2010, **49**, 10283.
144. J. D. Evans and F.-X. Coudert, *J. Am. Chem. Soc.*, 2016, **138**, 6131.
145. M. Fuentes-Cabrera, D. M. Nicholson and B. G. Sumpter, *J. Chem. Phys.*, 2005, **123**, 124713.

146. L.-M. Yang, P. Vajeeston, P. Ravindran, H. Fjellvåg and M. Tilset, *Phys. Chem. Chem. Phys.*, 2011, **13**, 10191.
147. L.-M. Yang, P. Ravindran, P. Vajeeston and M. Tilset, *RSC Adv.*, 2012, **2**, 1618.
148. L.-M. Yang, P. Ravindran, P. Vajeeston and M. Tilset, *Phys. Chem. Chem. Phys.*, 2012, **14**, 4713.
149. L.-M. Yang, P. Ravindran, P. Vajeeston and M. Tilset, *J. Mater. Chem.*, 2012, **22**, 16324.
150. L. Yang, M. P. Ravindran and M. Tilset, *Inorg. Chem.*, 2013, **52**, 4217.
151. L.-M. Yang, P. Ravindran, P. Vajeeston, S. Svelle and M. Tilset, *Microporous Mesoporous Mater.*, 2013, **175**, 50.
152. L.-M. Yang, *Microporous Mesoporous Mater.*, 2014, **183**, 218.
153. L.-M. Yang, G.-Y. Fang, J. Ma, E. Ganz and S. S. Han, *Cryst. Growth Des.*, 2014, **14**, 2532.
154. L.-M. Yang, G.-Y. Fang, J. Ma, R. Pushpa and E. Ganz, *Phys. Chem. Chem. Phys.*, 2016, **18**, 32319.
155. A. J. Graham, D. R. Allan, A. Muszkiewicz, C. A. Morrison and S. A. Moggach, *Angew. Chem., Int. Ed.*, 2011, **50**, 11138.
156. J. Joo, H. Kim and S. S. Han, *Phys. Chem. Chem. Phys.*, 2013, **15**, 18822.
157. M. R. Ryder, J. Maul, B. Civalleri and A. Erba, *Adv. Theory Simul.*, 2019, **2**, 1900093.
158. J. P. Perdew, A. Ruzsinszky, G. I. Csonka, O. A. Vydrov, G. E. Scuseria, L. A. Constantin, X. Zhou and K. Burke, *Phys. Rev. Lett.*, 2008, **100**, 136406.
159. R. Dovesi, R. Orlando, A. Erba, C. M. Zicovich-Wilson, B. Civalleri, S. Casassa, L. Maschio, M. Ferrabone, M. De La Pierre, P. D'Arco, Y. Noël, M. Causà, M. Rérat and B. Kirtman, *Int. J. Quantum Chem.*, 2014, **114**, 1287.
160. R. E. Allen and F. W. De Wette, *Phys. Rev.*, 1969, **179**, 873.
161. L. Wang, C. Wang, Y. Sun, K. Shi, S. Deng and H. Lu, *Mater. Chem. Phys.*, 2016, **175**, 138.
162. Y.-Q. Tian, C.-X. Cai, X.-M. Ren, C.-Y. Duan, Y. Xu, S. Gao and X.-Z. You, *Chem. – Eur. J.*, 2003, **9**, 5673.
163. K. S. Park, Z. Ni, A. P. Côté, J. Y. Choi, R. Huang, F. J. Uribe-Romo, H. K. Chae, M. O'Keeffe and O. M. Yaghi, *Proc. Natl. Acad. Sci. U. S. A.*, 2006, **103**, 10186.
164. T. D. Bennett, J.-C. Tan, S. A. Moggach, R. Galvelis, C. Mellot-Draznieks, B. A. Reisner, A. Thirumurugan, D. R. Allan and A. K. Cheetham, *Chem. Eur. J.*, 2010, **16**, 10684.
165. J.-C. Tan, B. Civalleri, C.-C. Lin, L. Valenzano, R. Galvelis, P.-F. Chen, T. D. Bennett, C. Mellot-Draznieks, C. M. Zicovich-Wilson and A. K. Cheetham, *Phys. Rev. Lett.*, 2012, **108**, 095502.
166. A. D. Becke, *J. Chem. Phys.*, 1993, **98**, 1372.
167. R. Dovesi, R. Orlando, B. Civalleri, C. Roetti, V. R. Saunders and C. M. Zicovich-Wilson, *Z. Kristallogr. Cryst. Mater.*, 2005, **220**, 571.
168. T. Weng and J. R. Schmidt, *J. Phys. Chem. A*, 2019, **123**, 3000.

169. W. Zhang, J. Maul, D. Vulpe, P. Z. Moghadam, D. Fairen-Jimenez, D. M. Mittleman, J. A. Zeitler, A. Erba and M. T. Ruggiero, *J. Phys. Chem. C*, 2018, **122**, 27442.
170. Y. Zhao and D. G. Truhlar, *Theor. Chem. Acc.*, 2008, **120**, 215.
171. H. Gao, W. Wei, L. Dong, G. Feng, X. Jiang, R. Wu, Z. Lin and W. Li, *Crystals*, 2017, 7, 99.
172. J.-C. Tan, B. Civalleri, A. Erba and E. Albanese, *CrystEngComm*, 2015, **17**, 375.
173. M. R. Ryder and J.-C. Tan, *Dalton Trans.*, 2016, **45**, 4154.
174. B. Zheng, Y. Zhu, F. Fu, L. L. Wang, J. Wang and H. Du, *RSC Adv.*, 2017, **66**, 41499.
175. S. M. Moosavi, P. G. Boyd, L. Sarkisov and B. Smit, *ACS Cent. Sci.*, 2018, **4**, 832.
176. L. Zhang, Z. Hu and J. Jiang, *J. Phys. Chem. C*, 2012, **116**, 19268.
177. H. Wu, T. Yildirim and W. Zhou, *J. Phys. Chem. Lett.*, 2013, **4**, 925.
178. C. L. Hobday, R. J. Marshall, C. F. Murphie, J. Sotelo, T. Richards, D. R. Allan, T. Düren, F.-X. Coudert, R. S. Forgan, C. A. Morrison, S. A. Moggach and T. D. Bennett, *Angew. Chem., Int. Ed.*, 2016, **55**, 2401.
179. L.-M. Yang, E. Ganz, S. Svelle and M. Tilset, *J. Mater. Chem. C*, 2014, **2**, 7111.
180. L. R. Redfern, M. Ducamp, M. C. Wasson, L. Robison, F. A. Son, F.-X. Coudert and O. K. Farha, *Chem. Mater.*, 2020, **32**, 5864.
181. M. Vandichel, J. Hajek, A. Ghysels, A. De Vos, M. Waroquier and V. Van Speybroeck, *CrystEngComm*, 2016, **18**, 7056.
182. S. M. J. Rogge, P. G. Yot, J. Jacobsen, F. Muniz-Miranda, S. Vandenbrande, J. Gosch, V. Ortiz, I. E. Collings, S. Devautour-Vinot, G. Maurin, N. Stock and V. Van Speybroeck, *ACS Mater. Lett.*, 2020, **2**, 438.
183. S. S.-Y. Chui, S. M.-F. Lo, J. P. H. Charmant, A. G. Orpen and I. D. Williams, *Science*, 1999, **283**, 1148.
184. J. P. Perdew and Y. Wang, *Phys. Rev. B*, 1992, **23**, 13244.
185. T. Watanabe and D. S. Sholl, *J. Chem. Phys.*, 2010, **133**, 094509.
186. M. Tafipolsky, S. Amirjalayer and R. Schmid, *J. Phys. Chem. C*, 2010, **114**, 14402.
187. L. Zhao, Q. Yang, Q. Ma, C. Zhong, J. Mi and D. Liu, *J. Mol. Model.*, 2011, **17**, 227.
188. M. R. Ryder, B. Civalleri, G. Cinque and J.-C. Tan, *CrystEngComm*, 2016, **18**, 4303.
189. K. Barthelet, J. Marrot, D. Riou and G. Férey, *Angew. Chem., Int. Ed.*, 2002, **41**, 281.
190. F. Millange, C. Serre and G. Férey, *Chem. Commun.*, 2002, **8**, 822.
191. P. G. Yot, Q. Ma, J. Haines, Q. Yang, A. Ghoufi, T. Devic, C. Serre, V. Dmitriev, G. Férey, C. Zhong and G. Maurin, *Chem. Sci.*, 2012, **3**, 1100.
192. I. Beurroies, M. Boulhout, P. L. Llewellyn, B. Kuchta, G. Férey, C. Serre and R. Denoyel, *Angew. Chem., Int. Ed.*, 2010, **49**, 7526.
193. G. Férey and C. Serre, *Chem. Soc. Rev.*, 2009, **38**, 1380.

194. R. Caruana-Gauci, E. P. Degabriele, D. Attard and J. N. Grima, *J. Mater. Sci.*, 2018, **53**, 5079.
195. J. M. Ogborn, I. E. Collings, S. A. Moggach, A. L. Thompson and A. L. Goodwin, *Chem. Sci.*, 2012, **3**, 3011.
196. A. U. Ortiz, A. Boutin, A. H. Fuchs and F.-X. Coudert, *Phys. Rev. Lett.*, 2012, **109**, 195502.
197. A. U. Ortiz, A. Boutin, A. H. Fuchs and F.-X. Coudert, *J. Chem. Phys.*, 2013, **138**, 174703.
198. J. Heinen and D. Dubbeldam, *Dalton Trans.*, 2016, **45**, 4309.
199. P. Ramaswamy, J. Wieme, E. Alvarez, L. Vanduyfhuys, J.-P. Itié, P. Fabry, V. Van Speybroeck, C. Serre, P. G. Yot and G. Maurin, *J. Mater. Chem. A*, 2017, **5**, 11047.
200. M. Wang, X. Zhang, Y. Chen and D. Li, *J. Phys. Chem. C*, 2016, **120**, 5059.
201. F. Mouhat, D. Bousquet, A. Boutin, L. Bouëssel du Bourg, F.-X. Coudert and A. H. Fuchs, *J. Phys. Chem. Lett.*, 2015, **6**, 4265.
202. Q. Ma, Q. Yang, A. Ghoufi, G. Férey, C. Zhong and G. Maurin, *Dalton Trans.*, 2012, **41**, 3915.
203. Y.-Y. Liu, S. Couck, M. Vandichel, M. Grzywa, K. Leus, S. Biswas, D. Volkmer, J. Gascon, F. Kapteijn, J. F. M. Denayer, M. Waroquier, V. Van Speybroeck and P. Van Der Voort, *Inorg. Chem.*, 2013, **52**, 113.
204. J. Wieme, L. Vanduyfhuys, S. M. J. Rogge, M. Waroquier and V. Van Speybroeck, *J. Phys. Chem. C*, 2016, **120**, 14934.
205. D. E. P. Vanpoucke, J. W. Jaeken, S. De Baerdemacker, K. Lejaeghere and V. Van Speybroeck, *Beilstein J. Nanotechnol.*, 2014, **5**, 1738.
206. A. E. J. Hoffman, J. Wieme, S. M. J. Rogge, L. Vanduyfhuys and V. Van Speybroeck, *Z. Kristallogr.*, 2019, **234**, 529.
207. C. Volkringer, T. Loiseau, N. Guillou, G. Férey and E. Elkaïm, *Solid State Sci.*, 2009, **11**, 1507.
208. V. Guillerm, F. Ragon, M. Dan-Hardi, T. Devic, M. Vishnuvarthan, B. Campo, A. Vimont, G. Clet, Q. Yang, G. Maurin, G. Férey, A. Vittadini, S. Gross and C. Serre, *Angew. Chem., Int. Ed.*, 2012, **51**, 9267.
209. M. R. Ryder, B. Civalleri and J.-C. Tan, *Phys. Chem. Chem. Phys.*, 2016, **18**, 9079.
210. H. C. Dong, H. L. Nguyen, H. M. Le, N. Thoai, Y. Kawazoe and D. Nguyen-Manh, *Sci. Rep.*, 2018, **8**, 16651.
211. N. Klein, I. Senkovska, I. A. Baburin, R. Grünker, U. Stoeck, M. Schlichtenmayer, B. Streppel, U. Mueller, S. Leoni, M. Hirscher and S. Kaskel, *Chem. – Eur. J.*, 2011, **17**, 13007.
212. Z. Wang, J. Liu, B. Lukose, Z. Gu, P. G. Weidler, H. Gliemann, T. Heine and C. Wöll, *Nano Lett.*, 2014, **14**, 1526.
213. N. Li, K. Blankson, Y. Yang, P. Zhang and X. Zhao, *APL Mater.*, 2020, **8**, 111101.
214. S. Krause, V. Bon, I. Senkovska, U. Stoeck, D. Wallacher, D. M. Többens, S. Zander, R. S. Pillai, G. Maurin, F.-X. Coudert and S. Kaskel, *Nature*, 2016, **532**, 348.

215. J. Wieme, S. M. J. Rogge, P. G. Yot, L. Vanduyfhuys, S.-K. Lee, J.-S. Chang, M. Waroquier, G. Maurin and V. Van Speybroeck, *J. Mater. Chem. A*, 2019, **7**, 22663.
216. A. U. Ortiz, A. Boutin, K. J. Gagnon, A. Clearfield and F.-X. Coudert, *J. Am. Chem. Soc.*, 2014, **136**, 11540.
217. X. Yang, L. Si, G. Xie, L. Zhang, D. Guo and J. Luo, *Comput. Mater. Sci.*, 2020, **177**, 109543.
218. Z. Zheng, X. Jiang and J. Zhao, *Chem. Phys. Lett.*, 2015, **628**, 76.
219. E. V. Alexandrov, A. V. Goltsev, R. A. Eremin and V. A. Blatov, *J. Phys. Chem. C*, 2019, **123**, 24651.
220. V. A. Saleev and A. V. Shipilova, *Russ. J. Phys. Chem. A*, 2018, **92**, 1940.
221. A. Gładysiak, K. S. Deeg, I. Dovgaliuk, A. Chidambaram, K. Ordiz, P. G. Boyd, S. M. Moosavi, D. Ongari, J. A. R. Navarro, B. Smit and K. C. Stylianou, *ACS Appl. Mater. Interfaces*, 2018, **10**, 36144.
222. P. Canepa, K. Tan, Y. Du, H. Lu, Y. J. Chabal and T. Thonhauser, *J. Mater. Chem. A*, 2015, **3**, 986.
223. M. Wittman, S. Ling, A. Gladysiak, K. C. Stylianou, B. Smit, B. Slater and M. Haranczyk, *J. Phys. Chem. C*, 2017, **121**, 1171.
224. A. U. Ortiz, A. Boutin and F.-X. Coudert, *Chem. Commun.*, 2014, **50**, 5867.
225. F. Niekiel, J. Lannoeye, H. Reinsch, A. S. Munn, A. Heerwig, I. Zizak, S. Kaskel, R. I. Walton, D. de Vos, P. Llewellyn, A. Lieb, G. Maurin and N. Stock, *Inorg. Chem.*, 2014, **53**, 4610.
226. R. Anderson and D. A. Gómez-Gualdrón, *CrystEngComm*, 2019, **21**, 1653.
227. G. Feng, X. Jiang, W. Wei, P. Gong, L. Kang, Z. Li, Y. Li, X. Li, X. Wu, Z. Lin, W. Li and P. Lu, *Dalton Trans.*, 2016, **45**, 4303.
228. I. M. Hönicke, I. Senkovska, V. Bon, I. A. Baburin, N. Bönisch, S. Raschke, J. D. Evans and S. Kaskel, *Angew. Chem., Int. Ed.*, 2018, **57**, 13780.
229. M. Parrinello and A. Rahman, *J. Chem. Phys.*, 1982, **76**, 2662.
230. J. R. Ray, *J. Appl. Phys.*, 1982, **53**, 6441.
231. J. R. Ray, M. C. Moody and A. Rahman, *Phys. Rev. B*, 1985, **32**, 733.
232. D. R. Squire, A. C. Holt and W. G. Hoover, *Physica*, 1969, **42**, 388.
233. K. Van Workum, K. Yoshimoto, J. J. de Pablo and J. F. Douglas, *Phys. Rev. E*, 2005, **71**, 061102.
234. K. Van Workum, G. Gao, J. D. Schall and J. A. Harrison, *J. Chem. Phys.*, 2006, **125**, 144506.
235. A. U. Ortiz, A. Boutin, A. H. Fuchs and F.-X. Coudert, *J. Phys. Chem. Lett.*, 2013, **4**, 1861.
236. P. Ying, J. Zhang, X. Zhang and Z. Zhong, *J. Phys. Chem. C*, 2020, **124**, 6274.
237. P. Ying, J. Zhang and Z. Zhong, *Microporous Mesoporous Mater.*, 2021, **312**, 110765.
238. J. Heinen, A. D. Ready, T. D. Bennett, D. Dubbeldam, R. W. Friddle and N. C. Burtch, *ACS Appl. Mater. Interfaces*, 2018, **10**, 21079.

239. P. Ying, J. Zhang and Z. Zhong, *J. Phys. Chem. C*, 2021, **125**, 12991.
240. F. Trousselet, A. Boutin and F.-X. Coudert, *Chem. Mater.*, 2015, **27**, 4422.
241. C. Zener, *Elasticity and Anelasticity of Metals*, University of Chicago Press, Chicago, 1948.
242. A. Schmitz, M. Chandrasekaran, G. GHosh and L. Delaey, *Acta Metall.*, 1989, **37**, 3151.
243. C.-S. Man, *J. Elastoplast.*, 1995, **39**, 165.
244. L. Sarkisov, R. L. Martin, M. Haranczyk and B. Smit, *J. Am. Chem. Soc.*, 2014, **136**, 2228.
245. P. M. Marcus, H. Ma and S. L. Qiu, *J. Phys.: Condens. Matter*, 2002, **14**, 525.
246. D. C. Wallace, *Phys. Rev.*, 1967, **162**, 776.
247. T. H. K. Barron and M. L. Klein, *Proc. Phys. Soc., London*, 1965, **85**, 523.
248. P. M. Marcus and S. L. Qiu, *J. Phys.: Condens. Matter*, 2009, **21**, 115401.
249. Z. Zhou and B. Joós, *Phys. Rev. B*, 2002, **66**, 054101.
250. Z. Zhou, *Chin. J. Phys.*, 2004, **42**, 21.
251. M. Born, *Math. Proc. Cambridge Philos. Soc.*, 1940, **36**, 160.
252. F. Mouhat and F.-X. Coudert, *Phys. Rev. B*, 2014, **90**, 224104.
253. J. Wang, S. Yip, S. R. Phillpot and D. Wolf, *Phys. Rev. Lett.*, 1993, **71**, 4182.
254. J. Wang, J. Li, S. P. S. Yip and D. Wolf, *Phys. Rev. B*, 1995, **52**, 12627.
255. Z. Zhou and B. Joós, *Phys. Rev. B*, 1996, **54**, 3841.
256. M. R. Ryder, B. Civalleri, T. D. Bennett, S. Henke, S. Rudić, G. Cinque, F. Fernandez-Alonso and J.-C. Tan, *Phys. Rev. Lett.*, 2014, **113**, 215502.
257. P. Zhao, G. I. Lampronti, G. O. Lloyd, M. T. Wharmby, S. Facq, A. K. Cheetham and S. A. T. Redfern, *Chem. Mater.*, 2014, **26**, 1767.
258. S. A. Moggach, T. D. Bennett and A. K. Cheetham, *Angew. Chem., Int. Ed.*, 2009, **48**, 7087.
259. H. Fang, M. T. Dove, L. H. N. Rimmer and A. J. Misquitta, *Phys. Rev. B*, 2013, **88**, 104306.
260. S. Krause, N. Hosono and S. Kitagawa, *Angew. Chem., Int. Ed.*, 2020, **59**, 15325.
261. A. Ghoufi, A. Subercaze, Q. Ma, P. G. Yot, Y. Ke, I. Puente-Orench, T. Devic, V. Guillerm, C. Zhong, C. Serre, G. Férey and G. Maurin, *J. Phys. Chem. C*, 2012, **116**, 13289.
262. Q. Ma, Q. Yang, A. Ghoufi, K. Yang, M. Lei, G. Férey, C. Zhong and G. Maurin, *J. Mater. Chem. A*, 2014, **2**, 9691.
263. A. V. Neimark, F. X. Coudert, A. Boutin and A. H. Fuchs, *J. Phys. Chem. Lett.*, 2010, **1**, 445.
264. C. Serre, S. Bourrelly, A. Vimont, N. A. Ramsahye, G. Maurin, P. L. Llewellyn, M. Daturi, Y. Filinchuk, O. Leynaud, P. Barnes and G. Férey, *Adv. Mater.*, 2007, **19**, 2246.
265. P. G. Yot, Z. Boudene, J. Macia, D. Granier, L. Vanduyfhuys, T. Verstraelen, V. Van Speybroeck, T. Devic, C. Serre, G. Férey, N. Stock and G. Maurin, *Chem. Commun.*, 2014, **50**, 9462.

266. V. Haigis, Y. Belkhodja, F.-X. Coudert, R. Vuilleumier and A. Boutin, *J. Chem. Phys.*, 2014, **141**, 064703.
267. L. D. Landau and E. M. Lifshitz, Course of Theoretical Physics, *Statistic Physics, Part 1*, Elsevier, Oxford, 2011, vol. 5.
268. E. Cockayne, *J. Phys. Chem. C*, 2017, **121**, 4312.
269. J. Wieme, K. Lejaeghere, G. Kresse and V. Van Speybroeck, *Nat. Commun.*, 2018, **9**, 4899.
270. P. G. Yot, L. Vanduyfhuys, E. Alvarez, J. Rodriguez, J.-P. Itié, P. Fabry, N. Guillou, T. Devic, I. Beurroies, P. L. Llewellyn, V. Van Speybroeck, C. Serre and G. Maurin, *Chem. Sci.*, 2016, **7**, 446.
271. R. Demuynck, S. M. J. Rogge, L. Vanduyfhuys, J. Wieme, M. Waroquier and V. Van Speybroeck, *J. Chem. Theory Comput.*, 2017, **13**, 5861.
272. P. Iacomi, J. S. Lee, L. Vanduyfhuys, K. H. Cho, P. Fertey, J. Wieme, D. Granier, G. Maurin, V. Van Speybroeck, J.-S. Chang and P. G. Yot, *Chem. Sci.*, 2021, **12**, 5682.
273. C. L. Hobday and G. Kieslich, *Dalton Trans.*, 2021, **50**, 3759.
274. L. Vanduyfhuys, S. M. J. Rogge, J. Wieme, S. Vandenbrande, G. Maurin, M. Waroquier and V. Van Speybroeck, *Nat. Commun.*, 2018, **9**, 204.
275. G. Bussi and A. Laio, *Nat. Rev. Phys.*, 2020, **2**, 200.
276. A. Laio and M. Parrinello, *Proc. Natl. Acad. Sci. U. S. A.*, 2002, **99**, 12562.
277. G. M. Torrie and J. P. Valleau, *J. Comput. Phys.*, 1977, **23**, 187.
278. M. R. Shirts and J. D. Chodera, *J. Chem. Phys.*, 2008, **129**, 124105.
279. J. G. Kirkwood, *J. Chem. Phys.*, 1935, **3**, 300.
280. R. W. Zwanzig, *J. Chem. Phys.*, 1954, **22**, 1420.
281. S. M. J. Rogge, S. Caroes, R. Demuynck, M. Waroquier, V. Van Speybroeck and A. Ghysels, *J. Chem. Theory Comput.*, 2018, **14**, 1186.
282. R. Demuynck, J. Wieme, S. M. J. Rogge, K. D. Dedecker, L. Vanduyfhuys, M. Waroquier and V. Van Speybroeck, *J. Chem. Theory Comput.*, 2018, **14**, 5511.
283. P. St. Petkov, V. Bon, C. L. Hobday, A. B. Kuc, P. Melix, S. Kaskel, T. Düren and T. Heine, *Phys. Chem. Chem. Phys.*, 2019, **21**, 674.
284. S. Henke, A. Schneemann, A. Wütscher and R. A. Fischer, *J. Am. Chem. Soc.*, 2012, **134**, 9464.
285. J. Keupp, J. P. Dürholt and R. Schmid, *Faraday Discuss.*, 2021, **225**, 324.
286. P. Vervoorts, J. Keupp, A. Schneemann, C. L. Hobday, D. Daisenberger, R. A. Fischer, R. Schmid and G. Kieslich, *Angew. Chem. Int. Ed.*, 2021, **60**, 787.
287. S. Krause, J. D. Evans, V. Bon, P. Senkovska, P. Iacomi, F. Kolbe, S. Ehrling, E. Troschke, J. Getzschmann, D. M. Többens, A. Franz, D. Wallacher, P. G. Yot, G. Maurin, E. Brunner, P. L. Llewellyn, F.-X. Coudert and S. Kaskel, *Nat. Commun.*, 2019, **10**, 3632.

288. S. Krause, J. D. Evans, V. Bon, I. Senkovska, S. Ehrling, P. Iacomi, D. M. Többens, D. Wallacher, M. S. Weiss, B. Zheng, P. G. Yot, G. Maurin, P. L. Llewellyn, F.-X. Coudert and S. Kaskel, *Chem. Sci.*, 2020, **11**, 9468.
289. C. Triguero, F.-X. Coudert, A. Boutin, A. H. Fuchs and A. V. Neimark, *J. Phys. Chem. Lett.*, 2011, **2**, 2033.
290. L. Schaper, J. Keupp and R. Schmid, *Front. Chem.*, 2021, **9**, 757680.
291. F.-X. Coudert, M. Jeffroy, A. H. Fuchs, A. Boutin and C. Mellot-Draznieks, *J. Am. Chem. Soc.*, 2008, **130**, 14294.
292. S. M. J. Rogge, R. Goeminne, R. Demuynck, J. J. Gutiérrez-Sevillano, S. Vandenbrande, L. Vanduyfhuys, M. Waroquier, T. Verstraelen and V. Van Speybroeck, *Adv. Theory Simul.*, 2019, **2**, 1800177.
293. R. Goeminne, S. Krause, S. Kaskel, T. Verstraelen and J. D. Evans, *J. Am. Chem. Soc.*, 2021, **143**, 4143.
294. T. D. Bennett and S. Horike, *Nat. Rev. Mater.*, 2018, **3**, 431.
295. M. Erkartal and M. Durandurdu, *ChemistrySelect*, 2018, **3**, 8056.
296. K. Banlusan, E. Antillon and A. Strachan, *J. Phys. Chem. C*, 2015, **119**, 25845.
297. A. C. T. van Duyn, S. Dasgupta, F. Lorant and W. A. Goddard, *J. Phys. Chem. A*, 2001, **105**, 9396.
298. B. Zheng, F. Fu, L. L. Wang, J. Wang, L. Du and H. Du, *J. Phys. Chem. C*, 2018, **122**, 4300.
299. K. Banlusan and A. Strachan, *J. Phys. Chem. C*, 2016, **120**, 12463.
300. R. Pallach, J. Keupp, K. Terlinden, L. Frentzel-Beyme, M. Kloß, A. Machalica, J. Kotschy, S. K. Vasa, P. A. Chater, C. Sternemann, M. T. Wharmby, R. Linser, R. Schmid and S. Henke, *Nat. Commun.*, 2021, **12**, 409.
301. V. I. Hegde, J.-C. Tan, U. V. Waghmare and A. K. Cheetham, *J. Phys. Chem. Lett.*, 2013, **4**, 3377.
302. L. Bouëssel du Bourg, A. U. Ortiz, A. Boutin and F.-X. Coudert, *APL Mater.*, 2014, **2**, 124110.
303. R. N. Widmer, G. I. Lampronti, S. Anzellini, R. Gaillac, S. Farsang, C. Zhou, A. M. Belenguer, C. W. Wilson, H. Palmer, A. K. Kleppe, M. T. Wharmby, X. Yu, S. M. Cohen, S. G. Telfer, S. A. T. Redfern, F.-X. Coudert, S. G. MacLeod and T. D. Bennett, *Nat. Mater.*, 2019, **18**, 370.
304. E. M. Reynolds, E. H. Wolpert, A. R. Overy, L. Mizzi, A. Simonov, J. N. Grima, S. Kaskel and A. L. Goodwin, *Faraday Discuss.*, 2021, **225**, 241.
305. B. B. Shah, T. Kundu and D. Zhao, *Top. Curr. Chem.*, 2019, **377**, 25.
306. F.-X. Coudert, A. H. Fuchs and A. V. Neimark, *Dalton Trans.*, 2016, **10**, 4136.
307. F.-X. Coudert, A. Boutin, A. H. Fuchs and A. V. Neimark, *J. Phys. Chem. Lett.*, 2013, **4**, 3198.
308. J. Wieme and V. Van Speybroeck, *J. Mater. Chem. A*, 2021, **9**, 4898.
309. D. Dubbeldam, A. Torres-Knoop and K. S. Walton, *Mol. Simul.*, 2013, **39**, 1253.
310. G. Steinle-Neumann and R. E. Cohen, *J. Phys.: Condens. Matter*, 2004, **16**, 8783.

311. P. M. Marcus and S. L. Qiu, *J. Phys.: Condens. Matter*, 2004, **16**, 8787.
312. I. T. Todorov, W. Smith, K. Trachenko and M. T. Dove, *J. Mater. Chem.*, 2006, **16**, 1911.
313. S. Plimpton, *J. Comput. Phys.*, 1995, **117**, 1.
314. J. VandeVondele, M. Krack, F. Mohamed, M. Parrinello, T. Chassaing and J. Hutter, *Comput. Phys. Commun.*, 2005, **167**, 103.

CHAPTER 4

High-pressure Mechanical Behaviour Under Hydrostatic Compression

STEPHEN A. MOGGACH*[a] AND GEMMA F. TURNER*[a]

[a]School of Molecular Sciences, The University of Western Australia, 35 Stirling Hwy, Crawley, Western Australia, Australia
*E-mail: stephen.moggach@uwa.edu.au, gemma.turner@uwa.edu.au

4.1 Introduction

Metal–organic frameworks (or MOFs) inherit structural flexibility from deformable metal nodes and conformationally flexible organic linkers, giving rise to structural and functional phenomena under applied hydrostatic pressure.[1,2] Structural characterisation of MOFs, in particular by *in situ* high-pressure single-crystal X-ray diffraction, provides a mechanistic insight into their pressure-dependent behaviour. Currently, <0.2% of the 103 000 MOF structures in the Cambridge Structural Database (CSD) have been studied at high-pressure.[†] Even so, high-pressure analysis of MOFs has led to the discovery of a collection of flexible structural responses to pressure, including 'breathing' of the pore volume,[3–8] 'gate-opening' rotation of the linkers,[9,10] negative linear compressibility (NLC),[11–14] Jahn–Teller switching,[13] or 'hyper-filling' of the pores with a guest species[15–17] (Figure 4.1). Structural

[†]161 structures contain a pressure descriptor out of the 102 583 structures in the MOF subset (65 are unique MOFs), CSD 2019.

Inorganic Materials Series No. 12
Mechanical Behaviour of Metal–Organic Framework Materials
Edited by Jin-Chong Tan

Published by the Royal Society of Chemistry, www.rsc.org

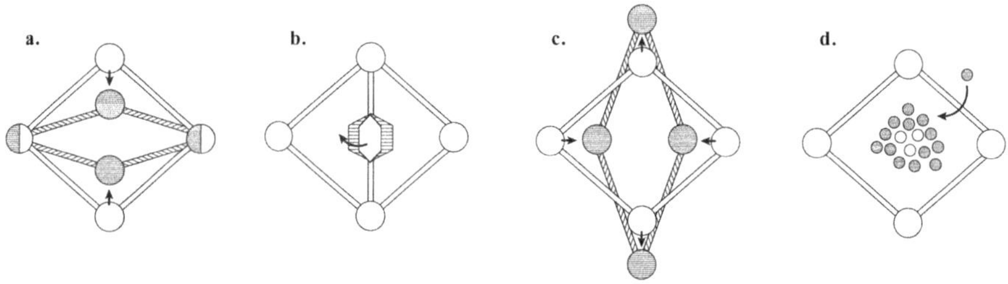

Figure 4.1 Illustration of flexible behaviour in MOFs. (a) Breathing. (b) Gate-opening linker rotation. (c) Negative linear compressibility. (d) 'Hyper-filling' of the framework pores with guest from a pressure-transmitting medium. White structures are at ambient pressure, or do not contain guest. Shaded structures are at high-pressure and/or contain guest.

changes are sometimes associated with functional behaviour, such as mechanical energy storage,[18,19] piezochromism,[13,20] bathochromism,[21] or piezoelectricity.[22,23] Such behaviour is relevant to their technological applications, which often depend upon the pore structure, mechanical stability, or functional properties of the framework.

High-pressure generally refers to the operating pressure of diamond anvil cells (DACs, 0.1 GPa to 400 GPa).[24–26] The minimum loading pressure of a DAC is equivalent to conditions in deep ocean trenches, while the maximum loading pressure is comparable to the Earth's core. Lower applied pressures are achieved in capillary cells (0.001 GPa to 0.15 GPa)[27] or gas cells ($<$0.001 GPa). Hydrostatic pressure (where pressure is applied equally in all directions) is generated in the cell by suspending the crystal or powder in a pressure-transmitting medium, such as a gas, liquid, or soft solid. Hydrostatic compression of MOFs elicits distinct structural behaviour compared with non-hydrostatic compression. Anisotropic compression generates deviatoric stresses (causing shear distortion) and depends upon the orientation of the crystal or crystallites, whilst hydrostatic compression is pressure that is applied evenly to the crystal, and is independent of the crystal orientation. Appropriate selection of the pressure-transmitting medium is important to the structural integrity and pressure-response of a given MOF. For porous MOFs, media composed of small molecules (*e.g.*, H_2O, short-chain alcohols and hydrocarbons) may enter the framework pores during compression, mediating the framework flexibility,[15–17] or promoting post-synthetic modification.[28] Direct compression of the framework therefore requires an inert, bulky pressure-transmitting medium that is too large to enter the pores. High-pressure cells are compatible with *in situ* X-ray and neutron diffraction, and spectroscopic characterisation methods, permitting analysis of pressure-induced dynamic structural and chemical processes in MOFs. This chapter provides an overview of the mechanical response of MOFs to hydrostatic pressure.

4.2 High-pressure Experimental Techniques

4.2.1 Hydrostatic Pressure Generation

4.2.1.1 DACs

Modern pressure cells incorporate diamond anvils[24–26] to generate hydrostatic pressure between 0.1 GPa and 400 GPa.[29,30] There are various designs of DACs, such as Mao–Bell,[31] Piermarini and Block,[32] double-stage,[33] or miniature Merrill–Bassett cells,[34] each of which uses a different mechanism to generate pressure. The Merrill–Bassett DAC[34] is the most commonly used for *in situ* single-crystal diffraction owing to its small size and versatility, which allows it to be mounted on a standard goniometer head and fit easily on laboratory diffractometers, though it can also be used for powder diffraction studies.

The miniature Merrill–Bassett DAC is composed of two triangular, stainless-steel or vascomax platens, each of which houses a diamond anvil (Figure 4.2).[34] The diamonds are positioned in a hole in the centre of each platen to form a window through the cell, and are supported by beryllium or tungsten carbide backing seats (Figure 4.2F).[35] The diamonds usually have

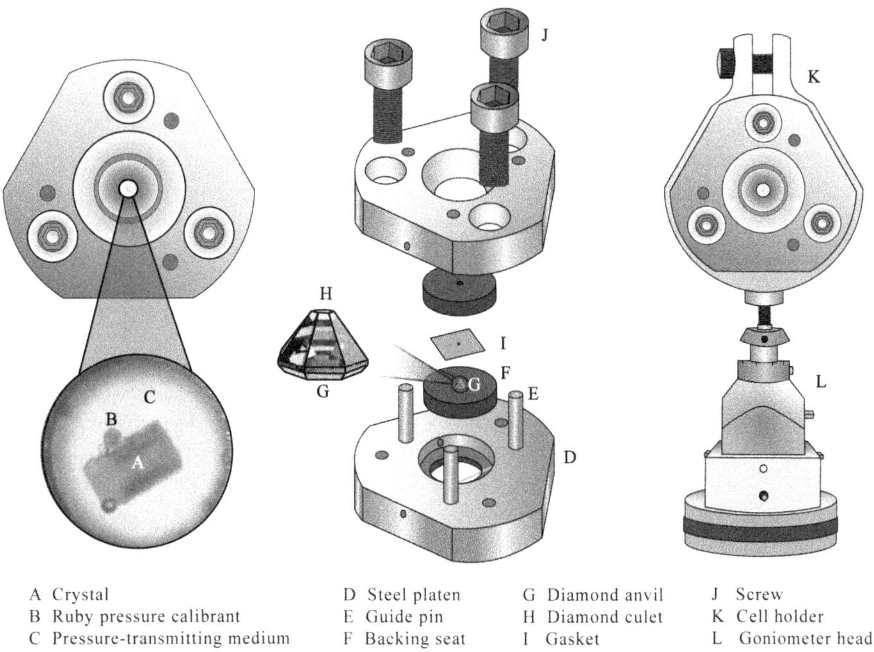

Figure 4.2 Illustrations of a miniature Merrill–Bassett diamond anvil, showing a top-down view of the cell (left), an exploded view. Centre, reproduced from ref. 35 with permission from International Union of Crystallography, Copyright 2008. The cell mounted on a goniometer head in a bespoke holder (right). Parts are labelled from A to L.

a flat surface, referred to as the culet faces of the diamond anvils. The faces must be perfectly aligned when the cell is closed to avoid fracturing the diamonds. The position of the anvils can be adjusted by screws connected to the backing seat, while the alignment of the platens is maintained by guide pins affixed to the platens. The sample chamber is formed from a cylindrical hole drilled into a metal gasket (usually composed of tungsten, ruthenium or steel), which sits between the diamond anvils (Figure 4.2I). The hole is usually <50% of the diameter of the diamond culets to ensure the sample chamber is sealed when the DAC is closed. The culet diameter, hole diameter and thickness of the gasket all determine the volume of the sample chamber, which restricts the size of the single crystal or volume of powder to be analysed. A pressure calibrant is placed alongside the analyte in the sample chamber to determine the pressure *in situ*, which is discussed in Section 4.2.1.1.4. The sample chamber is filled with a fluid pressure-transmitting medium to ensure hydrostatic pressure conditions. Liquid pressure-transmitting media solidify under sufficient pressure, generating non-hydrostatic conditions that can crush the crystal, which is discussed in Section 4.2.1.1.3. The cell is sealed and pressurised by tightening a set of screws that connect the platens. Central alignment of the sample chamber with the diamond culets allows the sample to be viewed in the cell (Figure 4.2A). Bespoke cell holders are required to incorporate the DAC on the diffractometer or spectroscopic instrument.

4.2.1.1.1 Diamond Anvils. The operational pressure range of the DAC is principally determined by the size and geometry of the diamond anvils. Conical diamond anvils are cut in brilliant, Drukker, or Boehlar–Almax geometries, which feature a large, flat table face at the top of the anvil (~1 mm to 3 mm) and a small, flat or bevelled culet face (~100 μm to 800 μm) at the base of the anvil, as illustrated in Figure 4.3. The hardness of the diamond and the large table to culet diameter ratio generates high-pressure with little force input. The distance between the table and culet, the angle of the pavilion (~60° to 130°), and the bevel angle also affect the maximum generated pressure (Figure 4.3). Opposed anvil apparatus are based upon the principle of massive support, which states that the massive support factor is proportional to the square of the ratio of the table diameter to the culet diameter. The massive support principle is expressed in eqn (4.1),[36] where P is the maximum generated pressure, S_A is the compressive strength of the anvil material, a is the culet radius, and b is related to the distance between the table and diameter and the bevel angle.

$$P = S_A \frac{b^2}{a^2} \tag{4.1}$$

To generate pressure of ~400 GPa, the table to culet ratio should be between 14 and 18, with a bevel angle of 8.5°.[37] Pressure up to 600 GPa can be achieved using a toroidal diamond tip, which permits smaller culet diameter without comprising the strength of the anvil.[38] In general, pressure below

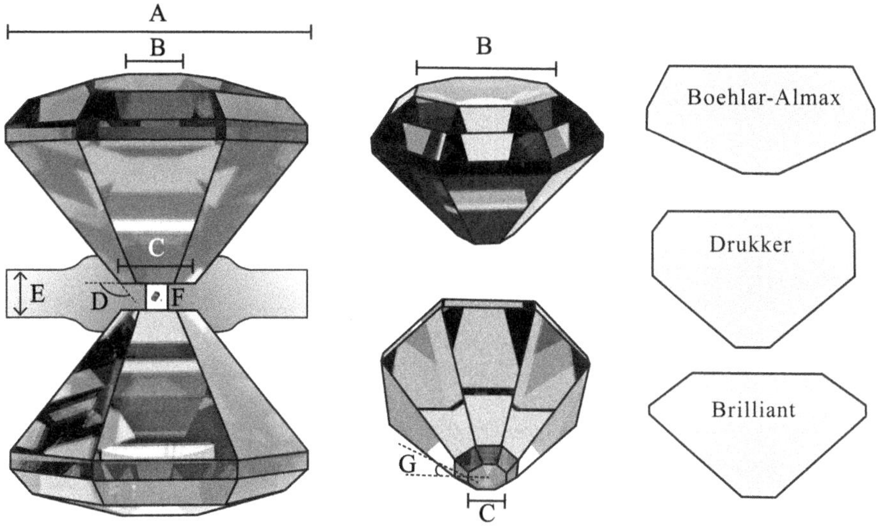

A Anvil diameter
B Table diameter
C Culet diameter
D Pavillion angle
E Gasket thickness
F Sample chamber
G Bevel angle

Figure 4.3 Illustrations of a gasketed opposed diamond anvil arrangement (left), brilliant cut diamond anvils showing the table and culet faces (centres), and Boehlar–Almax, Drukker and brilliant cut diamond geometries. Components and parameters are labelled from A–G. The figure is partly reproduced from ref. 39 with permission from International Union of Crystallography, Copyright 2019.

10 GPa is sufficient to characterise structural flexibility in MOFs, and in fact most other molecular materials, since crystallinity is usually lost well before 10 GPa is reached.[2]

4.2.1.1.2 Gasket Material. The generated pressure also depends upon the gasket material. The relationship between pressure and the gasket properties is expressed in eqn (4.2),[36] where A relates to the strength of the gasket material, μ is the friction coefficient between the anvil and gasket, h is the thickness of the gasket, and r is the radius of the sample chamber.

$$P = A\exp\left[\frac{2\mu}{h}(a-r)\right] \tag{4.2}$$

Gasket materials with high mechanical strength and rigidity, chemical inertness, and low absorption of X-rays are required to provide uniform stress in the sample chamber over a large pressure range with minimal pollution of the diffraction pattern. Materials with good X-ray transparency, such as boron and its composites, are vulnerable to mechanical failure, resulting in bursting of the gasket and total loss of pressure.[40–43] Therefore, strong and stiff materials, such as stainless steel,[44] rhenium,[45] and tungsten[46] are preferred since the cylindrical shape of the sample chamber and the integrity of the gasket is retained during compression. Powder rings in the X-ray diffraction images from the polycrystalline metal are manually excluded from the sample reflections during indexing and data reduction.

The mechanical stability of the sample chamber depends upon the thickness of the gasket and its pre-indentation.[47] Thick gaskets (>100 μm) provide a voluminous sample chamber and a linear relationship between the applied force and generated pressure, yet suffer from considerable deformation of the chamber during compression, which leads to non-uniform pressure application and potential leakage of the pressure-transmitting medium. Thin gaskets (≤100 μm) can access higher pressures, provide a better seal, and retain the integrity of the sample chamber, but the pressure–force relationship becomes non-linear and is therefore difficult to control. Thin gaskets also minimise the shaded area of the detector, increasing the available reciprocal space. In practise, a relatively thick gasket, between 200 and 300 μm is pre-indented between the diamond anvils to provide an area of thin gasket (~100 μm) in which the sample hole is drilled. Pre-indentation increases the massive support of the gasket and diamonds, increasing the maximum operating pressure of the DAC.[47]

4.2.1.1.3 Pressure-transmitting Medium. The pressure-transmitting medium must be suitably inert and maintain hydrostatic conditions in the desired pressure range. Liquid media, such as inert oils[48–51] and small organic molecules,[52–54] are the most commonly used, although gases[53,55,56] and soft solids are also suitable.[52,57–60] Solidification of the medium at a critical pressure introduces deviatoric stress around the sample, resulting in non-hydrostatic pressure conditions. Oils typically have a wide range of hydrostatic limits, with the perfluorinated hydrocarbons Fluorinert® FC-72 to FC-87[49,50] and Daphne 7474[51] remaining hydrostatic between 0.5 and 3.7 GPa, whilst silicone oil is hydrostatic up to 10 GPa.[48] Small molecules generally have higher hydrostatic limits. Mixtures of methanol, ethanol, and water remain hydrostatic in the range of 10–17 GPa,[53] while 2-propanol,[61] glycerol,[54] and pentanes have limits in the range of 4.2–7.4 GPa.[53] Gaseous media are typically noble gases, such as He, Ne, and Ar, with hydrostatic limits of 1.4 GPa,[53] 20 GPa,[55,56] and 11 GPa,[56] respectively. Solid media include mineral salts, such as NaCl, AgCl and MgO, which have limits in the range of 7–11 GPa,[52,59,60] and metals, such as Na or Al, which have limits of 69 GPa and 215 GPa, respectively.[57,58] For porous MOFs, the molecular size of the pressure-transmitting medium must be considered, as small molecules may enter the pores during compression, interacting dynamically with the sample, changing the composition, structure, and mechanical properties of the material.[2]

4.2.1.1.4 Pressure Determination. The pressure inside the sample chamber is measured using an internal standard with pressure-dependent properties, or with an established equation of state (EoS). Ruby (Al_2O_3: Cr^{3+}) is the most common pressure calibrant, which exhibits a linear relationship between applied pressure and its fluoresence emission wavelength upon iradiation with laser light up to 25 GPa.[62,63] The shift in the emission doublet with a maximum at ~694 nm upon compression is related to pressure by eqn (4.3), where P is the calculated applied pressure in GPa and $\Delta\lambda$ is the shift in the doublet maximum between ambient pressure and the given pressure.[62,63] The value of 2.746 is the gradient of the $\Delta\lambda$ *vs.* pressure plot up to 19.5 GPa.

Determination of the pressure using the equation of state of a standard is achieved by tracking changes in phase or volume of the calibrant.[64–66] Gold is one of the most commonly used pressure calibrants.[65–68] Its equation of state was first determined using the ruby fluorescence method and powder X-ray diffraction in a DAC up to 70 GPa, and was found to fit a third-order Eulerian finite strain.[65] Changes in the unit cell volume can be related to the applied pressure within an accuracy of 2%.[65]

$$P = 2.746(\Delta\lambda) \tag{4.3}$$

4.2.1.2 Capillary Pressure Cells

Flexible MOFs may undergo pronounced structural distortions under applied hydrostatic pressure much lower than the minimum loading pressure of a DAC (<0.1 GPa). Pressure cells that operate in the moderate pressure range between 5 MPa and 0.1 GPa are less common than DACs and implement a capillary to serve as the hydrostatic chamber.[27,69–75] The capillary is usually manufactured from X-ray transparent materials that can withstand moderate pressure without shattering, such as beryllium, sapphire, quartz, stainless steel, or glassy carbon. Capillary cells are advantageous over diamond anvils for materials that crystallise in low-symmetry (triclinic or monoclinic) systems since they do not impose restrictions on the available reciprocal space.

Early capillary cells employed beryllium due to its X-ray translucency (*ca.* transmission of Mo Kα ~ 70%, Cu Kα ~ 17%) and good tensile strength, which permit *in situ* X-ray diffraction to a maximum operating pressure of 0.1–0.4 GPa.[69–71] The beryllium capillary is affixed to a stainless-steel ring that can be mounted on a platform or stud, and affixed to a standard goniometer head. A single crystal is glued to a metal or glass fibre and is inserted into the bore of the capillary. The bore is filled with a pressure-transmitting medium and sealed at the base using either a metal-to-metal cone connection[69,71] or O-rings.[70] Hydrostatic pressure is generated by a mechanical hand pump to a precision of 0.0002 GPa.[69–71] The maximum operating pressure is determined by the thickness of the capillary, which varies between around 6 mm to 1 mm.[69–71] Thin capillaries suffer from fragility but can accommodate larger crystals in the bore. The operating pressure of thin capillaries can be increased by reinforcing the capillary with a copper–beryllium support, although absorption of X-rays by copper requires windows to be cut into the support.[73] Unlike DACs, (which due to the metal body restrict access to large regions of reciprocal space, otherwise known as 'shading') single-crystal X-ray diffraction in capillary cells is not restricted by shading of the detector. However, diffraction experiments are complicated by the inability to view the crystal in the capillary, preventing optical centring, and by the introduction of Debye–Scherrer rings in the diffraction pattern, which must be addressed during data processing.

Other capillary cells replace beryllium with quartz or sapphire to allow optical centring of the crystal.[27,72,74,75] Quartz is commonly used in flow cells, but is rarely implemented in liquid capillary cells due to its brittleness.[75] A quartz capillary cell with a maximum operating pressure of 0.1 GPa and an accuracy of 0.003 GPa has been reported, which features a valved, brass pressure line connected to the open end of the capillary.[75] The capillary is pressurised and the valve closed before mounting on the diffractometer.[75] Sapphire capillaries are more resilient and are operational up to 0.15 GPa in liquid or gaseous pressure-transmitting media, and have been designed and optimised for use at central facilities, which include synchrotron X-ray and neutron diffraction techniques.[27,72,74]

A sapphire capillary cell optimised for use on beamline I19-2 at Diamond Light Source in the United Kingdom uses a piston–pump mechanism to control the pressure of the liquid-filled chamber (Figure 4.4).[27] The single-crystal sapphire capillary is held in a stainless steel holder and is connected to a phase separator through a cone-type metal-to-metal seal.[27] The phase separator contains two liquid-filled chambers, separated by a moveable piston. One chamber contains the pressure-transmitting medium and is connected to the capillary, while the other chamber contains a hydraulic fluid and is connected to a hand-pump for pressure generation. Pressure control to within 0.0001 GPa is achieved using a SITEC valve, which decreases the volume of the phase separator. Tubing from the pressure cell to the hand-pump can be fed from the experimental hutch to the control room to allow remote modification of the pressure.[27] The capillary, which is between 1.0 mm and 1.4 mm thick, can be safely operated to 0.15 GPa.[27] Diffraction from the pressure cell is minimal with only broad reflections from the sapphire contaminating the diffraction pattern, which can be removed during processing.

Sapphire capillaries can also be implemented in moderate-pressure gas cells, and are particularly suited to examining hydrogen storage materials by neutron diffraction owing to the neutron transparency of sapphire and its inertness to hydrogen gas.[72,74] Maximum pressure ratings between 0.0275 GPa and 0.1 GPa are achieved using various cell configurations.[72,74] Such pressure conditions are comparable to hydrogen storage tanks (0.035 GPa to 0.07 GPa). A relatively low operating pressure of 0.0275 GPa is achieved using an open-ended sapphire capillary connected to a gas manifold by quick-connect fittings, mounted in a steel holder.[72] Open-ended capillaries allow gas delivery from both ends, minimising sample movement in the chamber. Higher operating pressures between 0.068 GPa and 0.07 GPa are achieved by using more robust connection fittings at the ends of the open-ended tube, such a metal-cone, support ring and flange joint,[74] or a Swagelok stainless

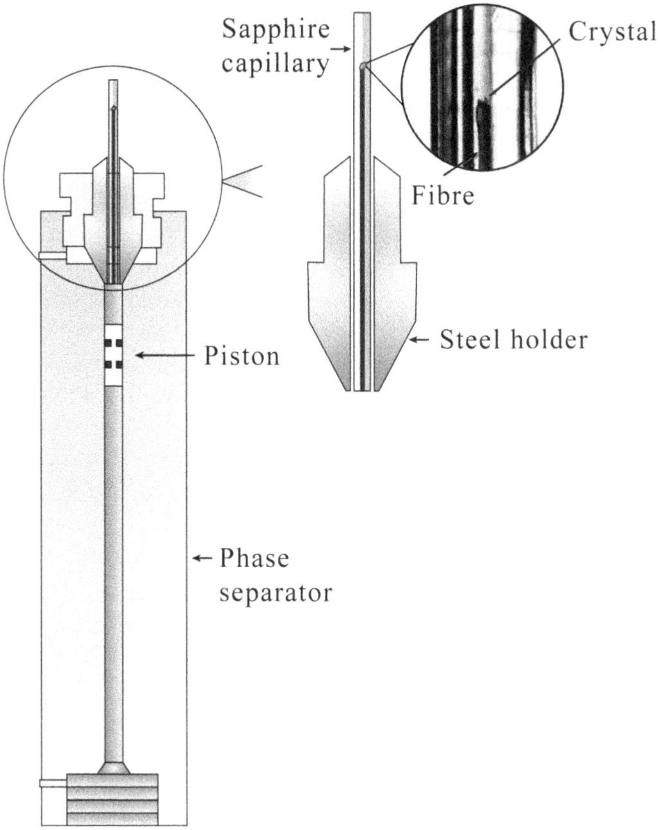

Figure 4.4 Illustration of a sapphire capillary cell. Reproduced from ref. 27 with permission from International Union of Crystallography, Copyright 2020. Hydrostatic pressure is generated in the liquid-filled capillary by a hand-pump that increases the pressure of the hydraulic fluid in the lower part of the phase separator, which operates the piston to increase the pressure in the capillary. The piston separates the hydraulic fluid from the pressure-transmitting medium.

steel isolation valve.[72] The pressure rating of the cell is determined by the resilience of the seals. Removing one of the seals by using a closed-ended capillary improves the pressure rating to 0.1 GPa.[72] In addition, the steel holder is replaced with a copper–beryllium alloy, and the capillary is sealed with brass and Teflon connections.[72] While sapphire capillary cells are suitable for both neutron and X-ray diffraction, their use in X-ray diffraction is limited to radiation with a wavelength shorter than 1 Å due to absorption of low-energy X-rays by sapphire.[72]

Capillary pressure cells, although limited in their operating pressure, allow much finer control of pressure than any other device and are optimal for measuring data on very soft condensed matter. As the systems that we study become ever more complex, we believe the use of capillary cells will become much more common in the years to come.

4.2.2 High-pressure X-ray Diffraction

4.2.2.1 Merrill–Bassett DAC

The Merrill–Bassett DAC is compatible with laboratory and synchrotron four-circle diffractometers with subtle modification to the instrument configuration.[76,77] The DAC can be mounted onto a standard goniometer head using a custom holder (Figure 4.2K). The large size of the cell is accommodated on the diffractometer by use of a shorter X-ray collimator, adjustment of the cryostream nozzle position, and by increasing the distance between the sample and detector during the data collection.

Enclosure of the single crystal within the pressure cell introduces a number of limitations to the diffraction experiment. Firstly, the cell body shades portions of the detector, restricting the accessible reciprocal space. This can be problematic for low-symmetry crystals (triclinic or monoclinic), where the data completeness may be as low as 20%. Secondly, diffraction from the cell components, including intense diamond and ruby reflections, and powder rings from the gasket or backing seat material lowers the signal-to-noise ratio of the sample reflections and can further decrease the data completeness if the sample and cell reflections overlap. For these reasons, together with the restricted size of the crystal due to the small sample chamber and the requirements of the diffraction experiment, high-pressure crystallography using a laboratory source is reserved for well-diffracting crystals.

The quality and completeness of the diffraction data is optimised by using a defined collection strategy. Diffraction data are collected in goniometer positions in which the detector is shaded least by the cell body. An optimised strategy for a DAC with a 80° opening angle has been devised, based on ω-scans in eight settings of φ (90° or 270°) and 2θ (±28°) (Table 4.1), though modern instruments usually allow an optimised strategy to be calculated based on the opening angle of the DAC and knowing its orientation with respect to the diffractometer circles.[76] In all positions, at least 20% of the

Table 4.1 Data collection strategy for a miniature Merrill–Bassett DAC with a 40° opening angle devised by Dawson *et al.*[76]

Run	ω Start/°	ω End/°	2θ/°	φ/°
1	−10	−40	−28	90
2	40	−25	28	90
3	−155	−220	−28	90
4	−140	−170	28	90
5	−155	−220	−28	270
6	−140	−170	28	270
7	−10	−40	−28	270
8	40	−25	28	270

detector is unshaded.[76] Data quality may also be improved by modifying the X-ray source or wavelength depending upon the intensity of the diffraction and the symmetry of the crystal. Short-wavelength radiation, such as from a silver anode (Ag Kα = 0.56 Å), increases the accessible reciprocal space but suffers from decreased intensity. Synchrotron radiation is therefore required for low-symmetry and weakly diffracting samples. Longer wavelength radiation, such as from a copper anode (Cu Kα = 1.54 Å) restricts the data completeness but provides strong diffraction intensity, however absorption from the diamond anvils at this wavelength is too high. For strongly diffracting, high-symmetry crystals, Mo Kα radiation (λ = 0.71 Å) on a laboratory source is sufficient, providing enough intensity and transmission through the DAC body.

The output diffraction images are polluted with reflections from components of the DAC, which must be manually removed before indexing the unit cell and integrating the data (Figure 4.5). Shaded regions of the detector are excluded from the background calculation by use of masks generated from the opening angle of the cell. Cell reflections persisting in the reduced data may be removed manually during refinement.

Crystallographic data may also be improved by modification of the DAC components or configuration. Low symmetry crystals with incomplete diffraction data benefit from large opening angles of the windows in the steel platens, which allow more reflections to exit the cell (Figure 4.5A). A standard Merrill–Bassett DAC employs an opening angle of 80°,[34] although opening angles up to 120° are viable.[78,79] Larger opening angles improve the

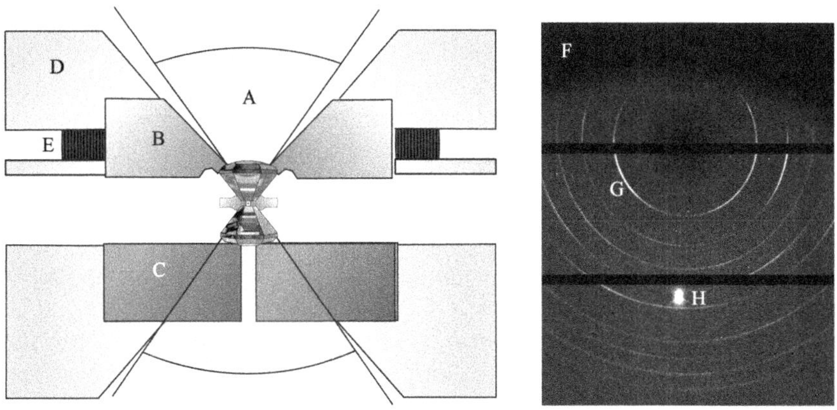

A Opening angle B Steel backing plate C Be or WC seat D Steel cell body E Anvil adjustment screw F Shadow from cell G Gasket powder ring H Diamond reflection

Figure 4.5 Cross-sectional diagram of a Merrill–Bassett DAC (left), reproduced from ref. 39 with permission from International Union of Crystallography, Copyright 2019. Components are labelled A–E. Typical diffraction image from a crystal loaded in a DAC (right), showing the shadow from the cell body, powder rings from the metal gasket (tungsten), and a bright reflection from the diamond anvils, labelled F–H.

completeness of the crystallographic data but decrease the maximum operating pressure of the cell. The backing seats supporting the diamond anvils may also be modified to reduce X-ray absorption,[35] or to increase the cell operating pressure.[80] For X-ray opaque materials, the angle of the confocal hole in the backing seat affects the opening angle of the cell. Tungsten carbide is one of the most commonly used support materials, which eliminates the diffraction halos from traditional beryllium seats.[35] Rarely, diamond backing seats are used to access higher pressure.[80]

4.2.3 High-pressure Neutron Diffraction

High-pressure neutron diffraction can be used to follow gas adsorption processes in MOFs, such as locating hydrogen adsorption sites,[81–83] or structural changes involving hydrogen atoms that are indiscernible by X-ray diffraction. Presently, analyses of MOFs by neutron diffraction have only been performed at gas loading pressure,[81,83] although it is possible to perform high-pressure neutron diffraction for these systems.

Neutron sources often produce low-flux radiation, requiring large sample volumes exceeding 1 mL (or ~1 g) to obtain sufficiently intense diffraction. Early pressure cells were therefore limited to a relatively low maximum operating pressure of 3 GPa to accommodate a more voluminous sample chamber.[84–86] Large sample volumes can be incorporated in multi-anvil pressure cells or opposed anvil cells with large anvils.[84–91] The anvil material must be neutron transparent and processible to achieve a culet dimeter on the millimetre-scale, such as zirconia,[85] sapphire[86,90] or moissanite.[91] Diamond anvils, although limited by their small size, can also be incorporated into large-volume cells by the addition of a cup to the culet face, such as in the Paris–Edinburgh cell.[88,89] The Paris–Edinburgh cell is among the most widely used for high-pressure powder neutron diffraction and is operational up to 40 GPa with a sample volume of between 25 mm^3 to 30 mm^3.[92] The anvils are cut to a toroidal geometry with sintered diamond tips, 1 mm in diameter. Polycrystalline diamond is used to minimise Bragg reflections from the anvils, which limits the intensity of the neutron beam at a particular wavelength.[92] Pressure is generated by an automated volumetric pump, which applies load to the anvils.

Gem-quality anvils of single-crystal sapphire are used in the Kurchatov–LLB cell.[90] The anvils are cut from a spherical, synthetic sapphire with a large culet diameter of between 1.5 mm and 8 mm, and a table-to-culet ratio of >15 mm, generating maximum pressure between 1 GPa and 15 GPa with sample volumes up to 35 mm^3. A pair of opposed anvils straddle a niobium, copper–nickel or copper–beryllium gasket. The anvils are housed in a steel or tungsten carbide cylinder and load is applied to the upper anvil by a tungsten carbide piston, which is moved by a pair of screws.[86,90] Tungsten carbide is a good thermal conductor, allowing low-temperature, high-pressure diffraction experiments to be performed. The effect of cooling on the cell can be further minimised by insertion of bevelled springs around the piston.[90]

Sapphire–anvil cells have been used to derive equations of state for molecular deuterium and metal-deuterides by neutron diffraction.[93,94]

Moissanite anvils are also implemented in opposed anvil cells for single crystal neutron diffraction.[91] The 'panoramic' piston-cylinder type cell employs smooth-cut conical anvils with a culet diameter of 1 mm and a sample volume of ~0.6 mm^3, allowing pressures of as high as 30 GPa to be reached.[91] Decreasing the culet diameter to 0.5 mm increases the maximum operating pressure to 38.5 GPa. Three windows (105° equatorial and 68° azimuthal opening angles) are cut into the wall of the cylindrical cell, allowing the diffracted neutrons to exit the cell largely unimpeded.[91] Neutron absorption mostly arises from the titanium–zirconium alloy gasket, which must be corrected for during data processing.[91]

Multi-anvil cells can accommodate up to 8 mm^3 of sample and operate to 8 GPa.[84,85,87] Anvils of zirconia ceramic are employed in a palm cubic anvil cell, which features six conical anvils about a cubic gasket of alloyed aluminium.[85] The sample is housed in a Teflon capsule within the gasket, which are held together in the cell by a guide block in the lower part of the cylindrical cell body. Load is applied to the anvils by a piston in the upper part of the cell body.[85] High thermal–conductivity of the cell components permits low-temperature, high-pressure neutron diffraction studies to be performed; an example of which is the single–crystal study of manganese phosphide to 7.5 GPa and <50 K, which was found to undergo a transition in its ferromagnetic structure to helical-*c* below 50 K and exhibited eventual loss of magnetism above 3 GPa.[85]

Advances in detector technology have improved the attainable spatial resolution of Laue diffraction to allow analysis of small sample quantities in DACs.[95–101] A miniature Merrill–Bassett DAC was first implemented for Laue diffraction at the OPAL reactor at ANSTO on the KOALA instrument, to examine single crystals of hexamethylenetetramine and L-arginine.[95] A data collection strategy based on twelve settings of ϕ with a step-size of 5° was employed for a cell with a half-opening angle of 39°.[95] Unlike X-ray diffraction, neutrons transmit the cell body to produce nearly complete diffraction for low-symmetry crystals. The intensity of neutrons transmitting the cell body are attenuated compared with those transmitting the anvils, and thus require a different absorption correction. Diffraction from the gasket and anvils must also be accounted for in the data processing. Manufacture of the cell body and diamond backing plates from a copper–beryllium alloy improves the thermal conductivity of the cell, offering the possibility for variable temperature, high-pressure diffraction.[95]

The sample may become activated in the neutron beam, posing a potential hazard when increasing the pressure by hand. Modification of the Merrill–Bassett DAC to include an inflatable helium-filled membrane mitigates this hazard.[101] The modified cell is constructed from circular copper–beryllium platens, connected by four guide pins, with a large half-opening angle of 80°.[101] The membrane is housed in a cup that is placed above one of the anvils. Pressure is generated by tightening a set of screws threaded through the platens, or by inflating the helium membrane. A diamond culet diameter of 0.6 mm can access 10 GPa, while a diameter of 1.0 mm can access 4 GPa. Components of the cell are coated in a gadolinium paint to minimise scattering from the cell.[101]

Replacing single-crystal diamond anvils with polycrystalline or nano-polycrystalline diamond can improve both the maximum operating pressure of the cell, and the quality of the neutron diffraction data.[96–100] As previously mentioned, polycrystalline diamond eliminates parasitic Bragg reflections from single-crystal anvils, permitting access to a larger portion of reciprocal space.[98] In addition, correction procedures for the attenuated intensity of neutrons that are diffracted through the anvil are simplified.[98]

Various piston–cylinder DACs are optimised for neutron diffraction, which generally use alloyed copper–beryllium for the major cell components and feature large cut-outs about the equator of the cylinder to provide wide opening angles between 60° and 135°.[96–100] Depending upon the cell components and configuration, maximum operating pressures between 40 GPa and 94 GPa are achieved with sample volumes between 0.02 mm^3 and 0.10 mm^3.[98] The highest operating pressure requires polycrystalline diamond backing seats to support the anvils,[98] while lower operating pressures are achieved using tungsten carbide or copper–beryllium.[96–100] Such pressures far exceed those required to analyse MOFs. To date, piston–cylinder DACs have been used to study minerals, such as manganese phosphide, sodium chloride and ice VII.[96,99]

4.2.4 High-pressure Spectroscopy

Dynamic chemical processes in MOFs under high-pressure or pressure-stimulated functional behaviour can be followed using *in situ* spectroscopic methods, including Raman, IR, NMR, UV-visible, and fluorescence emission spectroscopies. Changes in lattice vibrations resulting from phase transitions, guest adsorption, or local structural changes (*e.g.* intermolecular interactions) are detectable by vibrational and absorption/emission spectroscopies,[21,102–105] while distortion of hydrogen bonding networks may be monitored by ^1H NMR spectroscopy,[106] providing structural data for poorly diffracting MOFs and insights into chemical behaviour. Incorporation of pressure cells to spectrometers generally requires bespoke apparatus.

High-pressure Raman, IR and fluorescence spectroscopies involve a bespoke arrangement of the pressure cell, excitation laser, focussing and collecting lensing, turning mirrors, and the spectrometer.[21,105,107–111] Early use of DACs in Raman and IR spectroscopies involved directing the laser directly through the cell windows to the aperture in the spectrometer.[112] Although simple, this set-up permits only the forward scattering mode to be measured, restricting samples to transparent, neutral solids without polarisation effects and high vibrational frequencies.[107,108]

This problem is mitigated by modification of the cell design and experimental set-up to align the anvil faces parallel to the laser fringes.[108] Parallel alignment is retained under pressure by applying force normal to the diamond faces using a hemispherical bearing.[108] To avoid unintentional tilting of the anvils upon hand-tightening the screws to apply pressure, pressure is instead generated using a hydraulic ram, ensuring that the optical alignment is not offset between measurements.[108] Absorption and fluorescence from the diamond anvils pollute the sample spectrum, but are minimised by use of type II diamond anvils and by modification of the excitation wavelength.[108] Type I diamond anvils contain nitrogen platelet impurities, rendering them unsuitable for IR and Raman spectroscopies due to their strong fluorescence at all excitation wavelengths. By contrast, type II diamond anvils exhibit weak fluorescence below 600 cm^{-1} with an excitation wavelength of 632.8 nm.[108] The signal-to-noise ratio of the spectrum may be improved by replacing one of the anvils with tungsten carbide, or by using sapphire anvils, albeit at the expense of operating pressure.[108,109] Other types of cell configurations, such as piston-cylinder cells with sapphire windows may also be used for Raman, IR and fluorescence spectroscopies.[107,110] Characterisation of MOFs by high-pressure vibrational spectroscopy is common compared with other spectroscopies, and has been used to characterise phase transitions and guest adsorption in ZIF-8 (zeolitic imidazolate framework), $[Zn(mIm)_2]_n$ (mIm = 2-methylimidazolate)[19,113] (discussed in Section 4.4) and HKUST-1 (Hong Kong University of Science and Technology), $[Cu_3(btc)_2(H_2O)_3]_n$ (where btc^{3-} = 1,3,5-benzenetricarboxylate,[114] discussed in Section 4.11).

Fluorescence emission and UV-visible absorption spectroscopy follows a similar experimental set-up to vibrational spectroscopy.[21,105,111] The pressure cell must be optically pure to eliminate discrepancies between spectra depending upon the position of the sample in the sample chamber.[111] A bespoke set-up has been employed to examine the absorption and emission spectra of the Hf-based MOF, Hf-peb (where peb = 1,4-phenylene-bis(4-ethynylbenzoate), discussed in Section 4.13).[21] Light from a broadband lamp is passed through a bandpass filter to set the excitation wavelength, and then focussed through the windows of a Merrill–Bassett DAC using a reflective-objective lens.[111] The cell is fixed in position using a custom holder. Light transmitted through the sample is projected onto the detector for UV-visible spectroscopy by a second lens, or into the fibre optic of the detector by two additional lenses for fluorescence emission spectroscopy. Multiple crystals can be loaded in the sample chamber to increase the absorption or emission intensity. High-pressure absorption and emission spectroscopy of MOFs has only been reported for a handful of frameworks, including UiO-type (Universitetet i Oslo) frameworks,[21,105] CuPyr-I (where Pyr-I = bis(1-(4-pyridyl) butane-2,3-dione))[13] and AMU-1, $Co_2(bdc)_2(dabco)_4$ (where bdc = 1,4-benzenedicarboxylate and dabco = 1,4-diazabicyclo[2.2.2]octane)),[115] (discussed in Section 4.13).

High-pressure NMR spectroscopy is rare owing to the practical difficulties it poses.[106,116–119] Firstly, the cell must be non-magnetic, be small enough to fit within the superconducting magnet, and be capable of generating pressure of at least 10 GPa. Secondly, the cell must not shield the sample from the applied magnetic field. Merrill–Bassett DACs are suitable for NMR spectroscopy owing to their small size. The cell body, backing seats and gasket are composed from diamagnetic copper–beryllium or rhenium.[106,116,117,119] However, the sample chamber experiences very low magnetic flux owing to shielding of the magnetic field by the gasket material. A local magnetic field must therefore be generated inside the cell by inclusion of a Helmholtz coil within the cell configuration. The sample chamber itself is too small to accommodate the coil without displacing most of the analyte.[120] As a result, it is generally wound round the anvils and/or backing seat to produce a uniform magnetic field in the chamber.[106,116,117,119] Orientation of the magnetic field parallel to the gasket surface causes the field to effectively 'sink' into the sample chamber and is thus more effective for large-diameter, shallow gasket holes.[116,117] A wide sample chamber requires diamond anvils with a large culet diameter (~1 mm).[116] The magnetic field generated by the coil must be perpendicular to the static field of the cryomagnet.

The configuration of the Helmholtz coil in the cell should maximise exposure of the sample to the magnetic field of the coil whilst minimising its form factor. Smaller coils improve the sensitivity and spectral resolution of the NMR measurement.[121,122] Winding a coil round the conical anvils is space efficient but is ineffective at exposing the sample chamber to the magnetic field.[117] A split-pair configuration, whereby two coils are placed either side of the sample chamber orthogonal to the diamond anvils produces a stronger field in the sample chamber but is less space-efficient.[117] The split-pair coils are formed by first winding them around a frame, fixing their form by epoxy, and then removing the frame.

Introduction of an electromagnetic (Lenz) lens near the sample chamber can locally amplify the magnetic field, permitting the use of small coils without compromising the field experienced by the sample.[118,119] The Lenz lens serves as a magnetic flux transformer, which collects flux from a large area and deposits it over a small area (*i.e.* the sample chamber).[118,119] A torus-shaped Lenz lens wrapped in an excitation coil can be placed inside the sample chamber in a rhenium gasket.[118] The gasket is coated in alumina to insulate its contact with the lens. The sample is housed in the 230 pL chamber in the centre of lens. Amplification of the local magnetic field provides good-quality ^{1}H NMR for paraffin to 72 GPa, in spite of the small sample volume.[118] An even smaller sample volume of 100 pL can be analysed using a double-stage Lenz lens, with the excitation coil wrapped around the diamond backing seats.[119] One stage of the lens resides inside the gasket hole, as before, while the second stage is on the surface of the diamonds, as a 2 μm-thick copper coating deposited by physical vapor deposition.[119] High-pressure NMR spectroscopy of MOFs in DAC apparatus is currently unreported. However, adsorption of xenon in a series of DUT (Dresden University of Technology) MOFs has been examined by ^{129}Xe NMR spectroscopy at a gas loading pressure of ~0.002 GPa using a sapphire capillary gas rig (discussed in Section 4.10), demonstrating its utility in monitoring dynamic host–guest processes.[123,124]

4.3 Equations of State (EoS)

In undergraduate courses, EoS are introduced that relate thermodynamic state functions and include temperature (T), pressure (P), volume (V), internal energy (U), enthalpy (H), Gibbs free energy (G) and entropy (S). The simplest and most fundamental EoS is $PV = nRT$, the ideal gas law, which relates changes in volume to pressure and temperature for an ideal gas. In the compression of solids, several different equations of state are used to quantify the bulk compressibility, referred to as the bulk moduli (K or sometimes given as B). The simplest EoS for a solid is shown in eqn (4.4). Here, the variation of the volume of a solid ($\mathrm{d}V/V$) with hydrostatic pressures P, where the temperature is fixed, is termed its isothermal EoS. Hydrostatic pressures, here, is the term used to describe a special state in which all stresses are all equal (*i.e.* no shear stresses).

$$K = -V\left(\frac{\mathrm{d}P}{\mathrm{d}V}\right) \tag{4.4}$$

For MOFs, and other molecular solids, by far the most commonly used EoS to fit experimental volume *vs.* pressure data is the Birch–Murnaghan EoS.[125] For this particular EoS, how many volume derivatives you evaluate defines the 'order' of the EoS. Equations for a 1st and 3rd order Birch–Murnaghan EoS are shown in eqn (4.5) and (4.6), respectively, where V_0 is the reference volume, V is the deformed volume, K is the bulk modulus, and K' is the derivative of the bulk modulus with respect to pressure. The bulk modulus, K, and its derivative, K', are usually obtained from fits to experimental data.

$$P(V) = \frac{K}{K'}\left(\left(\frac{V_0}{V}\right)^{K'} - 1\right) \tag{4.5}$$

$$P(V) = \frac{3K_0}{2}\left[\left(\frac{V_0}{V}\right)^{\frac{7}{3}} - \left(\frac{V_0}{V}\right)^{\frac{5}{3}}\right]\left\{1 + \frac{3}{4}(K'-4)\left[\left(\frac{V_0}{V}\right)^{\frac{2}{3}} - 1\right]\right\} \tag{4.6}$$

All EoS used have assumptions, the validity of which can be judged by whether the derived EoS reproduces experimental data. It is therefore important when comparing the compressibility of solids that the same EoS is used and that the fit is sensible. For the compressibility of most organic and inorganic compounds, the bulk moduli (measured in GPa) usually lie in the range of 10–70 GPa, with ceramics and very hard solids in the 100 s of GPa. For MOFs, the bulk compressibility can vary wildly depending on how the material is compressed. For specific MOFs, this is discussed in detail in the individual sections that follow, though some common trends have been observed.

The bulk modulus depends on whether a penetrating or non-penetrating media is used. If the hydrostatic liquid is composed of molecules too large to enter the pores, direct compression occurs. In this way, the true extent to how robust the framework is to external pressure before failing can be tested. For MOFs, the bulk moduli tend to be <20 GPa before catastrophic failure of the framework occurs, and the crystalline MOF structure usually becomes amorphous. This would make MOFs relatively soft solids (see Section 1.8.1 of Chapter 1). However, when penetrating liquids are used, the bulk moduli can be in the 100s of GPa, behaving like a very hard ceramic. One noteable example which demonstrates this nicely is in the compressibility of the isostructural frameworks UiO-67 and UiO-adbc (adbc = 4,4′-azobenzene dicarboxylate, see also Section 4.5).[126] On increasing pressure in either Fluorinert® FC-70 or methanol, the bulk modulus varies between 15.2 GPa and 533 GPa, respectively. Note that this reflects whether the hydrostatic medium is penetrating (methanol) or non-penetrating (FC-70). This can be seen clearly in Figure 4.6, where the compressibility curve is almost flat for both frameworks in methanol compared to FC-70.

One other feature unique to MOFs and related to their bulk compressibility, is that on increasing pressure using a penetrating hydrostatic medium, the volume can increase with the pressure. This characteristic feature of MOFs can be seen in Figure 4.6 in the compression of both UiO frameworks in methanol. When calculating the bulk modulus of MOFs with a penetrating media, this is characterised by a negative K' value, the first derivative of the bulk modulus, and is characteristic of uptake of the hydrostatic media. This undoubtedly comes from the fact that on increasing pressure, the contents of the pores and therefore stoichiometry of the material change.

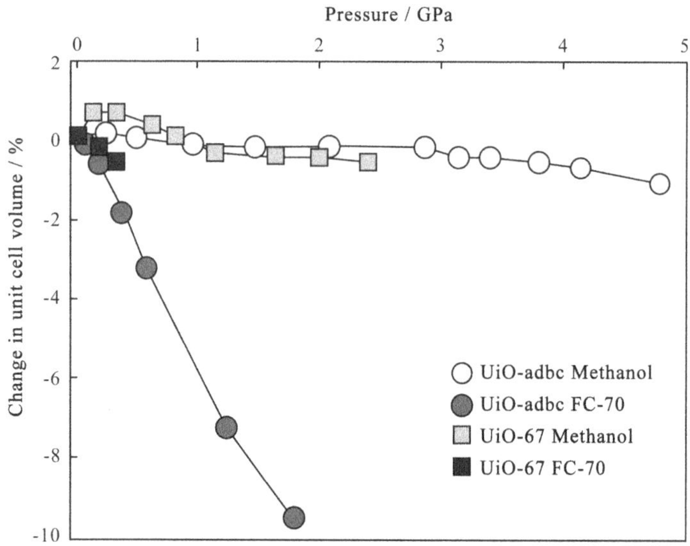

Figure 4.6 Percentage change in volume *vs.* pressure (GPa) for UiO-abdc in methanol (circles), FC-70 (diamonds), and UiO-67 in methanol (squares), FC-70 (triangles). Reproduced from ref. 126, https://doi.org/10.1002/anie.201509352, under the terms of the CC BY 4.0 license https://creativecommons.org/licenses/by/4.0/.

This phenomenon does present a unique opportunity though, as this means that the flexibility of the framework, how resistant and robust the crystalline phase is to external pressure, and how varying molecules interact with the pores, can be experimentally explored. This phenomenon was first noticed by Karena Chapman on a seminal piece of experimental work on the compression of HKUST-1 using high-pressure powder diffraction, using several different penetrating and non-penetrating hydrostatic media.[127] The negative K' value indicates that MOFs become more compressible on increasing pressure, something that is somewhat counterintuitive.

Very little experimental data have been obtained on MOFs by varying temperature and pressure, with the only two current studies found, both published on denser ZIFs (ZIF-62 and ZIF-4), where dense amorphous phases, liquid MOF phases and crystalline phases are mapped, highlighting the rich phase behaviour of these solids. These dense MOFs have bulk moduli of 4.7 GPa and 2.7 GPa for ZIF-62 and ZIF-4, respectively. However, this can vary on increasing the temperature to 600 °C and pressure to 8 GPa for ZIF-4, which undergoes four phase transitions, with bulk moduli of as high as 21.6(1) GPa reported for one of the high-temperature and pressure phases.[128]

If you are interested in general in understanding better the compression of solids and obtaining an overview of the background theory on EoSs used in quantifying the compressibility of solids, we encourage those of you interested to read the paper by Angel, Gonzalez-Platas and Alvaro.[129]

4.4 Zeolitic Imidazolate Frameworks (ZIFs)

The effect of pressure on MOFs was first investigated for ZIFs, which remain the most studied class of frameworks at high-pressure.[9,127] ZIFs are isomorphous with porous aluminosilicate zeolites. Tetrahedrally-coordinated transition-metal (*e.g.* Zn, Co or In) centres are bridged by imidazolate linkers to form a three-dimensional porous framework with an expanded zeolite network topology (*e.g.* **sod**, **lta**, **gme**). To date, approximately 70 unique native ZIF structures have been deposited in the Cambridge Structural Database (CSD).[130,131] Among them, ZIF-8, $[Zn(mIm)_2]_n$ (mIm = 2-methylimidazolate, Figure 4.7), has been the most studied at pressure.[9,127]

4.4.1 ZIF-8

The nanoporous framework of ZIF-8 forms colourless crystals in the cubic space group, $I\bar{4}3m$, with a sodalite, **sod**, framework topology. A large nanopore with a diameter of 11.6 Å ($V \sim 2500$ Å^3) centrally occupies the unit cell and is connected to symmetrically equivalent adjacent pores by eight six-membered ring (6MR) and six four-membered ring (4MR) windows with diameters of ~3.4 Å and ~0.8 Å, respectively. The narrow pore windows make ZIF-8 an effective material for gas filtration membranes.

Native ZIF-8 is compressible under applied hydrostatic pressure with an isothermal bulk modulus, K, between 6.5 GPa and 7.1 GPa from *in situ*

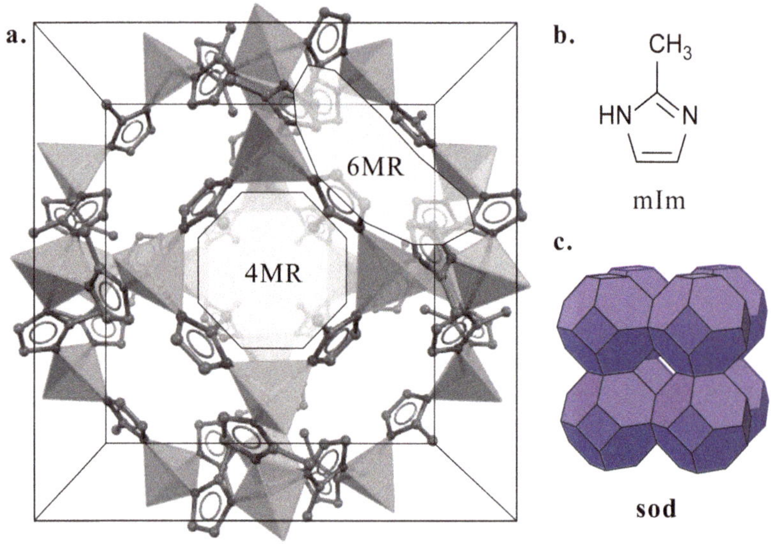

Figure 4.7 (a) Crystal structure of ZIF-8 under ambient conditions of pressure and temperature ($I\bar{4}3m$, CSD reference = 'TUDHUW').[9] The central pore is accessible from six four-membered ring windows, labelled 4MR, and eight six-membered ring windows, labelled 6MR. Hydrogen atoms are removed for clarity. (b) 2-Methylimidazole (mIm) ligand. (c) Illustration of the **sod** topology of ZIF-8.

powder X-ray diffraction.[132,133] As a polycrystalline powder, amorphisation occurs at 0.34 GPa due to inter-crystallite collisions,[132] while single crystals of ZIF-8 retain their structural integrity up to ~39 GPa with methanol as the pressure-transmitting medium.[132] Pressure-induced amorphisation is signalled by a large and irreversible decrease in the volume and porosity of the framework caused by partial collapse of the central nanopore. Recently, the amorphisation process of powdered ZIF-8 at 0.34 GPa has been re-examined by high-pressure *in situ* powder X-ray diffraction, revealing the occurrence of a reversible phase transition at 0.3 GPa, characterised by a sudden change in the unit cell volume.[133]

The amorphisation pressure is mediated by the presence of guest species into the framework pores; compression of native ZIF-8 in a non-penetrating pressure-transmitting medium of Fluorinert® FC-70 results in amorphisation at 0.34 GPa,[132] whilst ZIF-8 loaded with isopropanol, toluene, or *o*-xylene delays amorphisation to 1.15 GPa (11.5 kbar),[134] or to 7 GPa (70 kbar) loaded with argon.[135] This irreversible collapse of the pores can be exploited to trap the guest species, such as molecular iodine, to improve the storage and handling of hazardous compounds.[136] Shock-wave compression of ZIF-8 has demonstrated that the pore collapse is associated with absorption of 0.51 kJ g^{-1} of energy.[19,131] *Ex situ* Raman spectroscopy confirmed that energy dissipation occurred *via* scission of some of the zinc–imidazolate (Zn–N) bonds, leading to pore collapse with retention of the local coordination geometry.[19]

Single crystals of ZIF-8 have considerably delayed amorphisation pressures compared with its polycrystalline phase, allowing structural evolution of the framework under pressure to be followed by *in situ* single-crystal X-ray diffraction.[9] Gradual pressure-induced intrusion of guest into the framework pores from the pressure-transmitting medium prompts a structural phase transition at a critical pressure, characterised by rotation of the 2-methylimidazolate linkers to form a 'gate-opened' structure, in which the 4MR and 6MR windows partially open to connect adjacent nanopores (Figure 4.7). Intrusion of methanol into the framework pores prompts a sudden and reversible rotation of the linkers by 12° at 1.47 GPa, attributed to the increased quantity of adsorbed guest.[9] Inelastic neutron scattering has since shown that the phase transition is stimulated by adsorption of N_2 at a loading pressure of only 0.85 bar, indicating that the phase transition is driven by host–guest interactions rather than the mechanical perturbation of the framework by pressure.[137] Pressure-induced intrusion of the pressure-transmitting medium into the framework pores is common in MOFs, and is generally associated with an increase in the rigidity and structural stability of the framework due to the increased density of the guest-loaded material.[15–17] This 'gate-opening' phase transition has since been observed in powdered ZIF-8,[138] ZIF-7,[139] ZIF-90 and ZIF-65.[140] Notably the transitions were observed to be dependent upon both the crystallite size,[138] and the quantity and nature of the adsorbent.[137,141,142]

4.4.1.1 Guest Adsorption and Gate Opening in ZIF-8

Host–guest interactions between native ZIF-8 and an adsorbed guest species dictate the response of the framework to hydrostatic compression. Adsorption of weakly interacting guests, such as methane or water, stiffen and reinforce the framework,[143,144] while strongly interacting guests promote a 'gate-opening' structural phase transition.[9,140,142] The phase transition is driven by the maximisation of the host–guest interactions in the gate-opened structure.[140,142] The magnitude of the linker rotation is proportional to the strength of the host–guest interactions, which depends upon the nature and quantity of the adsorbent, and the functional group appended to the imidazolate linker.[140,142] Following the first observation of this phase transition upon pressure-induced adsorption of methanol in ZIF-8,[9] as previously discussed, the gate-opening effect has been followed experimentally by diffraction methods[137,141] and Brillouin spectroscopy,[113] and the energetic origin of this behaviour has been traced by theoretical simulations.[143,145]

Brillouin spectroscopy has been used to follow the effect of the phase transition on the acoustic phonons in ZIF-8 during hydrostatic compression in methanol and isopropanol media.[113] Only methanol is small enough to infiltrate the framework, causing an initial shift in the Brillouin peak to a higher frequency upon uptake. At 1.3 GPa, two longitudinal acoustic modes are formed, corresponding to both the ambient pressure phase (gate-closed) and the high-pressure phase (gate-opened), characteristic of a gradual phase transition. By 1.5 GPa, transformation to the gate-opened phase is complete, in agreement with earlier single-crystal X-ray diffraction results.[9]

The effect of the host–guest interaction strength on the magnitude of the imidazolate linker rotation has been examined for methanol adsorption in a series of isostructural ZIFs: ZIF-8, ZIF-90 and ZIF-65, which contain imidazolate linkers functionalised by methyl (MeIm), formyl (CHOIm) and nitro (NO_2Im) groups, respectively,[140] and in ZIF-8 using a series of liquefied gases, methane, argon, oxygen and nitrogen, as the guest species.[142] High-pressure single-crystal X-ray crystallography, and Grand Canonical Monte Carlo (GCMC) and density functional theory (DFT) simulations are used in combination to locate the position of the guest in the framework pore, and quantify the energy barrier to the imidazolate rotation. For the phase transition to occur, the adsorption of the guest must exceed the rotation barrier of the imidazolate linker.

Adsorption of methanol in ZIF-8, ZIF-90, and ZIF-65 becomes increasingly favourable across the series, lowering the energy barrier and activation pressure of the phase transition across the series (Table 4.2).[140] The stronger host–guest interaction between methanol and the formyl group in ZIF-90 compared with the methyl group in ZIF-8 increases the rotational angle of the imidazolate linker, θ (Table 4.2). A still greater rotation is expected for ZIF-65 on account of its strongly interacting nitrosyl group. However, intra-framework N···O interactions between the neighbouring NO_2Im linkers stabilise their conformation at ambient pressure, resulting in an unexpected, attenuated rotation.

For the adsorption of liquefied gases in ZIF-8, the interaction energy between the adsorbate and the methyl–imidazolate linker increases in the order of: methane, argon, nitrogen and oxygen.[142] Single-crystal X-ray crystallography and DFT adsorption energy calculations were used to locate the

Table 4.2 Activation pressure, linker rotation angle, energy barrier to linker rotation and adsorption energy of the most stable site in ZIFs upon undergoing the 'gate-opening' phase transition.[140,142]

	P/GPa	$\Delta\theta$/°[c]	ΔE_θ/kJ mol^{-1}[d]	Max. $E_{ads.}$/kJ mol^{-1}[e]
ZIF-8[a]	1.47	+12.0	+8.0	n/a
ZIF-90[a]	0.88	+57.3	+6.1	−30
ZIF-65[a]	0.73	−11.3	+5.6	−40
CH_4[b]	0.70	+21.5	+5.8	−13
Ar[b]	0.75	+21.6	n/a	−16
N_2[b]	0.21	+21.1	n/a	−21
O_2[b]	0.75	+22.1	n/a	−22

[a]Adsorption of methanol.
[b]Adsorption in ZIF-8.
[c]Rotation angle of imidazolate linker.
[d]Energy barrier to imidazolate linker rotation (ambient pressure $\theta_{ZIF\text{-}8}$ = 64.3°, $\theta_{ZIF\text{-}90}$ = 65.5°, $\theta_{ZIF\text{-}65}$ = 46.3°).
[e]Adsorption energy of the most stable site in the high-pressure, gate-opened phase.

adsorption sites of the guests in ZIF-8, and rank them by stability (Table 4.2). Six independent adsorption sites were identified for each argon, methane, nitrogen and oxygen. Nitrogen and oxygen are exceptional in that are positionally disordered. The most stable site is centralised in the larger, 6MR window, whilst the least stable site centrally occupies the nanopore. A more favourable adsorption energy is calculated for argon over methane due to its larger size, which enhances the dispersion interactions between the guest and the framework, although a similar rotation of the imidazolate linker by ~21.5° is observed in the region of 0.70 GPa to 0.75 GPa. Adsorption of oxygen is the most stable, resulting in the largest linker rotation of 22.1° at 0.75 GPa.

The adsorption energies of methanol and liquefied gases in the ambient and high-pressure phase ZIFs by DFT shows an increase in the stability of the adsorption sites by up to ~7 kJ mol^{-1} in the high-pressure phase, with the most interacting adsorbates becoming the most stabilised (Table 4.2). Stabilisation of the adsorbent in the gate-opened phase is a probable driving force for the phase transition, see Figure 4.8. Notably, maximisation of the host–guest interactions was only achieved by a forwards rotation of the imidazolate linker; simulated backwards rotation of the imidazolate linker in ZIF-8, ZIF-90 and ZIF-65 afforded higher rotation barriers (*ca.* ZIF-8 ΔE_θ = 160 kJ mol^{-1}) and less stabilised guest adsorption energies. Host–guest interactions in ZIF-8 therefore dictate both the occurrence of the phase transition and its dynamic structural properties.

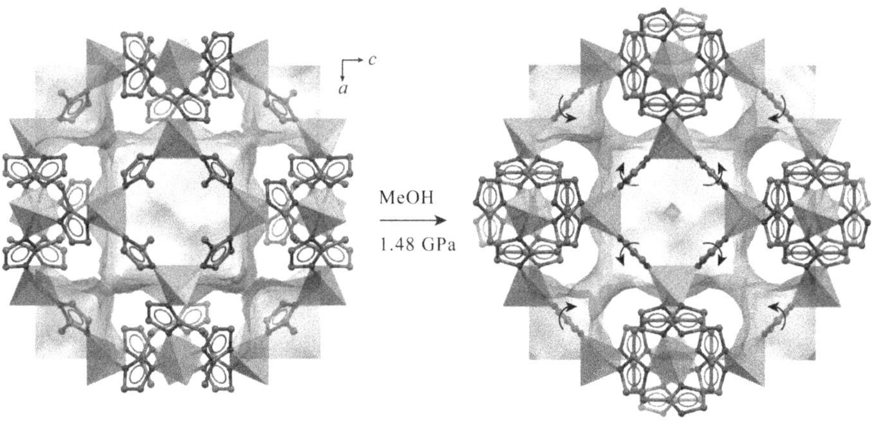

Figure 4.8 'Gate-opening' phase transition in ZIF-8 upon pressure-induced intrusion of methanol into the framework pores, characterised by a rotation of the 2-methylimidazole linker. The ambient pressure structure is from CSD reference: 'TUDHUW' and the structure at 1.48 GPa is from 'TUDJOS'.[9] The white region represents the solvent accessible volume. H-atoms are removed for clarity.

Guest-dependent 'gate-opening' is also observed in other classes of MOFs, such as the Sc-based framework, Sc_2BDC_3 (where BDC = 1,4-benzenedicaboxylate),[146] the Cu-paddlewheel based MOF with 5-(pyrimdin-5-yl)isophthalate ligands, NJU-Bai8,[10] and the Ce-based framework, MOF-76(Ce),[147] among others.[7,148]

4.4.2 Other ZIFs

Pressure-responsive structural behaviour is common among ZIFs owing to their high compressibility.[149–151] In general, the flexibility of native ZIFs increases with less sterically encumbered imidazolate-derived linkers and with increased electron density in the imidazolate ring.[102] One of the earliest high-pressure single-crystal X-ray diffraction experiments on MOFs was performed on $ZnIm_2$, $[Zn_2(imidazolate)_4]_n$, also denoted ZIF-zni.[152] Tetragonal ($I4_1cd$) crystals of $ZnIm_2$ in the α-phase undergo a rearrangement of the imidazolate bridges to form the lower symmetry β-phase ($I4_1$) during isotropic compression. The framework of ZIF-zni is less compressible than ZIF-8 ($K \sim 14$ GPa), owing to the inflexibility of the **zni** topology of the framework. Since then, various pressure-responsive behaviours, such as polymorphic phase transitions,[128,152,153] 'breathing' of the pore volume,[3,154,155] and modulation of the thermal properties, have been reported.[128]

Breathing behaviour in ZIFs was first observed in ZIF-7, $Zn(bIm)_2$ (where bIm = 2*H*-benzoimidazolate), upon adsorption and desorption cycles of carbon dioxide at gas-loading pressure.[141] Pressure-induced breathing was later observed in ZIF-4, $M(Im)_2$ ($M = Co^{II}$ or Zn^{II} and Im = imidazolate) upon hydrostatic compression to relatively low pressures between 28 and 50 MPa.[3,154,155] Purple, orthorhombic (*Pbca*) crystals of ZIF-4 form a microporous framework with a **cag** topology. Hydrostatic compression to 28 MPa for ZIF-4(Zn) or 50 MPa or ZIF-4(Co) prompts a polymorphic phase transition from *Pbca* to $P2_1/c$, concomitant with contraction of the pore volume by ~20%, from powder X-ray diffraction measurements.[3] Complete decompression of ZIF-4(Co) or heating of ZIF-4(Zn) to 130 °C returns them to the open-pore, *Pbca* phase. Temperature-dependent inelastic neutron scattering indicates that the closed-pore to open-pore transition is driven by an increase in vibrational entropy associated with rotation of the organic linkers.[155] Intuitively, the open-pore phase is more compressible than the closed-pore phase, with bulk moduli of 2.0 GPa and 4.4 GPa, respectively, according to *in situ* powder X-ray diffraction measurements using a plastic capillary pressure cell.[155]

4.5 UiO Frameworks

Isoreticular UiO-MOFs (Universitetet i Oslo) comprise hexanuclear metal nodes, $M^{III}{}_6O_4(OH)_4$ (where M = Zn, Hf, or Ce), linked by carboxylate bridging ligands. Currently, there are eleven unique native UiO-type structures in the CSD.[156] High-pressure studies have focussed on the isostructural

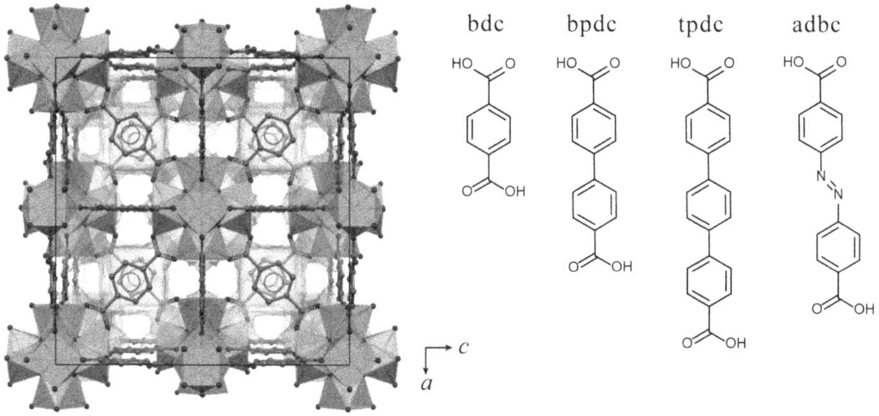

Figure 4.9 (Left) Crystal structure of UiO-66 under ambient conditions ($Fm\bar{3}m$, CSD reference = 'RUBTAK').[171] The white region represents the solvent accessible volume. H-atoms are removed for clarity. (Right) Protonated linker species in UiO-66 (1,4-benzenedicarboxylate, bdc), UiO-67 (biphenyl-4,4′-dicarboxylate, bpdc), UiO-68 (*p*-terphenyl-4,4′-dicarboxylic acid, tpdc), and UiO-adbc (4,4′-azobenzene dicarboxylate, adbc).

series UiO-66, UiO-67, UiO-68, and UiO-adbc,[126] which feature 1,4-benezendicarboxylate (bdc), biphenyl-4,4′-dicarboxylate (bpdc), *p*-terphenyl-4,4′-dicarboxylate (tpdc), and 4,4′-azobenzene dicarboxylate (adbc), respectively (Figure 4.9).

The cubic ($Fm\bar{3}m$) frameworks contain a large, central, octahedral pore that shares faces with eight smaller, tetrahedral voids. Thermal and mechanical stability is provided by the highly coordinated metal ions (MO_8) and the 12-connected paddlewheel nodes. This high connectivity limits the compressibility of UiO-frameworks, which have bulk moduli between 8 GPa and 65 GPa.[126,157–161] The combined mechanical stability and high porosity of these frameworks benefit practical applications in heterogeneous catalysis and gas storage.[162–164] The mechanical properties of UiO-frameworks can be tuned by modification of the metal or linker species,[126,157–161,165,166] or by the introduction of structural defects.[165,167–170]

Changing the metal species in the bdc-containing framework, UiO-66, $[M^{III}{}_6O_4(OH)_4(bdc)_6]_n$ (where M = Zr, Hf or Ce), alters the compressibility and stability of the material.[165] The Zr analogue is the most mechanically stable, remaining crystalline upon hydrostatic compression to 1.4 GPa at ambient temperature, whilst the Hf and Ce analogues have lower stability, becoming amorphous at 0.6 GPa and between 0.1 GPa and 0.4 GPa, respectively.[165,166] The comparative fragility of the Ce framework is ascribed to its weaker node–linker (Ce–O) coordination bonds, caused by the presence of Ce^{3+} in the material (10% Ce^{3+}, 90% Ce^{4+} according to XPS).[166] The larger ionic radius of Ce^{3+} is also likely to disrupt the structure of the Ce_6O_6 nodes, lowering the mechanical stability of the framework.

Although the Hf and Ce analogues of UiO-66 exhibit inferior mechanical properties compared with the Zr framework, they possess enhanced catalytic activity.[172] Combining Zr with Hf or Ce in bimetallic UiO-66(Hf : Zr) and UiO-66(Ce : Zr) imparts both catalytic activity and mechanical stability to the material.[165] The properties of the framework can be modified by adjusting the Zr to Hf/Ce ratio.[165] Increasing the Hf or Ce content to 50% (Zr_3Hf_3 or Zr_3Ce_3) decreases its mechanical stability by introducing structural defects to the framework, which lower the onset pressure of amorphisation from 1.4 GPa (Zr_6) to 0.6 GPa and 0.2 GPa for Zr_3Hf_3 and Zr_3Ce_3, respectively.[165]

Defect engineering is commonly used to control the mechanical properties and gas adsorption properties of UiO frameworks.[162,167] While UiO frameworks exhibit superior thermal and mechanical stability compared with other MOF subclasses, they have comparatively modest porosity and flexibility. Addition of an acidic modulator during the solvothermal synthesis introduces structural defects in the framework, either as 'missing linker' defects, or 'missing cluster' defects, which increase the porosity, compressibility, and chemical functionality of the framework.[165,167–170] Missing linker defects involve exchange of the bidentate ligand for a terminal capping ligand, such as acetate[162] or benzoate,[169] while missing cluster defects create a charge imbalance that is compensated by anions from the modulator.[168,170]

The number of defects is proportional to the number of stoichiometric equivalents of modulator used. In general, introducing defects to UiO-66 increases its compressibility, yet decreases its mechanical stability.[167] However, this relationship is non-linear since the addition of missing linker sites decreases the connectivity of the Zr nodes and thus changes the topology of the framework, complicating the structure–property relationships. For example, UiO-66 with 5 equivalents of modulator is more compressible than when 10 equivalents are used, since the topology of the framework transitions from 12-connected **fcu** to 8-connected **reo**.[167] The bulk moduli and amorphisation limits of a series of pristine and defective UiO-66 frameworks are given in Table 4.3.[167] Reported values for the bulk modulus of 'defect-free' UiO-66 show a large discrepancy, falling between 17 GPa and 38 GPa,[161,166,167] due to differences in synthetic routes in which defects may be introduced, unintentionally.

The mechanical properties of UiO frameworks are also dependent upon the length, conformational flexibility, and chemical functionality of the organic linkers.[126,157–161,173] In general, long, conformationally flexible linkers increase the compressibility of the framework at the expense of structural resilience.[160,173] The length of the linker increases across the series, UiO-66, UiO-67 and UiO-68, which feature 1,4-benzenedicarboxylate, biphenyl-4,4′-dicarboxylate and *p*-terphenyl-4,4″-dicarboxylate, respectively, associated with a decreasing bulk modulus from 22 GPa to 13 GPa to 8 GPa, and a decrease in the amorphisation pressure from 1.83 GPa to 0.45 GPa to 0.20 GPa, respectively.[160] Similarly, replacing the bdc^{2-} linkers of UiO-66 for conformationally flexible *trans*-1,4-cyclohexanedicarboxylate

Table 4.3 Mechanical properties of UiO frameworks under hydrostatic pressure in a non-penetrating pressure-transmitting medium, with varying metal ions, linker functionalisation, and equivalents (#eq) of modulator added during the synthesis.[167]

Material	Bulk modulus, *K*/GPa	Onset of amorphisation/GPa
UiO-66(Zr)	17,[a] 22,[d] 26,[b] 38[c]	1.4,[b] 1.83[d]
UiO-66(Hf)	37[c]	~1[c]
UiO-66(Ce)	17[c]	0.1–0.4[c]
UiO-66(Zr_3Hf_3)	n/a	0.6[e]
UiO-66(Zr_3Ce_3)	n/a	0.2[e]
UiO-66(Zr)-1eq	18.3[b]	n/a
UiO-66(Zr)-5eq	12.2[b]	n/a
UiO-66(Zr)-10eq	13.9[b]	n/a
UiO-66(Zr)-NH_2	25[a]	>2[a]
UiO-67(Zr)	13.3[d]	0.45[d]
UiO-68(Zr)	8.1[d]	0.2[d]
UiO-adbc	16.8[f]	1.3[f]

[a]Yot *et al.*[161]
[b]Dissegna *et al.*[167]
[c]Redfern *et al.*[166]
[d]Rogge *et al.*[160]
[e]Rogge *et al.*[165]
[f]Hobday *et al.*[126]

(cdc), denoted Zr-CDC, lowers the mechanical stability of the framework, while facilitating flexible breathing behaviour.[173]

However, a degree of linker flexibility is required to allow the framework to contract under applied pressure without losing its crystallinity.[126] High-pressure single-crystal X-ray diffraction coupled with molecular dynamics calculations of UiO-67 and UiO-adbc indicate that $bpdc^{2-}$ and $adbc^{2-}$ linkers bend by about 3° and 7°, respectively under an applied pressure of 1 GPa in a hydrostatic medium of Fluorinert® FC-70.[126] The larger degree of bending in UiO-adbc increases the compressibility of the framework compared with UiO-67, with respective *K* values of 16.8 GPa and 17.4 GPa, translating to a higher pressure of amorphisation of 1.8 GPa for UiO-adb, compared to 0.3 GPa for UiO-67.[126] Mechanical stabilisation is improved in both frameworks upon compression in a hydrostatic medium of methanol, which infiltrates the framework pores, delaying amorphisation to 2.4 GPa for UiO-67 and 4.8 GPa for UiO-adbc.[126]

The structural resilience of the framework can also be improved through the introduction of intra-framework interactions by functionalisation of the organic linker.[161] Hydrostatic compression of UiO-66 and its amine functionalised derivative, UiO-66-NH_2, yields respective bulk moduli of 17–25 GPa. Neither framework becomes amorphous in the measured pressure range up to 2 GPa, although the crystallinity of both gradually decreases during compression, with UiO-66-NH_2 maintaining a greater degree of crystallinity than UiO-66 at high pressure.[161] Resilience is provided by the formation of intra-framework hydrogen bonding interactions between the amino functional group and the O atom of the carboxylate in UiO-66-NH_2.

4.6 MIL Frameworks

Matériaux de l'Institut Lavoisier (MIL) MOFs are constructed from polymeric chains of corner-sharing metal-oxide octahedra, $M^{III}O_4(OH)_2$ (where M^{III} = Te, Al, Ga, In or Sc), which are linked by 1,4-benzenedicarboxylate (bdc) ligands to form a 'wine-rack' structure with a **cds** topology (Figure 4.10). One-dimensional, rhombic channels perforate the framework. The surfaces of the channels are decorated with bridging hydroxyl groups, μ_2-OH, from the metal–organic chains that stabilise the adsorption of polar guests, while the channels walls are formed from the phenyl rings of bdc, stabilising aromatic and non-polar guests. To date, approximately 60 unique native MIL-type structures have been deposited in the CSD,[130] all of which are derived from powder X-ray diffraction measurements owing to the difficulty in growing large single crystals. High-pressure structural analysis has focussed on the three-dimensional, microporous frameworks, MIL-53, $[Al^{III}(OH)(bdc)]\cdot 0.70(bdcH_2)$, and MIL-47, $[V^{III}(OH)(bdc)]\cdot 0.75(bdcH_2)$, which crystallise in the orthorhombic space groups, *Imma* and *Pnma*, respectively. Under ambient conditions, MIL-53 and MIL-47 are penetrated by channels with cross-sections of 7.3 Å × 7.7 Å and 7.9 Å × 12.0 Å, respectively, which house disordered 1,4-benzenedicarboxylic acid ($bdcH_2$) molecules in partially occupied adsorption sites.[174–177]

The elastic properties of MIL-frameworks are highly anisotropic on account of the network topology.[179] In grid-like networks with **cds**, **bik**, **crd**, **dia**, **gis**

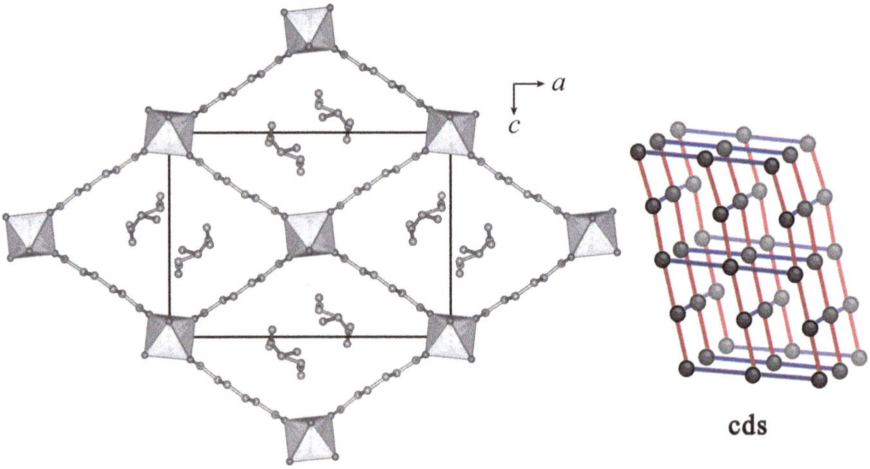

Figure 4.10 (Left) Crystal structure of MIL-53 under ambient conditions (*Pnma*, CSD reference = 'MINVOU').[178] H atoms are removed for clarity. (Right) Illustration of the **cds** topology of MIL-53.

and **mog** topologies, metal nodes are located at the intersecting points of the linker struts.[180] Under hydrostatic pressure, hinging motion of the linkers about the metal nodes imparts flexibility to the framework, affording low bulk moduli between 3 and 12 GPa and high onset pressures of amorphisation, typically exceeding 6 GPa.[181–183]

In wine-rack structures, the connectivity is such that orthogonal axes in the plane of the wine-rack are interdependent, with horizontal compression across the network prompting vertical extension, giving rise to negative linear compressibility (NLC, see Section 2.3.1 of Chapter 2), in which one or more axis of the unit cell *lengthens* under applied pressure. The softest direction corresponds to the diagonal of the rhombic channels, and the hardest direction coincides with the direction of the metal–linker chains.[177,180,184,185] Pronounced compression of the wine-rack is accommodated by narrowing of the rhombic channels, causing volumetric contraction of the pore volume by up to ~70%.[4,186–188] The expanded structure is returned upon releasing the pressure, corresponding to breathing behaviour. Breathing transitions and NLC benefit numerous practical applications, such as pressure sensors, shock-absorbing materials (Chapter 5), and actuators.

4.6.1 Breathing

Topological distortion results from rotation of the metal nodes or hinging motion of the organic linkers under hydrostatic pressure. Expansion and contraction of the wine-rack structure is associated with large volumetric changes in the pore and unit cell of the framework, characteristic of breathing behaviour.[186] Early reports of breathing behaviour in MIL frameworks observed that hydration and dehydration of the material stimulates pore expansion and contraction, respectively.[177,187] Adsorption and desorption of CO_2 was later found to induce the same effect.[188] The large-pore to narrow-pore phase transition is enabled by the rotation of phenyl rings of the bdc linkers about the C–C bond between the phenyl and carboxylate moieties.[4]

The first MIL-type MOF to be examined at high hydrostatic pressure was MIL-53, which undergoes a reversible breathing transition upon compression by mercury porosimetry.[189] Compression of powdered MIL-53(Al) and MIL-53(Cr) up to 55 MPa prompts a hysteretic phase transition from the large-pore *Imma* form to a dense *C*2/*c* phase with narrow pores, resulting in a decrease in the unit cell volume from 1490 Å^3 to 1020 Å^3. During a complete cycle of mercury intrusion and extrusion, the framework absorbs 12 J g^{-1} of energy, which is larger than commercially available aluminium foams (*ca.* 10 J g^{-1} at a mass density of 40 kg m^{-3}). Parallel behaviour is observed for powdered MIL-53(Al) upon hydrostatic compression in mineral oil, in which the large-pore phase (V = 1433 Å^3) is compressed to the narrow-pore phase (V = 1000 Å^3) at 2 GPa.[12] First-principles calculations for the breathing transition in analogous MIL-53(Cr) determine an energy barrier for the narrow-pore to large-pore transition of only 3 kJ mol^{-1} to 6 kJ mol^{-1}, accounting for the reversibility of the transition.[190] The large-pore phase is stabilised by an increase in vibrational entropy of the system, while the narrow-pore phase is stabilised by intra-framework van der Waals interactions between the phenyl rings of the bdc linkers.[190,191]

The magnitude of the breathing and the energy associated with the transition depends upon the hinging flexibility of framework. More rigid frameworks delay the large-pore to narrow-pore transition to a higher pressure, increasing the work energy.[5] For the Al-based derivative of MIL-53 with 2,5-thiophenedicarboxylate (tdc) linkers, denoted Al-MIL-53-tdc, $[Al(OH)(tdc)]_n$,[5] mercury intrusion and high-pressure powder X-ray diffraction experiments showed a reversible contraction of the unit cell volume from 1413 Å^3 to 1009 Å^3 at 0.275 GPa, with an associated work energy of 79 J g^{-1}. Rigidity may be introduced by using extended linkers, such as 4,4′-(ethyne-1,2-diyl)dibenzoate (edb), which forms a doubly interpenetrated framework with a MIL-53 topology, denoted GUF-1 (Glasgow University Framework, $[Sc^{III}(OH)(edb)]_n \cdot 2n$DMF]).[192] Interpenetration restricts the hinging motion of the edb linkers, leading to very subtle breathing upon guest exchange.

Pressure-induced breathing also occurs in MIL-47, which is isostructural to MIL-53.[4] Hydrostatic compression of the ambient pressure *Pmna* phase of MIL-47(V) to 0.18 GPa prompts a partial and reversible large-pore to narrow-pore transition, associated with contraction of the unit cell volume from 1528 Å^3 to 947 Å^3 and a phase transition to space group *C*2/*c*. The transition is hysteretic and separated by small energetic barriers of approximately 10 kJ mol^{-1} for the large-pore to narrow-pore transition, and 6 kJ mol^{-1} for the reverse, and is driven by free rotation of the bdc phenyl ring.[4] Molecular dynamic simulations and DFT replicate these results, with the formation of the narrow-pore phase occurring at 0.125 GPa, and the reverse at 0.056 GPa.[193] Symmetry-breaking during the breathing transition results from displacement of the V^{3+} nodes orthogonal to the direction of the metal–organic chains, which stabilises the narrow-pore structure by 20 kJ mol^{-1}.[193]

The mechanical properties and breathing transition are affected by the metal ion species in the nodes.[181] Hydrostatic compression of hydrated powders of MIL-53(M), where M is Al^{3+}, Cr^{3+}, Fe^{3+} or Sc^{3+}, up to 2 GPa show distinct breathing behaviours due to the increasing ionic radius of the metal (Table 4.4).[181] The Al, Cr and Fe analogues undergo a phase transition from a narrow-pore form (*C*2/*c*) to a very-narrow-pore form ($P\bar{1}$) at 1.36 GPa, 0.5 GPa and 0.4 GPa, respectively, associated with a decrease in the unit cell volume by ~50 Å^3. The framework with the most contracted metal, MIL-53(Al), forms the smallest pores, which allows formation of intra-framework π–π interactions between phenyl groups between the channel walls, leading to mechanical stabilisation of the narrow-pore (*C*2/*c*) phase over a larger pressure range than for either MIL-53(Cr) or MIL-53(Fe). This is also reflected in the comparative incompressibility of the material, with bulk moduli for the narrow-pore phase of 10.7 GPa for MIL-53(Al), 10.1 GPa for MIL-53(Fe) and 4.3 GPa for MIL-53(Cr). Breathing in the Cr and Fe analogues is generally similar owing to their comparable ionic radii, although the transition is more gradual for MIL-53(Fe). By contrast, the framework with the largest metal, MIL-53(Sc),

Table 4.4 Mechanical properties of breathing MIL-frameworks under hydrostatic pressure, with varying metal ions and linker functionalisation. 'Form' refers to the large-pore (lp), narrow-pore (np), or very-narrow-pore (vnp) phases. Where applicable, ionic radii are given for the low-spin state.

Material	*P*/GPa	Space group	Form	*K*/GPa
MIL-53(Cr)[a]	0	*C*2/*c*	np	4.3
Cr^{3+} $r = 75.5$ pm	0.5	$P\bar{1}$	vnp	3.4
MIL-53(Al)[a]	0	*C*2/*c*	np	10.7
Al^{3+} $r = 67.5$ pm	1.36	$P\bar{1}$	vnp	n/a
MIL-53(Fe)[a]	0	*C*2/*c*	np	10.1
Fe^{3+} $r = 78.5$ pm	0.4	$P\bar{1}$	vnp	7.4
MIL-53(Sc)[a]	0	$P\bar{1}$	vnp	10.3
Sc^{3+} $r = 88.5$ pm		$P\bar{1}$	vnp	
MIL-53(Cr)-Cl[a]	0	*C*2/*c*	np	12.4
	1.21	*C*2/*c*	np	
MIL-53(Cr)-CH_3[a]	0	*C*2/*c*	np	3.7
	2.54	*C*2/*c*	np	
MIL-53(Cr)-NO_2[a]	0	*C*2/*c*	np	7.3
	2.84	*C*2/*c*	np	
MIL-53(Al)-NH_2[c]	0	*Imma*	lp	7.4
	2.0	*Imma*	np	71
MIL-47(V)[b]	0	*Pnma*	lp	6.09
	0.125	*C*2/*c*	np	2.84

[a]Yot *et al.*[181]
[b]Theoretical, Vanpoucke *et al.*[200]
[c]Serra-Crespo *et al.*[12]

exists in the very-narrow-pore form under ambient conditions, and persists in this phase to at least 2 GPa.

Addition of functional head groups to the bdc linkers may also modify the breathing response of the framework under pressure.[181] In hydrated MIL-53(Cr), appendage of chloro, methyl, or nitrosyl groups to the bdc linker (MIL-53-X, where X is H, Cl, CH_3 or NO_2) 'switches off' the breathing behaviour by introducing steric bulk to the channel surface, which destabilises the less voluminous, very-narrow-pore phase.[181] The frameworks are instead gradually compressed to 2 GPa. The order of framework compressibility is informative of an interplay between steric bulk and intra-framework interactions, and is as follows: X = CH_3 > H > NO_2 > Cl. The low compressibility of the chloro- and nitrosyl-functionalised frameworks originates from an intra-framework hydrogen bond between the pendant functional group on bdc and the μ_2-OH group at the channel surface, which is absent in the methyl-functionalised and pristine materials. Unusually, the nitrosyl derivative is more compressible than the chloro-material, in spite of its greater steric hindrance, which is mirrored in the methyl and pristine analogues. This is attributed to the expanded structures of the nitrosyl and methyl derivatives under ambient conditions.[181]

Functionalisation of MIL-47(V) with bromo or trifluoromethyl pendant groups, in the form of MIL-47(V)-Br and MIL-47(V)-CF_3, draws a distinct structural response.[181] Unlike MIL-53 derivatives, sterically bulky groups do not inhibit the breathing transition in MIL-47, since the framework adopts an expanded, large-pore structure under ambient conditions. However, steric effects remain important in tuning the breathing transition. Mercury intrusion in MIL-47(V)-Br prompts a large-pore (*Pmcn* $V = 1481$ Å^3) to narrow-pore transition ($C2/c$ $V = 1126$ Å^3) at 0.085 GPa, associated with a work energy of 20 J g^{-1}.[181] Increasing the bulkiness of the pendant group to MIL-47(V)-CF_3 delays the transition to between 0.75 GPa and 0.85 GPa, resulting in larger volumetric contraction from 1600 Å^3 to 1145 Å^3 and pronounced increase in the work energy to between 160 J g^{-1} and 190 J g^{-1}.[181]

Guest-inclusion in MIL frameworks can also modulate the pressure-induced breathing regime.[12,175] Breathing may occur under ambient pressure conditions upon the reversible adsorption and desorption of guest due to modification of the host–guest interactions in the system and changes to the density of the material.[177,187,188,194,195] Pressure-induced adsorption of guest from the pressure-transmitting medium into the framework channels can therefore affect the guest-mediated breathing. The framework does not necessarily require evacuation prior to guest loading; in MIL-47 loaded with $bdcH_2$, hydrostatic compression in methanol or water promotes guest exchange at 0.3 GPa or 1.0 GPa, respectively.[175]

The magnitude and activation pressure of the breathing transition is dependent upon the size, quantity, and nature of the guest. Sterically bulky or densely-packed guest in the framework channels may ‘switch-off’ breathing by inhibiting contraction of the framework,[12] similar to materials with bulky pendant groups. For instance, compression of the amino-functionalised framework, MIL-53(Al)-NH_2, in a non-penetrating pressure-transmitting medium of mineral oil promotes a large-pore to narrow-pore transition at 2 GPa, while breathing is inhibited when ethanol is used as the hydrostatic medium, according to powder X-ray diffraction measurements.[12] Intrusion of ethanol into the framework channels during compression leads to the formation of a dense, incompressible phase with a bulk modulus of 85 GPa, while mineral oil is largely prevented from entering the pores, exhibiting a low bulk modulus of 7.4 GPa for the large-pore phase and 71 GPa for the narrow-pore phase.[12]

In some cases, adsorption of guests that interact strongly with the host framework may actually stabilise the narrow-pore phase, promoting breathing.[176,196] For example, adsorption of *n*-hexane or *n*-nonane in MIL-53(Cr) leads to the formation of the narrow-pore phase under ambient pressure due to the addition of host–guest CH–π interactions.[176] This transition does not occur in MIL-47(V) owing to its relative inflexibility.[176] Similarly, DFT indicates that adsorption of H_2O in MIL-53(Al) stabilises the narrow-pore phase through the introduction of host–guest hydrogen bonds.[196]

Guest-loaded frameworks possess distinct mechanical properties compared with their native analogues, leading to complicated breathing regimes and guest-mediated structural behaviour.[175,197] Infiltration of methanol in MIL-47(V) leads to the formation of an additional high-pressure phase during isotropic compression up to 2.3 GPa.[175] The primitive orthorhombic (*Pmna*) ambient phase becomes an expanded phase with body-centred symmetry (*Imcm*) at 0.3 GPa. The transition is accompanied by an increase in the volume of the unit cell by 6.3%, which remains constant to 1.5 GPa, although the framework channels become flattened. Further compression to 1.9 GPa affords a second expanded phase in which the flattened channels re-open, corresponding to breathing behaviour. When water is used as the guest species, only a single high-pressure is observed. The reversible narrow pore to large pore transition begins at 0.7 GPa and is complete by 1.4 GPa. The large pore phase is retained to 2.9 GPa before the framework becomes amorphous.

As is typical for MOFs, structural resilience is improved by pressure-induced intrusion of guest into the framework pores.[197] For example, hydrostatic compression of the mesoporous framework, MIL-101, $[Cr_3X(H_2O)_2O(bdc)_3]_n$ (where X = F or OH), which features quasi-spherical pores with a cross-section of 29 Å × 34 Å, in a solid pressure-transmitting medium of NaCl results in amorphisation of the framework at 0.4 GPa, from powder X-ray diffraction measurements.[197] However, amorphisation is delayed to 7 GPa in a medium of polydimethylsiloxane silicone oil, indicative of guest uptake.[197] This is surprising, considering the large size of the pressure-transmitting medium.

In MIL-53 and MIL-47, and their functionalised or guest-included derivatives, breathing occurs about a single hinge motif. However, incorporation of a dual hinge pivot motif in the Co or Mg-based MOF, CUK-1 (Cambridge University-Korean Research Institute of Chemical Technology) which is composed of 2,4-pyridinedicarboxylate and 6-mercapto-3-pyridinecarboxylate ligands in a wine-rack structure, facilitates a spring-back mechanical breathing distortion.[198,199] Upon compression of 0.5 GPa, the framework abruptly contracts from a large-pore form to a narrow-pore form, with NLC along the crystallographic *a*-axis.

4.6.2 Negative Linear Compressibility (NLC)

The volume of a MOF decreases under applied hydrostatic pressure when a non-penetrating hydrostatic medium is used. However, for structures with anisotropic elastic properties, the material may expand in one or more direction during compression, providing a positive bulk compressibility (β_V) but at least one direction of NLC (β_l, see Section 2.3.1 of Chapter 2). MIL frameworks with a trellis-like structure commonly exhibit NLC along the direction of the vertical or horizontal diameter of their rhombic channels. NLC has been reported for a number of MOF systems, including $[NH_4][Zn(HCOO)_3]$,[11] ZAGs,[201] Ag(meIm),[202] and metallocyanide frameworks,[203] and usually relies upon a topological distortion of wine-rack, honeycomb, scissor and butterfly network structures. However, it may also occur *via* local structural mechanisms, such as torsional flexibility about metal centres,[204] extension along the direction of empty channels,[205] or slippage in layered frameworks.[206]

For MIL frameworks, NLC was first reported in MIL-53(Al).[12,184] Compression of the framework in a hydrostatic medium of mineral oil up to 3

GPa yielded NLC of $\beta_l = -28$ TPa^{-1} along the *b*-axis of the unit cell, which is among the largest reported for any framework material (*ca.* $\beta_l = -45$ TPa^{-1} for $Zn[Au(CN)_2]_2$[75] and $\beta_l = -12$ TPa^{-1} for $KMn[Ag(CN)_2]_3$).[207] Above 3 GPa, the pressure is sufficient to override the NLC, causing the unit cell to contract in all directions before amorphisation at 6 GPa. Equivalent behaviour is exhibited by the amino-functionalised framework, NH_2-MIL-53(Al), with $\beta_l = -27$ TPa^{-1} along the *b*-axis up to 2 GPa.[12,208] In both cases, the NLC arises from hinging motion of the linkers in the wine-rack net.

The magnitude of the NLC is dependent upon the topological and local flexibility of the framework.[206,209,210] Large NLC generally requires short, rigid linkers with flexible metal coordination centres that permit hinging motion about nodes without deformation of the network struts (linkers). Modification of the metal ion affects the bulk compressibility of the material, altering the NLC. For example, replacing the Zr^{IV} metal centre in MFM-133(M) ($[M_6(\mu_3\text{-}OH)_8(OH)_8(tchb)_8]$ (where M = Hf^{IV} or Zr^{IV} and tchb =3,3′,5,5′-tetrakis(4-carboxyphenyl)-2,2′,4,4′,6,6′-hexamethyl-1,1′-biphenyl)) for Hf^{IV} decreases the compressibility of the framework, leading to attenuated NLC along the channel diameter (*c*-axis) from $\beta_l = -5.1$ TPa^{-1} for Zr^{IV} and $\beta_l = -7.9$ TPa^{-1} for the Hf^{IV} analogue.[206,209,210]

Modification of the linker species also changes the NLC properties. Rigid linkers provide a greater degree of interdependence of the orthogonal linear compressibility of the wine-rack network. This interdependency is further enhanced by interpenetration of the flexible networks. For example, the β-quartz-like framework, $InH(bdc)_2$[206,210] and the Mn-based 3-(pyridine-4-yl) benzoate (pba) framework, MCF-34 (MCF = metal carboxylate framework), $[Mn(pba)_2]_n$,[211] which are constructed from interpenetrated wine-rack subnets with rigid linkers, both exhibit large NLC along the channel diameter of $\beta_l = -62.4$ TPa^{-1} and $\beta_l = -47.3$ TPa^{-1}, respectively. Compared with MCF-34, the shorter, more rigid bdc^{2-} linkers of $InH(bdc)_2$ promote larger NLC.

Frameworks with conformationally flexible linkers exhibit more subtle NLC but allow the elastic properties of the framework to be controlled by pressure–guest-mediated conformational isomerism. For instance, the Al-based 1,4-cyclohexanedicarboxylate (cdc) framework, Al-CAU-13 (CAU = Christian Albrecht University), $[Al_2(cdc)_3]_n$, which shares its topology with MIL-88, shows NLC along the *b*-axis only after the framework is dehydrated and filled with tetramethylpyrazine guest, which causes the cdc linkers to isomerise from an axial–axial configuration to an equatorial–equatorial configuration to adopt a more open wine-rack structure.[212] The NLC is mediated by the pressure-transmitting medium; compression of Al-CAU-13 in silicone oil up to 4.19 GPa causes the *b*-axis to extend by +2.1%, while compression in He up to 11 GPa eliminates the NLC due to intrusion of He into the channels, which decreases the framework compressibility ($K = 8.9$ GPa in silicone oil, $K = 25.1$ GPa in He).[212] Inclusion of He in the framework at high-pressure stiffens the material along the channel direction (*a*-axis), resulting in near zero linear compressibility (ZLC) in this direction.

Rarely, topological distortion of the framework competes with distortion of local features to afford ZLC.[213] This was demonstrated in a lanthanide-based framework with 1,4,5,8-naphthalene tetradentate carboxylate (ntc) linkers, MIL-122, in which the *a*-axis maintains a constant length during compression up to 10.5 GPa, as a result of simultaneous topological expansion along the *a*-axis direction with a counteractive shortening of the Ln–ntc–Ln distance.[213]

4.7 Sc_2bdc_3

The three-dimensional, nanoporous framework of Sc_2bdc_3, $[Sc^{III}_2(bdc)_3]_n$, crystallises as colourless shards in the orthorhombic space group, *Fddd*.[214,215] Octahedral ScO_6 nodes are connected by bdc linkers to form a wine-rack-type structure with the bdc linkers cross-linking the channels (Figure 4.11). Narrow, triangular channels, approximately 4 Å in diameter, perforate the framework and are extended in three-dimensions by apertures between the bdc linkers. The channels are vacant upon crystallisation, and are a suitable size for molecular sieving applications. The bdc linkers occupy two symmetry-independent sites, and are denoted as Group 1 and Group 2. Linkers in Group 1 are positioned with the benzyl ring on an inversion centre, while linkers in Group 2 are bisected by two orthogonal two-fold rotation axes. High-pressure analysis of Sc_2bdc_3 is restricted to the guest-loaded structure, since the evacuated framework is structurally unstable near the initial loading pressure of a DAC.[216]

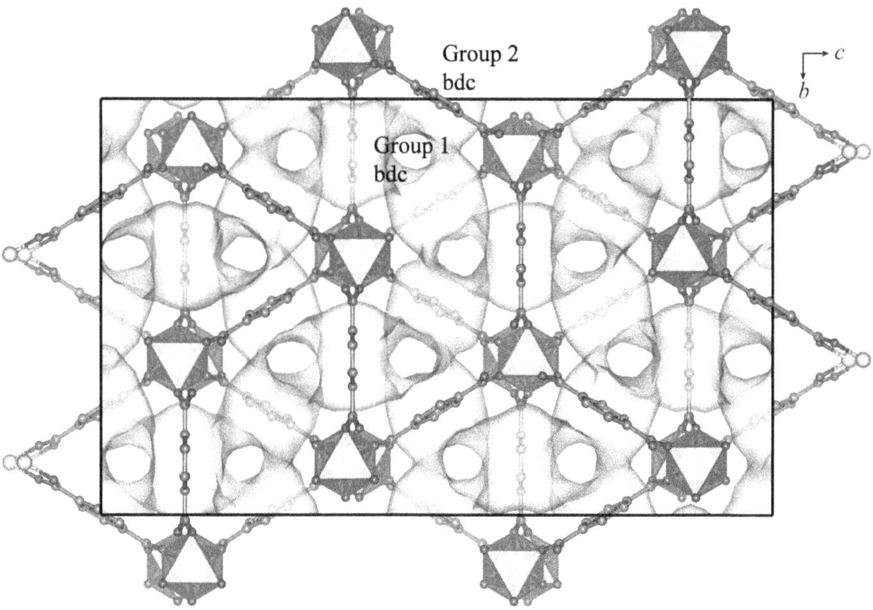

Figure 4.11 Crystal structure of Sc_2bdc_3 under ambient conditions (*Fddd*, CSD reference = 'HAZGOF'),[217] with the Group 1 and Group 2 bdc^{2-} linkers labelled. The white region represents the solvent accessible volume. H atoms are removed for clarity.

4.7.1 Guest Intrusion at High Pressure

The evacuated framework of Sc_2bdc_3 is brittle, becoming reversibly amorphous at approximately 0.4 GPa of hydrostatic pressure in a non-penetrating medium of Fluorinert® FC-77.[216] Inclusion of guest in the framework channels imparts structural resilience to the material and, depending upon the nature and quantity of the guest species, can promote or inhibit polymorphic and gate-opening phase transitions.[146,215,216,218,219]

Upon compression of a single crystal of Sc_2bdc_3 in a penetrating medium of methanol, the onset pressure of amorphisation is delayed from 0.4 GPa to at least 3 GPa.[216] Pressure causes gradual infiltration of methanol into the channels, which can be followed directly by single-crystal X-ray diffraction measurements. The triangular geometry and narrow diameter of the channels creates a confined space conducive to the formation of host–guest interactions, leading to adsorption of guest in ordered sites.[215,216] Soaking the crystal in methanol under ambient conditions promotes adsorption of methanol in two distinct, partially-occupied sites along the triangular channels, forming a stacking motif.[216] Under an applied pressure of 1.6 GPa, the channels reach their capacity, with the occupancy of the respective adsorption sites increasing from 50% and 39% to 100% and 80% between ambient pressure and 1.6 GPa, whereafter the methanol content remains constant up to 2.3 GPa.[216] The framework retains its *Fddd* symmetry in the measured pressure range (10^5 Pa–3 GPa).

Modification of the host–guest interactions can promote structural changes in MOFs, leading to guest-mediated phase transitions.[146,215,218,219] Adsorption of CO_2 in Sc_2bdc_3 at a gas loading pressure of only 1 bar (10^{-4} GPa) promotes a host–guest hydrogen bonding interaction between the O atom of CO_2 and an H atom of a bdc linker (O···H = 2.87–2.98 Å), resulting in a phase transition from *Fddd* to *C2/c* caused by tilting of the bdc linker.[215] The transition creates two symmetry-independent channels, one of which contains one fully-occupied CO_2 adsorption site, while the other contains two partially occupied adsorption sites, which have an overall adsorption energy of 20 kJ mol^{-1}. Adsorption of CO_2 in Sc_2bdc_3 at high pressure is achieved by using a hydrostatic medium of either liquefied gas[218] or CO_2 dissolved in Fluorinert® FC-70.[219] Upon compression to 0.2 GPa, an additional adsorption site is modelled in the crystal structure, concomitant with disordering of the Group 1 bdc ligands, leading to an increased maximum adsorption capacity of 6.2 mmol g^{-1}[218] (*ca.* 3.4 mmol g^{-1} at 1 bar).[215] The third site is located in the same channel as the ordered site, and is disordered across two orientations about a two-fold rotation axis.[218] The *C2/c* symmetry of the framework is retained in the hyperfilled phase.

More weakly interacting adsorbents, such as CH_4, require high pressure to stimulate considerable uptake into the channels of Sc_2bdc_3.[215,218] Dosing between 1 bar and 50 bar introduces CH_4 in two, low-occupancy sites in the channel in the *Fddd* phase, corresponding to a loading in the range of 0.1–0.5 mmol g^{-1}.[215] Dispersion interactions stabilise the CH_4 guest, with an adsorption energy between 17 kJ mol^{-1} and 14 kJ mol^{-1}, as measured by calorimetry.[215] Compression of the framework in liquefied CH_4 between 0.3 GPa and 2.5 GPa promotes hyperfilling, affording almost full occupancy of the adsorption sites.[218] Gradual intrusion of CH_4 into the channels during compression increases the libration of the bdc linkers and causes adjacent Sc_6O nodes to rotate in opposite directions, resulting in a phase transition from *Fddd* to *Fdd*2 at 1.3 GPa. Further rotation of the ScO_6 units upon compression to 2.5 GPa results in a symmetry-forbidden phase transition to a dense $P2_1/c$ phase, which must proceed *via* an unobserved intermediate.[218] Contraction of the pores in the $P2_1/c$ phase is associated with the desorption of CH_4.[218]

Bulky guest species that are too large to enter the narrow channels under ambient conditions can be forced into the pores under applied hydrostatic pressure, resulting in a gate-opening rotation of the bdc linkers.[146] Compression of a single crystal of Sc_2bdc_3 in linear and branched alkanes (C_3–C_7), which have kinetic diameters exceeding the framework channel diameter ($d_{kinetic}$ = 4.3 Å to 6.2 Å), are adsorbed between 0.2 GPa and 0.7 GPa, causing the Group 1 bdc linkers to rotate by between 10.9° and 51.4°.[146] The magnitude of the rotation is proportional to the quantity of the guest adsorbed, with linear alkanes (*n*-pentane, *n*-heptane) and shorter chain species (*i*-pentane) showing the greatest uptake and largest distortions.[146] Pronounced rotation of the bdc linker causes pairs of adjacent triangular channels to conjoin, forming a more voluminous, rhombic channel. The linker rotation is driven by the minimisation of steric contacts between the alkane guest and the channel walls.[146]

Functionalisation of the bdc linkers in Sc_2bdc_3 may also affect the structural response of the framework to hydrostatic pressure, as the pendant group can act as an immobilised guest.[216] Appendage of a bulky nitrosyl group to the bdc linkers increases the mechanical stability of the framework, as the functional groups occupy space in the channel, increasing the density of the material.[216] Native Sc_2bdc_3–NO_2 remains crystalline up to 2.6 GPa in a hydrostatic medium of Fluorinert® FC-77, before undergoing irreversible amorphisation.[216] Gradual rotation of the ScO_6 nodes during compression prompts a polymorphic phase transition from *Fddd* to *C*2/*c* at 0.8 GPa, characterised by distortion and collapse of the triangular channel.[216] The phase transition is inhibited when Sc_2bdc_3–NO_2 is compressed in a hydrostatic medium of methanol, due to the increased rigidity of the framework.[216] Methanol intrusion is inferred from the slight expansion of the channel direction (*a*-axis), although the quantity of methanol adsorbed is considerably lower than for Sc_2bdc_3.

4.8 Pillared–Layered MOFs

Pillared–layered MOFs are three-dimensional, microporous frameworks composed of two-dimensional grids formed from metal–ligand coordination polymers that are pillared in the third dimension by a bidentate ligand between the metal axial coordination sites with the general formula: $[M^{II}_2(L)_2P]_n$ (where M^{II} = Zn, Co, Ni or Cu, L = bidentate ligand, and P = bidentate pillaring ligand). The metals are arranged in dinuclear paddlewheel building units. Square, rectangular or rhombic-shaped channels perforate the framework parallel to the pillars.[220] Flexibility within the paddlewheel units facilitate guest-mediated breathing behaviour,[6,221–223] NLC, and flexible phase transitions.[224]

4.8.1 Guest-mediated Flexibility

The structure and flexibility of pillared-layered MOFs are sensitive to host–guest interactions. The framework, DMOF-1, $[Zn_2(bdc)_2(dabco)]_n{\cdot}x$(Guest), adopts different structures under ambient conditions depending upon the guest species occupying the channels (Figure 4.12).[220] A tetragonal (*P*4/*mmm*) framework with square channels is formed when guest is absent from the pores. The structure is unaffected by adsorption of ethanol upon soaking, indicative of weak host–guest interactions, while adsorption of *N*,*N*-dimethylformamide prompts a phase change to *I*4/*mcm* with rectangular channels, and adsorption of benzene affords *C*2/*mmm* symmetry with rhombus-shaped

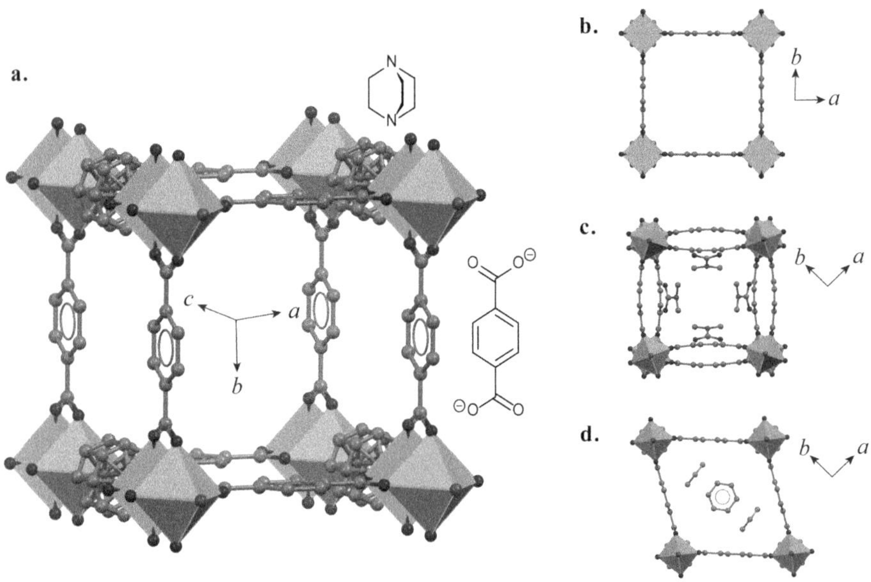

Figure 4.12 (a) Crystal structure of DMOF-1 under ambient conditions with evacuated channels (*P*4/*mmm*, CSD reference = 'WAFKEU').[220] (b) Square-shaped channels of DMOF-1·6(EtOH) (*P*4/*mmm*).[220] (c) Rectangular-shaped channels of DMOF-1·4(*N*,*N*-dimethylformamide) (*I*4/*mcm*, CSD reference = 'WAFKAQ').[220] (d) Rhombus-shaped channels of DMOF-1·3(benzene) (*Cmmm*, CSD reference = 'WAFKIY').[220] In all structures, H-atoms are removed for clarity.

channels.[220] Hydrostatic compression of a single crystal of each guest-loaded framework in a pressure-transmitting medium of the guest species elicits a unique structural response.[224] Pressure-induced intrusion of ethanol into the square channels of DMOF-1·6(EtOH) results in hyperfilling of the channel, which is accommodated by a change in the channel geometry from square-shaped to rhombus-shaped, concomitant with a phase transition from *P*4/*mmm* to *C*2/*m*, above 0.68 GPa. By contrast, the guest content and pore geometry of DMOF-1·4(*N*,*N*-dimethylformamide) is unchanged during compression in *N*,*N*-dimethylformamide, with an *I*4/*mcm* to *P*4/*mmm* transition at 0.32 GPa caused by disordering of the adsorbent. The benzene loaded framework is largely unaffected by hydrostatic pressure.

Hinging motion of the linkers about the metal paddlewheels in the two-dimensional grid layers in pillared-layered MOFs fosters breathing behaviour, reminiscent of the MIL-series. DMOF-1 and its functionalised derivatives undergo guest and pressure-dependent breathing, the magnitude of which can be controlled by modification of metal,[221] linker,[222] or crystallite size.[225] Small crystallites are generally more inflexible, attenuating the breathing magnitude, while large crystallites permit larger breathing and greater cyclability.[225] The high flexibility of pillared-layered MOFs allows breathing to occur at a low gas-loading pressure between only 1 bar and 20 bar.[221,225,226]

Pendant functional groups on the bdc linkers introduce steric hindrance or host–guest interactions in the framework channels that suppress or stimulate the breathing transition.[221,222] Upon adsorption of CO_2, DMOF-1 remains in the large-pore form.[220] However, addition of a polar side-chain to the bdc linker to form $[M_2(\text{bme–bdc})_2\text{dabco}]_n$ (where M^{II} = Co, Ni, Cu, or Zn, bme–bdc = 2,5-bis(2-methoxyethoxy)benzene dicarboxylate) stimulates a narrow-pore (*P*4/*mmm*) to large-pore ($P2_1/m$ or *C*2/*m*) phase transition upon adsorption of CO_2 up to 1 bar due to the addition of host–guest interactions with the polar adsorbent.[221]

The structural mechanism of the breathing is governed by the metal centre.[221] The Cu and Zn analogues show a sudden and complete transition, while the Ni and Co analogues gradually swell to form the eventual large-pore form, as observed from powder X-ray diffraction measurements. The magnitude and enthalpy of the breathing transition decreases across the series in the order of: Co ≈ Zn > Ni > Cu.[221] Electron paramagnetic resonance (EPR) spectroscopy measurements on a Zn framework doped with Cu, $[Zn_{1.9}Cu_{0.1}(\text{bme–bdc})_2\text{dabco}]_n$, indicate that the narrow-pore to large-pore transition increases the equatorial ligand field of the metal centres and decreases the rhombic distortion of the paddlewheel units.[227] Theoretical calculations indicate that the Cu and Zn analogues of DMOF-1 can absorb up to 119 J g^{-1} and 130 J g^{-1} during the breathing transition, respectively, in spite of their low activation pressure.[223]

Combined high-pressure powder diffraction experiments and molecular dynamics simulations of $[Cu_2(\text{bdc})_2\text{dabco}]_n$ and $[Cu_2(\text{db–bdc})_2\text{dabco}]_n$ (where db = 2,5-dibutoxy), highlight the effect that the configurational entropy of the functional side chain has on the breathing transition. Both frameworks adopt a large-pore regime under ambient conditions.[6] Under applied hydrostatic pressure to 0.40 GPa, $[Cu_2(bdc)_2dabco]_n$ is rigid and remains in the large-pore form. However, addition of the 2,5-dibutoxy side chain in $[Cu_2(db\text{-}bdc)_2dabco]_n$ stimulates a large-pore to narrow-pore transition at 0.15 GPa. The breathing is driven by the change in configurational entropy of the side chain, which gradually decreases as the framework is compressed under pressure due to its increasingly limited freedom in the confined channel.[6]

In addition to breathing, pillared-layered MOFs may also undergo 'gate-opening' linker rotations.[226] The Co-based MOF, $[Co_2(ndc)_2(bpy)]_n$ (where ndc = 2,6-napthalene and bpy = 4,4′-bipyridyl), which adopts an interpenetrated structure to form isolated pores rather than the one-dimensional channels common to other pillared-layered frameworks, exhibits a rotation of the ndc linker by ~130° upon adsorption of propane or butane.[226] The crystal structure of the framework shows that the propane or butane guest occupies an ordered adsorption site in each pore.[226] Molecular mechanics simulations indicate that guest is transported between adjacent pores *via* a 'turnstile' mechanism, involving concerted rotation of both the ndc and bpy linkers.[226]

4.9 Zinc Alkyl Gate Frameworks

A subclass of MOFs with structures that resemble a collapsible safety gate, denoted zinc alkyl gate (ZAG) frameworks, $[Zn(a\text{-}pa)\cdot 2(2H_2O)]_n$ (where a-pa = alkylbis(phosphonate)), are structurally resilient and highly flexible under applied pressure.[201,228] The 3-D, porous framework is constructed from tetrahedral ZnO_4 centres linked by bidentate alkyl phosphonate ligands in a wine-rack structure. Inorganic chains formed from eight-membered rings of Zn–O–P–O units are cross-linked by alkyl phosphonate ligands and are further stabilised by hydrogen bonds between neighbouring chains. The chain length of the alkyl linker is eponymous in the framework nomenclature, with ZAG-4 featuring 1,4-butanebisphosphonate linkers and ZAG-6 containing the hexane analogue (Figure 4.13). Guest H_2O is adsorbed in the rhombic channels upon synthesis, which are stabilised by three host–guest hydrogen bonding interactions between adsorbent H_2O, and three phosphonate groups.[201,228] The most commonly studied framework in this class, ZAG-4, has a bulk modulus between 11.7 GPa and 15.7 GPa, from hydrostatic compression of its single crystal to 7.2 GPa.[201,228] Its structure is highly resilient, with no discernible amorphisation up to 9.9 GPa according to *in situ* single-crystal X-ray diffraction measurements, owing to the ability of the alkyl chains to serve as a 'spring-like cushion', which structurally stabilises the framework.[201]

Under applied pressure, ZAG-4 exhibits modest NLC along the diameter of the rhombic channels, parallel to the $H_2O\cdots PO_3H$ hydrogen bonding network (b-axis), above 2 GPa ($\beta_{NLC} = -18\ TPa^{-1}$).[201,228] Between ambient pressure and 2 GPa, all axes have positive linear compressibility. The NLC at high-pressure

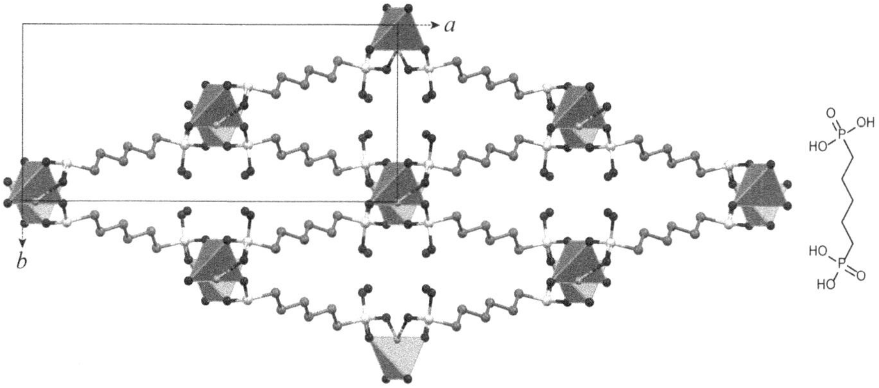

Figure 4.13 Crystal structure of ZAG-6 under ambient conditions (*C*2/*c*, CSD reference = 'VONBUX').[201] H-atoms are removed for clarity. The 1,6-hexanebiphosphate linker is illustrated to the right of the crystal structure.

partly originates from the wine-rack topology of ZAG-4, although is primarily caused by a structural transition at 3.8 GPa involving proton-exchange from a phosphonate group to adsorbed H_2O, forming H_3O^+ and R-POC$^-$ with retention of the *C*2/*c* symmetry of the framework.[228,229] The reaction is driven by shortening of the interaction distance between the host and guest as the H_3O^+ and R-POC$^-$ counterions are formed, which causes the unit cell volume to contract.[228] Quantum mechanical calculations indicate that the counterionic phase at 5 GPa is stabilised by 14 kJ mol^{-1} compared with its neutral (H_2O/PO_3H) counterpart.[228] Local flexibility in the inorganic chains of ZAG-4 also accounts for NLC along the *b*-axis.[201] Between ambient pressure and 7.2 GPa, the unit cell volume of ZAG-4 decreases by 27% simultaneously with an increase in the β angle of the unit cell by approximately 10°.[201] In this pressure range, the O–Zn–O angle of the inorganic chains gradually opens, while the intra-Zn distances gradually decreases.[201] Overall, the chain extends by ~0.2 Å along the *b* axis direction up to 7.2 GPa.[201]

Hydrostatic compression of a single crystal of isostructural ZAG-6 draws similar structural response to that of ZAG-4, exhibiting NLC along the *b* axis as a result of pressure-activated proton-exchange.[228] However, the longer, more flexible alkyl chains of the 1,6-hexanebisphosphonate linker undergo a 'coiling' transition at 6.9 GPa, not observed for ZAG-4. The transition is characterised by a decrease in the dihedral angles of the hexane from 177.1(2)° (∠P–C1–C2–C3) and 173.7(3)° (∠C1–C2–C3–C4) at ambient pressure to 170.2(6)° and 55.3(10)°, respectively, at 6.9 GPa.[228]

4.10 DUT Frameworks

A class of MOFs first developed at Dresden University of Technology, denoted DUT-frameworks, exhibit unusual pressure-induced negative gas adsorption, where gas is spontaneously desorbed from the framework pores under applied pressure, causing a sudden breathing-like transition from mesoporosity to microporosity.[230,231] The release of gas in a closed, pressurised system leads to pressure amplification, which benefits applications such as microfluidic devices, micropumps or artificial lungs. At a constant temperature, the magnitude of the breathing and force amplification is dependent upon the adsorbate species,[124,232] crystallite size,[233–235] number of defects within the framework,[235] and temperature.[124,231]

Negative gas adsorption was first reported for the Cu-based framework, DUT-49 (Figure 4.14), which is constructed from dinuclear Cu paddlewheel units with 9,9′-([1,1′-biphenyl]-4,4′-diyl)bis(9*H*-carbazole-3,6-dicarboxylate) (bbdc) linkers, connected in an **fcu** topology.[230] Cubic close packing of CuO_6 octahedra leave octahedral, tetrahedral, and cuboctahedral voids with respective diameters of 24 Å, 17 Å and 10 Å.[124,230] Dosing a crystalline powder of DUT-49 with methane causes initial adsorption of methane into the pores, followed by sudden desorption that is equivalent to −8.62 mmol g^{-1} at 10 kPa. Desorption of methane is associated with a reversible transition from an open-pore to closed-pore phase, corresponding to a decrease in the unit cell volume from 100 072 $Å^3$ to 47 282 $Å^3$, contraction of the octahedral and tetrahedral pore diameters to 11 Å and 8 Å and pressure amplification of ~2.3 kPa.[230] A similar structural response also occurs upon adsorption of butane to 30 kPa.[230]

Powder X-ray diffraction and extended X-ray absorption fine structure (EXAFS) measurements indicate that structural contraction originates from concerted rotation of CuO_6 units about a three-fold axis, which causes the

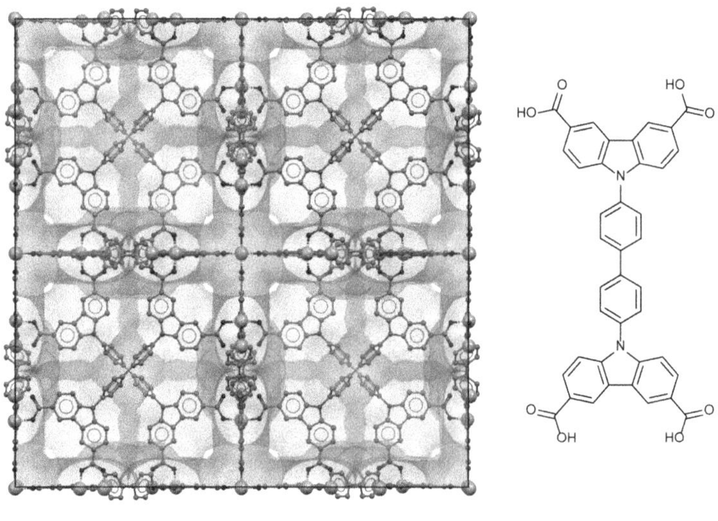

Figure 4.14 Crystal structure of DUT-49 under ambient conditions (*F*23, CSD reference = 'OKUSOE'),[230] showing the 9,9′-([1,1′-biphenyl]-4,4′-diyl)bis(9*H*-carbazole-3,6-dicarboxylic acid ($bbdcH_2$) linker. The white region represents the solvent accessible volume. H-atoms are removed for clarity.

bddc linkers to buckle, with retention of the framework connectivity.[230,231] The rotation is driven by enhancement of the host–guest interactions in the closed-pore form. Upon adsorption of *n*-butane, the adsorption enthalpy increases from 29 kJ mol^{-1} in the open-pore form to 50 kJ mol^{-1} in the closed-pore form.[230,231] Structural flexibility in the linker is necessary for negative gas adsorption to occur.[231] Frameworks with short, rigid linkers that are isostructural to DUT-49, such as DUT-48 and DUT-46, which have 9,9'-(1,4-phenylene) bis-carbazole-3,6-dicarboxylate and 9-(6-carbazol-9-ylnaphthalen-2-yl)carbazole-dicarboxylate linkers, respectively, are too stiff to accommodate the breathing transition, eliminating negative gas adsorption behaviour.[124,231]

The interplay of framework flexibility and maximisation of host–guest interactions is demonstrated by xenon adsorption at 200 K in a series of frameworks with increasing linker length: DUT-48, DUT-46, DUT-50 (9,9'-([1,1′-triphenyl]-4,4′-diyl)bis(9*H*-carbazole-3,6-dicarboxylate, tbdc)), and DUT-151 (9,9'-([1,1′-tetraphenyl]-4,4′-diyl)bis(9*H*-carbazole-3,6-dicarboxylate tdbdc)).[124] High-pressure ^{129}Xe NMR spectroscopy can be used to track the chemical shift of the adsorbed xenon as a function of applied pressure and pore size.[124] In the rigid frameworks DUT-48 and DUT-46, gradual uptake of xenon is evidenced by a gradual shift in the ^{129}Xe NMR signal to higher field as the xenon-framework interactions increase; the shift is continuous, indicating that no breathing transition occurs.[124] By contrast, the longer, flexible linkers of DUT-50 facilitate the open-pore to closed-pore transition above a relative applied pressure, p/p_0, of 0.4. The transition is accompanied by a sudden shift in the ^{129}Xe signal from 150 ppm at $p/p_0 = 0.4$ to 230 ppm at $p/p_0 = 0.5$, as the strength of the host–guest interactions increase in the confined pore.[124] Further increasing the length of the linkers to DUT-151 results in the formation of an interpenetrated framework, reversing the breathing to classical positive gas adsorption.[124] Under ambient conditions, evacuated DUT-151 adopts a closed-pore regime, which gradually transitions to the open-pore phase with increasing xenon pressure.[124] Frameworks with conformational flexibility in the organic linkers undergo breathing *via* conformational isomerism.[233] Adsorption of *n*-butane in DUT-13, $[Zn_4O(benztb)_{1.5}]_n$ (where benztb = *N*,*N*,*N*′,*N*′-benzidine tetrabenzoate), stimulates a transition from the open-pore to closed-pore phase by transformation of the benztb linker from a staggered to an eclipsed conformation, which requires contraction of the pore.[233]

Flexible breathing in DUT frameworks is not only mediated by gas adsorption and desorption, but also occurs under direct mechanical pressure.[231] Hydrostatic compression of DUT-49 and DUT-48 by mercury porosimetry up to 0.4 GPa, coupled with DFT calculations, revealed a stepwise contraction of the frameworks due to buckling of the bbdc and tbdc linkers under pressure.[231] Compression of DUT-48 results in compression of the unit cell by 64% in a two-step regime, while compression of DUT-49 follows a four-step regime to a total contraction of 72%.[231] This is distinct from guest-mediated breathing, which only occurs for DUT-49 in a single-step process. Buckling of the linker occurs at 0.15 GPa for DUT-48 compared with only 0.06 GPa for DUT-49. The pronounced volumetric contraction is associated with a large work energy of 85 J g^{-1} and 106 J g^{-1} for DUT-48 and DUT-49, respectively, indicating the promise of DUT-frameworks for use in shock-absorbing applications.[231]

4.11 Inflexible MOFs

Rigid frameworks (see Section 2.4 of Chapter 2 for discussion on framework flexibility) that retain their structure, porosity and crystallinity under applied pressure benefit catalytic, storage and templating applications, owing to their stable structures and predictable properties. The archetypal MOFs, MOF-5, $[ZnO_4(bdc)_3]_n$, and HKUST-1, $[Cu_3(btc)_2(H_2O)_3]_n$ (where btc = 1,3,5-benzenetricarboxylate) (Figure 4.15), are stiff when compared with other porous frameworks, with bulk moduli, K, of 16.5 GPa for MOF-5[15] and between 29.5 GPa and 30.7 GPa for HKUST-1,[127] based on their hydrostatic compression in a non-penetrating medium. Both MOF-5 and HKUST-1 are formed from three-dimensional, cubic scaffolds.[236–238] In MOF-5, metal clusters, $[Zn_4O]_6$, are linked to eight bdc linkers in an octahedral array, forming cubic voids.[236,237] In HKUST-1, dinuclear Cu paddlewheels are linked by four btc linkers to form a network punctuated by three distinct cavities, including two cuboctahedral cavities and an interstitial octahedral cavity.[238]

Local rigidity in the framework results in minimal perturbation of the structures under applied hydrostatic pressure.[15,16,114,127] High-pressure *in situ* X-ray diffraction of MOF-5 and HKUST-1 in non-penetrating pressure-transmitting media indicates gradual, direct compression of the unit cell volume until eventual amorphisation. Volumetric contraction originates from compression of the metal–linker coordination bonds.[15,16] Irreversible amorphisation

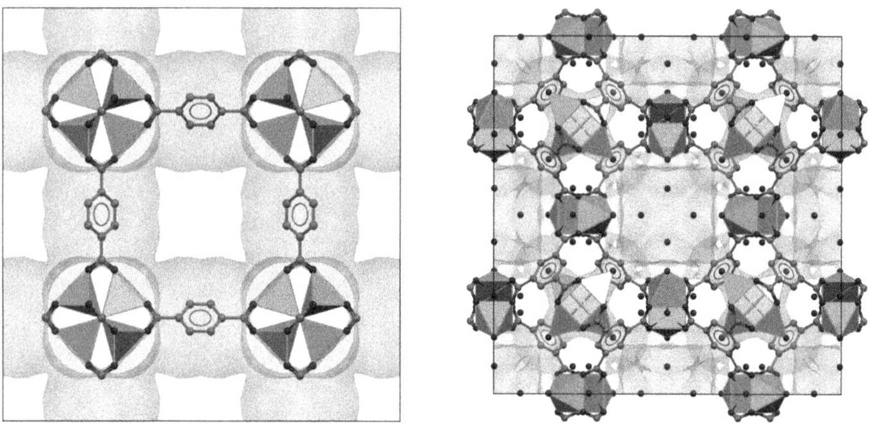

Figure 4.15 Crystal structure of MOF-5 (right, $Fm\bar{3}m$, CSD reference = 'SAHYIK')[239] and HKUST-1 (left, $Fm\bar{3}m$ CSD reference = 'FIQCEN').[240] The white region represents the solvent accessible volume. H atoms are removed for clarity.

of MOF-5 occurs at relatively low applied pressure of 3.5 MPa.[241] By contrast, HKUST-1 remains crystalline to at least 8 GPa.[127]

The compressibility of the MOF-5 and HKUST-1 is mediated by adsorption or desorption of guest.[15,127] As previously discussed, forcing inert guest into porous MOFs at high pressure can enhance the mechanical stability of the framework by rigidifying the pore region. Compression of a single crystal of MOF-5 in a penetrating hydrostatic medium of *N*,*N*-diethylformamide delays the pressure of amorphisation from 3.5 MPa to >3.2 GPa.[15] Pressure-induced adsorption of *N*,*N*-diethylformamide is followed by an increase in the disordered electron density in the framework pore between 0.33 GPa and 0.78 GPa by the equivalent of nine *N*,*N*-diethylformamide molecules, concomitant with swelling of the pore and unit cell volume. In this pressure region, the material is highly incompressible with a bulk modulus of 242 GPa.[15] Increasing the pressure to 1.35 GPa prompts sudden contraction of the framework, leading to desorption of fifty-four guest molecules. Further compression to 2.35 GPa causes gradual contraction of the framework and further desorption of seven *N*,*N*-diethylformamide molecules, corresponding to a bulk modulus of between 20 GPa and 40 GPa.[15] Pressure-induced desorption of guest is uncommon in MOFs, as the framework typically undergoes a phase transition to accommodate the increased quantity of adsorbent (*e.g.* ZIF-8, Sc_2bdc_3). However, the inflexibility of MOF-5 prevents such a transition from occurring, resulting in the desorption of guest as the pores are directly compressed.[15]

Although the $Fm\bar{3}m$ symmetry of MOF-5 is retained in the high-pressure crystallographic study,[15] quasi-harmonic lattice dynamics calculations of MOF-5 suggest that the crystal undergoes a phase transition to $Fm\bar{3}$ when the volumetric contraction exceeds 0.8%, due to softening of the A_{2g} phonon during compression.[242] The phase transition is accompanied by an increase in the compressibility of the framework, with K decreasing from 17 GPa ($Fm\bar{3}m$) to 14 GPa ($Fm\bar{3}$).

A similar structural response to guest-mediated compression is observed in HKUST-1. Compression of HKUST-1 in a hydrostatic medium of either a methanol–ethanol–water mixture or isopropanol from ambient pressure to 8 GPa elicits a sudden change in the framework compressibility from hard to soft above a threshold pressure, with no change in the symmetry or structure of the framework.[127] Initial compression to 0.8 GPa in isopropanol or 2.2 GPa in methanol–ethanol–water is associated with adsorption of the medium into the framework pores,[16] corresponding to a high K of 117 GPa or 116 GPa, respectively.[127] In these pressure regions, the framework swells due to lengthening of the coordination bond between Cu and an axially bound water ligand oriented in the pore, Cu–O(H_2), which becomes involved in hydrogen bonding with the incoming guest, thus weakening the Cu–O bond.[16] Compression above 0.8 GPa or 2.2 GPa in respective isopropanol or methanol–ethanol–water to 8 GPa prompts a sudden transition to a soft phase, with a decrease in K to 25.9 GPa for isopropanol and 33.6 GPa for methanol–ethanol–water.[127] In contrast to MOF-5, the sudden increase in compressibility occurs with only a small degree of desorption (16%) of guest from the pores between 3.9 GPa and 5.0 GPa, although the reason for this is ambiguous.[16]

Spectroscopic analysis of HKUST-1 at high-pressure informs of chemical changes that are imperceptible from diffraction methods.[114] Raman and infrared spectroscopy of HKUST-1 upon between ambient pressure and 15 GPa without a pressure-transmitting medium imply that at least one chemical phase transition may occur between 4.1 GPa and 5.1 GPa and/or at 9 GPa.[114] Two new vibrational peaks associated with the benzene C–H stretch of btc (600 cm^{-1} to 1200 cm^{-1}) form in the Raman spectrum at >9 GPa, while new peaks in the infrared spectrum are observed at >4.1 GPa.

Frameworks with native flexibility can become rigid by molecular 'retrofitting', where an additional linker is added across the pores to inhibit contraction under pressure.[243] The ability to modulate the flexibility of MOFs with retention of the original architecture is necessary to optimise their performance in industrial or technological applications. Post-synthetic modification of MOF-520, $[Al_8(\mu\text{-}OH)_8(HCOO)_4(btb)_4]_n$ (where btb = 1,3,5-benzenetribenzoate),[244] with 4,4′-biphenyldicarboxylate (bpdc) across the pores enhances the mechanical stability of the framework compared with the pristine material.[243] MOF-520 crystallises in the tetragonal space group, $P4_22_12_1$, and is constructed from octametallic rings of $Al_8(\mu\text{-}OH)_8(HCOO)_4(btb)_4$, connected through twelve btb linkers in a (12,3)-connected **fon** topology. The framework is punctuated by one-dimensional channels approximately 9.5 Å in diameter. Functionalisation across the pores with bpdc linkers forms a (16,3)-connected **skl** net, in which the octametallic rings are linked through four additional bpdc ligands. The original architecture and porosity of the framework is retained, although the onset pressure of amorphisation is delayed from 3 GPa to 5.5 GPa.[243] The stiffness of the framework is also increased, contracting by 0.3% in unit cell volume at 1 GPa compared with 1.2% at 0.9 GPa for the pristine framework. Upon adsorption of methanol, the usual volumetric expansion of the pores is inhibited by the rigid bpdc 'girders', which are equivalent in length to the original pore diameter of MOF-520.

4.12 Natural Product MOFs

Metal–organic frameworks derived from natural products, such as sugars or peptides, are uniquely suited to biological or medicinal applications, such as drug delivery or medical imaging. Conformational flexibility within the framework backbone and the formation of intra-framework hydrogen bonding networks provide structural resilience and flexibility under applied pressure.

To date, the homochiral material, $Zn(GlyTyr)_2$ (where GlyTyr = glycyl-L-tyrosine) is the only peptidic MOF that has been examined at high pressure.[245] The framework is composed of two-dimensional layers of Zn^{2+} coordinated to the carboxylate of tyrosine and the amine of glycyl in the GlyTyr linkers, which are extended in three dimensions by the GlyTyr pillars. A network of hydrogen bonds between hydroxyl and amino residues forms between the layers. Hydrostatic compression to 4 GPa generates strain within the peptide linkers, triggering cooperative rotation of the amino groups and closing of the dihedral angle of the peptide backbone from 164° to 146°.[245] Compression of the peptide backbone leads to isotropic contraction of the two-dimensional layers and shortening of the inter-layer distance, which strengthens the inter-layer hydrogen bonds, leading to propagation of the phase transition through the crystal.

Porous frameworks derived from γ-cyclodextrin exhibit unusual pressure inverse solubility and pressure-induced polymorphism.[246] The framework, CD-MOF-1, which is composed of γ-cyclodextrin and K^+ cations crystallises in one of two polymorphs under ambient conditions: a cubic *I*432 phase (denoted CD-MOF-1-α) or a trigonal *R*32 phase (CD-MOF-1-β). Under applied pressure of 0.1 MPa in a pressure-transmitting medium of isopropanol, the β-phase transitions to the α-phase, where it is retained upon compression to 1.36 GPa.[246] A crystal of CD-MOF-1-α, exclusively, may also be grown *in situ* in the DAC. Upon compression to 2.1 GPa in a medium of methanol and water, the crystal of CD-MOF-1-α completely dissolves, and returns to hexahedral crystals upon lowering the pressure to 0.53 GPa.[246] This is unusual, and is thought to result from the pressure-induced adsorption of polar medium into the framework pores.

4.13 Piezoresponsive Functional MOFs

Piezoresponsive materials exhibit changes in their functional properties under a mechanical stimulus. Piezoelectric, piezomagnetic and piezochromic behaviour in MOFs is relatively uncommon. However, functional MOFs are promising candidates for gas/pressure sensors and energy harvesting.

Piezochromism is the most commonly reported functional behaviour in MOFs, and was first noted for $[Co(bdc)(dabco)]_n{\cdot}x(H_2O)$ (AMU-1, Adam Mickiewicz University),[115] $Co_2(bdc)_2(dabco)_4$[246] and the coordination polymer, $CoCl_2bpp$ (where bpp = 1,3-bis(4-pyridyl)propane).[247] Red crystals of AMU-1 gradually become yellow under increasing hydrostatic pressure to 3.5 GPa as the octahedral coordination environment of the divalent cobalt centres is strained. The colour change results from shortening and strengthening of intra-framework hydrogen bonds between ligated H_2O and bdc linkers at elevated pressure, which distorts the Co^{2+} coordination centre. Similarly, compression of $CoCl_2bpp$ causes reversible piezochromism from blue/green to colourless as a result of distortion of the Co^{2+} octahedral coordination sphere and the framework topology under applied pressure.[247] In $Co_2(bdc)_2(dabco)_4$, blue to purple piezochromism results from twisting and shearing of the Co paddlewheel secondary building units.[246] In all cases, pressure-induced strain at the metal centres results in a blue-shift in the colour of the crystal.

Piezochromism in the Cu-based framework, CuPyr-I (where Pyr-I = bis(1-(4-pyridyl)butane-2,3-dione)) from green to red during hydrostatic compression originates from pressure-induced and guest-mediated Jahn–Teller switching of the octahedral Cu^{2+} centres.[13] Compression to 3.34 GPa promotes gradual distortion of the Cu centres from an elongated to compressed geometry without a structural phase transition. The pressure at which the Jahn–Teller switch occurs depends upon the pressure-transmitting medium used, occurring at 0.57 GPa in Fluorinert® FC-70 and 1.28 GPa in methanol. As has been described for other MOFs, intrusion of methanol into the framework pores under pressure reduces the compressibility of the framework, delaying the Jahn–Teller switch.[13] Unusually, this framework also exhibits various other structural phenomena, including pressure and temperature induced phase transitions, and negative linear compressibility.[13]

Piezochromism and bathochromism may also result from conformational flexibility of the organic ligands. Hydrostatic compression of a single crystal of a UiO framework, Hf-peb (where peb = 1,4-phenylene-bis(4-ethynlbenzoate)) up to 2.1 GPa prompts a red-shift in both the fluorescence emission and UV-visible absorption energy as a result of a pressure-induced rotation of the central phenyl ring in the peb linker, from a coplanar to a twisted conformation.[21] Under ambient conditions, the central phenyl ring of the peb linker is disordered, occupying the planar conformation in 73% of sites, and the twisted conformation in the remainder, with an energy barrier to interconversion of 2.5 kJ mol^{-1}.[21] High-pressure single-crystal X-ray diffraction coupled with high-pressure UV-visible and fluorescence spectroscopy shows an increase in the population of the twisted peb conformation from 27% at ambient pressure to 100% at 2.1 GPa is associated with a red-shift in the fluorescence emission from 424 nm for the planar form to 450 nm for the twisted form. Isolation of the peb fluorophore in the framework allows its luminescent properties to be studied in the absence of intermolecular interactions, offering unique insight into the effect of intramolecular flexibility on fluorescence.[21]

Piezoelectricity originates from the perturbation of polarised charge in non-centrosymmetric space groups under mechanical stress. Piezoelectric MOFs include MOF-1-Cd(imazethapyr)$_2$ (where imazethapyr = (2-(4,5-dihydro-4-methyl-4-(1-methylethyl)-5-oxo-1*H*-imidazol-2-yl)-5-ethyl-3-pyridinecarboxylic acid))[248] and perovskite frameworks.[22] Structures that crystallise in non-centrosymmetric space groups give rise to permanent polarisation of charge, inducing ferroelectricity. Changes to the polarisation under applied stress, or conversely, the generation of strain due to changes in the polarisation, afford piezoelectricity.

The herbicide-based framework, MOF-1-Cd(imazethapyr)$_2$, forms a dense, three-dimensional diamondoid structure, crystallising in the space group *Fdd*2.[248] Bulky isopropyl and methyl groups protrude into the framework cavities, eliminating porosity. The diamond structure is conducive to good ferroelectric behaviour, with a remnant polarisation of 0.006 μC cm^{-2} and a coercive field (electric field strength at which the charge polarisation is zero) of 0.9 kV cm^{-1} to 1.1 kV cm^{-1}.[248] Deformation of the structure under applied pressure produces a small piezoelectric coefficient, defined as the polarisation of the material under applied stress, of 60.1 pC N^{-1}, compared with 300 pC N^{-1} to 2500 pC N^{-1} for $BaTiO_3$.[249]

Pressure-induced phase transitions between non-centrosymmetric and centrosymmetric space groups can lead to on/off switching of ferroelectricity.[22] The heterometallic perovskite frameworks, $EtAM^aM^bX_3$ (where EtA^- = ethylammonium, $M^a = Na^+$, $M^b = Al^+$, X = $HCOO^-$), show ferroelectric to non-ferroelectric switching behaviour under applied hydrostatic pressure due to a phase transition from *Pn* to $P2_1/n$ at 3.6 GPa.[22] Compression of the polar *Pn* phase to 3.6 GPa causes the ferroelectric polarisation to increase from 0.9 μC cm^{-2} at ambient pressure to 1.1 μC cm^{-2} at 3.6 GPa, before polarisation is lost after the phase transition.[22] Structural characterisation by high-pressure X-ray diffraction and Raman spectroscopy indicates that the Na_6O subnetwork is compressed and distorted at elevated pressure, leading to a symmetry-breaking change in the configuration of the ethylammonium cations that occupy the framework pores.[22]

Electrical conductivity in MOFs is generally poor, limiting their use as semi-conductor components in electronic devices. However, the conductivity of MOFs can be enhanced under applied hydrostatic pressure.[250] Structural deformation of the semi-conducting, monolayer MOF, Ni_3(hexaiminotriphenylene)$_2$ under a negative hydrostatic pressure of −0.001 GPa (*i.e.* stretching the framework) causes a semi-conductor to metallic transition.[250] In the analogous framework Ni_3(hexaiminobenzene)$_2$, stretching of the Ni centres to −0.0011 GPa results in piezoreduction of the metal with extension of the Ni–N bond length by 10%, leading to pressure-induced magnetisation of the framework.[250] Enhancement of the electronic properties and installation of functional properties in MOFs under pressure benefits their implementation in technological devices and sensors.

4.14 Pressure-stimulated Post-synthetic Modification

Post-synthetic modification of MOFs allows the composition of a framework to be changed with retention of its crystallinity and parent topology, permitting incorporation of functional groups or guest species that are inaccessible through *de novo* routes.[251–253] There are three main methods to achieve post-synthetic modification: (*i*) covalent modification, in which pendant functional groups on the organic linker react with an added guest species, (*ii*) coordinate modification, in which the metal coordination sphere is altered by addition, removal or exchange of a ligand, and (*iii*) metal or ligand exchange, in which a competitive metal ion or ligand fully or partially exchanges with that of the parent framework. Modification of the ligand can be used to control the geometry and internal surface of the framework pores, whilst modification of the metal can alter the thermal stability and catalytic properties of the material.

Post-synthetic modification may occur upon soaking a solid sample of a MOF in a suitable reactant under ambient conditions. In porous frameworks, an appreciable quantity of reactant must infiltrate the pores to access the reaction sites, rendering many MOF/reactant systems inert due to size-exclusion effects. High pressure can promote post-synthetic modification in otherwise inert systems by forcing the reactant inside the framework pores. Pressure-stimulated post-synthetic modification has been observed in the Cu-framework, STAM-1 (St Andrews MOF), $[Cu_3(btc)_3]_n \cdot 5n(H_2O)$,[28] by compression of a single crystal in a hydrostatic medium of the liquid reactant species in DAC apparatus. Presently, this is the only example of post-synthetic modifications in a MOF at high pressure.

STAM-1 is a two-dimensional, layered framework with an amphiphilic pore regime.[28] Monomethyl-esterified benzenetricarboxylate (btc) linkers connect five-coordinate dinuclear Cu paddlewheels to form interdigitated layers with two types of channels. One channel is lined by methyl groups and is hydrophobic, and the other is decorated by H_2O ligands axially bound to Cu^{2+} and is thus hydrophilic. The H_2O ligand is labile to post-synthetic ligand exchange upon soaking in a range of small organic molecules, including methanol, methylamine, and ethylamine under ambient conditions, without the need for prior activation.[28] However, more bulky acetonitrile does not undergo exchange under ambient conditions due to minimal diffusion into the hydrophilic channel.[28] Upon compression to at least 0.3 GPa, partial ligand exchange of H_2O for acetonitrile is observed by high-pressure *in situ* single-crystal X-ray diffraction due to pressure-induced intrusion of the hydrostatic medium into the framework channels.[28] The reaction occurs in a single-crystal-to-single-crystal process, with retention of the $P\bar{3}m1$ symmetry and minimal structural distortion. Addition of alkyl-bearing ligands at the surface of the hydrophilic pores eliminates the hydrophilicity of the framework and decreases the solvent-accessible volume of the channels, significantly changing its properties.

4.15 Concluding Remarks

Since the first high-pressure structural studies began on MOFs over 10 years ago, the range of physical and structural phenomena observed has grown substantially. The mechanical response of MOFs to high hydrostatic pressure is informative of their structural stability and flexibility. From the small number of MOFs characterised at high-pressure, some general structural features guiding flexible behaviour have been identified, providing a basis for rational MOF design. By combining high-pressure diffraction, with other techniques, including NMR, UV-vis, Raman, infrared and fluorescence spectroscopies, pressure has been used as a fantastic tool to probe structure–property relationships in a range of MOFs, which include some of the largest and most complex systems ever to be studied at pressure. What is clear, is that we are starting to get to grips with how to build more chemically and mechanically resistant frameworks, or at least control the stability of the crystalline phases of MOFs. The effect hydrostatic compression has on frameworks, depending on whether a penetrating or non-penetrating medium is used, is beginning to become somewhat predicable, though the bulk moduli, and changes in physical properties (and chemical reactivity), still need more study to better evaluate and design frameworks in the future. One of the biggest questions to answer, is how and why amorphous phases occurs at elevated pressure (which in almost all high-pressure studies occurs eventually), and how we characterise them (and the resulting physical properties), is a challenging area, and one that is ripe for exploration.

References

1. S. A. Moggach and I. D. Oswald, *Crystallography Under High Pressures*, Springer, Cham, 2020.
2. S. C. McKellar and S. A. Moggach, *Acta Crystallogr.*, 2015, **B71**, 587.
3. S. Henke, M. T. Wharmby, G. Kieslich, I. Hante, A. Schneemann, Y. Wu, D. Daisenberger and A. K. Cheetham, *Chem. Sci.*, 2018, **9**, 1654.
4. P. G. Yot, Q. Ma, J. Haines, Q. Yang, A. Ghoufi, T. Devic, C. Serre, V. Dmitriev, G. Férey, C. Zhong and G. Maurin, *Chem. Sci.*, 2012, **3**, 1100.
5. M. Wahiduzzaman, N. Reimer, J.-P. Itié, N. Stock, G. Maurin and P. G. Yot, *Polyhedron*, 2018, **155**, 144.
6. P. Vervoorts, J. Keupp, A. Schneemann, C. L. Hobday, D. Daisenberger, R. A. Fischer, R. Schmid and G. Kieslich, *Angew. Chem., Int. Ed.*, 2021, **60**, 787.
7. H. Yang, F. Guo, P. Lama, W. Y. Gao, H. Wu, L. J. Barbour, W. Zhou, J. Zhang, B. Aguila and S. Ma, *ACS Cent. Sci.*, 2018, **4**, 1194.
8. N. Chanut, A. Ghoufi, M. V. Coulet, S. Bourrelly, B. Kuchta, G. Maurin and P. L. Llewellyn, *Nat. Commun.*, 2020, **11**, 1216.
9. S. A. Moggach, T. D. Bennett and A. K. Cheetham, *Angew. Chem., Int. Ed.*, 2009, **48**, 7087.
10. X. Wang, R. Krishna, L. Li, B. Wang, T. He, Y.-Z. Zhang, J.-R. Li and J. Li, *Chem. Eng. J.*, 2018, **346**, 489.
11. W. Li, M. R. Probert, M. Kosa, T. D. Bennett, A. Thirumurugan, R. P. Burwood, M. Parinello, J. A. Howard and A. K. Cheetham, *J. Am. Chem. Soc.*, 2012, **134**, 11940.
12. P. Serra-Crespo, A. Dikhtiarenko, E. Stavitski, J. Juan-Alcaniz, F. Kapteijn, F. X. Coudert and J. Gascon, *CrystEngComm*, 2015, **17**, 276.
13. C. J. McMonagle, P. Comar, G. S. Nichol, D. R. Allan, J. González, J. A. Barreda-Argüeso, F. Rodríguez, R. Valiente, G. F. Turner, E. K. Brechin and S. A. Moggach, *Chem. Sci.*, 2020, **11**, 8793.
14. Z. Chen, B. Xu, Q. Li, Y. Meng, Z. Quan and B. Zou, *Inorg. Chem.*, 2020, **59**, 1715.
15. A. J. Graham, D. R. Allan, A. Muszkiewicz, C. A. Morrison and S. A. Moggach, *Angew. Chem., Int. Ed.*, 2011, **50**, 11138.

16. A. J. Graham, J. C. Tan, D. R. Allan and S. A. Moggach, *Chem. Commun.*, 2012, **48**, 1535.
17. A. J. Graham, A. M. Banu, T. Duren, A. Greenaway, S. C. McKellar, J. P. Mowat, K. Ward, P. A. Wright and S. A. Moggach, *J. Am. Chem. Soc.*, 2014, **136**, 8606.
18. P. G. Yot, M. Wahiduzzaman, E. Elkaim, P. Fertey, P. Fabry, C. Serre and G. Maurin, *Dalton Trans.*, 2019, **48**, 1656.
19. X. Zhou, Y. R. Miao, W. L. Shaw, K. S. Suslick and D. D. Dlott, *J. Am. Chem. Soc.*, 2019, **141**, 2220.
20. T. Gong, Q. Sui, P. Li, X. F. Meng, L. J. Zhou, J. Chen, J. Xu, L. Wang and E. Q. Gao, *Small*, 2019, **15**, e1803468.
21. A. Sussardi, C. L. Hobday, R. J. Marshall, R. S. Forgan, A. C. Jones and S. A. Moggach, *Angew. Chem., Int. Ed.*, 2020, **59**, 8118.
22. M. Ptak, I. E. Collings, K. L. Svane, A. Sieradzki, W. Paraguassu and M. Mączka, *J. Mater. Chem. C*, 2019, **7**, 8660.
23. A. S. Babal, L. Donà, M. R. Ryder, K. Titov, A. K. Chaudhari, Z. Zeng, C. S. Kelley, M. D. Frogley, G. Cinque, B. Civalleri and J.-C. Tan, *J. Phys. Chem. C*, 2019, **123**, 29427.
24. A. W. Lawson and T. Y. Tang, *Rev. Sci. Instrum.*, 1950, **21**, 815.
25. J. C. Jamieson, A. W. Lawson and N. D. Nachtrieb, *Rev. Sci. Instrum.*, 1959, **30**, 1016.
26. G. J. Piermarini and C. E. Weir, *J. Res. Natl. Inst. Stand. Technol.*, 1962, **66**, 325.
27. C. J. McMonagle, D. R. Allan, M. R. Warren, K. V. Kamenev, G. F. Turner and S. A. Moggach, *J. Appl. Crystallogr.*, 2020, **53**, 1519.
28. S. C. McKellar, A. J. Graham, D. R. Allan, M. I. Mohideen, R. E. Morris and S. A. Moggach, *Nanoscale*, 2014, **6**, 4163.
29. A. Katrusiak, *Acta Crystallogr.*, 2008, **A64**, 135.
30. A. Jayaraman, *Rev. Sci. Instrum.*, 1986, **57**, 1013.
31. H. K. Mao, *Science*, 1978, **200**, 1145.
32. G. J. Piermarini and S. Block, *Rev. Sci. Instrum.*, 1975, **46**, 973.
33. S. S. Lobanov, V. B. Prakapenka, C. Prescher, Z. Konôpková, H.-P. Liermann, K. L. Crispin, C. Zhang and F. Goncharov, *J. Appl. Phys.*, 2015, **118**, 035905.
34. L. Merrill and W. A. Bassett, *Rev. Sci. Instrum.*, 1974, **45**, 290.
35. S. A. Moggach, D. R. Allan, S. Parsons and J. E. Warren, *J. Appl. Crystallogr.*, 2008, **41**, 249.
36. M. Yousuf and G. K. Rajan, *Pramana*, 1982, **18**, 1.
37. E. F. O'Bannon 3rd, Z. Jenei, H. Cynn, M. J. Lipp and J. R. Jeffries, *Rev. Sci. Instrum.*, 2018, **89**, 111501.
38. A. Dewaele, P. Loubeyre, F. Occelli, O. Marie and M. Mezouar, *Nat. Commun.*, 2018, **9**, 2913.
39. A. Katrusiak, *Int. Tables Crystallogr.*, 2019, **H**, 159.
40. J.-F. Lin, J. Shu, H.-K. Mao, R. J. Hemley and G. Shen, *Rev. Sci. Instrum.*, 2003, **74**, 4732.
41. N. Funamori and T. Sato, *Rev. Sci. Instrum.*, 2008, **79**, 053903.

42. G. Shen, V. B. Prakapenka, M. L. Rivers and S. R. Sutton, *Rev. Sci. Instrum.*, 2003, **74**, 3021.
43. S. Merkel and T. Yagi, *Rev. Sci. Instrum.*, 2005, **76**, 046109–046112.
44. H. Miyaji, *J. Phys. Soc. Jpn.*, 1975, **39**, 1346.
45. Y. K. Vohra, S. J. Duclos and A. L. Ruoff, *Phys. Rev. B*, 1987, **36**, 9790.
46. R. J. Nelmes, J. S. Loveday, E. Kaldis and J. Karpinksi, *Phys. C*, 1990, **172**, 311.
47. D. J. Dunstan, *Rev. Sci. Instrum.*, 1989, **60**, 3789.
48. Y. Shen, R. S. Kumar, M. Pravica and M. F. Nicol, *Rev. Sci. Instrum.*, 2004, **75**, 4450.
49. T. Varga, A. P. Wilkinson and R. J. Angel, *Rev. Sci. Instrum.*, 2003, **74**, 4564.
50. V. A. Sidorov and R. A. Sadykov, *J. Condens. Matter Phys.*, 2005, **17**, S3005.
51. K. Murata, K. Yokogawa, H. Yoshino, S. Klotz, P. Munsch, A. Irizawa, M. Nishiyama, K. Iizuka, T. Nanba, T. Okada, Y. Shiraga and S. Aoyama, *Rev. Sci. Instrum.*, 2008, **79**, 085101.
52. G. J. Piermarini, S. Block and J. D. Barnett, *J. Appl. Phys.*, 1973, **44**, 5377.
53. S. Klotz, J. C. Chervin, P. Munsch and G. Le Marchand, *J. Phys. D: Appl. Phys.*, 2009, **42**, 075413.
54. S. Klotz, K. Takemura, T. Strassle and T. Hansen, *J. Phys.: Condens. Matter*, 2012, **24**, 325103.
55. A. Dewaele and P. Loubeyre, *High Pressure Res.*, 2007, **27**, 419.
56. R. J. Angel, D. R. Allan, R. Miletich and L. W. Finger, *J. Appl. Crystallogr.*, 1999, **30**, 461.
57. A. K. Singh, H.-P. Liermann, Y. Akahama and H. Kawamura, *J. Appl. Phys.*, 2007, **101**, 123526.
58. M. Hanfland, K. Syassen and J. Köhler, *J. Appl. Phys.*, 2002, **91**, 4143.
59. A. K. Singh and G. C. Kennedy, *J. Appl. Phys.*, 1977, **48**, 3362.
60. W. A. Bassett, M. S. Weathers, T. C. Wu and T. Holmquist, *J. Appl. Phys.*, 1993, **74**, 3824.
61. R. J. Angel, M. Bujak, J. Zhao, G. D. Gatta and S. D. Jacobsen, *J. Appl. Crystallogr.*, 2007, **40**, 26.
62. R. A. Forman, G. J. Piermarini, J. D. Barnett and S. Block, *Science*, 1972, **176**, 284–285.
63. J. D. Barnett, S. Block and G. J. Piermarini, *Rev. Sci. Instrum.*, 1973, **44**, 1.
64. A. H. Shen, W. A. Bassett and I.-M. Chou, in *High Pressure Research*, ed. Y. Syono and M. H. Manghnani, Terra Scientific Publishing Company, Tokyo, 1992, vol. 67, pp. 61–68.
65. D. L. Heinz and R. Jeanloz, *J. Appl. Phys.*, 1984, **55**, 885.
66. Y. Meng and D. J. Weidner, *Geophys. Res. Lett.*, 1983, **20**, 1147.
67. D. Fan, J. Xu, Y. Kuang, X. Li, Y. Li and H. Xie, *Phys. Chem. Miner.*, 2015, **42**, 529.
68. G. Fiquet and B. Reynard, *Am. Mineral.*, 1999, **84**, 856.
69. B. Morosin and J. E. Schirber, *J. Appl. Crystallogr.*, 1974, **7**, 295.
70. C. E. Kundrot and F. M. Richards, *J. Appl. Crystallogr.*, 1986, **19**, 208.
71. P. Urayama, G. N. Phillips and S. M. Gruner, *Structure*, 2002, **10**, 51.

72. B. R. S. Hansen, K. T. Møller, M. Paskevicius, A.-C. Dippel, P. Walter, C. J. Webb, C. Pistidda, N. Bergemann, M. Dornheim, T. Klassen, J.-E. Jørgensen and T. R. Jensen, *J. Appl. Crystallogr.*, 2015, **48**, 1234.
73. N. Hamaya, Y. Kuroiwa and Y. Fujii, *Nucl. Instrum. Methods Phys. Res., Sect. B*, 1987, **29**, 537.
74. R. Finger, N. Kurtzemann, T. C. Hansen and H. Kohlmann, *J. Appl. Crystallogr.*, 2021, **54**, 839.
75. D. S. Yufit and J. A. K. Howard, *J. Appl. Crystallogr.*, 2005, **38**, 583.
76. A. Dawson, D. R. Allan, S. Parsons and M. Ruf, *J. Appl. Crystallogr.*, 2004, **37**, 410.
77. H. Nowell, S. A. Barnett, K. E. Christensen, S. J. Teat and D. R. Allan, *J. Synchrotron Radiat.*, 2012, **19**, 435.
78. I. Kantor, V. Prakapenka, A. Kantor, P. Dera, A. Kurnosov, S. Sinogeikin, N. Dubrovinskaia and L. Dubrovinsky, *Rev. Sci. Instrum.*, 2012, **83**, 125102.
79. C. Perreault, Y. K. Vohra, A. M. dos Santos, J. J. Molaison and R. Boehler, *High Pressure Res.*, 2019, **39**, 588.
80. G. Krauss, H. Reifler and W. Steurer, *Rev. Sci. Instrum.*, 2005, **76**, 105104–105109.
81. J. Getzschmann, I. Senkovska, D. Wallacher, M. Tovar, D. Fairen-Jimenez, T. Düren, J. M. van Baten, R. Krishna and S. Kaskel, *Microporous Mesoporous Mater.*, 2010, **136**, 50.
82. C. M. Brown, Y. Liu and D. A. Neumann, *Pramana*, 2008, **71**, 755.
83. K. Sumida, C. M. Brown, Z. R. Herm, S. Chavan, S. Bordiga and J. R. Long, *Chem. Commun.*, 2011, **47**, 1157.
84. Y. Uwatoko, K. Matsubayashi, T. Matsumoto, N. Aso, M. Nishi, T. Fujiwara, M. Hedo, S. Tabata and K. Takagi, *Rev. High Pressure Sci. Technol.*, 2008, **18**, 230.
85. S. E. Dissanayake, M. Matsuda, K. Munakata, H. Kagi, J. Gouchi and Y. Uwatoko, *J. Phys.: Condens. Matter*, 2019, **31**, 384001.
86. T. Okuchi, M. Yoshida, Y. Ohno, N. Tomioka, N. Purevjav, T. Osakabe, S. Harjo, J. Abe, K. Aizawa and S. Sasaki, *High Pressure Res.*, 2013, **33**, 777.
87. A. Sano-Furukawa, T. Hattori, H. Arima, A. Yamada, S. Tabata, M. Kondo, A. Nakamura, H. Kagi and T. Yagi, *Rev. Sci. Instrum.*, 2014, **85**, 113905.
88. W. G. Marshall and D. J. Francis, *J. Appl. Crystallogr.*, 2002, **35**, 122.
89. S. Klotz, J. Besson, G. Hamel, R. Nelmes, J. Loveday and W. Marshall, *High Pressure Res.*, 1996, **14**, 249.
90. I. N. Goncharenko, *High Pressure Res.*, 2004, **24**, 193.
91. J.-A. Xu, H.-K. Mao, R. J. Hemley and E. Hines, *J. Phys.: Condens. Matter*, 2002, **14**, 11543.
92. T. Hattori, A. Sano-Furukawa, S. Machida, J. Abe, K. Funakoshi, H. Arima and N. Okazaki, *High Pressure Res.*, 2019, **39**, 417.
93. V. Glazkov, S. Besedin, I. Goncharenko, A. Irodova, I. Makarenko, V. Somenkov, S. Stishov and S. S. Shil'Shteĭn, *Pis'ma Zh. Eksp. Teor. Fiz.*, 1988, **47**, 661.
94. I. N. Goncharenko, V. P. Glazkov, A. V. Irodova and V. A. Somenkov, *Phys. B*, 1991, **174**, 117.

95. J. Binns, K. V. Kamenev, G. J. McIntyre, S. A. Moggach and S. Parsons, *IUCrJ*, 2016, **3**, 168.
96. B. Haberl, S. Dissanayake, Y. Wu, D. A. A. Myles, A. M. Dos Santos, M. Loguillo, G. M. Rucker, D. P. Armitage, M. Cochran, K. M. Andrews, C. Hoffmann, H. Cao, M. Matsuda, F. Meilleur, F. Ye, J. J. Molaison and R. Boehler, *Rev. Sci. Instrum.*, 2018, **89**, 092902.
97. B. Haberl, S. Dissanayake, F. Ye, L. L. Daemen, Y. Cheng, C. W. Li, A. J. Ramirez-Cuesta, M. Matsuda, J. J. Molaison and R. Boehler, *High Pressure Res.*, 2017, **37**, 495.
98. R. Boehler, M. Guthrie, J. J. Molaison, A. M. dos Santos, S. Sinogeikin, S. Machida, N. Pradhan and C. A. Tulk, *High Pressure Res.*, 2013, **33**, 546.
99. K. Komatsu, S. Klotz, S. Nakano, S. Machida, T. Hattori, A. Sano-Furukawa, K. Yamashita and T. Irifune, *High Pressure Res.*, 2020, **40**, 184.
100. K. Yamashita, K. Komatsu, S. Klotz, M. T. Fernández-Díaz, O. Fabelo, T. Irifune, K. Sugiyama, T. Kawamata and H. Kagi, *High Pressure Res.*, 2019, **40**, 88.
101. A. Grzechnik, M. Meven, C. Paulmann and K. Friese, *J. Appl. Crystallogr.*, 2020, **53**, 9.
102. P. Usov, C. F. Leong, B. Chan, M. Hayashi, H. Kitagawa, J. J. Sutton, K. C. Gordon, I. Hod, O. K. Farha, J. T. Hupp, M. Addicoat, A. B. Kuc, T. Heine and D. M. D'Alessandro, *Phys. Chem. Chem. Phys.*, 2018, **20**, 25772.
103. N. Nijem, P. Thissen, Y. Yao, R. C. Longo, K. Roodenko, H. Wu, Y. Zhao, K. Cho, J. Li, D. C. Langreth and Y. J. Chabal, *J. Am. Chem. Soc.*, 2011, **133**, 12849.
104. M. Zhou, K. Wang, Z. Men, C. Sun, Z. Li, B. Liu, G. Zou and B. Zou, *CrystEngComm*, 2014, **16**, 4084.
105. A. Sussardi, R. J. Marshall, S. A. Moggach, A. C. Jones and R. S. Forgan, *Chem. – Eur. J.*, 2021, **27**, 14871.
106. T. Okuchi, G. D. Cody, H. K. Mao and R. J. Hemley, *J. Chem. Phys.*, 2005, **122**, 244509.
107. O. Brafman, S. S. Mitra, R. K. Crawford, W. B. Daniels, C. Postmus and J. R. Ferraro, *Solid State Commun.*, 1969, **7**, 449.
108. D. M. Adams and S. J. Payne, *Appl. Spectrosc.*, 1973, **27**, 377–381.
109. R. S. Hawke, K. Syassen and W. B. Holzapfel, *Rev. Sci. Instrum.*, 1974, **45**, 1598.
110. D. B. Tefelski, C. Jastrzębski, M. Wierzbicki, R. M. Siegoczyński, A. J. Rostocki, K. Wieja and R. Kościesza, *High Pressure Res.*, 2010, **30**, 124.
111. A. N. P. Sussardi, Doctor of Philosophy, The University of Edinburgh, 2021.
112. C. Postmus, V. A. Maroni, J. R. Ferraro and S. S. Mitra, *J. Inorg. Nucl.*, 1968, **4**, 269.
113. D. Radhakrishnan and C. Narayana, *J. Chem. Phys.*, 2016, **144**, 134704.
114. Z. Dong, Z. Mi, W. Shi, H. Jiang, Y. Zheng and K. Yang, *RSC Adv.*, 2017, **7**, 55504.
115. M. Andrzejewski and A. Katrusiak, *J. Phys. Chem. Lett.*, 2017, **8**, 279.

116. S.-H. Lee, K. Luszczynski and R. E. Norberg, *Rev. Sci. Instrum.*, 1987, **58**, 415.
117. R. Bertani, M. Mali, J. Roos and D. Brinkmann, *Rev. Sci. Instrum.*, 1992, **63**, 3303.
118. T. Meier, N. Wang, D. Mager, J. G. Korvink, S. Petitgirard and L. Dubrovinsky, *Sci. Adv.*, 2017, **3**, 1.
119. T. Meier, S. Khandarkhaeva, S. Petitgirard, T. Korber, A. Lauerer, E. Rossler and L. Dubrovinsky, *J. Magn. Reson.*, 2018, **292**, 44.
120. P. L. Walling and J. R. Ferraro, *Rev. Sci. Instrum.*, 1978, **49**, 1557.
121. T. L. Peck, R. L. Magin and P. C. Lauterbur, *J. Magn. Reson.*, 1995, **B108**, 114.
122. M. E. Lacey, R. Subramanian, D. L. Olson, A. G. Webb and J. V. Sweedler, *Chem. Rev.*, 1999, **99**, 3133.
123. H. C. Hoffmann, B. Assfour, F. Epperlein, N. Klein, S. Paasch, I. Senkovska, S. Kaskel, G. Seifert and E. Brunner, *J. Am. Chem. Soc.*, 2011, **133**, 8681.
124. F. Kolbe, S. Krause, V. Bon, I. Senkovska, S. Kaskel and E. Brunner, *Chem. Mater.*, 2019, **31**, 6193.
125. F. Birch, *Phys. Rev.*, 1947, **71**, 809.
126. C. L. Hobday, R. J. Marshall, C. F. Murphie, J. Sotelo, T. Richards, D. R. Allan, T. Duren, F.-X. Coudert, R. S. Forgan, C. A. Morrison and S. A. Moggach, *Angew. Chem., Int. Ed.*, 2016, **55**, 2401.
127. K. W. Chapman, G. J. Halder and P. J. Chupas, *J. Am. Chem. Soc.*, 2008, **130**, 10524.
128. R. N. Widmer, G. I. Lampronti, S. Anzellini, R. Gaillac, S. Farsang, C. Zhou, A. M. Belenguer, C. W. Wilson, H. Palmer, A. K. Kleppe, M. T. Wharmby, X. Yu, S. M. Cohen, S. G. Telfer, S. A. T. Redfern, F.-X. Coudert, S. G. MacLeod and T. D. Bennett, *Nat. Mater.*, 2019, **18**, 370.
129. R. J. Angel, M. Alvaro, J. Gonzalez-Platas and Z. Kristallogr, *Cryst. Mater.*, 2014, **229**, 405.
130. C. R. Groom, I. J. Bruno, M. P. Lightfoot and S. C. Ward, *Acta Crystallogr.*, 2016, **B72**, 171.
131. P. Z. Moghadam, A. Li, S. B. Wiggin, A. Tao, A. G. P. Maloney, P. A. Wood, S. C. Ward and D. Fairen-Jimenez, *Chem. Mater.*, 2017, **29**, 2618.
132. K. W. Chapman, G. J. Halder and P. J. Chupas, *J. Am. Chem. Soc.*, 2009, **131**, 17546.
133. P. Vervoorts, S. Burger, K. Hemmer and G. Kieslich, ChemRxiv, 2020.
134. A. S. Poryvaev, D. M. Polyukhov and M. V. Fedin, *ACS Appl. Mater. Interfaces*, 2020, **12**, 16655.
135. Z. Su, W. L. Shaw, Y. R. Miao, S. You, D. D. Dlott and K. S. Suslick, *J. Am. Chem. Soc.*, 2017, **139**, 4619.
136. T. D. Bennett, P. J. Saines, D. A. Keen, J. C. Tan and A. K. Cheetham, *Chem. – Eur. J.*, 2013, **19**, 7049.
137. M. E. Casco, Y. Q. Cheng, L. L. Daemen, D. Fairen-Jimenez, E. V. Ramos-Fernandez, A. J. Ramirez-Cuesta and J. Silvestre-Albero, *Chem. Commun.*, 2016, **52**, 3639.

138. T. Tian, M. T. Wharmby, J. B. Parra, C. O. Ania and D. Fairen-Jimenez, *Dalton Trans.*, 2016, **45**, 6893.
139. C. Gücüyener, J. van den Bergh, J. Gascon and F. Kapteijn, *J. Am. Chem. Soc.*, 2010, **132**, 17704.
140. C. L. Hobday, T. D. Bennett, D. Fairen-Jimenez, A. J. Graham, C. A. Morrison, D. R. Allan, T. Duren and S. A. Moggach, *J. Am. Chem. Soc.*, 2018, **140**, 382.
141. S. Aguado, G. Bergeret, M. P. Titus, V. Moizan, C. Nieto-Draghi, N. Bats and D. Farrusseng, *New J. Chem.*, 2011, **35**, 546.
142. C. L. Hobday, C. H. Woodall, M. J. Lennox, M. Frost, K. Kamenev, T. Duren, C. A. Morrison and S. A. Moggach, *Nat. Commun.*, 2018, **9**, 1429.
143. D. Fairen-Jimenez, R. Galvelis, A. Torrisi, A. D. Gellan, M. T. Wharmby, P. A. Wright, C. Mellot-Draznieks and T. Duren, *Dalton Trans.*, 2012, **41**, 10752.
144. G. Ortiz, H. Nouali, C. Marichal, G. Chaplais and J. Patarin, *Phys. Chem. Chem. Phys.*, 2013, **15**, 4888.
145. D. Fairen-Jimenez, S. A. Moggach, M. T. Wharmby, P. A. Wright, S. Parsons and T. Duren, *J. Am. Chem. Soc.*, 2011, **133**, 8900.
146. S. C. McKellar, J. Sotelo, A. Greenaway, J. P. S. Mowat, O. Kvam, C. A. Morrison, P. A. Wright and S. A. Moggach, *Chem. Mater.*, 2016, **28**, 466.
147. J. Ethiraj, F. Bonino, J. G. Vitillo, K. A. Lomachenko, C. Lamberti, H. Reinsch, K. P. Lillerud and S. Bordiga, *ChemSusChem*, 2016, **9**, 713.
148. Y. Ying, Z. Zhang, S. B. Peh, A. Karmakar, Y. Cheng, J. Zhang, L. Xi, C. Boothroyd, Y. M. Lam, C. Zhong and D. Zhao, *Angew. Chem., Int. Ed.*, 2021, **60**, 11318.
149. J. C. Tan, T. D. Bennett and A. K. Cheetham, *Proc. Natl. Acad. Sci. U. S. A.*, 2010, **107**, 9938.
150. T. D. Bennett, J. C. Tan, S. A. Moggach, R. Galvelis, C. Mellot-Draznieks, B. A. Reisner, A. Thirumurugan, D. R. Allan and A. K. Cheetham, *Chem. – Eur. J.*, 2010, **16**, 10684.
151. P. Adhikari, N. Li, P. Rulis and W.-Y. Ching, *Phys. Chem. Chem. Phys.*, 2018, **20**, 29001.
152. E. C. Spencer, R. J. Angel, N. L. Ross, B. E. Hanson and J. A. K. Howard, *J. Am. Chem. Soc.*, 2009, **131**, 4022.
153. R. N. Widmer, G. I. Lampronti, S. Chibani, C. W. Wilson, S. Anzellini, S. Farsang, A. K. Kleppe, N. P. M. Casati, S. G. MacLeod, S. A. T. Redfern, F. X. Coudert and T. D. Bennett, *J. Am. Chem. Soc.*, 2019, **141**, 9330.
154. P. Vervoorts, C. L. Hobday, M. G. Ehrenreich, D. Daisenberger and G. Kieslich, *Z. Anorg. Allg. Chem.*, 2019, **645**, 970.
155. K. T. Butler, P. Vervoorts, M. G. Ehrenreich, J. Armstrong, J. M. Skelton and G. Kieslich, *Chem. Mater.*, 2019, **31**, 8366.
156. C. R. Groom, I. J. Bruno, M. P. Lightfoot and S. C. Ward, *Acta Crystallogr., Sect. B*, 2016, **72**, 171.
157. G. W. Peterson, J. B. Decoste, F. Fatollahi-Fard and D. K. Britt, *Ind. Eng. Chem. Res.*, 2014, **53**, 701.
158. G. W. Peterson, J. B. Decoste, T. G. Glover, Y. Huang, H. Jasuja and K. S. Walton, *Microporous Mesoporous Mater.*, 2013, **179**, 48.

159. Y.-R. Miao, Z. Su and K. S. Suslick, *J. Am. Chem. Soc.*, 2017, **139**, 4667.
160. S. M. J. Rogge, J. Wieme, L. Vanduyfhuys, S. Vandenbrande, G. Maurin, T. Verstraelen, M. Waroquier and V. Van Speybroeck, *Chem. Mater.*, 2016, **28**, 5721.
161. P. G. Yot, K. Yang, F. Ragon, V. Dmitriev, T. Devic, P. Horcajada, C. Serre and G. Maurin, *Dalton Trans.*, 2016, **45**, 4283.
162. N. Liu, L. Shi and X. Meng, *J. Chem. Sci.*, 2019, **131**, 51.
163. M. Lammert, M. T. Wharmby, S. Smolders, B. Bueken, A. Lieb, K. A. Lomachenko, D. D. Vos and N. Stock, *Chem. Commun.*, 2015, **51**, 12578.
164. L. Li, S. Tang, C. Wang, X. Lv, M. Jiang, H. Wu and X. Zhao, *Chem. Commun.*, 2014, **50**, 2304.
165. S. M. J. Rogge, P. G. Yot, J. Jacobsen, F. Muniz-Miranda, S. Vandenbrande, J. Gosch, V. Ortiz, I. E. Collings, S. Devautour-Vinot, G. Maurin, N. Stock and V. Van Speybroeck, *ACS Mater. Lett.*, 2020, **2**, 438.
166. L. R. Redfern, M. Ducamp, M. C. Wasson, L. Robison, F. A. Son, F.-X. Coudert and O. K. Farha, *Chem. Mater.*, 2020, **32**, 5864.
167. S. Dissegna, P. Vervoorts, C. L. Hobday, T. Duren, D. Daisenberger, A. J. Smith, R. A. Fischer and G. Kieslich, *J. Am. Chem. Soc.*, 2018, **140**, 11581.
168. M. J. Cliffe, W. Wan, X. Zou, P. A. Chater, A. K. Kleppe, M. G. Tucker, H. Wilhelm, N. P. Funnell, F. X. Coudert and A. L. Goodwin, *Nat. Commun.*, 2014, **5**, 4176.
169. C. Atzori, G. C. Shearer, L. Maschio, B. Civalleri, F. Bonino, C. Lamberti, S. Svelle, K. P. Lillerud and S. Bordiga, *J. Phys. Chem. C*, 2017, **121**, 9312.
170. G. C. Shearer, S. Chavan, S. Bordiga, S. Svelle, U. Olsbye and K. P. Lillerud, *Chem. Mater.*, 2016, **28**, 3749.
171. J. H. Cavka, S. Jakobsen, U. Olsbye, N. Guillou, C. Lamberti, S. Bordiga and K. P. Lillerud, *J. Am. Chem. Soc.*, 2008, **130**, 13850.
172. V. R. Bakuru, S. R. Churipard, S. P. Maradur and S. B. Kalidindi, *Dalton Trans.*, 2019, **48**, 843.
173. B. Bueken, F. Vermoortele, M. J. Cliffe, M. T. Wharmby, D. Foucher, J. Wieme, L. Vanduyfhuys, C. Martineau, N. Stock, F. Taulelle, V. Van Speybroeck, A. L. Goodwin and D. De Vos, *Chem. – Eur. J.*, 2016, **22**, 3264.
174. K. Barthelet, J. Marrot, D. Riou and G. Férey, *Angew. Chem., Int. Ed.*, 2002, **114**, 291.
175. J. Im, D. Seoung, G. C. Hwang, J. W. Jun, S. H. Jhung, C.-C. Kao, T. Vogt and Y. Lee, *Chem. Mater.*, 2016, **28**, 5336.
176. T. K. Trung, I. Déroche, A. Rivera, Q. Yang, P. Yot, N. Ramsahye, S. D. Vinot, T. Devic, P. Horcajada, C. Serre, G. Maurin and P. Trens, *Microporous Mesoporous Mater.*, 2011, **140**, 114.
177. T. Loiseau, C. Serre, C. Huguenard, G. Fink, F. Taulelle, M. Henry, T. Bataille and G. Ferey, *Chem. – Eur. J.*, 2004, **10**, 1373.
178. F. Millange, C. Serre and G. Ferey, *Chem. Commun.*, 2002, 822.
179. M. R. Ryder, B. Civalleri and J.-C. Tan, *Phys. Chem. Chem. Phys.*, 2016, **18**, 9079.

180. E. V. Alexandrov, A. V. Goltsev, R. A. Eremin and V. A. Blatov, *J. Phys. Chem. C*, 2019, **123**, 24651.
181. P. G. Yot, K. Yang, V. Guillerm, F. Ragon, V. Dmitriev, P. Parisiades, E. Elkaïm, T. Devic, P. Horcajada, C. Serre, N. Stock, J. P. S. Mowat, P. A. Wright, G. Férey and G. Maurin, *Eur. J. Inorg. Chem.*, 2016, **2016**, 4424.
182. S. C. McKellar and S. A. Moggach, *Acta Crystallogr., Sect. B*, 2015, **71**, 587.
183. N. C. Burtch, J. Heinen, T. D. Bennett, D. Dubbeldam and M. D. Allendorf, *Adv. Mater.*, 2018, **30**, 1704124.
184. A. U. Ortiz, A. Boutin, A. H. Fuchs and F. X. Coudert, *Phys. Rev. Lett.*, 2012, **109**, 195502.
185. A. U. Ortiz, A. Boutin, A. H. Fuchs and F. X. Coudert, *J. Chem. Phys.*, 2013, **138**, 174703.
186. F.-X. Coudert, A. Boutin and A. H. Fuchs, *Mol. Phys.*, 2014, **112**, 1257.
187. C. Serre, F. Millange, C. Thouvenot, M. Noguès, G. Marsolier, D. Louer and G. Ferey, *J. Am. Chem. Soc.*, 2002, **124**, 13519.
188. S. Bourrelly, P. L. Llewellyn, C. Serre, F. Millange, T. Loiseau and G. Ferey, *J. Am. Chem. Soc.*, 2005, **127**, 13519.
189. A. V. Neimark, F. X. Coudert, C. Triguero, A. Boutin, A. H. Fuchs, I. Beurroies and R. Denoyel, *Langmuir*, 2011, **27**, 4734.
190. E. Cockayne, *J. Phys. Chem. C*, 2017, **121**, 4312.
191. A. M. Walker, B. Civalleri, B. Slater, C. Mellot-Draznieks, F. Cora, C. M. Zicovich-Wilson, G. Roman-Perez, J. M. Soler and J. D. Gale, *Angew. Chem., Int. Ed.*, 2010, **49**, 7501.
192. A. Thom, D. G. Madden, R. Bueno-Perez, A. A. N. Al Shakhs, C. T. Lennon, R. J. Marshall, C. W. Wilson, C. Murray, S. Thompson, G. F. Turner, D. Bara, S. A. Moggach, D. Fairen-Jimenez and R. S. Forgan, unpublished, 2021.
193. J. Wieme, L. Vanduyfhuys, S. M. J. Rogge, M. Waroquier and V. Van Speybroeck, *J. Phys. Chem. C*, 2016, **120**, 14934.
194. T. Devic, P. Horcajada, C. Serre, F. Salles, G. Maurin, B. Moulin, D. Heurtaux, G. Clet, A. Vimont, J.-M. Grenèche, B. Le Quay, F. Moreau, E. Magnier, Y. Flinchuk, J. Marrot, J.-C. Lavalley, M. Daturi and G. Férey, *J. Am. Chem. Soc.*, 2010, **132**, 1127.
195. M. Maes, F. Vermoortertele, L. Alaerts, S. Couck, C. E. Kirschhock, J. F. M. Denayer and D. C. de Vos, *J. Am. Chem. Soc.*, 2010, **132**, 15277.
196. M. Wang, X. Zhang, Y. Chen and D. Li, *J. Phys. Chem. C*, 2016, **120**, 5059.
197. A. Celeste, A. Paolone, J. P. Itie, F. Borondics, B. Joseph, O. Grad, G. Blanita, C. Zlotea and F. Capitani, *J. Am. Chem. Soc.*, 2020, **142**, 15012.
198. S. M. Humphrey, J. S. Chang, S. H. Jhung, J. W. Yoon and P. T. Wood, *Angew. Chem., Int. Ed.*, 2007, **46**, 272.
199. P. Iacomi, J. S. Lee, L. Vanduyfhuys, K. H. Cho, P. Fertey, J. Wieme, D. Granier, G. Maurin, V. Van Speybroeck, J.-S. Chang and P. G. Yot, *Chem. Sci.*, 2021, **12**, 5682.
200. D. E. P. Vanpoucke, K. Lejaeghere, V. Van Speybroeck, M. Waroquier and A. Ghysels, *J. Phys. Chem. C*, 2015, **119**, 23752.

201. K. J. Gagnon, C. M. Beavers and A. Clearfield, *J. Am. Chem. Soc.*, 2013, **135**, 1252.
202. J. M. Ogborn, I. E. Collings, S. A. Moggach, A. L. Thompson and A. L. Goodwin, *Chem. Sci.*, 2012, **3**, 3011.
203. A. B. Cairns, J. Catafesta, P. Hermet, J. Rouquette, C. Levelut, D. Maurin, A. van der Lee, V. Dmitriev, J. L. Bantignies, A. L. Goodwin and J. Haines, *J. Phys. Chem. C*, 2020, **124**, 6896.
204. J. Binns, K. V. Kamenev, K. E. Marriott, G. J. McIntyre, S. A. Moggach, M. Murrie and S. Parsons, *Chem. Commun.*, 2016, **52**, 7486.
205. F. Colmenero, *Phys. Chem. Chem. Phys.*, 2021, **23**, 8508.
206. Q. Zeng, K. Wang, Y. Qiao, X. Li and B. Zou, *J. Phys. Chem. Lett.*, 2017, **8**, 1436.
207. A. B. Cairns, A. L. Thompson, M. G. Tucker, J. Haines and A. L. Goodwin, *J. Am. Chem. Soc.*, 2012, **134**, 4454.
208. P. Serra-Crespo, E. Gobechiya, E. V. Ramos-Fernandez, J. Juan-Alcaniz, A. Martinez-Joaristi, E. Stavitski, C. E. Kirschhock, J. A. Martens, F. Kapteijn and J. Gascon, *Langmuir*, 2012, **28**, 12916.
209. Y. Yan, A. E. O'Connor, G. Kanthasamy, G. Atkinson, D. R. Allan, A. J. Blake and M. Schroder, *J. Am. Chem. Soc.*, 2018, **140**, 3952.
210. I. E. Collings, M. G. Tucker, D. A. Keen and A. L. Goodwin, *CrystEngComm*, 2014, **16**, 3498.
211. Q. Zeng, K. Wang and B. Zou, *ACS Mater. Lett.*, 2020, **2**, 291.
212. M. T. Wharmby, F. Niekiel, J. Benecke, S. Waitschat, H. Reinsch, D. Daisenberger, N. Stock and P. G. Yot, *Nanomaterials*, 2020, **10**, 1698.
213. Y. Yu, Q. Zeng, Y. Chen, L. Jiang, K. Wang and B. Zou, *ACS Mater. Lett.*, 2020, **2**, 519.
214. S. R. Miller, P. A. Wright, C. Serre, T. Loiseau, J. Marrot and G. Ferey, *Chem. Commun.*, 2005, 3850.
215. S. R. Miller, P. A. Wright, T. Devic, C. Serre, G. Ferey, P. L. Llewellyn, R. Denoyel, L. Gaberova and Y. Filinchuk, *Langmuir*, 2009, **25**, 3618.
216. A. J. Graham, A. M. Banu, T. Duren, A. Greenaway, S. C. McKellar, J. P. Mowat, K. Ward, P. A. Wright and S. A. Moggach, *J. Am. Chem. Soc.*, 2014, **136**, 8606.
217. J. P. Mowat, S. R. Miller, J. M. Griffin, V. R. Seymour, S. E. Ashbrook, S. P. Thompson, D. Fairen-Jimenez, A. M. Banu, T. Duren and P. A. Wright, *Inorg. Chem.*, 2011, **50**, 10844.
218. J. Sotelo, C. H. Woodall, D. R. Allan, E. Gregoryanz, R. T. Howie, K. V. Kamenev, M. R. Probert, P. A. Wright and S. A. Moggach, *Angew. Chem., Int. Ed.*, 2015, **54**, 13332.
219. S. C. McKellar, J. Sotelo, J. P. S. Mowat, P. A. Wright and S. A. Moggach, *CrystEngComm*, 2016, **18**, 1273.
220. D. N. Dybtsev, H. Chun and K. Kim, *Angew. Chem., Int. Ed.*, 2004, **116**, 5143.
221. A. Schneemann, P. Vervoorts, I. Hante, M. Tu, S. Wannapaiboon, C. Sternemann, M. Paulus, D. C. F. Wieland, S. Henke and R. A. Fischer, *Chem. Mater.*, 2018, **30**, 1667.

222. S. Henke, A. Schneemann, A. Wutscher and R. A. Fischer, *J. Am. Chem. Soc.*, 2012, **134**, 9464.
223. J. Wieme, S. M. J. Rogge, P. G. Yot, L. Vanduyfhuys, S.-K. Lee, J.-S. Chang, M. Waroquier, G. Maurin and V. Van Speybroeck, *J. Mater. Chem. A*, 2019, 7, 22663.
224. G. F. Turner, S. C. McKellar, D. R. Allan, A. K. Cheetham, S. Henke and S. A. Moggach, *Chem. Sci.*, 2021, **12**, 13793.
225. S. Wannapaiboon, A. Schneemann, I. Hante, M. Tu, K. Epp, A. L. Semrau, C. Sternemann, M. Paulus, S. J. Baxter, G. Kieslich and R. A. Fischer, *Nat. Commun.*, 2019, **10**, 1.
226. D. P. van Heerden, V. J. Smith, H. Aggarwal and L. J. Barbour, *Angew. Chem., Int. Ed.*, 2021, **60**, 13430.
227. M. Mendt, P. Vervoorts, A. Schneemann, R. A. Fischer and A. Pöppl, *J. Phys. Chem. C*, 2019, **123**, 2940.
228. A. U. Ortiz, A. Boutin, K. J. Gagnon, A. Clearfield and F. X. Coudert, *J. Am. Chem. Soc.*, 2014, **136**, 11540.
229. H. Wang, M. Feng, Y. F. Wang and Z. Y. Gu, *Sci. Rep.*, 2016, **6**, 26015.
230. S. Krause, V. Bon, I. Senkovska, U. Stoeck, D. Wallacher, D. M. Tobbens, S. Zander, R. S. Pillai, G. Maurin, F. X. Coudert and S. Kaskel, *Nature*, 2016, **532**, 348.
231. S. Krause, J. D. Evans, V. Bon, I. Senkovska, S. Ehrling, U. Stoeck, P. G. Yot, P. Iacomi, P. Llewellyn, G. Maurin, F.-X. Coudert and S. Kaskel, *J. Phys. Chem. C*, 2018, **122**, 19171.
232. S. Krause, J. D. Evans, V. Bon, I. Senkovska, F. X. Coudert, D. M. Tobbens, D. Wallacher, N. Grimm and S. Kaskel, *Faraday Discuss.*, 2021, **225**, 168.
233. B. Felsner, V. Bon, J. D. Evans, F. Schwotzer, R. Grünker, I. Senkovska and S. Kaskel, *Chem. Eng. J.*, 2021, **27**, 9708.
234. S. Krause, V. Bon, I. Senkovska, D. M. Tobbens, D. Wallacher, R. S. Pillai, G. Maurin and S. Kaskel, *Nat. Commun.*, 2018, **9**, 1573.
235. S. Krause, F. S. Reuter, S. Ehrling, V. Bon, I. Senkovska, S. Kaskel and E. Brunner, *Chem. Mater.*, 2020, **32**, 4641.
236. M. Eddaoudi, J. Kim, N. Rosi, D. J. Vodak, J. Wachter, M. O'Keeffe and O. M. Yaghi, *Science*, 2002, **295**, 469.
237. N. L. Rosi, J. Eckert, M. Eddaoudi, D. T. Vodak, J. Kim, M. O'Keeffe and O. M. Yaghi, *Science*, 2003, **300**, 1127.
238. S. S. Y. Chui, S. M.-F. Le, J. P. H. Charmant, A. G. Orpen and I. D. Williams, *Science*, 1999, **283**, 1148.
239. H. Li, M. Eddaoudi, M. O'Keefe and O. M. Yaghi, *Nature*, 1999, **402**, 276.
240. S. S. Y. Chui, S. M.-F. Lo, J. P. H. Charmant, A. G. Orpen and I. D. Williams, *Science*, 1999, **283**, 1148.
241. Y. H. Hu and L. Zhang, *Phys. Rev. B*, 2010, **81**, 174103.
242. M. R. Ryder, J. Maul, B. Civalleri and A. Erba, *Adv. Theory Simul.*, 2019, **2**, 1900093.
243. E. A. Kapustin, S. Lee, A. S. Alshammari and O. M. Yaghi, *ACS Cent. Sci.*, 2017, **3**, 662.

244. S. Lee, E. A. Kapustin and O. M. Yaghi, *Science*, 2016, **353**, 808.
245. J. Navarro-Sanchez, I. Mullor-Ruiz, C. Popescu, D. Santamaria-Perez, A. Segura, D. Errandonea, J. Gonzalez-Platas and C. Marti-Gastaldo, *Dalton Trans.*, 2018, **47**, 10654.
246. M. Andrzejewski, N. Casati and A. Katrusiak, *Dalton Trans.*, 2017, **46**, 14795.
247. M. Andrzejewski and A. Katrusiak, *J. Phys. Chem. Lett.*, 2017, **8**, 929.
248. D. W. Fu, W. Zhang and R. G. Xiong, *Dalton Trans.*, 2008, 3946.
249. Y. Saito, H. Takao, T. Tani, T. Nonoyama, K. Takatori, T. Homma, T. Nagaya and M. Nakamura, *Nature*, 2004, **432**, 84.
250. K. N. Le and C. H. Hendon, *Phys. Chem. Chem. Phys.*, 2019, **21**, 25773.
251. K. K. Tanabe and S. M. Cohen, *Chem. Soc. Rev.*, 2011, **40**, 498.
252. Z. Yin, S. Wan, J. Yang, M. Kurmoo and M.-H. Zeng, *Coord. Chem. Rev.*, 2019, **378**, 500.
253. S. Mandal, S. Natarajan, P. Mani and A. Pankajakshan, *Adv. Funct. Mater.*, 2020, **31**, 2006291.

CHAPTER 5

Mechanical Energy Absorption of Metal–Organic Frameworks

YUETING SUN*[a] AND HEBIN JIANG[a]

[a]School of Engineering, University of Birmingham, Edgbaston, Birmingham, United Kingdom
*E-mail: y.sun.9@bham.ac.uk

5.1 Introduction: Energy Absorption

Development of energy absorption materials and structures are required for improved protection from damages and injuries associated with mechanical impact, vibration, or explosion. These can be seen in many fields such as vehicle crash safety, anti-blast body armour or armoured vehicles, vibration-proof precision equipment, earthquake-resistant building structures, and sports products such as helmets. The high surface area and porosity offered by metal–organic frameworks (MOFs) can be exploited to develop efficient energy absorption materials based on solid–liquid interactions or framework deformations under mechanical pressure. In these processes, the energy absorption can be amplified by the internal surface area and porosity of MOFs and therefore holds great potential and has attracted increasing interest over the past decade. For example, the pressurised intrusion of non-wetting liquid into MOF nanopores can absorb mechanical energy and mitigate impact by generating a large solid–liquid interface.[1,2]

This chapter will start by giving the reader an overview of energy absorption and current energy absorption materials (Section 5.1). Subsequently,

Inorganic Materials Series No. 12
Mechanical Behaviour of Metal–Organic Framework Materials
Edited by Jin-Chong Tan

Published by the Royal Society of Chemistry, www.rsc.org

the concept of nanofluidic energy absorption and notable developments in the field are presented (Section 5.2). This will be followed by extensive discussions on the liquid intrusion of MOFs from quasi-static studies (Section 5.3) to recent development in the dynamic regime (Section 5.4). Finally, we will conclude this chapter with two non-intrusion-based energy absorption mechanisms of MOFs: structural transition (Section 5.5) and plastic deformation (Section 5.6).

5.1.1 Concept of Mechanical Energy Absorption

Energy absorption is the process of absorbing mechanical energy through the deformation of materials or structures. Figure 5.1 shows representative behaviour of energy absorption materials. These materials usually start with linear elastic compression, followed by a plateau during which a large deformation is gained within a small increase of pressure. There are different mechanisms by which to obtain such a plateau, for example, the buckling of thin-walled structures, crushing of porous structures, and liquid intrusion of nanopores. Undergoing a plateau plays a key role in energy absorption materials for protection applications as the likelihood of injuries is associated with the peak force or the integration of deceleration within a short duration of time during an impact, *e.g.*, the widely used head injury criterion

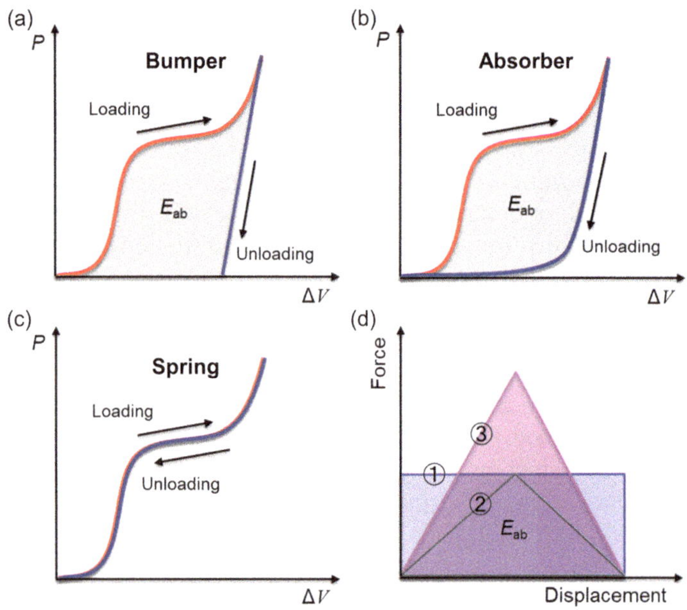

Figure 5.1 Typical mechanical behaviour of energy absorption materials: (a) bumper, (b) absorber, and (c) spring, plotted as pressure–volume change (P–ΔV) with the energy absorption denoted as E_{ab}. (d) Three different cases of force–displacement relationship.

(HIC).[3] The design target is to absorb enough energy within an acceptable peak force, so that less energy is transmitted to the protected object and the peak force is also kept under a reasonable threshold to prevent injuries. A plateau can serve this purpose well: as shown in Figure 5.1(d), where case 1 is the best as more energy is absorbed than in case 2 under the same peak force and it has a lower peak force than case 3 with the same energy absorption. The length of the plateau affects energy absorption capacity and depends on how the energy is absorbed. For porous materials, the plateau length is usually determined by their porosity.

When the plateau comes to an end, there is another linear elastic deformation stage. The amount of energy absorbed, E_{ab}, can be calculated from the area of the hysteresis loop enclosed by the loading and unloading curves. Figure 5.1(a) shows a linear unloading behaviour, like a bumper, which means that the material does not recover after deformation, *i.e.*, the plateau is underpinned by a permanent deformation. Figure 5.1(b) represents a reusable energy absorber, with its deformation recovered during the unloading process while still absorbing a substantial amount of energy. In Figure 5.1(c), the unloading curve overlaps with the loading curve, resulting in a spring-like behaviour with no energy absorption. This kind of spring may find its application in areas such as energy storage, where minimal dissipation is required. Energy absorption density is commonly used to evaluate these materials. This can be expressed in terms of the amount of energy absorbed per unit mass, *i.e.*, specific energy absorption, or per unit volume, *i.e.*, volumetric energy absorption capacity. One can choose between the two, depending on whether mass saving (*i.e.*, lightweight) or space saving is more important for a particular application. Some other evaluation criteria, such as crush force efficiency (CFE), are also used, which is defined as the ratio between the mean force and peak force during the loading process, so it is still about the plateau: CFE equals 1 when a perfect plateau or square wave is obtained.

5.1.2 Energy Absorption Materials

Conventional energy absorption mechanisms include plastic deformation of ductile materials, structure buckling, polymer damping, *etc.*, with typical examples including thin-walled tube structures,[4] polymer composites,[5,6] and cellular solids.[7,8] Thin-walled tubes absorb energy through a buckling process, which produces folded patterns and a plateau, and as such they are widely used in vehicles, with active research on increasing the energy absorption density by adding specific internal architectures or filling-materials like foams,[9,10] and improving the CFE by buckling triggers,[11] origami patterns,[12] or liquid fillers.[13,14] Polymer composites made of two or more constituents with distinct properties are also commonly used for energy absorption, such as carbon or glass fibre reinforced polymers, as they absorb energy *via* a variety of mechanisms, including matrix cracking, matrix–fibre debonding, fibre breakage, delamination, *etc.*[15–17] Cellular solids such as foams, honeycombs, or architected materials absorb energy by crushing their porous structures. Due to their random porous geometry, foams have the limitation of coupled plateau stress and length: a higher plateau stress means a reduced porosity and a shorter plateau length. Architected materials can address this limitation by topological design and optimisation,[18] and potentially maintain high strength and porosity simultaneously. This is because architected designs allow the constituent materials to be engaged in the energy absorption process in the most efficient way (*e.g.*, through a tensile mode rather than bending mode).

Despite the wide variety of energy absorption materials and structures currently being used, there are critical challenges to be addressed. For example, how to improve the energy absorption density beyond the order of 1–10 J g^{-1}, how to obtain reusability to withstand more than one impact, how to increase the response rate to be effective at high speed, *etc.* Some materials perform well under static conditions but poorly at high rates due to insufficient response time or stress localisation. Therefore, novel energy absorption materials have been developed over the past few decades, *e.g.*, graphene has been demonstrated to exhibit outstanding energy absorption under supersonic micro-projectile impacts;[19] some glassy polymer thin films presented surprisingly high energy absorption due to localised heating in high-rate deformation.[20] Among these materials innovations, nanoporous materials have also attracted substantial interest due to their high porosity and internal surface area. As reviewed in previous articles,[2,21] there are different mechanisms *via* which MOFs absorb mechanical energy, this chapter will mainly discuss the pressurised liquid intrusion of MOFs, but further considers other mechanisms such as the structural transition of flexible MOFs under hydrostatic pressure in pressure transmitting media,[22] and the plastic deformation of MOFs under uniaxial force that may involve nanopore collapse and chemical bond breakage.[21,23,24]

5.2 Nanofluidic Energy Absorption

5.2.1 Concept of Nanofluidic Energy Absorption

Conventional hydraulic shock absorbers make use of viscous liquid flow inside a small channel, either through a valve on the piston or between the piston and cylinder wall, to mitigate shock or vibration.[25] The energy absorption capacity of such mechanical systems can be characterised by the shear stress of the fluid and the contact area between the fluid and the solid structure. One can maximise energy absorption by amplifying the solid–liquid contact area, *i.e.*, by making the channels extremely small. This offers nanoporous materials a unique opportunity. Materials such as MOFs, zeolites, and silica have extremely small pore sizes down to the nanometre, which are comparable to the size of liquid molecules. Therefore, pressurised liquid intrusion into their nanopores can convert bulk liquid into a large solid–liquid interface, which promises high energy absorption density.

To enable such an energy absorption process, several criteria need to be met. In addition to a large internal surface area, the nanopores should be bigger in size than liquid molecules, and they should be chemically compatible with the liquid. Moreover, the nanopores must be lyophobic, *i.e.*, the liquid must be non-wetting to the nanopores so that liquid intrusion only takes place when external pressure is applied, otherwise the nanopores would absorb the liquid spontaneously at ambient pressure. For this reason, the water intrusion of hydrophobic nanopores is the most studied system.

Figure 5.2(a) shows a schematic overview of the pressurised water intrusion and extrusion process. A typical P–ΔV relationship is shown in Figure 5.2(b) using the water intrusion of ZIF-71 as an example, measured by the setup in Figure 5.2(c). Both intrusion and extrusion involve three stages. With an increase in the applied hydrostatic pressure on the water suspension of zeolitic imidazolate frameworks (ZIFs), as shown in Figure 5.2(c), the material starts with a linear elastic compression, designated as stage 1 in Figure 5.2(a and b). Once the pressure increases to a threshold value at which the water molecules start to penetrate the nanopores, a plateau is attained

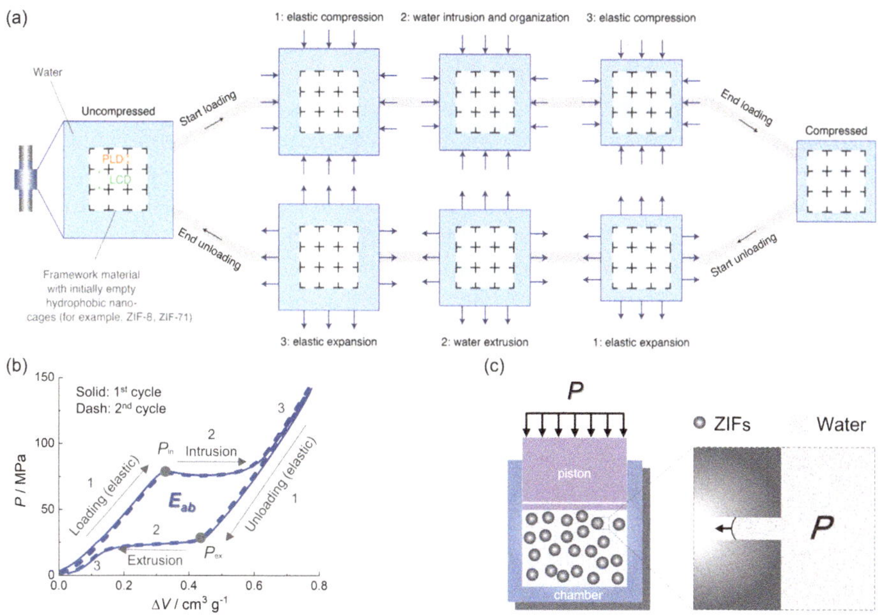

Figure 5.2 Pressure-induced water intrusion for energy absorption. (a) Schematic overview of the liquid intrusion and extrusion process. The framework in the middle with white background represents the nanopores, and the surrounding blue part represents the liquid water. Adapted from ref. 1 with permission from Springer Nature, Copyright 2021. (b) A typical P–ΔV curve, using the water intrusion of ZIF-71 as an example. (c) Schematic of the liquid intrusion setup for measuring the P–ΔV curves. Adapted from ref. 26 with permission from the Royal Society of Chemistry.

during which the mechanical energy is stored through a nanoscale liquid intrusion process, different from conventional energy absorption mechanisms. Such an intrusion process is designated as stage 2 in Figure 5.2(a and b) and the hydrostatic pressure at the onset of this process is called 'intrusion pressure' or 'infiltration pressure', denoted as P_{in} in Figure 5.2(b). Once all the nanopores are filled up by the liquid, another linear compression stage is obtained, *i.e.*, stage 3 (Figure 5.2(a and b)) as no more pores are available. As illustrated in Figure 5.2(a), the elastic compression in stages 1 and 3 is different, with the former applied on the guest-free nanoporous material and liquid, while the latter is on the liquid-filled nanoporous materials and liquid.

When the pressure is reduced during the unloading process, there is an elastic expansion first to recover the compression in stage 3 of the loading process. Once the external pressure is reduced to a certain level, the intruded water molecules start to extrude from the nanopores. Such a pressure is denoted as the extrusion pressure P_{ex} in Figure 5.2(b). Note that in some works, P_{in} and P_{ex} are also taken as the average pressure of the intrusion and extrusion plateau, respectively. In this example, water intrusion of ZIF-71 absorbs around 20 J g^{-1} of energy under quasi-static conditions. Figure 5.2(b) also shows that once the pressure is removed all water molecules escape from the nanopores and the material returns to its initial volume, this means that it is reusable, which is further validated by its performance in the second cycle. A bumper or a spring behaviour can also be attained *via* a liquid intrusion mechanism, where for a bumper system intruded water molecules are trapped inside the nanopores, and for a spring system P_{ex} equals P_{in} with no hysteresis. The performance of liquid intrusion systems depends on the nanoporous materials and the type of liquids used, so the highly modular architecture and physicochemical properties of MOFs promise diverse and tuneable energy absorption behavior, which can be customised for target applications.

5.2.2 The Emergence of the Field

Here, we provide a brief overview on how the research on the nanofluidic energy phenomenon of porous solids emerged from the mid-1990s to the mid-2000s. The field has developed extensively since 2006 in terms of various porous materials, which will be discussed in later sections. The earliest research on non-wetting liquid intrusion of porous solids originated from mercury porosimetry, which has been widely used to characterise the pore size of materials containing macropores (*i.e.*, pore diameter > 50 nm).[27,28] Different from capillary condensation, where fluid wets the pores, mercury has a high surface tension and does not wet most solid surfaces (*i.e.*, contact angle > 90°), so external pressure is required to force mercury into the pores. Such a mercury intrusion process, in which a progressive increase in hydrostatic pressure forces the mercury into the pores starting from the larger ones to the smaller ones, can be used to assess the size distribution ranging from 3 nm to *ca.* 400 μm.[29] The inverse relationship between the applied pressure P and the pore diameter d_p can be described by the modified Young–Laplace equation for cylindrical pores, known as the Washburn's equation, $d_p = -(4\gamma/P)\cos\theta$, where γ is the surface tension of mercury and θ is the contact angle between the solid sample and mercury.[30,31] However, due to the environmental and health concerns of mercury as well as its high working pressure of up to 400 MPa, which is too high for some materials,[29] studies have been carried out to replace mercury with less hazardous non-wetting liquids such as other liquid metals and alloys.[27] Water porosimetry has also been proposed as an alternative solution, although applicable only to hydrophobic materials, and this drew on the research interests into the science of water intrusion in hydrophobic nanopores.

Early works in this field started in the mid-1990s on the water intrusion of hydrophobic porous silica. Fadeev and Eroshenko employed a specially designed water porosimeter to investigate the penetration of water into hydrophobised porous silica.[32] In their system, water and silica are sealed inside a flexible envelope and subjected to hydrostatic pressure transmitted by oil. The pressurised intrusion of water into silica allows the envelope to shrink and a differential transformer and pressure gauge can be used to capture the volume change and pressure, respectively, during the process. They studied mesoporous grafted silicas and proved that Washburn's equation, which is in the context of continuum fluid mechanics, is still valid in the mesoporous range (2–50 nm), allowing use of a water porosimeter to characterise mesoporous and macroporous materials. Nowadays, there are commercial water porosimeters based on this principle. Some works at this stage also incorporated calorimetry into the system to investigate the thermodynamics during the water intrusion process of porous silica.[33,34]

It was not long before groups of researchers discovered that the non-wetting intrusion phenomenon behind water porosimetry can also be used to accumulate mechanical energy. In their early work,[35] Bogomolov made a brief comment that liquid metal confined in zeolite contains a great amount of surface energy and can potentially be used to construct mechanical energy accumulators. This was an extension from the well-studied concept of using zeolites as molecular sieves for wetting liquids, but the use of mercury for nonwetting liquid intrusion of zeolites was rare and indicated a new application of such materials systems. Seminal work in this field was reported by Eroshenko *et al.*,[36] which focused on the energetic applications of hydrophobic zeolites to accumulate, restore, and dissipate mechanical energy through a water intrusion process. This work inspired the exploration of more hydrophobic materials over the past two decades. Microporous siliceous zeolites (zeosils) were studied in this work, but other porous solids and liquids were proposed in earlier patents.[37]

Similar concept was implemented on hydrophobic porous silica in the early 2000s. Borman *et al.* studied the mechanical energy accumulation of silica with aqueous ethylene glycol solutions,[38,39] backed by the percolation theory they developed earlier for the intrusion of liquid metal and alloy.[40,41] Around the same time, Martin *et al.* and Lefevre *et al.* studied the water intrusion and extrusion of a mesoporous silica MCM-41.[42,43] MCM-41 is made up of parallel cylindrical pores with a narrow pore size distribution, so once grafted (to be hydrophobic) it is an ideal candidate for developing theoretical models or controlling intrusion behaviour. Their theoretical analysis of experimental results found that the standard capillarity theory can be used to describe the intrusion process well, but for the extrusion process the nucleation barrier of a vapour bubble, including the line tension effect, should be considered rather than using a receding contact angle, and it was also predicted that the intrusion–extrusion hysteresis should vanish as the pore becomes smaller,[43] consistent with more recent simulation work.[44]

In terms of experimental techniques, these early works used a water porosimeter or similar metallic chamber apparatus to supply hydrostatic pressure to nanoporous solids and nonwetting liquids, during which the volume change was measured as a function of the applied pressure. The porous solids and liquids are either directly placed into the chamber or encapsulated in a flexible cell surrounded by pressure transmitting fluid inside the chamber—just like the design of a water porosimeter.[36,43] The volume change is usually measured by the displacement of a piston that enters the chamber, and the pressure is usually measured by a pressure gauge. As the field was moving towards energy absorption applications, mechanical testing techniques started to enter this field, which enabled more diverse testing conditions. For example, since a damping application was first proposed by Eroshenko,[37,45–47] Suciu *et al.* investigated the concept of a colloidal damper using water–silica systems. Static tests were driven by a hand pump[48–50] and sinusoidal dynamic tests up to 10 Hz were driven by a vibrator,[51] with the pressure, stroke, and temperature measured by a high-pressure gauge, displacement sensor, and thermometer, respectively. They compared the single-cylinder and double-cylinder damper configurations,[51] carried out endurance studies,[52,53] investigated thermal effects,[54–56] and explored its application in vehicle suspensions.[57,58] Li *et al.* investigated a damping structure using magnetized multiwalled carbon nanotubes.[59] The Qiao group used a universal Instron machine to achieve cyclic loading on mesoporous silica with water and aqueous solutions of sodium chloride,[60] ethanol,[61] and sarcosyl,[62] and tested liquid intrusion at different temperatures[63] and at dynamic loading using the Hopkinson bar.[64]

Since 2006, there have been extensive investigations in this field on various porous solids and non-wetting liquids under different conditions. The porous solids investigated to date include silica,[65,66] zeolites,[36,67] MOFs,[68,69] polymers,[70] porous carbon,[71,72] carbon nanotubes,[59,73] and mesoporous organosilicas,[74] amongst others. The non-wetting liquids used include water,[1] aqueous solutions of salts[66] and alcohols,[26] non-aqueous liquids such as nonpolar hydrocarbons like paraxylene,[73] cyclohexane,[72] and diethyl ether,[75] and liquid metals such as mercury,[65] gallium,[71] and Wood's alloy.[41] Depending on the pore system of the solids (*e.g.*, pore size and dimensionality) as well as liquid properties, these materials systems present different energy absorption behaviour in their intrusion–extrusion process. Research on the liquid intrusion of MOFs is still in its early stages. The first publication dates back to 2013, by Ortiz *et al.* who reported the water intrusion of ZIF-8 under quasi-static conditions using a modified mercury porosimeter,[68] following earlier work from the Patarin group on various zeolites using a similar method.[76,77] In their experiments, the sample was contained in a polypropylene cylinder sealed by a mobile piston and then placed in another glass cell of the porosimeter filled with mercury. This work focused on energy storage, highlighting the ability of ZIF-8 to store 13.3 J g^{-1} of energy at a relatively low pressure (<30 MPa), but from the results one can also see its potential for energy absorption applications. Since then, a group of ZIFs have been reported for this application, which will be introduced in Section 5.3.

5.2.3 Water Intrusion of Microporous Zeolites

Pressurised water intrusion was first demonstrated on inorganic materials such as zeolites and silica before being extended to MOFs. Zeolites are microporous aluminosilicate minerals, and in the pioneering work by Eroshenko *et al.* the water intrusion characteristics of three hydrophobic pure silica zeolites were reported, including Silicalite-1 (F^-) and Silicalite-1 (OH^-) of the MFI type synthesised in hydroxide medium (OH^-) and fluoride medium (F^-) respectively, as well as Zeolite β (F^-) of the BEA type.[36] As shown in Figure 5.3, these three zeolites exhibit water intrusion at 99 MPa (curve 3), 81 MPa (curve 2), and 57 MPa (curve 1), respectively. Such behaviour contrasts with the more hydrophilic commercial material Na-ZSM-5 of the MFI type (curve 4), the volume of which varies linearly with pressure without any intrusion plateau.

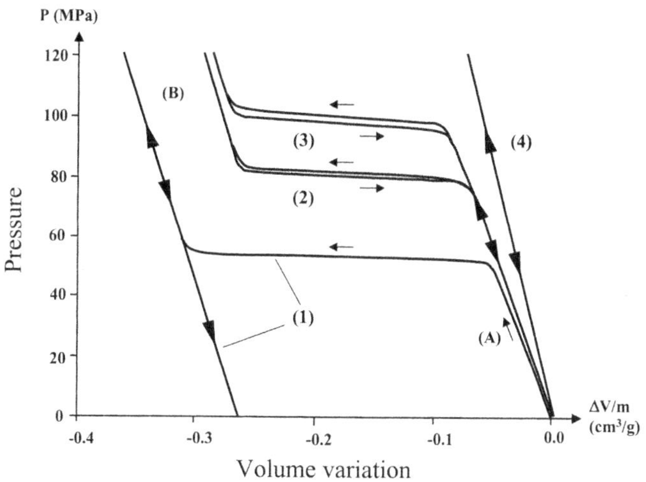

Figure 5.3 *P*–Δ*V* curves of water intrusion in (1) Zeolite β, (2) Silicalite-1 (OH^-), (3) Silicalite-1 (F^-), and (4) Na-ZSM-5. Reproduced from ref. 36 with permission from American Chemical Society, Copyright 2001.

The water intrusion of the two Silicalite-1 samples is reversible, and their extrusion pressures are almost identical to the intrusion pressures. Such molecular spring behaviour has also been observed by the same group in another hydrophobic pure silica zeolite, decadodecasil 3R of the DDR type.[67] In fact, most of the pure silica zeolites display spring behaviour, as shown in Table 5.1. A later study examined these two types of Silicalite-1 samples further and found that Silicalite-1 (OH^-) contains a small number of silanol groups while Silicalite-1 (F^-) does not, which explains why Silicalite-1 (OH^-) has a relatively lower intrusion pressure, reduced intrusion volume, and a more gradual onset or rounding of the intrusion plateau.[78]

Zeolite β (curve 1) behaves differently from Silicalite-1, behaving as an irreversible bumper with no extrusion when the pressure is removed. This irreversibility can be attributed to the presence of silanol defects at the interface of the polytypes of Zeolite β: these defects become accessible to water molecules during the intrusion step and prevent water molecules from being expelled due to their strong interactions.[36] However, a later study on Zeolite β, *via* more comprehensive structural characterisation of the intruded sample such as nuclear magnetic resonance (NMR) measurements, revealed that these hydrophilic silanol defects are created by the water intrusion process and are not present in the sample before the intrusion experiments.[79] Intrusion-induced silanol defects have also been observed in other zeolites, an interesting example of which is the intrusion of high-concentration sodium hydroxide into an MFI-type zeolite called ZSM-5, which effectively tunes the performance from spring to bumper.[80] It is also worth noting that a reversible performance of Zeolite β can be achieved under certain conditions, *e.g.*, at a higher intrusion rate,[81] with added ions in water.[79,82]

Table 5.1 summarises the quasi-static water intrusion behaviour of the pure-silica zeolites reported so far, with some of them also listed in earlier works.[2,84,87,100,101] Table 5.1 is not an exhaustive list of all zeolite materials that have been tested, as the same zeolite framework can be synthesised to yield materials with different Si/Al ratios or defects resulting in different performances, but Table 5.1 does represent the most important features of each framework using pure-silica zeolites. These zeolites are hydrophobic as synthesised, so most of them have substantial water intrusion pressure. Some zeolites such as OKO,[102] STO and MWW[87] exhibit spontaneous water intrusions (*i.e.*, P_{in} = 0 MPa) probably due to their high concentrations of silanol groups. These zeolites are not listed here, although it is still possible for them to absorb energy if more repellent liquids are used, such as concentrated salt solutions.[102] For hydrophilic aluminosilicate zeolites, post-synthesis heating treatment can be applied to increase their Si/Al ratio through dealumination. This has been proven to be effective in enhancing hydrophobicity and enabling energy absorption. For example, a heating process of 1000 °C for 3 h can promote mordenite and zeolite β to attain water intrusion plateaus, and it has also been found that a higher heating temperature can increase the water intrusion pressure and yield a higher intrusion volume.[1,103,104]

Table 5.1 Energy absorption of hydrophobic pure-silica zeolites under quasi-static water intrusion. Pore structure information is from the International Zeolite Association (IZA), including crystallographic direction, membered-ring numbers (in bold), pore diameter, and asterisk showing the dimensionality, more details from ref. 97. The pore limiting diameter (PLD) and largest cavity diameter (LCD) are from ref. 98. ΔV is the intruded pore volume, often slightly smaller than the theoretical pore volume, probably due to the different liquid density under nanoconfinement.[99] The energy absorption density $E_{ab} = E_{in} - E_{ex}$, with E_{in} being the energy stored during the intrusion process and E_{ex} being the energy recovered during the extrusion process. Unless accompanied by superscripts (1) or (2) meaning the data are from the first and second cycle respectively, all other data correspond to the first cycle. For spring systems, their following cycles have the same performance as the first cycle; for bumper systems, their following cycles do not have any intrusion process (P_{ex} = 0 MPa); for absorbers, the behaviour of the two cycles are shown in the table. '—' means data unavailable.

Framework	Material	Pore structure/Å	PLD/Å	LCD/Å	ΔV/cm^3 g^{-1}	P_{in}/MPa	P_{ex}/MPa	E_{in}/J g^{-1}	E_{ex}/J g^{-1}	E_{ab}/J g^{-1}	Behaviour	Ref.
AFI	SSZ-24	[001] **12** 7.3 × 7.3*	8.1	8.1	0.102	57	55	5.8	5.6	0.2	Spring	83 and 84
MTW	ZSM-12	[010] **12** 5.6 × 6.0*	6.3	6.3	0.114	132	126	15.0	14.4	0.6	Spring	83 and 84
TON	ZSM-22	[001] **10** 4.6 × 5.7*	5.7	5.7	0.075	186	172	14.0	12.7	1.3	Spring	83 and 84
CFI	—	[010] **14** 7.2 × 7.5*	7.4	7.5	0.08	75	75	6.0	6.0	0	Spring	85
DON	—	[010] **14** 8.1 × 8.2*	8.7	8.7	0.04	26	21	1.0	0.8	0.2	Spring	85
MTT	—	[001] **10** 4.5 × 5.2*	5.7	5.7	0.03	176	174	7	5.3	1.7	Spring	86
IFR	ITQ-4	[001] **12** 6.2 × 7.2*	6.3	6.3	0.136	42	0	5.7	0	5.7	Bumper	77 and 87
MTF	—	[001] **8** 3.6 × 3.9*	4.6	6.9	0.008	125	125	1.0	1.0	0	Spring	87
STF	—	[001] **10** 5.4 × 5.7*	6.0	8.3	0.04$^{(1)}$ 0.03$^{(2)}$	51$^{(1)}$ 38$^{(2)}$	37$^{(1)}$ 37$^{(2)}$	2.04$^{(1)}$ 1.15$^{(2)}$	1.1$^{(1)}$ 1.1$^{(2)}$	0.94$^{(1)}$ 0.05$^{(2)}$	Absorber$^{(1)}$ Spring$^{(2)}$	86 and 87
CDO	—	[010] **8** 3.1 × 4.7* ↔ [001] **8** 2.5 × 4.2*	3.9	6.4	0.03	210	180	6.3	5.4	0.9	Spring	87
FER	Ferrierite	[001] **10** 4.2 × 5.4* ↔ [010] **8** 3.5 × 4.8*	5.3	7.0	0.102	147	142	15	14.5	0.5	Spring	84 and 88
RRO	RUB-41	[100] **10** 4.0 × 6.5* ↔ [001] **8** 2.7 × 5.0*	4.4	4.4	0.15$^{(1)}$ 0.15$^{(2)}$	1.5$^{(1)}$ 1.2$^{(2)}$	0.5$^{(1)}$ 0.5$^{(2)}$	0.37$^{(1)}$ 0.30$^{(2)}$	0.18$^{(1)}$ 0.18$^{(2)}$	0.19$^{(1)}$ 0.12$^{(2)}$	Absorber$^{(1)}$ Absorber$^{(2)}$	89
MFI	Silicalite-1	[100] **10** 5.1 × 5.5 ↔ [010] **10** 5.3 × 5.6***	5.0	7.0	0.110	96	91	10.6	10	0.6	Spring	36 and 84
MEL	Silicalite-2	⟨100⟩ **10** 5.3 × 5.4***	5.7	8.4	0.103	63	58	6.5	5.9	0.6	Spring	84

(continued)

Table 5.1 (*continued*)

Framework	Material	Pore structure/Å	PLD/Å	LCD/Å	ΔV/cm^3 g^{-1}	P_{in}/MPa	P_{ex}/MPa	E_{in}/J g^{-1}	E_{ex}/J g^{-1}	E_{ab}/J g^{-1}	Behaviour	Ref.
ISV	ITQ-7	⟨100⟩ **12** 6.1 × 6.5** ↔ [001] **12** 5.9 × 6.6*	6.2	6.9	0.21$^{(1)}$ 0.22$^{(2)}$	30$^{(1)}$ 0.45$^{(2)}$	0.25$^{(1)}$ 0.23$^{(2)}$	4.3$^{(1)}$ 1.1$^{(2)}$	0.5$^{(1)}$ 0.7$^{(2)}$	3.8$^{(1)}$ 0.4$^{(2)}$	Absorber$^{(1)}$ Spring$^{(2)}$	90
BEC	—	[001] **12** 6.3 × 7.5* ↔ ⟨100⟩ **12** 6.0 × 6.9**	7.2	7.2	0.08	41	0	3.3	0	3.3	Bumper	91
BEA	Zeolite β	⟨100⟩ **12** 6.6 × 6.7** ↔ [001] **12** 5.6 × 5.6*	6.7	6.9	0.14	53	0	8.3	0	8.3	Bumper	36 and 79
ITH	—	[001] **10** 4.8 × 5.3* ↔ [010] **10** 4.8 × 5.1* ↔ [100] **9** 4.0 × 4.8*	5.6	7.4	0.08	82	0	6.6	0	6.6	Bumper	92
ITW	ITQ-12	[100] **8** 2.4 × 5.4* ↔ [001] **8** 3.9 × 4.2*	3.4	5.4	0.047	172	172	8.08	8.08	0	Spring	93
DDR	Decadodecasil 3R	⊥[001] **8** 3.6 × 4.4**	3.3	8.3	0.112	60	51	6.7	5.7	1	Spring	67 and 84
ITE	—	[010] **8** 3.8 × 4.3* ↔ [001] **8** 2.7 × 5.8*	4.9	8.9	0.10	32	32	3.2	3.2	0	Spring	87
STT	SSZ-23	[101] **9** 3.7 × 5.3* ↔ [001] **7** 2.4 × 3.5*	4.4	7.7	0.135$^{(1)}$ 0.114$^{(2)}$	40$^{(1)}$ 36$^{(2)}$	33$^{(1)}$ 32$^{(2)}$	7.0$^{(1)}$ 5.7$^{(2)}$	5.6$^{(1)}$ 4.7$^{(2)}$	1.4$^{(1)}$ 1.0$^{(2)}$	Spring$^{(1)}$ Spring$^{(2)}$	84 and 94
CHA	Chabazite	⊥[001] **8** 3.8 × 3.8***	4.2	8.0	0.15$^{(1)}$ 0.13$^{(2)}$	29$^{(1)}$ 22$^{(2)}$	20$^{(1)}$ 20$^{(2)}$	4.4$^{(1)}$ 2.9$^{(2)}$	2.6$^{(1)}$ 2.6$^{(2)}$	1.8$^{(1)}$ 0.3$^{(2)}$	Absorber$^{(1)}$ Spring$^{(2)}$	76 and 95
LTA	—	⟨100⟩ **8** 4.1 × 4.1***	4.9	11.7	0.17	20	0	3.4	0	3.4	Bumper	96

As shown in Table 5.1, most of the hydrophobic zeolites exhibit reversible spring behaviour with small hysteresis (<20%), which can be quantified by $(E_{in} - E_{ex})/E_{in}$, or $(P_{in} - P_{ex})/P_{in}$ in an ideal case. These materials are suitable for energy storage rather than energy absorption. Zeolites usually have a relatively smaller pore volume (*ca.* 0.1 $cm^3\ g^{-1}$) than MOFs and silica, thereby limiting the energy absorption density (E_{ab}) of zeolites, mostly to around 1 $J\ g^{-1}$, as seen in Table 5.1. It is worth noting that the intruded volume (ΔV) in Table 5.1 is usually lower than the theoretical pore volume or the value measured by gas sorption due to the lower water density in hydrophobic zeolite frameworks compared to the bulk, *e.g.*, the mass density of water in MFI zeolite is determined to be approximately 0.6 $cm^3\ g^{-1}$,[99] and 0.8–0.9 $cm^3\ g^{-1}$ in LTA,[105] IFR and TON zeolites.[106] There are a few irreversible bumpers in Table 5.1, including IFR,[74,82] BEC,[91] BEA,[36,79] ITH,[92] and LTA,[96] which absorb a slightly higher amount of energy (*ca.* 5 $J\ g^{-1}$) than the springs, but these zeolites can only function for one cycle, probably due to the formation of silanol groups during water intrusion. It has not yet been possible to use water–zeolite systems to make energy absorbers that are reusable and exhibit consistent performance for multiple cycles. There are a few zeolites that absorb energy and exhibit complete (*e.g.*, RRO,[89] ISV[90]) or partial (*e.g.*, STF,[86] CHA[76,95]) extrusions in their first cycle, broadly defined as 'absorbers' in Table 5.1, but their following cycles do not reproduce their first-cycle performance. Their intrusion pressure becomes substantially lower in subsequent cycles but without significant change in extrusion pressures, which reduces the energy absorption (*e.g.*, RRO[89]) or even turns the system into a spring (*e.g.*, ISV,[90] STF,[86] CHA[76,95]) for subsequent cycles.

The change in behaviour from the first cycle to the second cycle can be explained by the formation of silanol groups in the first cycle, which essentially makes the material less hydrophobic and thus lowers the intrusion pressure or retains some water in the framework. Although an ideal pure-silica zeolite can be free of defects, the water intrusion process can generate new defects by breaking the siloxane bridge (Si–O–Si bonds). Such defect formation seems to be more significant for the bumper and absorber types of zeolites mentioned above, but can also be present in spring-type zeolites. For example, STT retains its spring behaviour for multiple water intrusion cycles but the intrusion volume reduces gradually after each cycle due to the increase in the number of defects.[94] However, some zeolites can be recovered through calcination after intrusion tests and regain their initial performance, *e.g.*, the ISV-type zeolite ITQ-7 has a drop in intrusion pressure from 30 MPa in its first cycle to almost zero in its second cycle, but heating the zeolite at 600 °C for 4 h in air can regain its performance shown in the first cycle.[90]

Table 5.1 motivates the need to understand the correlation between framework structure and intrusion performance, as attempted in ref. 84. Figure 5.4(a) shows a plot of the intrusion pressure P_{in} over the pore limiting diameter (PLD), with the information of framework type, LCD/PLD ratio, and largest cavity diameter (LCD) values labelled on each data point, and colour coded by the dimensionality of the pore systems. The LCD/PLD ratio is a

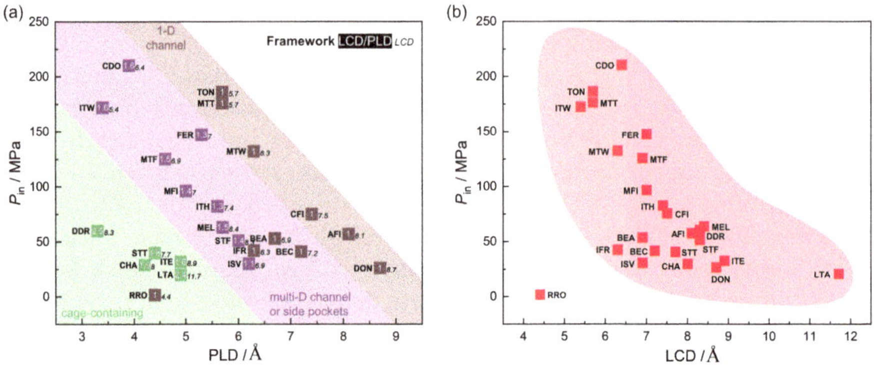

Figure 5.4 The intrusion pressure of various zeolite frameworks in Table 5.1, plotted in relation to (a) PLD and (b) LCD, respectively. In (a), each data point is labelled with the framework name, the LCD/PLD ratio, and the value of LCD; the data points are distributed in three different zones that correspond to different pore systems and different levels of LCD/PLD ratio. RRO is an outlier in both (a) and (b).

useful indicator of the pore geometry and dimensionality of the pore system. In Figure 5.4(a), the frameworks located in the top right zone are the unidimensional (1-D) channel systems with LCD/PLD equalling 1, including AFI, MTW, TON, CFI, DON, and MTT, which are listed at the top of Table 5.1. The frameworks found in the middle zone of Figure 5.4(a) are multi-dimensional channel systems, including CDO, FER, MFI, MEL, ISV, BEC, BEA, ITH, and ITW, as well as three 1-D channel frameworks that have side pockets including IFR, MTF, and STF. These frameworks have slightly higher LCD/PLD ratios in the range of 1–1.6 and are listed in the middle of Table 5.1. The bottom left corner of Figure 5.4(a) shows the frameworks containing a cage geometry with a LCD/PLD ratio of around 2, including DDR, ITE, STT, CHA, and LTA, the detailed information of which is listed at the bottom of Table 5.1.

Figure 5.4(a) shows that the intrusion pressure of zeolite frameworks is not only dependent on the PLD, but also related to LCD or the geometry and dimensionality of the pore system. For the frameworks with the same type of pore system, *i.e.*, very similar LCD/PLD ratio and dimensionalities, P_{in} decreases with the increase in PLD or LCD. This is in line with the Laplace–Washburn relation, although it is not applicable for describing microporous intrusion.[43] For a comparison across the three different types of pore systems, 1-D channel systems with the lowest LCD/PLD ratio have the highest intrusion pressure, cage-containing zeolites with the highest LCD/PLD ratio have the lowest intrusion pressure, and multi-dimensional channel systems and 1-D channel systems with side pockets are found to have an intrusion pressure between the two. These findings reveal the important role of pore geometry and dimensionality. Basically, with the same PLD, P_{in} decreases with the increase in LCD. In contrast, if comparing frameworks with the same LCD across different types of pore systems, PLD does not seem to have a significant influence on the intrusion pressure. For example, if a group of frameworks have a similar LCD of around 8 Å, including DDR, CHA, STT, ITE, MEL, STF, AFI, DON, their intrusion pressures are evenly distributed within the range of 26–63 MPa with no correlation to PLD. This analysis demonstrates that the intrusion pressure of zeolite frameworks is more closely related to LCD rather than PLD. Therefore, the plots in Figure 5.4(b) of P_{in} against LCD show an inverse relationship between the two parameters, whereas other factors have less of an effect on the distribution of the data points. Therefore, LCD should be regarded as a major design factor in the intrusion pressure. The outlier in Figure 5.4, RRO, has an exceptionally low intrusion pressure, most likely due to the presence of structural defects.

From a thermodynamic point of view, Bushuev *et al.* investigated different zeolite structures to estimate the water intrusion pressure associated with the configurational entropy of confined water.[106] Recently, they further demonstrated that the presence of secondary porosity, such as lateral windows on the main channels, can promote the water intrusion of microporous zeolites.[107] They established a correlation between the intrusion pressure P_{in} and the ratio of the accessible pore surface area to total pore volume (A/V): at the same A/V, a 1-D system has the highest P_{in} compared to multi-dimensional or cage systems.[108] These findings agree with our discussion above on the effect of pore geometry and dimensionality. Despite the interesting observations we established above on the intrusion pressure, less insight can be obtained from the data of the extrusion pressure and energy absorption, which seem to be predominantly affected by structural defects rather than geometric features. Nevertheless, the intrusion pressure does have energetic implications: it determines the energy storage of spring systems ($P_{ex} = P_{in}$) and the energy absorption of bumper systems (P_{ex} = 0 MPa). To obtain a spring with the highest energy storage or a bumper with the highest energy absorption, it follows that unidimensional channel systems with small pore diameters should be chosen.

5.2.4 Water Intrusion of Mesoporous and Macroporous Silica

Mesoporous and macroporous silica are another group of materials studied in this area. Their much bigger nanopores allow them to be made hydrophobic by the covalent grafting of nonpolar organic chains on the interior of their pores through a surface treatment process.[109] This approach provides the opportunity to select suitable modifiers to tailor the pore size and surface properties, and hence attain different intrusion pressures and extrusion performances.[110] For example, longer chains can effectively reduce the pore size and increase the intrusion pressure.[32] Note that classical theories such as the Laplace–Washburn equation may still work at meso and macro levels where the pore diameter is larger than ten water molecules and the confined water dynamics retain some continuum features, so the water intrusion performance of silica is more predictable than microporous materials although their extrusion is complex.[111]

Table 5.2 shows data of some representative hydrophobic silicas determined by quasi-static water intrusion. Compared to microporous zeolites, the water intrusion of silicas exhibits a significantly larger hysteresis area, which is good for energy absorption applications. Due to the big pore size, the intrusion pressures are not very high, but their extrusion pressures are even lower, thereby resulting in significant hysteresis between intrusion and extrusion. As silicas also have large pore volumes, their water intrusion provides promising energy absorption capacity. Table 5.2 shows that among the listed materials, Fluka 100 C8 exhibits the highest energy absorption density of up to 11 J g^{-1}.[64] However, in many cases their extremely low extrusion pressures result in irreversible bumper behaviour (when P_{ex} = 0 MPa), although the use of chemical admixtures such as sodium chloride can promote water extrusion and improve reusability.[60] Recent studies on water–silica systems have also focused on the outflow behaviour, including the effect of gas phase, surface properties, pore size, pore morphology, *etc.*,[110,112–118] with the aim of controlling the extrusion and reusability through materials design. Overall, it seems that for water intrusion, the spring behaviour dominates microporous zeolites, while the bumper behaviour dominates mesoporous and macroporous silicas. As will be shown in the following section, cage-type ZIFs can exhibit reusable absorber behaviour with a good combination of reusability and energy absorption.

Table 5.2 Energy absorption of selected hydrophobic silicas under quasi-static water intrusion. Unless accompanied by superscripts (1) or (2) meaning the data are from the first and second cycle, respectively, all other data correspond to the first cycle. For bumper systems, their following cycles do not have any intrusion process (P_{ex} = 0 MPa); for absorbers if the behaviours in two cycles are substantially different then both cycles are shown in the table. ΔV is the intruded pore volume. In this table, P_{in} is taken as the onset of the plateau while P_{ex} is taken as the end of the plateau. Data with asterisks are values estimated from testing curves.

Silica	Pore size/nm	ΔV/cm^3 g^{-1}	P_{in}/MPa	P_{ex}/MPa	E_{ab}/J g^{-1}	Behaviour	Ref.
TMS-PhSBA-1 (2:1)	2.1	0.07*	15	0	1.8	Bumper	119
MCM-41	2.4	0.17*	40	15	4.5	Absorber	42
KIT-6CF	6.3	0.27$^{(1)}$	31.5$^{(1)}$	6.5$^{(1)}$	6.9$^{(1)}$	Absorber	120
		0.25$^{(2)}$	29.5$^{(2)}$	6.5$^{(2)}$	5.8$^{(2)}$		
MSU-H	6.6	0.7	4.5	0	5.1	Bumper	121
Fluka 100 C8	7.8	0.5	16	0	11	Bumper	13 and 60
Zeoflo-TL	100	1.7*	1.6	0	4.5	Bumper	13

5.3 Liquid Intrusion of MOFs

5.3.1 Intrusion of Water

As a subfamily of MOFs, ZIFs have zeolitic topologies constructed from divalent cations connected through imidazolate-based ligands. ZIFs have been the main materials explored to date for water intrusion applications, due to their hydrothermal stability[122] and the availability of hydrophobic ZIFs.[123] Table 5.3 lists all the MOF materials with measured water intrusion behaviour, including ZIF-8, ZIF-67, ZIF-7, ZIF-9, ZIF-71, ZIF-11, ZIF-12, MAF-6, and Cu_2(tebpz). All of them belong to the ZIF family, with the exception of Cu_2(tebpz). Due to MOFs being a large family of materials (*ca.* 70 000),[124] there is substantial scope to explore other MOF materials for this application.

Figure 5.5 presents the quasi-static water intrusion performance of various ZIFs. A wide range of behaviour can be observed, with different intrusion and extrusion pressures, hysteresis areas, and reversibility. This is an indication of the diverse energy absorption performance that ZIFs can exhibit through the liquid intrusion mechanism, which stems from their highly modular architecture, tuneable geometry, and physicochemical properties. In principle, one can select suitable chemical moieties to construct desirable nanoporous structures for target applications. Among these materials, three of them have reversible responses, as evidenced by their extrusion plateaus and second loading cycles, including ZIF-8, ZIF-67, and ZIF-71, while other ZIFs have a linear unloading curve with no water extrusion after the experiment. Therefore, ZIF-8, ZIF-67, and ZIF-71 are reusable energy absorbers, while ZIF-7, ZIF-9, ZIF-11, ZIF-12 are irreversible bumpers like MAF-6.[127] The other MOF in Table 5.3, Cu_2(tebpz), behaves differently, it is a spring that does not absorb energy.[100]

ZIF-8 was the first MOF material investigated for water intrusion, although when it was first discovered the focus was on mechanical energy storage rather than energy absorption.[68] ZIF-67, the structural analogue of ZIF-8, was later demonstrated to have very similar performance.[69,128] ZIF-8 and ZIF-67 only absorb around 3 J g^{-1} of energy under quasi-static conditions, which implies that they exhibit spring performance due to their limited hysteresis loop, although they are classified as absorbers in Table 5.3 due to their enhanced performance under dynamic conditions (see Section 5.4). The water intrusion phenomenon of ZIF-71 was observed in 2014, with it showing reusable shock absorber behaviour.[126] The substantial hysteresis between its intrusion and the extrusion pressure leads to it exhibiting the highest energy absorption density amongst all the materials shown in Table 5.3. Compared with ZIF-8 and ZIF-67, ZIF-71 has a much higher intrusion pressure but a similar level of extrusion pressure, therefore it absorbs a much higher amount of energy. The fact that ZIF-71 has a higher intrusion pressure than ZIF-8 is counterintuitive because its PLD and LCD are larger than those of ZIF-8. This could potentially be explained by the different flexibilities of their structures related to the twist of the linkers:[126] the mIm linker of ZIF-8 has a higher capacity to twist compared to the dcIm linker of ZIF-71.[129]

Table 5.3 Energy absorption of hydrophobic MOFs under quasi-static water intrusion. mIm = 2-methylimidazolate, bIm = benzimidazolate, dcIm = 4,5-dichloroimidazolate. eIm = 2-ethylimidazolate, tebpz = 3,3′,5,5′-tetraethyl-4,4′-bipyrazolate. ΔV is the intruded pore volume. The energy absorption density $E_{ab} = E_{in} - E_{ex}$, with E_{in} being the energy stored during the intrusion process and E_{ex} being the energy recovered during the extrusion process. All data are obtained in the first cycle. For spring and absorber systems, their following cycles have the same performance as the first cycle; for bumper systems, their following cycles do not have any intrusion process ($P_{ex} = 0$ MPa). ZIF-8 and ZIF-67, with a behaviour between absorber and spring at quasi-static conditions, are listed as absorber in the table due to their higher energy absorption at dynamic conditions (see Section 5.4). For the various ZIFs from ZIF-8 to ZIF-71, the data are obtained from ref. 1. Due to the very short intrusion plateau of ZIF-7, ZIF-9, ZIF-11, and ZIF-12, the P_{in} and E_{ab} values of these materials are of relatively low accuracy. For example, the measured E_{ab} of ZIF-7 3.3 J g^{-1} is also contributed to by the penetration of water into the space between nanoparticles at a very low pressure which absorbs a specific energy of 1.2 J g^{-1}, and the energy absorption from the nanoporous water intrusion process is around 2.1 J g^{-1}. Data with asterisks are values estimated from testing curves in other references and '—' means data unavailable.

ZIFs	Topology	Metal	Linker	PLD/Å	LCD/Å	ΔV/cm^3 g^{-1}	P_{in}/MPa	P_{ex}/MPa	E_{in}/J g^{-1}	E_{ex}/J g^{-1}	E_{ab}/J g^{-1}	Behaviour	Ref.
ZIF-8	**sod**	Zn	mIm	3.4	11.4	0.35	24.9	21.5	16.5	13.6	2.9	Absorber	1 and 68
ZIF-67	**sod**	Co	mIm	3.4	11.4	0.40	16.3	14.7	14.0	11.1	3.0	Absorber	1 and 69
ZIF-7	**sod**	Zn	bIm	2.4	5.6	0.04	60.0	0	19.7	16.3	3.3	Bumper	1 and 125
ZIF-9	**sod**	Co	bIm	2.4	5.6	0.07	46.1	0	20.1	16.7	3.4	Bumper	1
ZIF-11	**rho**	Zn	bIm	3.0	14.6	0.07	23.7	0	20.8	17.2	3.6	Bumper	1
ZIF-12	**rho**	Co	bIm	3.0	14.6	0.04	26.9	0	19.6	17.1	2.5	Bumper	1
ZIF-71	**rho**	Zn	dcIm	5.5	17.0	0.32	78.8	23.7	53.3	34.3	19.0	Absorber	1 and 126
MAF-6	**rho**	Zn	eIm	7.6	18.1	0.52	25	<5	13*	0*	13*	Bumper	127
Cu_2(tebpz)	—	Cu	tebpz	Three 1D channels (5.94, 6.67, 11.6)		0.12	35.7	35.4	4.3	4.3	0*	Spring	100

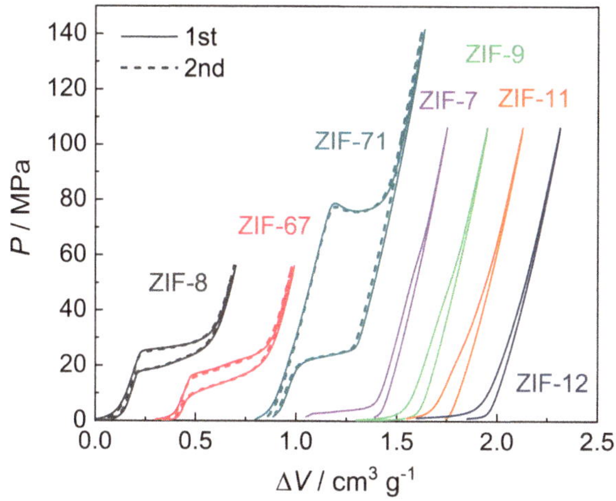

Figure 5.5 Quasi-static water intrusion of ZIF-8, ZIF-67, ZIF-7, ZIF-9, ZIF-71, ZIF-11, and ZIF-12, using data obtained from ref. 1. The second cycles of ZIF-8, ZIF-67, and ZIF-71 are shown as dashed lines, indicating the same responses as the first cycles.

ZIF-7 and ZIF-11 were first tested by Khay *et al.* in 2016 but no water intrusion was detected,[69] however, subsequent experiments by Sun *et al.* employing much smaller crystals revealed their water intrusion phenomena.[1,125] They exhibit bumper behaviour with no extrusion process, whereby all the intruded water molecules are trapped inside the framework. Their structural analogues ZIF-9 and ZIF-12 exhibit very similar performances to ZIF-7 and ZIF-11, respectively. It is important to recognise that ZIF-7 can adopt multiple phases and only the second phase (ZIF-7-II) can be intruded by water molecules, as its first phase (ZIF-7-I) is occupied by solvent molecules such as dimethylformamide (DMF), and the third phase (ZIF-7-III) features a two-dimensional layered architecture (see Figure 1.1(b and c) in Chapter 1).[125] Since water molecules are trapped within ZIF-11 and ZIF-12 (PLD = 3.0 Å) but flow out of ZIF-8 and ZIF-67 (PLD = 3.4 Å) when the pressure is removed, the threshold of PLD to enable water extrusion is determined to be around 3 Å, *i.e.*, a material must have a PLD of larger than 3.0 Å to exhibit reusable absorber behaviour.

In terms of the three pairs of structural analogues shown in Table 5.3 (ZIF-8 *vs.* ZIF-67, ZIF-7 *vs.* ZIF-9, ZIF-11 *vs.* ZIF-12), it has been demonstrated that the metal cation has very limited influence on the water intrusion behaviour. However, the linker plays a significant role, which can be revealed by comparison between MAF-6 and ZIF-71. Like the other ZIFs shown in Table 5.3, MAF-6 is also a hydrophobic ZIF known for its low water adsorption.[130] This material possesses the same topology and metal cation as ZIF-71 but has a different type of linker. In contrast to the reusable absorber performance of ZIF-71, MAF-6 exhibits irreversible behaviour, which, as will be discussed further in Section 5.3.5, lies in its partial structural collapse under pressurised water intrusion, although thermogravimetric analysis (TGA) seems to suggest that the intruded water molecules in the remaining pores come out of the framework.[127]

The chemical functionalisation of the linker has a significant impact on the pore hydrophobicity:[123] ZIFs comprising organic linkers bearing polar substituents (*e.g.*, ZIF-90) are capable of adsorbing water at low pressures or in the absence of pressure,[69,131] and not favourable for water intrusion applications. Therefore a recent study by Sun *et al.* proposed that hydrophobicity is used as the first design rule of constructing water-intrusion based shock absorbers.[1] There is also scope to design or modify linkers chemically to yield different water intrusion performances. For example, in addition to traditional ZIF-8_CH_3, its halogenated derivatives ZIF-8_Cl and ZIF-8_Br can also be intruded by water but exhibit different behaviour. ZIF-8_Br and ZIF-8_Cl exhibit lower water intrusion pressure than ZIF-8_CH_3, with ZIF-8_Br being <5 MPa and ZIF-8_Cl being 22 MPa; the intruded volume of ZIF-8_Br is also substantially lower than that of ZIF-8_CH_3 and ZIF-8_Cl, which might be related to the swing of imidazolate linkers.[132] It should be noted that all three ZIF-8 are hydrophobic due to the nonpolar methyl, bromine, and chlorine substituents borne by the imidazolate ligand.

The ZIFs in Table 5.3 are of different chemical compositions and topologies but most of them are constructed from nanocages connected through narrow window apertures, apart from Cu_2(tebpz), which is a channel-type MOF. Cu_2(tebpz) proves to be highly stable under water intrusion, including at high pressure and high temperature. It has three different channels and behaves as a molecular spring,[100] like many of the zeolites shown in Table 5.1. Apparently, channel-type frameworks usually have small hysteresis and energy absorption, probably due to the lower water transport resistance than observed in cage-type frameworks, although more investigation is required.[1]

Despite the limited number of MOF materials studied to date (Table 5.3), the overview in this section has demonstrated the potential in the rational design of MOF materials both chemically (*i.e.*, metal, linker, functionalisation) and structurally (*i.e.*, dimension, geometry, topology) for various behaviours and engineering performances.

5.3.2 Intrusion of Electrolyte Solutions

One advantage of exploiting the liquid intrusion mechanism for energy absorption applications is that its performance can be tuned by controlling the liquid properties, and therefore versatile systems can be developed using the same MOF material. It can be as easy as replacing the pure water with aqueous electrolyte solutions. It is found that adding some ions in water such as sodium chloride can effectively increase the intrusion pressure and promote the extrusion process. Such effect can potentially improve the energy absorption and reusability of the system, so the intrusion of electrolytes has attracted great interest in this field. Figure 5.6(a) gives an example of such an effect using ZIF-7 in a lithium chloride (LiCl) solution.[125] At 10 M

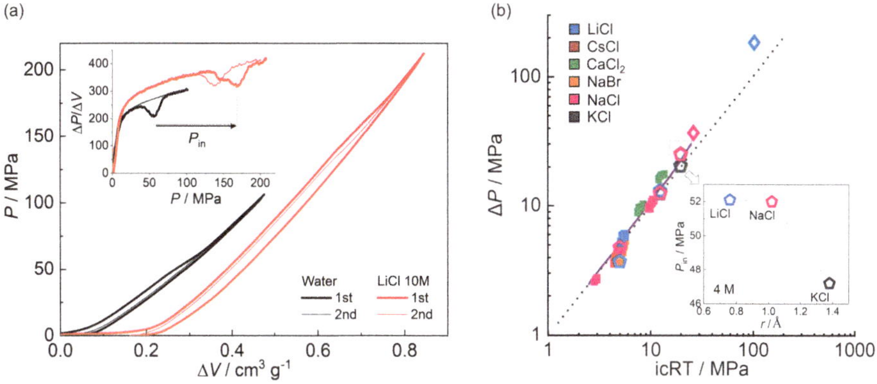

Figure 5.6 Effect of ions on water intrusion. (a) P–ΔV curves of ZIF-7-II nanocrystals (phase II) under the intrusion of water and 10 M LiCl solution for two consecutive cycles. The inset shows that with the addition of LiCl there is an increase in intrusion pressure and re-intrusion in the 2nd cycle. Adapted from ref. 125 with permission from American Chemical Society, Copyright 2018. (b) Excess intrusion and extrusion pressures of electrolyte solutions with respect to pure water intrusion, with data of ZIF-8 from ref. 133 (solid square) and ref. 134 (open pentagon) and Silicate-1 from ref. 135 (open rhombus). The dotted line is the van't Hoff osmotic pressure *icRT*, with *i* being the number of ions per salt molecule, *c* being the salt concentration, *R* being the perfect gas constant, and *T* being the temperature. The solid line is the simulated osmotic pressure of NaCl solutions from ref. 136, which is in agreement with the experimental data from ref. 134 and 135 at high concentrations, where there are measurable deviations from the van't Hoff law. The inset is a magnification of the ZIF-8 data at 4 M from ref. 134 plotted as the intrusion pressure P_{in} against the radius r of the cation. Adapted from ref. 133 with permission from American Physical Society, Copyright 2015.

(mol kg_{H2O}^{-1}), LiCl increases the intrusion pressure from 57 to 168 MPa and the associated energy absorption is tripled from 2.1 to 6.7 J g^{-1}. Importantly, LiCl not only serves as an intrusion inhibitor but also as an extrusion promoter: the intruded liquid flows out of the framework upon unloading as opposed to the non-extrusion behaviour of the pure water system. Therefore, intrusion plateaus are also observed in subsequent loading cycles, *i.e.*, LiCl not only allows ZIF-7 to absorb more energy, but also turns it from an irreversible bumper into a reusable absorber.

The effects of electrolytes on water intrusion have been observed for other nanoporous materials, including zeolites, silica, and MOFs. For example, the aqueous solutions of chloride salts (such as LiCl, NaCl, KCl, and $MgCl_2$) have been proven to increase the intrusion pressure of various pure silica zeolites,[101] such as MFI,[135,137–139] DDR,[140] ITH,[92] CFI,[85] DON,[85] FER,[141,142] OKO,[102] BEA,[79,143] CHA,[95,144] LTA,[96] MTF,[87] CDO,[87] FAU,[143] and BEC,[91] aluminosilicates such as ZSM-5[145,146] and Zeolite Y,[147,148] mesoporous silica,[120,149–151] and ZIFs such as ZIF-8[133,134] and its derivatives,[132] ZIF-7,[125] ZIF-71,[126,127] and MAF-6.[127]

In all cases, the higher the concentration of electrolyte solutions, the higher the intrusion pressure. Intriguingly, for some hydrophilic materials with spontaneous water intrusion, using electrolytes can potentially enable their energy absorption, as has been demonstrated for commercial aluminosilicates that are less hydrophobic than pure silica zeolites, including Zeolyst CBV780 (FAU, Si/Al = 40),[143] Zeolyst CBV-901 (FAU, Si/Al = 80),[147,152] and Clariant H-BEA-150 (BEA, Si/Al = 90).[143] Similarly, the extrusion promoting effect of electrolytes has also been demonstrated for different materials, including mesoporous silica,[149,150] ZIF-7,[125] and various pure-silica zeolites (such as BEA,[79] BEC,[91] LTA,[96] and CHA),[95] where LiCl is believed to protect zeolites from forming hydrophilic silanol defects that appear after pure water intrusion. For some zeolites, such as MFI,[137] DDR,[140] and MTF,[87] concentrated LiCl solutions (15–20 M) are found to increase their hysteresis and energy absorption as opposed to a spring behaviour under water intrusion.

The detailed mechanisms of how ions increase water intrusion pressure are not yet fully understood. Several hypotheses have been proposed, mainly including the increase in liquid–solid interfacial tension,[129] the osmotic phenomenon,[133] the confinement effect of nanopore walls,[147] and ion desolvation.[122,124] At the length scale of mesopores and macropores, the Laplace–Washburn equation may still apply, and the increased intrusion pressure of electrolyte solutions can be explained by the higher surface tension of the liquid compared to that of pure water.[149,153] Although this hypothesis can be used for mesoporous or macroporous silica, it is not valid for the molecular-sized confinement in microporous MOFs and zeolites.

Michelin-Jamois *et al.* proposed the use of osmotic pressure to describe the increased intrusion pressure associated with the ion exclusion process.[133] As shown in Figure 5.6(b), the van't Hoff law of osmotic pressure ($icRT$) works well in describing the excess intrusion and extrusion pressure (ΔP) of various electrolyte solutions in microporous ZIF-8, indicating that the increased intrusion pressure with the electrolyte solution comes from the osmotic pressure, *i.e.*, the additional pressure required to extract pure water from the surrounding ions before the water molecules can enter the ZIF-8 pores, also the pressure difference between the pure water inside and the salt solution outside the framework. However, although ZIF-8 proves to be a good sieve for water desalination (*i.e.*, Na^+ and Cl^- ions),[154,155] the assumption of ion exclusion, *i.e.*, only water molecules pass through the ZIF-8 apertures leaving ions outside in the bulk liquid, can be problematic. For example, there are deviations from the van't Hoff law for the intrusion of NaI and LiI solutions, explained by the possible penetration of ions into ZIF-8.[133] There is also experimental evidence of ion penetration into microporous zeolites that have similar pore sizes to the apertures in ZIF-8: *in situ* X-ray diffraction (XRD) studies have proven that at high pressures $MgCl_2$ can enter FER zeolite,[122] NaCl, NaBr, and $CaCl_2$ can enter CHA zeolites,[124] NaCl and $CaCl_2$ can enter LTA zeolites.[156] Moreover, the use of the van't Hoff law of osmotic pressure indicates that the increased intrusion pressure of electrolyte solutions is independent of the chemical identity of ions. However, as shown in Figure 5.6(b), different salts can have different values of intrusion pressure (or ΔP) at the same concentrations (or $icRT$). It seems that smaller cations have higher intrusion pressures,[134] which is also observed for various microporous materials, *e.g.*, ZIF-8 derivatives,[132] Silicalite-1,[135] Zeolite Y,[147] and mesoporous MCM-41.[151] This might be related to the fact that smaller cations have a higher solvation energy,[157] and tend to penetrate nanopores together with the neighbouring water molecules rather than separately,[144,156,158] therefore resulting in a higher free energy barrier for the intrusion process to occur.[158]

The above findings suggest a more complex phenomenon that underpins the increase in intrusion pressure with the addition of ions. The key is to understand the nature and dynamics of intruded species (water, ions) which can be associated with the hydration structure of ions under confinement. This requires experiments[142,144] or molecular simulations[158–160] to provide microscopic insight. *In situ* high-pressure XRD measurements of FER, CHA, and LTA zeolites have proved that both water and salt molecules enter micropores, and that the composition of the intruded species is different from that of the initial solution: the concentrations of cations and anions in the framework are substantially higher than those in the bulk liquid outside the framework indicating the partial desolvation of ions during the intrusion process. For example, with the intrusion of $MgCl_2$ solution in Ferrierite at high pressure, the liquid in the framework becomes $MgCl_2 \cdot 10H_2O$, a solvation degree that is close to a saturated solution and lower than that of the bulk liquid $MgCl_2 \cdot 21H_2O$.[142] Similarly, with the intrusion of NaCl and $CaCl_2$ solutions into Chabazite and LTA zeolite,[144,156] the intruded solution has a higher concentration than that of the initial bulk solution or even the saturated solution. Notably, the composition of the intruded phase can also vary with the pressure applied, indicating a sequence for water and ions to penetrate the pores. For example, for Chabazite under the intrusion of NaCl or $CaCl_2$ solutions, ions are only detected inside the framework when the pressure rises beyond 200 MPa, at a lower pressure only water molecules penetrate the Chabazite pores.[144] Similarly, for LTA zeolite, its intrusion by NaCl solution also starts with water molecules followed by ions at a higher pressure, but its intrusion by $CaCl_2$ solution leads to both water and ions being accepted at the same pressure as solvated ions,[156] probably due to the relatively higher solvation energy of Ca^{2+} compared to that of Na^{+}.[157]

Molecular simulations have been conducted on the intrusion of electrolytes in simple nanochannels as well as framework materials. Liu *et al.* simulated the pressurised intrusion of chloride salts with different cation sizes (LiCl, NaCl, KCl and CsCl) in a rigid silicon dioxide nanotube with a diameter of 7.4 Å.[147] It was found that such confinement prevented ion solvation and the intruded water molecules and ions formed a quasi 1-D chain with periodical cation–anion couples separated by water molecules. Different from pure water intrusion, there is an additional energy barrier for the intrusion of electrolytes due to the nonbonded interaction between the ion couples and the rest of the system, so the continuous liquid transport requires additional external work. The distance between the two ions in a couple and the distance between two ion couples are smaller for cations of smaller sizes, and since a smaller spacing between ion couples means a better chance for them to enter the nanotube, with smaller cations requiring more external work to drive the liquid transport, as a higher number of ion couples are present although the energy consumption of each ion couple is lower for smaller cations. This explains the higher intrusion pressure of smaller cations, as has been reported in experimental work.

Fraux *et al.* simulated the intrusion of LiCl solution in ZIF-8.[158] Ion desolvations were observed in their simulations, but it was found that only chlorine anions had a significant change in the number of water neighbours in the first solvation shell due to their weaker solvation,[157] while lithium cations penetrate ZIF-8 pores as a whole solvation sphere. As expected, ion desolvation favours the occurrence of anion–cation pairs with a higher number compared to the bulk. There are free energy barriers for the entry of ions but not for water molecules, and the free energy profiles of Li and Cl ions were found to be different. Compared to chlorine ions and water molecules, lithium exhibits weaker interactions with aromatic linkers and a more even distribution inside ZIF-8. With framework confinement and the presence of ions, the dynamics and diffusion of water will slow down, and ions cannot diffuse through ZIF-8 apertures due to energy barriers. Interestingly, the presence of water can deform the ZIF-8 structure, including linker rotation, while ions cannot. These findings highlight the need to investigate both cations and anions of different sizes and nature. In fact, the anion effect of zeolites has been experimentally studied, for Silicalite-1,[138] ZSM-5,[161] and Zeolite Y.[152,162] To exclude the influence from cations, anions with different charges (NaCl, CH_3COONa, Na_2SO_4 and Na_3PO_4) have been used to prove that anions can also increase the intrusion pressure when the cation concentration is unchanged.[161] Moreover, like the size effect of cations, smaller anions exhibit higher intrusion pressures, as demonstrated on Zeolite Y (P_{in}: NaF > NaCl > NaBr)[152,162] and ZSM-5 (P_{in}: NaCl > CH_3COONa).[161] It has also been discussed in both experiments and simulations that anions can have strong guest–host interactions with frameworks, including the aromatic linkers of ZIF-8,[158] and the silanol defects of zeolites.[156]

In terms of extrusion behaviour, although electrolytes promote water extrusion of most nanoporous materials, they can act in the opposite way in some cases. For example, ZIF-8 and ZIF-71 behave like reusable absorbers under water intrusion but adding ions can to some extent reduce their reversibility. This behaviour contrasts with the more general extrusion promoting effect of electrolytes that has been observed on other nanoporous materials, and provides another indication that the intrusion of electrolytes is not merely a size exclusion process. Specifically, with the intrusion of LiCl solution of ZIF-8 at a concentration of 4 M, some of the intruded liquid is trapped inside the framework (~22% according to gas adsorption) though the framework is preserved.[134] Such effect is most significant for a system consisting of LiCl and ZIF-8_CH_3, in comparison to other electrolytes (*e.g.*, NaCl, KCl) or ZIF-8 derivatives (*e.g.*, ZIF-8_Cl, ZIF-8_Br).[132,134] This discrepancy can be explained by the strongest interaction between the lithium ion and the imidazolate linker of ZIF-8_CH_3, compared to the interactions between other cations and imidazolate linker[163] or between lithium and the halogenated linkers. The latter is due to the electron-withdrawing effect of Cl and Br on the imidazolate linker in contrast to the electron-donating effect of the methyl group.[132] The different extrusion performance between cations might also be related to their different degrees of solvation due to ionic radius and solvation energy discussed earlier. For ZIF-71, a transition from absorber to bumper is observed when water is replaced with concentrated KCl[126] or LiCl[127] solutions. But different from ZIF-8, such phenomenon of ZIF-71 comes from framework collapse evidenced by its reduced pore volume and crystallinity after the intrusion of electrolyte. Importantly, the framework collapse proves to be a combined effect of pressure and liquid, the ZIF-71 framework can survive mechanical pressure or liquid immersion, but it collapses when subjected to the pressurised intrusion of concentrated electrolyte solutions.[127] The significant influence of electrolytes on the extrusion of ZIF-8 and ZIF-71 indicates the inclusion of ions in their framework, although more direct evidence is needed to understand the mechanism involved.

5.3.3 Intrusion of Alcohol Solutions

Alcohol is also a good additive that can be used to tune the performance of MOF liquid intrusion systems. As will be discussed below, most research on alcohol intrusion is on mesoporous materials, with the study by Sun *et al.* on ZIF-8 being the first experiment on the intrusion of alcohols into micropores.[26] As shown in Figure 5.7(a), the results demonstrate that alcohols can either decrease or increase the water intrusion pressure depending on the size of the alcohol molecules. Smaller alcohol molecules that can possibly penetrate the nanopores (*e.g.*, ethylene glycol) can lower the intrusion pressure, becoming lower than that of the pure water intrusion. Larger alcohol molecules, which are too bulky to enter the nanopores, will increase the intrusion pressure (*e.g.*, erythritol, xylitol), making it higher than that of pure water intrusion. The extent of such a notable change in the intrusion pressure can be controlled by the concentration of the aqueous solutions of the alcohols.[26]

For the intrusion of glycerol solution, since glycerol is 3.6 Å in the least extended dimension,[164] comparable with the theoretical limiting diameter of ZIF-8 (3.4 Å), the first intrusion cycle shows an increase in intrusion pressure with the addition of glycerol, but after a few cycles the intrusion pressure becomes lower than that of pure water intrusion (Figure 5.7(a and b)). Besides this, in the extrusion of pure water systems, nearly all the water molecules flow out of the ZIF-8 framework, although a tiny amount of residual water does exist, which leads to a small decline in intrusion pressure (*ca.* 1 MPa) in cyclic tests. However, much more liquid is trapped inside the framework with the addition of glycerol, as evidenced by the larger residual

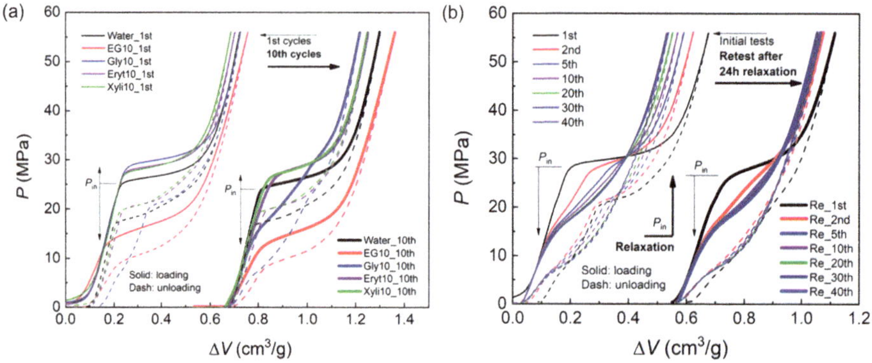

Figure 5.7 P–ΔV curves of cyclic intrusion of alcohol aqueous solutions into the ZIF-8 framework. Extrusion curves are shown as dotted lines and retested results are shifted horizontally for clarity. (a) Different alcohols at the same concentration of 10 wt%. (b) Cyclic intrusion of the 10 wt% glycerol aqueous solution. Two sets of 40-cycle measurements were performed with a relaxation time of 24 h in between them. Reproduced from ref. 26 with permission from the Royal Society of Chemistry.

deformation. Further experiments of repeated intrusion tests show that such residual deformation is recoverable after 24 h of relaxation (Figure 5.7(b)), and another test at a higher intrusion rate but with a holding of the system under the peak pressure[26] further demonstrated the time dependence of glycerol intrusion. It seems that glycerol molecules cannot be forced into the nanopore until the pressure is applied for an extended period, and it cannot leave the nanopore until the system has been at the ambient pressure for long enough. Such time dependence is not obvious for small liquid molecules but becomes significant for glycerol, which is comparable to the pore aperture size and thus has a very low diffusion rate.

The effect of alcohol on ZIF-7 has also been investigated, for which ethanol decreases its intrusion pressure significantly:[125] a spontaneous intrusion is reached (*i.e.*, P_{in} = 0 MPa) when the ethanol concentration is increased to 10 wt%. These two reports on ZIF-8 and ZIF-7 are the only MOFs or any microporous materials that have been studied under alcohol intrusion. There have been attempts of intruding zeolite ZSM-5 with glycerol,[165,166] but the authors believed that glycerol molecules are not small enough to enter the 5 Å micropores[166] and that the observed intrusion phenomenon is probably a result of glycerol intrusion into the mesoporous defects of the ZSM-5 crystals (*ca.* 2 nm). Therefore, the intrusion performance is very similar to that of mesoporous silica, characterised by a relatively low intrusion pressure (10–20 MPa) and limited extrusion.[167]

In comparison, there has been more extensive research on the alcohol intrusion of mesoporous materials, *e.g.*, the intrusion of ethanol,[61] ethylene glycol,[168] and glycerol[165–167,169,170] in hydrophobic mesoporous silica (pore size 7.8 nm), and porous carbon.[170] These alcohols are smaller in size than mesopore openings and will reduce the intrusion pressure.[61,165,166,168]

It is energetically favourable for these alcohol molecules to get close to the pores and meanwhile form strong bonds with water molecules, promoting water intrusion. Further, it is believed that alcohol molecules can potentially form an interface layer on the pore wall with a higher concentration than in the bulk phase.[61] Such an intrusion promoting effect can also be achieved by adding surfactants such as *N*-lauroylsarcosine sodium salt (sarcosyl), sodium cholate hydrate (SCH), ammonium 8-anilino-1-naphthalenesulfonate (1,8-ANS-NH_4), and 8-anilino-1-naphthalenesulfonic acid ammonium salt (ANSA).[62,171–173] Their molecular dimensions are close to the mesopore size but much bigger than the alcohols discussed earlier, therefore there is partial inclusion and exclusion due to the broad range of pore sizes found in silica, resulting in a two-staged intrusion plateau and selective energy absorption performance. This behaviour was also observed in the intrusion of polyethylene glycol (PEG) into silica, where the result was used to estimate the size of confined PEG at different molecular weights and concentrations in water.[174] In terms of extrusion, hydrophilic promoters inhibit mesoporous extrusion, as demonstrated by glycerol,[167] ethylene glycol,[168] and PEG,[174] although silica–water systems already retain most of the intruded water. Elevated temperature promotes extrusion for both water intrusion and alcohol intrusion.[165,174]

The intrusion of viscous alcohols such as glycerol at different flow rates and temperatures also provides insight into the effective viscosity of confined liquid inside mesopores.[167,169,170] It was found that the effective liquid viscosity confined in hydrophobic mesopores is orders of magnitude lower than its bulk counterpart, and the flow resistance is determined by the shear liquid viscosity force typically present in conventional laminar flow[175] as well as the nanoscale velocity slip at the solid–liquid interface.[176,177] Mesoscale fluids essentially lie between the continuum fluid and nanoscale transport: the effective liquid viscosity of the confined liquid is dependent on pore size and flow rate instead of being a materials constant,[170] it is influenced by the bulk liquid viscosity, but to a very limited extent due to the confinement effect.[167]

Although there is limited research on the pressurised alcohol intrusion of hydrophobic microporous solids, there are two research areas that are highly relevant. The first is the use of alcohol solutions as penetrating pressure transmitting media (PTM) to study the mechanical response of microporous materials *via* high-pressure synchrotron X-ray diffraction (Chapter 4). A mixture of methanol : ethanol : water (16 : 3 : 1, m.e.w.) is commonly used in these experiments and the research is mainly focused on the pressure induced hydration (PIH) phenomenon, *i.e.*, the penetration of additional water molecules into the framework under a pressure on the order of GPa. For example, Arletti and Lotti *et al.* studied pure-silica FER zeolite ferrierite using standard m.e.w. solution,[178] ethanol : water (1 : 3, e.w.) solution,[179] and ethylene glycol and 2-methyl-2-propen-1-ol.[180] Under compression with m.e.w. solution, more water molecules penetrated the 2-D channels of FER, while no evidence of methanol or ethanol penetration was observed. However, under compression with e.w. solution, both water and ethanol penetrated FER channels and were found in distinct channels with different arrangements: water was present in the 6-membered ring (6MR) channels in tetramer squares while ethanol formed linear dimers in 10MR channels. The reason for the distinct results from using different solutions is not well understood yet, but it is also possible that some alcohol molecules have disordered distributions that have not been observed in these experiments. Similar work has been conducted on the MFI zeolites H-ZSM-5 and Na-ZSM-5 using m.e.w. solution,[181–183] the MOR zeolite mordenite using m.e.w. solution, e.w. solution, and ethylene glycol,[184] as well as a group of MOF materials such as CuBTC (HKUST-1),[185] ZIF-8,[186,187] MOF-5,[188] Sc_2BDC_3 and $Sc_2(NO_2\text{-}BDC)_3$,[189] and UiO-67 and UiO-abdc,[190] using methanol, ethanol, methanol : ethanol solution (4 : 1), m.e.w. solution, isopropyl alcohol, and diethyl formamide, as well as non-penetrating PTM such as Fluorinert® and silicone oil (see further details in Chapter 4). The second relevant research area is the use of microporous solids for separating water–alcohol mixtures including gas adsorption,[191–194] membrane separations,[195–197] and pervaporation,[197] *etc.* The foregoing areas provide additional motivation to understand the intrusion of alcohol solutions into hydrophobic microporous solids.

In terms of practical applications, electrolytes and alcohol solutions provide good flexibility for designing versatile liquid intrusion performances targeting different applications. Some even bring additional benefit, such as the anti-freezing function of glycerol. From a fundamental point of view, further investigations are needed to understand the details about the intrusion of liquid mixture, *e.g.*, what determines the inclusion and exclusion of ionic or alcohol species, and then use these fundamental understandings to guide the rational design of engineering applications. A combination of liquid intrusion experiments, molecular simulations, and high-pressure XRD experiments is needed. There have been some studies on the intrusion of non-aqueous liquids such as paraxylene,[73] cyclohexane,[72] and diethyl ether,[75] as introduced in Section 5.2.2, but due to their relatively large molecular sizes such studies have not yet been attempted for MOFs.

5.3.4 Effect of Crystal Size and Other Design Considerations

In addition to framework design (Section 5.3.1) and liquid design (Sections 5.3.2–5.3.3), crystal size and shape can also be controlled to tune water intrusion performance. For example, Sun *et al.* discovered that the crystal size of ZIF-7 plays a role in enabling the water intrusion process: due to its small aperture size (PLD) which is smaller than water molecules, only nanocrystals of ZIF-7 can be intruded by water molecules, while bigger crystals on the micrometre scale show no water intrusion plateau, indicating ultra-high intrusion pressure beyond the experimental range.[125] Khay *et al.* compared different crystal sizes of ZIF-8 and found that bigger ZIF-8 crystals have a slightly higher water intrusion pressure than smaller crystals,[198] which is in agreement with the results of ZIF-7. Similar observations have been made on ZIF-8 at various intrusion rates.[1,199] It was found that under same loading condition, larger crystals exhibit slightly higher intrusion and extrusion pressure and energy absorption, although their overall behaviour is similar across different crystal sizes (from 100 nm to 1 μm), including the strain rate effect, as will be discussed in detail in Section 5.4.[1] Similar crystal size effect has also been observed on the quasi-static water intrusion of zeolite ZSM-5.[103]

Such a crystal size effect can be explained by the possible presence of disordered structures near the outer shell of small crystals.[198,200] Since the surface of a crystal can be more flexible than its core, it can potentially undergo gate-opening that facilitates water intrusion. Therefore, nanocrystals can effectively exhibit high structural flexibility due to their high surface area that interacts directly with liquid molecules in the intrusion process. The cage-by-cage water flow in ZIF-8 can also be interrupted by the crystal surfaces so smaller crystals seem to exhibit lower energy absorption density.[125] A similar crystal size effect has also been observed on the CO_2 adsorption of ZIF-7 with different crystal sizes, ascribed to their different degrees of structural flexibility.[200] The increase in flexibility by crystal downsizing has also been demonstrated by the Young's moduli of micro- and nano-sized ZIF-8 crystals.[201] Recently, Möslein *et al.* employed a combination of nearfield infrared spectroscopy (nanoFTIR) and density functional theory (DFT) simulations to elucidate how the structural defects of ZIF-8 evolve with crystallisation time: finer nanocrystals (1 min growth time) feature considerable surface defects and thus lower Young's modulus than less-defective crystals harvested after 60 min (see Figure 1.11 in Chapter 1).[202] Further investigation is thus needed to reveal the underlying mechanisms of the crystal size effect and to enable rational design.

There is also scope to design crystal shapes. Khay *et al.* compared different shapes of ZIF-8 and found that cubic-shaped crystals attain higher water intrusion pressure than rhombic dodecahedra, which can be explained by the different crystal facets and cage apertures that are exposed to water.[198] Zajdel *et al.* recently investigated the effect of macroscopic morphology and found that turning ZIF-8 from a fine powder into a dense monolith can substantially increase the energy absorption due to the different external surface areas of the forms that come into contact with the bulk liquid.[199] More complex design of crystals is possible but has only been demonstrated on zeolites. For example, Huve *et al.* designed Silicalite-1 nanoboxes, which contain mesoporous cavities and therefore reach a mesoporous intrusion plateau in addition to a relatively high microporous intrusion plateau. Like the performance of mesoporous silica described in Section 5.2.4, additional mesoporous intrusion is irreversible and has a low intrusion pressure: water intrusion is spontaneous and 20 M LiCl solution intrudes at 98 MPa.[139] Trzpit *et al.* also introduced mesopores into Silicalite-1 using carbon black or surfactant [3-(trimethoxysilyl)propyl]hexadecyldimethylammonium chloride as porogen and templating agents respectively, synthesised in both hydroxide[203] and fluoride media.[204] Different from the nanobox structure that reduces the micropore volume, using carbon black confers additional micropores to Silicalite-1. These methods to combine mesopores and micropores in a single material offer two separate liquid intrusion plateaus and can be beneficial to some energy absorption applications. For example, when a single energy absorption system is designed to protect against different levels of mechanical impact, its lower plateau can work with modest impact while its higher plateau can address more aggressive impact.

We have demonstrated the substantial scope to design MOF materials for water intrusion energy absorptions, from framework design (*e.g.*, metal, linker, structure), to liquid properties (*e.g.*, electrolytes, alcohols), and crystal size and shape. There is also potential to use external stimuli such as temperature and electric field to control liquid intrusion performances, related to the concept of thermowetting and electrowetting phenomena.[205,206] As will be discussed shortly, elevated temperature may cause ZIFs to degrade in water, but experiments on zeolites[165,207–211] and silica[63,167,174,212,213] have shown that an increase in temperature will decrease the liquid intrusion pressure, which can be explained by the lower solid–liquid interface tension or the reduced interaction between water molecules. Electric field has also been used to tune the intrusion pressure of salt solutions, including mesoporous silica,[214–217] zeolites,[218] and carbon.[219] For example, an applied voltage reduces the intrusion pressure of KCl solution into mesoporous silica,[214,217] but increases the intrusion pressure of LiCl aqueous solution into zeolite Y.[218]

5.3.5 Materials Stability Under Liquid Intrusion

The stability of MOFs in the presence of water is an important topic for its deployment in practical applications. Water, as a polar molecule, can form metal–water interactions with MOFs, especially those with open metal sites[220,221] which are accessible for water to form new chemical bonds.[222,223] Such interactions can potentially compete with and disrupt existing coordination bonds and breakdown the framework.[159,224,225] These effects can be less prominent for hydrophobic MOFs due to the weak interaction between the framework and the intruded water molecules. For example, the recent simulations by Sun *et al.* demonstrated that intruded water molecules tend to agglomerate around the cage centre and avoid the hydrophobic 6MR apertures of ZIF-8.[1]

Even though ZIF-8 was initially regarded to be stable in water,[122] recent studies reported that ZIF-8 can actually decompose into zinc and imidazolate ions *via* framework dissolution to form new substances in water.[226–229] Materials characterisation (*e.g.*, scanning electron microscopy (SEM), XRD, FTIR, gas adsorption) as well as repeated water intrusion can be used to examine materials degradation after water intrusion experiments. Sun *et al.* recently conducted a cyclic water intrusion test of ZIF-8 for 1000 cycles, showing its highly consistent performance (Figure 5.8(a)). This indicates that its molecular structure remains intact, which is also supported by the consistent XRD patterns before and after the intrusion tests.[1] Figure 5.8(b and c) shows the

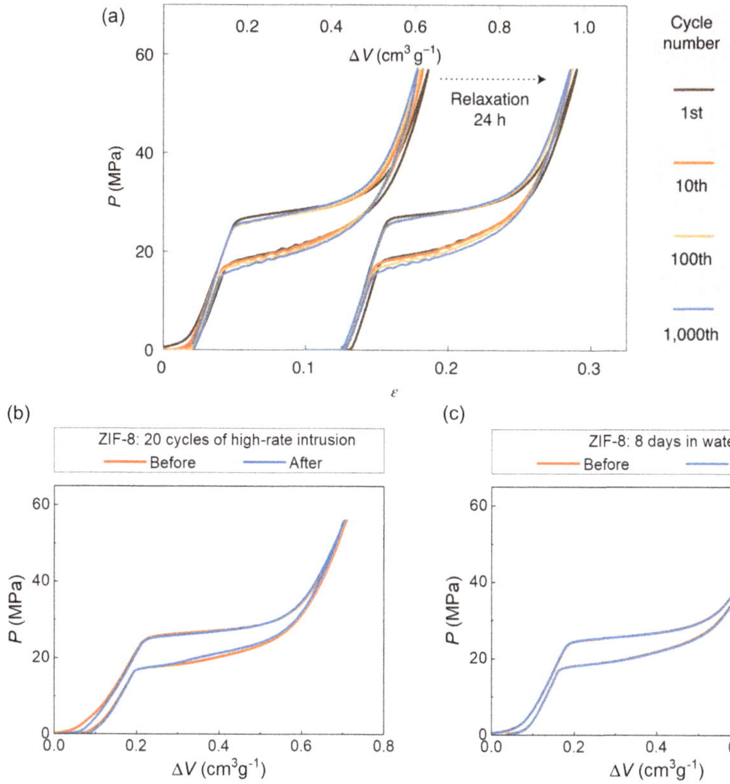

Figure 5.8 Cyclic water intrusion of ZIF-8 to confirm materials stability. (a) P–ΔV and stress–strain curves for two sets of 1000 cycles with a 24 h relaxation period between them. (b) P–ΔV curves before and after 20 cycles of impact-driven high-rate water intrusion. (c) P–ΔV curves before and after 8 days of immersion of ZIF-8 in water. Reproduced from ref. 1 with permission from Springer Nature, Copyright 2021.

water intrusion measurements after multiple impact-driven water intrusion tests and after being immersed in water for 8 days, respectively. No appreciable change of behaviour was observed, which also demonstrated the stability of ZIF-8 under such conditions. Specifically, Figure 5.8(b) shows a comparison before and after a set of twenty high-rate tests, which will be discussed in detail in Section 5.4. Figure 5.8(c) shows a comparison to examine the long-term water stability of ZIF-8 samples, which were not subjected to mechanical pressure beforehand. Such quasi-static water intrusion experiments are similar to the concept of water porosimetry,[27,32] in which the length of the intrusion plateau of the P–ΔV curves represents the total pore volume of the material, and therefore can be used to identify any potential change in the porosity and to evaluate materials stability.

The ZIF material that is known to be less stable under water intrusion is MAF-6. It undergoes substantial phase transformation during water intrusion and becomes a non-porous phase.[127] Such transformation comes from the synergetic effect of pressure and water, *i.e.*, the structure collapses during the penetration of water molecules. Therefore, the 'intrusion plateau' observed in its $P–\Delta V$ curve corresponds to the water intrusion of porosity, which is believed to be reversible from TGA results, as well as the partial phase transformation of the MAF-6 structure, which is irreversible. ZIF-7 is another ZIF material that presents identified stability issues, it is adequately stable against water intrusion, but since porous ZIF-7-II will transform into nonporous ZIF-7-III upon prolonged immersion in water for around one week, its long-term durability needs to improved.[125]

All the experiments above were conducted at room temperature and it should be noted that materials stability can be temperature dependent. ZIF-8 becomes less stable under high-temperature water intrusion. Grosu *et al.* demonstrated an irreversible structural transition of ZIF-8 from cubic to orthorhombic symmetry, which only occurs during water intrusion at a temperature of up to around 90 °C.[230] Such structural degradation has been proven to be a synergetic effect of high temperature and high pressure. Similarly, ZIF-67 exhibits reversible water intrusion at room temperature, however, this becomes irreversible at a high temperature[128] due to destruction of its framework structure, as revealed by XRD and FTIR spectroscopy measurements. All these results show that temperature is an important factor for water intrusion-based energy absorption applications, which should be considered together with pressure. Understanding thermomechanical effects is important because it is the process of water molecules penetrating ZIF pores that facilitates the decline of the stability of the ZIF under high pressure and temperature.

The water stability of ZIFs is related to the mass ratio between solids and liquids.[227] Water intrusion applications usually use a high mass ratio because a higher amount of solid can increase the total porosity and energy absorption capacity of a system, but it also helps to enhance materials stability as an additional benefit. For example, the results shown in Figure 5.8 involve the use of 25 wt% as the mass ratio of ZIF-8 to water (*i.e.*, 25 mg of ZIF-8 in 0.1 mL of water), which is much higher than the value in traditional water stability studies typically in the range of 0.2–6 wt%.[227] A higher mass ratio of ZIF-8 to water can inhibit the dissolution of ZIF-8 as the high-concentration solution can be easily saturated with inhibitive imidazolate ligands.[226,227]

Moreover, the stability of ZIFs is certainly also dependent on the nature of the liquid. As mentioned in Section 5.3.2, the intrusion of highly concentrated electrolyte solutions can result in structural collapse, *e.g.*, the ZIF-71 structure is well preserved under water intrusion but collapses under the intrusion of concentrated KCl and LiCl solutions.[126,127] MAF-6 collapses partially under the intrusion of either LiCl solution or pure water.[127] This is in contrast to ZIF-7, which survives and gains reversibility under the intrusion of concentrated LiCl solutions.[125]

There is a growing interest on the water stability of MOFs, as reviewed in ref. 231, due to the importance of materials stability for various applications. Some approaches have been proposed to improve the hydrothermal stability of ZIFs, such as surface ligand exchange or partial carbonisation,[232–234] and there have also been fundamental simulation studies to understand confined water in hydrophobic MOFs and the mechanism of instability.[159,231,235,236] However, for water intrusion energy absorption applications, more work needs to be done to understand and establish the long-term stability of MOFs in water and under cyclic mechanical pressure, including the coupled effects of temperature.

5.3.6 Thermal Effects of Liquid Intrusion

Calorimetric measurements have been carried out on some nanoporous liquid intrusion systems to better understand thermodynamics and energy conversion, with the results summarised in Table 5.4. These measurements were conducted under isothermal conditions using high-pressure calorimetric devices, with some studies also measuring temperature variations of the liquid intrusion process that are not included in the table.[56,146,237] The two MOFs reported to date, ZIF-8 and Cu_2(tebpz), exhibit endothermic water intrusion and exothermic water extrusion behaviour. The MFI-type pure silica zeolite Silicalite-1 exhibits the same type of thermal behaviour, but the CHA-type Chabazite behaves differently, exhibiting exothermic water intrusion and endothermic water extrusion. A number of mesoporous silicas have been tested but have shown inconsistent performance under water intrusion, with some of them being exothermic (PEP-100/300, EVA, KSK-G) while others are endothermic (WC8, CF3). More studies are thus needed to provide a better thermodynamic understanding of these results.

The results in Table 5.4 demonstrate that the thermal behaviour of nanoporous liquid intrusion is related to the pore systems.[88,244,245] An interesting comparison is Silicalite-1 (F^-) *vs.* Chabazite (F^-), which have similar surface chemistries but different pore geometries, MFI features channels while CHA is made up of cages. Karbowiak *et al.* carried out water intrusion calorimetric measurements and found that MFI is endothermic and CHA is exothermic under water intrusion and both have an opposite sign in extrusions.[244] Cailliez *et al.* suggested that the internal energy of intrusion can be seen as the sum of the energy of interaction between water and a zeolite surface, which is exothermic, and the energy due to the change in the number of hydrogen bonds between water molecules, which is endothermic or close to zero depending on the confinement.[88] Compared to CHA cages that can accommodate big water clusters (up to around 11 molecules), MFI channels provide a stronger confinement of water molecules that is close to a single-chain arrangement. Such strong confinement and associated reduction in coordination number counterbalance the exothermic water–surface interaction and lead to an overall endothermic intrusion process. In comparison,

Table 5.4 Thermal characteristics of MOFs and other nanoporous materials under quasi-static liquid intrusion. Thermal energy variation within an intrusion–extrusion cycle is $Q_{cycle} = Q_{in} + Q_{ex}$, where Q_{in} and Q_{ex} are the thermal energy during intrusion and extrusion, respectively. The sign '+' means endothermic and '–' means exothermic. Unless accompanied by superscripts (1) or (2) meaning the data are from the first and second cycles, respectively, all other data correspond to the first cycle. Data with asterisks are values estimated from testing curves and '—' means data unavailable.

	Material	Pore system and size/nm	Liquid	Temperature/K	Q_{in}/J g^{-1}	Q_{ex}/J g^{-1}	Q_{cycle}/J g^{-1}	Ref.
MOFs	ZIF-8	SOD cage 0.34–1.16	Water	300	+4.4	–	+	230 and 238
				310	+7.4	−5.4	+2.0	
				320	+9.4	−6.3	+3.1	
				330	+14.4	−12.7	+1.7	
				340	+13	−12.9	+0.1	
				350	+26.3	−25.7	+0.6	
	Cu_2(tebpz)	Channel ~1	Water	303	+2*	−0.5*	+1.5*	239
				323	+5*	−2.5*	+2.5*	
				343	+8.4*	−6*	+2.4*	
				363	+11.9*	−7.8*	+4.1*	
				360	+9.6$^{(1)}$	−7.2$^{(1)}$	+2.4$^{(1)}$	100
					+7.6$^{(2)}$	−7.2$^{(2)}$	+0.4$^{(2)}$	
Pure silica zeolites	Silicalite-1 (F^-)	MFI channel ~0.6	Water	283	+	–	+	240
				298	+7.8	−1.56	+6.24	241
				298	+7.8$^{(1)}$	−1.5$^{(1)}$	+6.3$^{(1)}$	242–244
					+3.5$^{(2)}$	−1.6$^{(2)}$	+1.9$^{(2)}$	
				303	+7.5	−7.3	+0.2	245
	Silicalite-1 (OH^-)	MFI channel ~0.6	Water	303	+5.7	−5	+0.7	245
	Chabazite (F^-)	CHA cage 0.38–0.74	Water	298	−7.85$^{(1)}$	+7.43$^{(1)}$	−0.42$^{(1)}$	244
					−7.6$^{(2)}$	+6.74$^{(2)}$	−0.86$^{(2)}$	
	Faujasite (simulation)	FAU cage 0.74–1.11	Water	300	–	—	—	88

Mesoporous silica	PEP-100	10	Water	298	−5.2*	—	—	34
	PEP-300	30	Water	298	−1.6*	—	—	34
	EVA	8	Water	303	−6.5	−1.2	−7.7	245
	KSK-G	13	Water	308	–	–	–	33
	WC8	8.4	Water	330	+10.9	−2.1	+8.8	238
			55% sucrose	370	Insignificant	—	—	246
			12% 1,3-dimethylimidazolium glutamate ionic	370	–	—	—	246
	CF3	10	Water	338	+11.7*	—	—	247
			5% ethanol	338	+9.9*	—	—	
			10% ethanol	338	+9.4*	—	—	
			15% ethanol	338	+7.2*	—	—	

water clusters in CHA cages are closer to the bulk phase, so the intrusion process of Chabazite is dominated by the exothermic water–surface interaction.[244] Such geometry effect is not observed between Cu_2(tebpz) and ZIF-8, both MOFs exhibit endothermic intrusion and exothermic extrusion processes despite their different geometries. However, more work needs to be done to fully understand the underlying mechanisms.

In terms of pore size effect, PEP-100 and PEP-300, two C18 grafted silicas, can be compared. Comparisons across different materials groups (*e.g.*, the 3-D channel-type MOF Cu_2(tebpz) *vs.* zeolite MFI *vs.* silica) are not appropriate due to their different chemistries. As shown in Table 5.4, both PEP-100 and PEP-300 exhibit exothermic water intrusion, which is consistent with the discussion on Chabazite above. These two mesoporous silicas have a pore size in the range of 10–30 nm, and at such a length scale, the confined water is close to a bulk and continuum phase,[248] with the thermal response in the intrusion process dominated by solid–liquid interaction effects. As PEP-100 has a larger pore volume (0.71 *vs.* 0.60 $cm^3\ g^{-1}$) and a substantially higher surface area (132 *vs.* 50.2 $m^2\ g^{-1}$), PEP-100 exhibits a relatively stronger exothermic effect under water intrusion.[34]

The surface chemistry of a material also plays an important role. This was first demonstrated by comparing Silicalite-1 (F^-) and Silicalite-1 (OH^-). Silicalite-1 (F^-) synthesised in a fluoride medium is more hydrophobic and contains fewer hydrophilic silanol defects than Silicalite-1 (OH^-). Table 5.4 shows that of the two materials, Silicalite-1 (F^-) exhibits a slightly more endothermic intrusion and exothermic extrusion process at the same temperature of 303 K, probably due to its relatively larger surface area and pore volume accessible for the hydrophobic intrusion process.[245] Another observation related to surface chemistry is the multi-cycle calorimetric measurements of Silicalite-1 (F^-) and Chabazite (F^-) carried out by Karbowiak *et al.*[244] and those of Cu_2(tebpz) by Grosu *et al.*,[100] as shown in Table 5.4. Due to the formation of hydrophilic silanol defects during cycling, the calorimetric heat of Silicalite-1 (F^-) decreases substantially after the first cycle. In addition to the solid–liquid interaction and confined water organisation mentioned above, there is an additional exothermic phenomenon associated with the bond-breakage of siloxane and silanol defect formation, which is strongest in the first intrusion.[241,242,244,249] Such a multi-cycle effect is not as prominent in Chabazite (F^-) due to it having fewer silanol defects compared to Silicalite-1 (F^-),[244] but is significant in Cu_2(tebpz) although ascribed to a different possible reason associated with the presence of air molecules and its influence on the confined water and thermal effect.[100] For all three materials, Silicalite-1 (F^-), Chabazite (F^-) and Cu_2(tebpz), their extrusion isotherms are similar across different cycles, as what is involved during the extrusion process is mainly the transition from confined water to bulk water.

Isothermal calorimetric measurements are carried out at a defined temperature. The results of ZIF-8 and Cu_2(tebpz) at different temperatures show that both Q_{in} and Q_{ex} increase at higher temperature. This can be understood using the macroscopic Gibbs heat of the solid–liquid interface development–reduction $Q = T[\mathrm{d}(\gamma\cos\theta)/\mathrm{d}T]\Omega$, where γ is the surface tension, θ is the contact angle, and Ω is the interface area, although its applicability to such a small length scale is questionable.[230,247,250] At a higher temperature, the Q_{cycle} of ZIF-8 decreases while the Q_{cycle} of Cu_2(tebpz) increases, indicating that for an athermal damping application, ZIF-8 is more suitable to be used at a relatively high temperature while Cu_2(tebpz) should be used at a low temperature.

Some studies were extended from the intrusion of pure water to aqueous solutions to investigate how liquid properties affect the thermal output. Like studies on temperature effects, investigations on liquid properties inevitably employed macroscopic concepts such as viscosity, contact angle, and surface entropy, and the pore systems investigated were mostly on the mesoscale.[247] For example, three liquids with different viscosities (water, 55% sucrose solution, and 12% 1,3-dimethylimidazolium glutamate ionic liquid solution) were used to intrude the mesoporous silica C8.[246] Since water has the lowest viscosity and the ionic solution has the highest, the results in Table 5.4 suggest that a more viscous liquid tends to exhibit exothermic intrusion while a less viscous liquid can have an endothermic intrusion.[246] This is in line with the concept of viscous dissipation because the internal shearing as for a continuum fluid is still present at the mesoscale. Table 5.4 shows that the intrusion of ethanol solutions in silica CF3 is endothermic but that such effect becomes smaller with an increase in ethanol concentration, which is consistent with the observation above and can also be explained by the decrease in surface entropy $[\mathrm{d}(\gamma\cos\theta)/\mathrm{d}T]$ with the variation in contact angle and surface tension.[247] Macroscopic descriptions such as surface entropy and Gibbs heat of interface prove to work well for rigid mesoporous silica and some microporous zeolites when the temperature dependence of the contact angle is considered, but not as good when applied to MOFs.[247]

It is also worth noting that in some systems, the simultaneous measurement of mechanical work and heat within an intrusion–extrusion cycle reveals an energy imbalance. Grosu *et al.* introduced the measurement of electricity generation during the water intrusion process and discovered an interesting electrification effect for the ZIF-8 and WC8 silica systems.[238] This not only implied a new application of materials systems in energy harvesting,[251–253] but also improved the energy balance of the system by compensating for mechanical and thermal energy hysteresis using an additional electrical energy term.[238]

5.4 Dynamic Liquid Intrusion

Water intrusion experiments were mostly performed under quasi-static conditions, but the target applications of mechanical energy absorption are under dynamic loading conditions, so there is a pressing need to study the liquid intrusion phenomenon at medium and high loading rates, which are relevant to practical applications. To enable such experiments, experimental techniques associated with mechanical vibrations and impacts have recently been introduced in this field.

5.4.1 High-rate Water Intrusion of ZIFs

The high-rate water intrusion of hydrophobic ZIFs was reported recently, which demonstrated their realistic behaviour under mechanical impact conditions.[1] The experimental method employed is called the split Hopkinson pressure bar (SHPB),[254] an impact mechanics technique used to measure the high-rate mechanical properties of materials, usually at a strain rate of 10^3 s^{-1}. As illustrated in Figure 5.9, there is a gas gun to the left of the setup that launches the striker at a high speed of up to tens of m s^{-1}. With the impact of the striker on the incident bar, there will be a stress wave generated inside the incident bar. This stress wave, with a certain amount being reflected, propagates through the sample, and enters another bar on the right called the transmitted bar. The sample, *e.g.*, a water suspension of ZIF-8 (Figure 5.9(c)), is sandwiched between the incident and transmitted bar, sealed inside a piston cylinder setup. There are strain gauges on the incident and transmitted bars that measure stress wave profiles, which can be used to calculate the forces and displacements at the sample–bar interfaces and produce P–ΔV curves for the sample.[254] The strain rate of the deformation can also be obtained from the recorded reflected wave.

The SHPB technique enables the measurement of the high-rate nanofluidic energy absorption of various nanoporous materials.[1] Remarkably, the results show that some materials exhibit very different performance from their quasi-static behaviour under such high-rate conditions. Figure 5.10(a) shows the water intrusion of ZIF-8 at different rates from hydrostatic compression to high-speed impact. The increase in strain rate comes with a sharp rise in intrusion pressure and slight drop in extrusion pressure, and these give rise

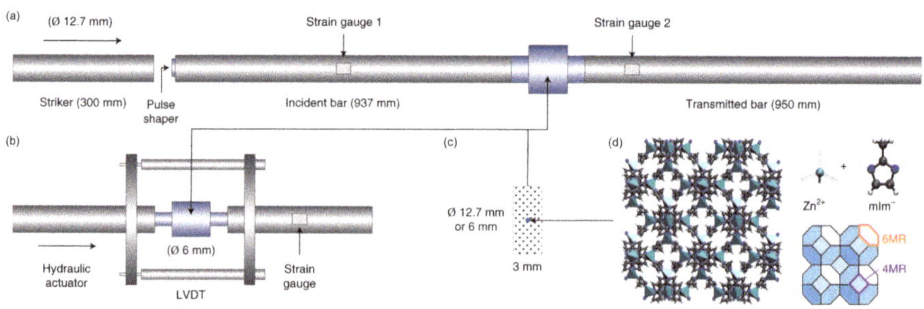

Figure 5.9 Dynamic water intrusion testing setup. (a) High-rate SHPB setup (10^3 s^{-1}). (b) Medium-rate hydraulic setup (1–10^2 s^{-1}). (c) As an example of the test sample, a water suspension of ZIF-8 sealed in a stainless-steel chamber. (d) Nanoporous framework structure, building blocks and sodalite (**sod**) topology of ZIF-8. Reproduced from ref. 1 with permission from Springer Nature, Copyright 2021.

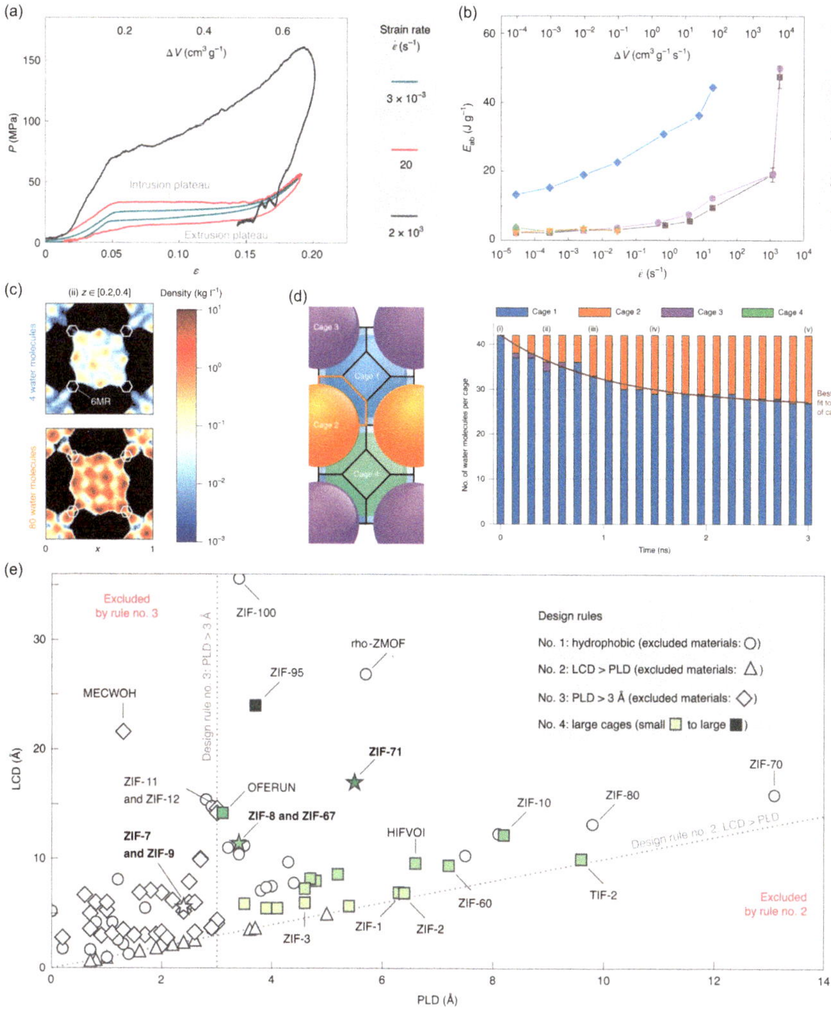

Figure 5.10 Dynamic water intrusion of ZIFs. (a) Compressive stress–strain curves of ZIF-8 at three different strain rates. (b) Energy absorption densities of various hydrophobic cage-type ZIFs as a function of the strain rate. The error bars represent the uncertainty due to the incomplete unloading curves. (c) Simulated water distribution in ZIF-8: symmetrised water density from a 2 × 2 × 2 ZIF-8 simulation cell as obtained from canonical Monte Carlo simulations at 298 K with either 4 or 80 water molecules per unit cell in the ambient-pressure (AP) phase. The axes are defined in fractional coordinates with one slab of the unit cell [0.2,0.4] shown. (d) The intrinsic timescale for water mobility in ZIF-8 according to non-equilibrium molecular dynamics (MD) simulations: evolution of the number of water molecules per cage in a 1 × 1 × 2 ZIF-8 supercell at 300 K and 0 MPa when starting from 42 water molecules in cage 1 and all other cages are initially empty. The best exponential fit of $a(1 - e^{-t/\tau})$ to the filling of cage 2 (red line) yields a time constant

(continued)

to substantial, up to 17-fold, enhancement in energy absorption, *i.e.*, much more energy can be absorbed under realistic impact conditions compared to quasi-static compression. Importantly, such rate-dependent water intrusion behaviour is also observed in other ZIF materials constructed from nanocages connected *via* hydrophobic narrow apertures, such as ZIF-67, ZIF-7, ZIF-9, and ZIF-71 (Figure 5.10(b)). Structural analogues, including ZIF-8 *vs.* ZIF-67, and ZIF-7 *vs.* ZIF-9, exhibit very similar dynamic performances, indicating the limited influence of metal cations.

MD simulations reveal that the rate effect of water intrusion into cage-type zeolitic materials originates from the intrinsic timescale for water clusters to nucleate inside the nanocages, which expedites the water transport across the cages and throughout the framework.[1] ZIF-8 consists of nanocages connected by small hydrophobic apertures; the simulated water distribution in ZIF-8 shows that water molecules tend to agglomerate at the cage centre and avoid hydrophobic apertures (Figure 5.10(c)). There is a substantial energy barrier for water molecules to transport from one cage to another, especially when the new cage is initially empty or only has very few water molecules. As shown in Figure 5.10(d), this gives an intrinsic timescale to form critically sized and stable hydrogen-bonded clusters in the neighbouring cage, which is on the nanosecond scale and can be translated into an intrinsic strain rate for ZIF-8 crystals that is comparable with the strain rate in SHPB experiments. Therefore, during slow compression, water clusters have enough time to nucleate inside cages, which will facilitate the intrusion of more water molecules, but when the loading rate is too high and water clusters cannot organize in time, additional external pressure and energy need to be applied to overcome the energy barrier for water transport.[1]

These fundamental understandings not only identified cage-type hydrophobic ZIFs as a unique class of energy-absorbing materials, but also led to four design rules by which to construct efficient and reusable impact energy absorbers *via* the pressurised liquid intrusion mechanism:[1] (1) the material should be hydrophobic; (2) the material should consist of nanocages, that is, LCD > PLD; (3) the apertures connecting the nanocages should be sufficiently large to allow liquid outflow and ensure reusability, *e.g.*, ZIF-8 has been demonstrated to be able to sustain repeated impacts or 1000 loading cycles with consistent performance (Figure 5.8); (4) large nanocages can accommodate large water clusters and hence increase the energy absorption density at high strain rates. Using these design rules, twenty ZIFs materials have been identified as potential materials, as depicted in Figure 5.10(e).[1]

Figure 5.10 *(continued)* $t \cong 1$ ns. (e) Materials selection map of the 105 ZIFs tabulated in ref. 255 according to their PLD and LCD, showing 20 promising materials fulfilling the four design rules (□ to ■); others are either not hydrophobic (as determined by their linkers, shown with circles), not cage-type (triangles) or have too small a PLD (diamonds). Experimentally validated materials are bold-faced (stars). Adapted from ref. 1 with permission from Springer Nature, Copyright 2021.

5.4.2 Dynamic Liquid Intrusion of Other Nanoporous Solids

High-rate water intrusion experiments have also been carried out on a group of hydrophobic zeolites.[1] Like the observations on ZIF pore geometries, the strain rate dependence is absent in channel-containing zeolites such as ZSM-5, zeolite β, and mordenite, but cage-type zeolites such as Chabazite can achieve enhanced energy absorption as the strain rate increases. Apart from SHPB, there are other dynamic testing methods available to characterise medium- and high-rate liquid intrusion behaviour. The medium-rate result at 20 s^{-1} in Figure 5.10(a) was obtained using the hydraulic setup illustrated in Figure 5.9(b), which consists of a hydraulic actuator, a strain-gauge based force transducer, and linear variable differential transformers (LVDT). As expected, medium-rate behaviour sits between static and high-rate performances and can be exploited for applications such as the vibration attenuations of vehicles and infrastructures. Static machines such as screw-driven mechanical testers or mercury porosimeters have also been used to explore the rate effect of ZIFs,[26,128,256] but they are limited to the low-strain rate range.[72,112,234]

For zeolite–water systems, Sun *et al.* used a drop-weight tower to study the water intrusion of Zeolite β, with the force detected by an accelerometer on the hammer and the displacement recorded by a high-speed camera (Figure 5.11(a)).[104] This is a medium-rate impact from a 200 kg hammer at a height of 0.2 m, giving an initial impact velocity of 1.98 m s^{-1} and strain rate of 27.50 s^{-1}, similar to the strain rate from a hydraulic machine. The reduced

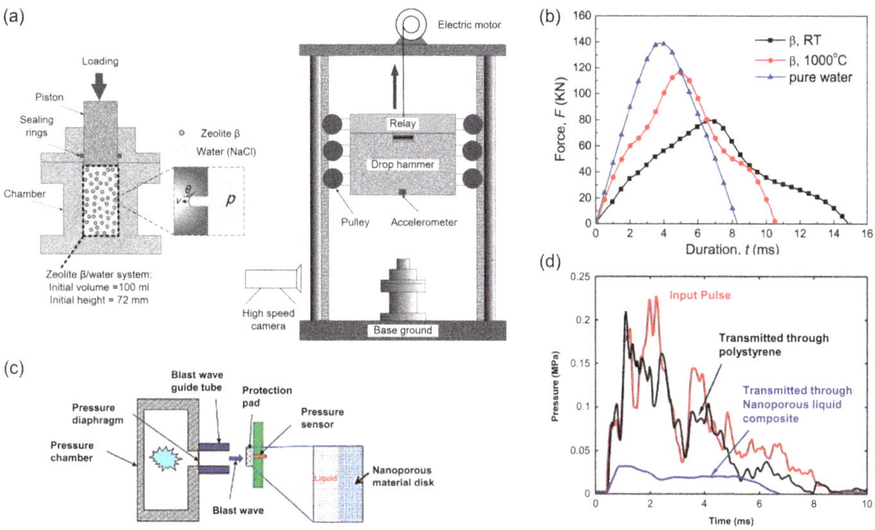

Figure 5.11 Dynamic testing of the water intrusion of zeolites.[104,257] (a and b) Drop-weight tower tests of Zeolite β with different heat treatment in comparison with pure water. Reproduced from ref. 86 with permission from Elsevier, Copyright 2015. (c and d) Blast chamber tests on a mesoporous zeolite from Zeolyst. Reproduced from ref. 238 with permission from Elsevier, Copyright 2014.

peak force compared with pure water demonstrated the cushioning effect of the water intrusion process (Figure 5.11(b)). In another study, Xu *et al.* tested a mesoporous zeolite using a blast chamber setup (Figure 5.11(c)), in which the pressure is built up by an air compressor until the vinyl diaphragm breaks and a Friedlander wave is generated that propagates onto the sample through a guide tube.[257] The input and transmitted pulses recorded by pressure transducers (Figure 5.11(d)) show that the zeolite–water system exhibits a better performance than the reference material expanded polystyrene (EPS) in attenuating the transmitted pressure. One limitation of drop-weight and blast tests is that they cannot provide a constant strain rate during the water intrusion process, therefore they are not the best choice to study the fundamental strain rate effect but are useful in evaluating the potential for related practical applications.

Some dynamic experiments have also been carried out on hydrophobic mesoporous and macroporous silicas. For example, drop-weight testing has been used to investigate various types of silicas, such as mesoporous Libersorb 23 and Polysorb-1,[66,70] and macroporous SP-1000-20.[258] SHPB has been used to test a mesoporous silica called Fluka 100 C8, which has a high energy absorption density of up to 41 J g^{-1} at high rates.[64] A custom-designed device has been used to test three mesoporous templated silicas (MCM-41, SBA-15, and HMS), which have been proven to have very limited rate effect in the medium-rate range.[259]

5.5 Structural Transition of Flexible MOFs for Energy Absorption

Apart from the liquid intrusion mechanism, MOFs can also absorb mechanical energy through structural transition under hydrostatic pressure,[22] or plastic deformation under uniaxial force.[21,23] These two mechanisms are phenomenologically similar to the crushing of metallic and polymeric foams,[4,7] where mechanical energy is absorbed by the collapse of pore volume.

5.5.1 MOF Structural Transition for Energy Absorption

Some MOFs can experience large structural transformation in response to external chemical or physical stimuli, referred to as 'flexible' MOFs.[260] Such dynamic features of framework materials led to the concept of soft porous crystals (SPCs) or 3rd generation porous coordination polymers,[261,262] which possess both highly ordered networks and structural transformability. This has been an active research field over the past two decades, representing compelling knowledge gaps and applications.[263–265] The structural transformation of flexible MOFs upon adsorption and desorption typically includes breathing, swelling, linker rotation, and subnetwork displacement.[266–268] The breathing behaviour is essentially a structural transition between different phases of the material, *e.g.*, from the open pore (OP) phase to the closed pore (CP) phase, or from the large pore (LP) phase to the narrow pore (NP) phase. Due to the pore volume reduction involved in such a process, if the structural transition can be provoked by an external pressure, it can enable energy absorption. The interested reader may consult Section 2.4 of Chapter 2 for the competing concepts and discussions on MOF flexibility.

MIL (Materials Institute Lavoisier) systems such as MIL-53 and MIL-47 were the first group of materials investigated in this area. These materials have a wine-rack-like structure with lozenge-shaped channels, constructed from infinite metal–hydroxide or metal–oxide chains crosslinked by linear dicarboxylates. Beurroies *et al.* pioneered this research by investigating the structural transition of MIL-53(Cr) between the LP and NP phases (Figure 5.12).[22] Mercury was used as the PTM, so the first step of volume contraction at around 1 MPa corresponded to the intrusion of mercury into the interparticle space, while the second step at around 55 MPa arose from the cell volume reduction in phase transition. Note that such a materials response is solely due to mechanical pressure without any adsorption effect, different from previous gas adsorption studies.[269–271] The system behaves as a reversible energy absorber, with substantial hysteresis and a very similar performance in its second cycle. The energy absorption density of this MIL-53(Cr)–mercury system was found to be 12 J g^{-1}, with a volume reduction of 0.25 cm^3 g^{-1}. There are two competing energetic contributions during this process: the deformation of bond angles and the distance between the ligands. The energy of the crystal increases when the bonds are deformed (which stabilises the LP phase), but decreases when the distance between the benzene rings decreases (which stabilises the NP phase).[22] The competition between these two stabilises one of the two phases, and there is an energy

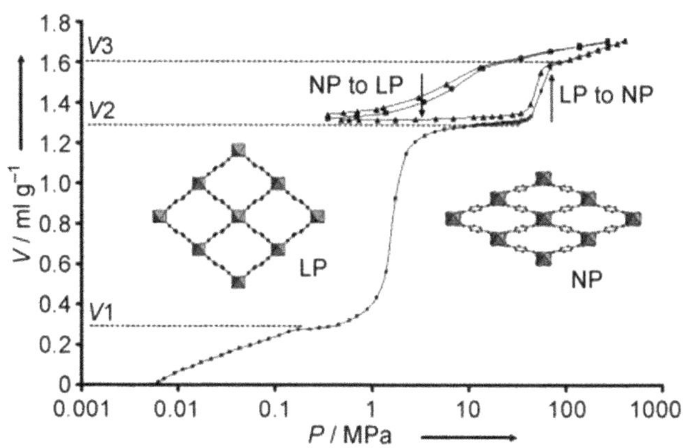

Figure 5.12 The pressure–volume change during the compression of MIL-53 (Cr) with mercury. The arrows show the transition between the LP and NP phases. The first cycle is represented by squares and the second cycle by triangles. Reproduced from ref. 22 with permission from John Wiley & Sons, Copyright © 2010 WILEY-VCH Verlag GmbH & Co. KGaA, Weinheim.

barrier between them which results in the hysteresis and energy absorption. Similar discussions have also been presented on other isostructural MIL structures.[272]

PTM is required to impose hydrostatic pressure onto the material in a controlled manner, but it is different from the liquid intrusion process as the liquid in this case should always stay outside the material and not enter the nanopores. Mercury and silicone oil are the main PTMs that have been used to date, as they can sustain high hydrostatic pressure and do not penetrate micropores within the range of pressure used in this research. Mercury porosimetry is often used to obtain the gross volume contraction of flexible MOFs in relation to the hydrostatic pressure,[22] in a similar way to liquid intrusion study. High-pressure X-ray diffraction measurements facilitated by a diamond anvil cell (DAC) can be used to characterise the structural transition of flexible MOFs at different pressures (see Section 4.2 of Chapter 4), to permit measurements of crystal parameters such as the unit cell volume rather than the gross volume contraction.[273] High-pressure microcalorimetry has also been used to measure the heat generated in the structural transition process.[274–277] These are three main techniques used in this area. However, some materials characterisation techniques that have already been used to examine adsorption-induced structural transitions can potentially also be considered for future pressure-induced structural transition studies with modification and development, *e.g.*, environmental scanning electron microscopy,[278] NMR,[279] UV–vis–NIR and IR spectroscopies,[280] and Raman spectroscopy.[281]

To date, the research on the pressurised structural transition of flexible MOFs has been limited to quasi-static equilibrium studies. There have been a few reports using different pressurisation rates to investigate the nonequilibrium transformations of solid materials,[185,282,283] although not necessarily looking at materials systems relevant to energy absorption applications. It has been found that the pressurisation rate, *i.e.*, the rate at which mechanical pressure is applied, plays an important role in the compressibility change,[185] and rapid pressurisation may result in different phases that are otherwise inaccessible under quasi-static conditions.[282,283] This provides the motivation to study dynamic structural transitions. In fact, the temporal dimension represents an underdeveloped perspective of MOF materials in general, with which one can view MOF properties more as a time-resolved dynamic process as opposed to steady-state properties. Evans *et al.* put forward the concept of 'four-dimensional MOFs' with the vision for a rational design of switchable MOFs on the time-axis to achieve deliberate and predictable time-dependent properties.[284] This will have a positive impact on applications such as impact mitigation, which requires a short response time in the spatiotemporal evolution of MOFs. Understanding and controlling the intrinsic timescale of framework transformations is an emerging field with outstanding challenges, with Cerasale *et al.* having recently reviewed the time-resolved experimental techniques that can be used to monitor the temporal evolution of MOF structures and their properties.[285]

5.5.2 Materials Design for Structural Transition

Currently, switchable MOFs explored for energy absorption applications mainly include MIL, ZIF, and DUT (Dresden University of Technology) systems. Their structural transitions lie in the flexibility of the wine-rack structure of the MIL system, the paddlewheel of the DUT system, and the cage structure of ZIFs. DMOF-1 and CUK-1 are the other two materials that have been investigated, which have pillared-layered and wine-rack structures, respectively. Table 5.5 summarises all the flexible MOF materials whose pressurised structural transitions have shown potential for energy absorption. Most of the MIL systems, including MIL-53, MIL-47 and their derivates, are reusable absorbers, apart from MIL-53(Al)-BDC and MIL-53(Ga)-FA, which behave as a bumper (with mercury) associated with their specific energy profiles during structural transition.[272,286] However, all the DUT materials behave as irreversible bumpers, although they absorb a substantially higher amount of energy due to their larger pore volume change compared with other groups of materials. ZIF systems exhibit the smallest volume changes and can behave as either absorbers or bumpers. For the other two materials, DMOF-1 exhibits similar bumper behaviour to that of DUT but operates at a relatively higher pressure and lower volume change; CUK-1 exhibits spring behaviour that has not been achieved by other materials in the table. It should be noted that the transition from NP to very narrow pore (VNP) observed in materials such as MIL-53 is not included in Table 5.5, as such a process has limited volume change and its working pressure is too high, therefore making these materials less promising for use in energy absorption applications.[287] The composition of a flexible MOF as well as the PTM can have a significant impact on its structural transition behaviour. As an example, Yot *et al.* investigated the influence of metal centre and linker functionalisation on the transition of MIL-53 from the monoclinic NP to the triclinic VNP phase.[287] In this section we will discuss the various factors that influence the structural transition behaviour of flexible MOFs based on the data collected in Table 5.5.

Metal ions that constitute the secondary building units (SBUs) of flexible MOFs can have a significant impact on structural transition behaviour, which is linked to the differential modulus or rigidity of the resultant framework (see Section 2.4 of Chapter 2). Table 5.5 contains five pairs of structural analogues, including MIL-53(Cr *vs.* Al)-BDC, MIL-53(Al *vs.* Ga)-FA, DMOF-1(Cu *vs.* Zn), ZIF-4(Co *vs.* Zn), and CUK-1(Co *vs.* Mg), the data of which are plotted in Figure 5.13(a), together with their bulk moduli at the bottom of the image. It can be seen that the analogue with a high bulk modulus tends to exhibit high structural transition pressure and small volume change upon compression, as well as a high structure recovery pressure associated with the reversibility of the structural transition. For example, MIL-53(Cr)-BDC exhibits a higher bulk modulus than its aluminium counterpart, so it also proves to have a higher structural transition pressure and a smaller volume change compared to MIL-53(Al)-BDC with either mercury or silicone oil as the PTM.[272] Having a

Table 5.5 Energy absorption of flexible MOFs under quasi-static structural transition. Hg is mercury, and oil is silicone oil such as AP 100. P_{com} and P_{dec} are the structural transition and recovery pressures, respectively. ΔV is the measured gross pore volume change and Δv is the unit cell volume change, in percentage. The energy absorption density $E_{ab} = E_{com} - E_{dec}$, with E_{com} being the energy stored during the compression process and E_{dec} being the energy recovered during the decompression process. Unless accompanied by superscripts (1) or (2) the data are from the first and second cycles, respectively, all other data correspond to the first cycle. For spring systems, their following cycles exhibit the same performance as the first cycle; for bumper systems, their following cycles do not indicate any structural transition (P_{dec} = 0 MPa); for absorbers, if the behaviour in two cycles is substantially different both cycles are shown in the table. Data with asterisks are values estimated from testing curves (apart from Δv, the asterisk associated with which represents simulated data), and '—' means data unavailable. P_{com} is taken as the onset of the compression plateau while P_{dec} is taken as the midpoint of the decompression plateau.

Material	PTM	P_{com}/MPa	P_{dec}/MPa	ΔV/cm^3 g^{-1}	Δv/%	E_{com}/J g^{-1}	E_{dec}/J g^{-1}	E_{ab}/J g^{-1}	Behaviour	Ref.
MIL-53(Cr)-BDC	Hg	55	10	0.25	29	13.75	−2.5	11.25	Absorber	22 and 275
MIL-53(Cr)-BDC	Oil	55	37*	0.2	35	9.5	−5.1	4.4	Absorber	276
MIL-53(Al)-BDC	Hg	13	0	0.35*	35	6.6	0	6.6	Bumper	272
MIL-53(Al)-BDC	Oil	30	10*	0.3$^{(1)}$ 0.23$^{(2)}$	35	7.8$^{(1)}$ 6.3$^{(2)}$	−1.1*$^{(1)}$ −1.7$^{(2)}$	6.7*$^{(1)}$ 4.6$^{(2)}$	Absorber	276
		33	7	0.23$^{(1)}$ 0.17*$^{(2)}$	—	7.8$^{(1)}$ 7$^{(2)}$	−1$^{(1)}$ −1.1$^{(2)}$	6.8$^{(1)}$ 5.9$^{(2)}$	Absorber	275
MIL-53(Al)-FA	Hg	110	65*	0.25	—	60	−17*	43*	Absorber	288
MIL-53(Al)-FA	Oil	100$^{(1)}$ 72$^{(2)}$	50*	0.25$^{(1)}$ 0.20*$^{(2)}$	—	41.7$^{(1)}$ 22.9$^{(2)}$	−10.8$^{(1)}$ −8$^{(2)}$	30.9$^{(1)}$ 14.9$^{(2)}$	Absorber	288
MIL-53(Ga)-FA	Hg	85	0	0.26*	14	65*	0	65*	Bumper	286
MIL-53(Al)-BDC@acetonitrile	Hg	60	32*	0.41	37	—	—	25	Absorber	289
MIL-53 (Al)-TDC	Hg	275	200	0.29	28.1	79	−40*	39*	Absorber	290
MIL-47(V)-BDC	Hg	85	75	0.38*	43	33	−28*	5*	Absorber	273
MIL-47(V)-BDC	Oil	86	76*	0.35	—	35.2	−23.8	11.4	Absorber	276
MIL-47(V)-BDC_Br	Hg	85	67*	0.16*	17	29*	−9*	20	Absorber	291

MIL-47(V)-BDC_ CF_3	Oil	750–850	—	—	28	160–190	—	—	Absorber	291
DUT-46	Hg	50	0	2.01	47.2*	101*	0	101*	Bumper	292
DUT-48	Hg	65	0	1.5	44.0*	85	0	85	Bumper	293
DUT-49	Hg	35	0	2.3	51.4*	106	0	106	Bumper	293
DUT-50	Hg	24	0	2.78	59.2*	67*	0	67*	Bumper	292
DUT-147	Hg	35	0	1.82*	52*	64*	0	64*	Bumper	294
DUT-148	Hg	32	0	2.65*	51*	85*	0	85*	Bumper	294
DUT-151	Hg	29	0	1.53	27.9*	44*	0	44*	Bumper	292
DUT-160	Hg	24	0	2.88*	53*	69*	0	69*	Bumper	294
DUT-161	Hg	24	0	2.82*	56*	68*	0	68*	Bumper	294
DUT-162	Hg	8*	0	0.82*	61*	6.6*	0	6.6*	Bumper	294
ZIF-4(Co)	Hg	50	10*	0.15	19.2	7.3	−1.5*	5.8*	Absorber	295
ZIF-4(Zn)	Hg	28	0	0.16	20.6	4.4	0	4.4	Bumper	295
ZIF-4(Zn)	Oil	50	—	—	24*	—	—	—	Absorber	296
ZIF-7-I	Hg	150	8*	0.05*	—	5.6	—	—	—	297
DMOF-1(Zn)	Hg	51	0	0.585	48	30	0	30	Bumper	298
DMOF-1(Cu)	Hg	193	0	0.404	34	80	0	80	Bumper	298
CUK-1(Co)	Hg	281	281	0.143	20.9	40.3	−40.3	0	Spring	299
CUK-1(Mg)	Hg	288	288	0.151	17.9	43.5	−43.5	0	Spring	299

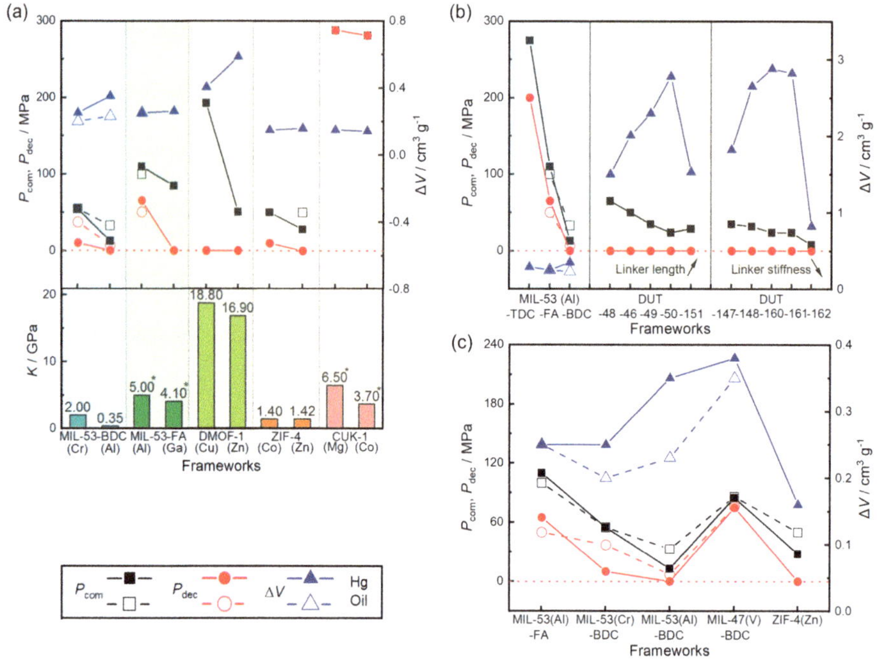

Figure 5.13 Selected experimental results from Table 5.5 on the structural transition of flexible MOFs. (a) Comparison of different metal sites, including their bulk moduli (MIL-53(Cr *vs.* Al)-BDC from ref. 272 and 304, MIL-53(Al *vs.* Ga)-FA simulated data from ref. 286, DMOF-1(Cu *vs.* Zn) from ref. 298, ZIF-4(Co *vs.* Zn) from ref. 295, CUK-1(Co *vs.* Mg) from ref. 299 estimated from testing curves). (b) Comparison of different linkers for the MIL-53 and DUT series, including those with different linker lengths (DUT-48 < DUT-46 < DUT-49 < DUT-50 < DUT-151) and linker stiffnesses (DUT-147 > DUT-148 > DUT-160 ~ DUT-161 > DUT-162), (c) Comparison between mercury and silicone oil as the PTM. The solid lines and filled symbols are results obtained using mercury as the PTM, dashed lines and hollow symbols are results using silicone oil as the PTM. The red dotted lines (P_{dec} = 0 MPa) indicate that the material exhibits irreversible bumper behaviour. All data correspond to the first cycle of structural transition.

high structural recovery pressure also means that the material can exhibit a reversible transition, in contrast to the irreversible transition of MIL-53(Al)-BDC when mercury is used as the PTM (*i.e.*, P_{dec} = 0 MPa in Figure 5.13(a)). Note that both analogues of DMOF-1 are irreversible bumpers due to their amorphisation under compression, but DFT calculations reveal reversible transition in DMOF-1(Cu) and irreversible transition in DMOF-1(Zn), consistent with the other materials.[298] As discussed for a multiferroic MOF material in Chapter 1 (Figure 1.6),[300] there is no clear correlation between bulk modulus and cation radius. For example, in Figure 5.13(a), the ionic radii of tetrahedrally coordinated Zn^{2+} and Co^{2+} are very similar in ZIF-4,[301] likewise for the ionic radius and metal–oxygen distances of Co^{2+} and Mg^{2+} in CUK-1.[302] Henke *et al.* interpreted the different responses of the two ZIF-4 analogues according to their different electronegativities (1.88 for Co *vs.* 1.65 for Zn)[303] and electron configurations ($3d^7$ for Co^{2+} *vs.* $3d^{10}$ for Zn^{2+}),[295] because a higher electronegativity means more covalent (directional) ligand-to-metal bonding as opposed to ionic (non-directional) bonding, and the $3d^7$ electron configuration of Co^{2+} indicates additional ligand-to-metal π bonding that is absent in the $3d^{10}$ of Zn^{2+}. More research is thus required to gain a better understanding of how metal ions influence the structural transition process.

Linkers and their functionalisation can also affect the structural transition performance. Figure 5.13(b) shows that among the three MIL-53(Al) materials incorporating different linkers -BDC, -FA, and -TDC, MIL-53(Al)-TDC exhibits the highest transition pressure and MIL-53(Al)-BDC exhibits the lowest and becomes irreversible when using mercury as the PTM. This is in line with their different behaviour upon gas absorption.[305] DUT systems reveal the important role of the length and stiffness of linkers, as shown in Figure 5.13(b). Frameworks with shorter and stiffer linkers can exhibit higher structural transition pressures.[292,294] For example, the main differences in the DUT-46, -48, -49, -50, and -151 frameworks lies in their linker length (DUT-48 < DUT-46 < DUT-49 < DUT-50 < DUT-151), which results in their different pore size, elasticity,[292] and structural transition pressure, with DUT-48 being the highest. Similarly, with different stiffness of linkers (DUT-147 > DUT-148 > DUT-160 ~ DUT-161 > DUT-162),[294] DUT-147 attains the highest structural transition pressure among this family of materials. The results thus reveal that it is possible to design linkers to achieve differential stiffness and structural transition performance, *e.g.*, using a short and bulky ligand backbone to achieve higher stiffness, or including acetylene, ethylene, and sp^3 hybridised groups to achieve lower stiffness.[292,294] Ligand functionalisation often results in a smaller pore volume change, ΔV, as the bulky functional groups can limit the structure contraction, allowing its CP phase to retain substantial pore volume. Therefore, Table 5.5 shows that MIL-47(V)-BDC_Br and MIL-47(V)-BDC_CF_3 exhibit a much smaller volume change than MIL-47(V)-BDC, and MIL-47(V)-BDC_CF_3 exhibits a much higher transition pressure than MIL-47(V)-BDC.[291]

The type of PTM, for instance mercury or silicone oil, has a significant effect on the materials performance. Unlike mercury, which requires external pressure for interparticle penetrations, silicone oil is a wetting fluid that can spontaneously occupy the interparticle domain.[288] Their differences in structural transition and energy absorption mainly lie in the fact that the molecular chain of silicone oil can partially penetrate the micropores on the crystal surface, which may provide additional resistance to structural transition and reduce its volume change during structural transition.[276] Therefore, as shown in Figure 5.13(c), in most cases, using silicone oil results in a higher structural transition pressure, P_{com}, than mercury (*e.g.*, MIL-53(Al)-BDC, MIL-47(V)-BDC, and ZIF-4(Zn)), and this also comes with a higher structure recovery pressure, P_{dec}, (*e.g.*, MIL-53(Cr)-BDC and MIL-47(V)-BDC) which in some cases can even turn bumpers into reusable absorbers by replacing mercury with silicone oil (*e.g.*, MIL-53(Al)-BDC and ZIF-4(Zn)). However, due to the partial penetration effect, silicone oil usually exhibits a slightly smaller volume change than mercury, as observed in MIL-53(Cr)-BDC, MIL-53(Al)-BDC, and MIL-47(V)-BDC. MIL-53(Al)-FA is an exception to the observations above: it has a lower transition pressure when using silicone oil, which might be related to its slightly smaller pore size compared to the other MILs (MIL-53(Al)-FA ~5.7–6.0 Å,[288] MIL-47(V)-BDC ~7.6 Å,[273] MIL-53(Cr)-BDC ~8 Å,[306] MIL-53(Al)-BDC ~8.5 Å[288]), because pore size can affect how the silicone oil penetrates and contributes to the transition pressure.[276] Moreover, both MIL-53(Al)-FA and MIL-53(Al)-BDC show a different performance in their first and second cycles when silicone oil is used as the PTM, although their structural transitions are reversible. This non-reproducibility can also be explained by the presence of silicone oil at the pore aperture on the crystal surface.[288] To fully utilise the structural transition mechanism for energy absorption, it is important to make sure that the PTM does not completely enter the framework within the pressure range of the experiments.[307] Most mercury porosimetry experiments are performed below 450 MPa, where either mercury or silicone oil is used, while high-pressure XRD experiments can be performed at a higher pressure, where usually silicone oil is used as the PTM.[272,275,287,288]

For the two isostructural MIL systems, *i.e.*, MIL-47 *vs.* MIL-53,[272] MIL-47 exhibits higher structural transition pressures in both compression and decompression, which lies in the fact that MIL-47 is more rigid while MIL-53 is highly flexible due to the presence of the –OH groups shared by its metallic octahedra.[273,308] Such observation is consistent with their different responses in gas adsorption: when including guest molecules, MIL-47(V) does not show any significant change in its crystal structure[309,310] or unit cell parameters.[311] It is also worth noting that the solvent used in the synthesis can also influence the materials performance. When acetonitrile is used to make MIL-53(Al)-BDC as opposed to the standard option of DMF, the structural transition occurs at a higher pressure and energy absorption, which can be explained by the stabilisation of the large-pore phase due to the presence of defects within the framework, and is potentially also contributed to by the small size and intergrowth of crystals.[289]

5.5.3 Thermal Effects of Structural Transition

Beurroies *et al.* pioneered the thermodynamic investigations of flexible MOFs using high-pressure calorimetry, which allows the measurement of mechanical and thermal energy at the same time.[274,275,277,288] They started from MIL-53(Al)[274] and found that its thermal energy during structural transition is of the same order of magnitude as the mechanical energy, where the temperature change can be around 1 °C.[275] In terms of practical application, a constant temperature is good for damping applications, but a large thermal energy variation can be also utilised to develop novel mechanical driven heating–cooling systems.

Table 5.6 and Figure 5.14 present the thermal response of flexible MOF transition published thus far. MIL-53(Cr)-BDC and MIL-53(Al)-BDC exhibit very similar thermal responses upon structural transition, with a thermal energy variation, Q_{cycle}, of around −5 J g^{-1}.[288] However, MIL-53(Al)-FA exhibits a much higher Q_{cycle} of up to −18 J g^{-1} in its first cycle.[288] These findings suggest that the linker has a much stronger effect than the metal on the thermal response of MIL-53, which is consistent with their effect on mechanical responses, as discussed in Figure 5.13(a and b). It is also shown that MIL-47 exhibits a higher thermal response than MIL-53, consistent with their different mechanical responses. In fact, among the four materials reported, those with stronger mechanical responses upon compression, *i.e.*, higher structural transition pressure and mechanical energy storage, also tend to exhibit more substantial thermal effect upon compression. And since these materials have similar thermal responses upon decompression, there also seems to be a correlation between the mechanical energy storage, E_{com}, and thermal energy variation within a cycle, Q_{cycle}. For example, MIL-53(Al)-FA has the smallest pore size in the range of 5.7–6.0 Å, highest transition pressure, and greatest mechanical energy storage, which intriguingly also leads

Table 5.6 Energy absorption and thermal response of flexible MOFs under quasi-static structural transition. The mechanical energy absorption density $E_{ab} = E_{com} + E_{dec}$ with E_{com} being the mechanical energy stored during the compression process and E_{dec} being the mechanical energy recovered during the decompression process. Here, E_{dec} is defined as negative to be consistent with the thermal energies. The thermal energy variation within a cycle $Q_{cycle} = Q_{com} + Q_{dec}$ with Q_{com} being the thermal energy during the compression process and Q_{dec} being the thermal energy during the decompression process. Internal energy variation within a cycle $\Delta U_{cycle} = E_{ab} + Q_{cycle}$. The superscript (#) denotes the cycle number. Silicone oil is used as the PTM in these experiments.

Material	E_{com}/J g^{-1}	E_{dec}/J g^{-1}	E_{ab}/J g^{-1}	Q_{com}/J g^{-1}	Q_{dec}/J g^{-1}	Q_{cycle}/J g^{-1}	ΔU_{cycle}/J g^{-1}	Ref.
MIL-53(Al)-BDC	7.8$^{(1)}$	−1$^{(1)}$	6.8$^{(1)}$	−10.8$^{(1)}$	2.0$^{(1)}$	−8.8$^{(1)}$	−2$^{(1)}$	275
	7$^{(2)}$	−1.1$^{(2)}$	5.9$^{(2)}$	−7.9$^{(2)}$	2.1$^{(2)}$	−5.8$^{(2)}$	0.1$^{(2)}$	
	6.0$^{(3)}$	−1.4$^{(3)}$	4.6$^{(3)}$	−6.7$^{(3)}$	2.0$^{(3)}$	−4.7$^{(3)}$	−0.1$^{(3)}$	
MIL-53(Al)-BDC	7.8$^{(1)}$	—	—	−9.9$^{(1)}$	—	—	—	276
	6.3$^{(2)}$	−1.7$^{(2)}$	4.6$^{(2)}$	−9.4$^{(2)}$	5.7$^{(2)}$	−3.7$^{(2)}$	0.9$^{(2)}$	
	5.9$^{(3)}$	−2.0$^{(3)}$	3.9$^{(3)}$	−8.7$^{(3)}$	4.5$^{(3)}$	−4.2$^{(3)}$	−0.3$^{(3)}$	
	5.9$^{(4)}$	−2.1$^{(4)}$	3.8$^{(4)}$	−8.2$^{(4)}$	4.9$^{(4)}$	−3.3$^{(4)}$	0.5$^{(4)}$	
MIL-53(Cr)-BDC	9.5$^{(1)}$	−5.1$^{(1)}$	4.4$^{(1)}$	−9.1$^{(1)}$	4.1$^{(1)}$	−5.0$^{(1)}$	−0.6$^{(1)}$	276
	9.4$^{(2)}$	−5.4$^{(2)}$	4.0$^{(2)}$	−8.1$^{(2)}$	4.5$^{(2)}$	−3.6$^{(2)}$	0.4$^{(2)}$	
	8.7$^{(3)}$	−5.5$^{(3)}$	3.2$^{(3)}$	−7.8$^{(3)}$	4.7$^{(3)}$	−3.1$^{(3)}$	0.1$^{(3)}$	
	9.2$^{(4)}$	−5.0$^{(4)}$	4.2$^{(4)}$	−8.3$^{(4)}$	4.7$^{(4)}$	−3.6$^{(4)}$	0.6$^{(4)}$	
MIL-53(Al)-FA	41.7$^{(1)}$	−10.8$^{(1)}$	30.9$^{(1)}$	−25.1$^{(1)}$	6.4$^{(1)}$	−18.7$^{(1)}$	12.2$^{(1)}$	288
	22.9$^{(2)}$	−8$^{(2)}$	14.9$^{(2)}$	−18.7$^{(2)}$	6.3$^{(2)}$	−12.4$^{(2)}$	2.5$^{(2)}$	
	22.2$^{(3)}$	−8.8$^{(3)}$	13.4$^{(3)}$	−18.2$^{(3)}$	6.5$^{(3)}$	−11.7$^{(3)}$	1.7$^{(3)}$	
MIL-47(V)-BDC	35.2$^{(1)}$	−23.8$^{(1)}$	11.4$^{(1)}$	−15.6$^{(1)}$	5.2$^{(1)}$	−10.4$^{(1)}$	1.0$^{(1)}$	276
	33.9$^{(2)}$	−22.0$^{(2)}$	11.9$^{(2)}$	−13.0$^{(2)}$	4.8$^{(2)}$	−8.2$^{(2)}$	3.7$^{(2)}$	
	33.1$^{(3)}$	−22.9$^{(3)}$	10.2$^{(3)}$	−12.9$^{(3)}$	4.9$^{(3)}$	−8.0$^{(3)}$	2.2$^{(3)}$	

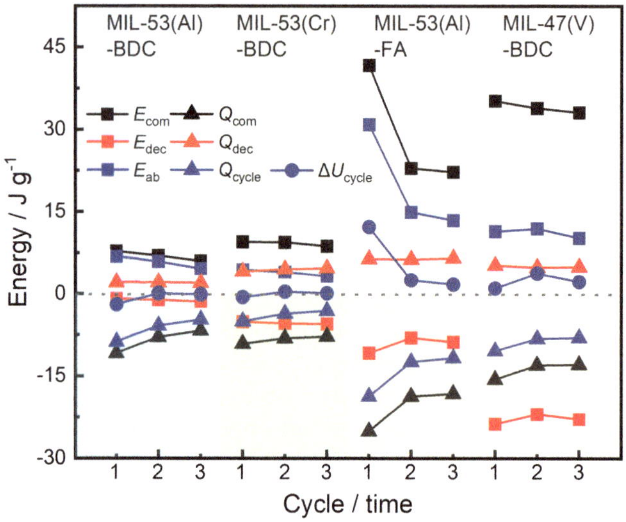

Figure 5.14 Mechanical and thermal responses in the structural transition of flexible MOFs, using selected data from Table 5.6. The data of MIL-53(Al)-BDC were taken from ref. 275.

to it having the highest thermal effects, Q_{in} and Q_{cycle}. The reversibility of structural transition can be revealed by multi-cycle responses as well as the internal energy variation, ΔU_{cycle}. For a reversible process, a similar materials response should be expected in different cycles and the internal energy variation, ΔU_{cycle}, should be close to zero.[276] It can be observed from Figure 5.14 that MIL-53(Al)-FA is far from being reversible in its first cycle, and MIL-53(Cr)-BDC and MIL-47(V)-BDC are almost entirely reversible.

5.6 Plastic Deformation of MOFs for Energy Absorption

The plastic deformation of MOF materials under uniaxial force can result in nanopore volume collapse and potentially mechanochemical reactions, providing another mechanism for MOFs to absorb mechanical energy, similar to foam crushing where the porous structure fails under compression or impact. There are two comprehensive reviews on this topic by the Suslick group.[21,23] Here, we provide a brief overview of the experimental methods and materials performance by putting them in the context of energy absorption and the other mechanisms discussed above. The experiments reported so far mainly include the quasi-static compression of single crystals and shock experiments on films, with the testing conditions and main results summarised in Table 5.7. Different from the liquid intrusion and structural transition mechanisms described in the previous sections, the experiments here do not require a PTM.

Quasi-static compression tests have been carried out on ZIF-8[312] and UiO materials[313] by recording the load–displacement curve of an individual MOF crystal under the uniaxial compression of a flat-punch inside a transmission electron microscope (TEM), as shown in Figure 5.15(a and b). It is found that the presence of solvent as guest molecules inside ZIF-8 cages shatter the crystal completely at a low pressure (*ca.* 0.8 GPa) while guest-free crystals are tough and absorb energy,[312] as demonstrated by the results of the two ZIF-8_hexagonal (d = 1.2 μm) samples shown in Table 5.7. The data in Table 5.7 also demonstrate that ZIF-8 experiences amorphisation at a lower pressure during powder compaction than during single-crystal compression

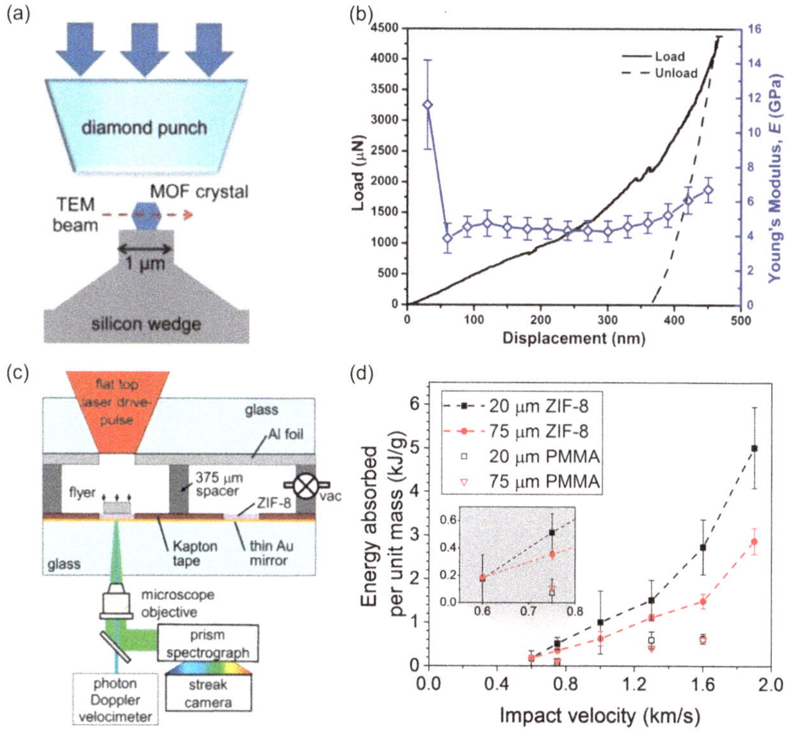

Figure 5.15 Mechanical tests of ZIF-8. (a) Schematic of the nanocompression apparatus on an individual crystal within a TEM. Loading force is measured by a piezo-actuator pressing the diamond punch against the crystal. (b) Representative load–displacement curve of a ZIF-8 crystal (1.2 μm, hexagonal projection) using the nanocompression setup. The loading Young's modulus is shown on the right axis (blue), with error bars being the standard deviation from around ten independent measurements. (c) Schematic of the table-top laser-driven flyer plate apparatus, which generates a planar shock on the ZIF-8 sample at km s^{-1}. (d) Shock wave energy absorbed per unit mass: comparison between ZIF-8 and PMMA, with error bars representing one standard deviation. Reproduced from ref. 312, 313, 317 and 318 with permissions from American Chemical Society, Copyright 2015, 2017, 2017, 2019.

Table 5.7 Energy absorption of MOFs through plastic deformation under uniaxial quasi-static and shock compressions, including the key parameters: crystal size d, compression pressure P, energy absorption density E_{ab}. For quasi-static single-crystal experiments, the values of loading modulus E_{load} and unloading modulus E_{unload} are presented. For shock experiments, the impact speed v and sample thickness t are also included. ZIF-8 single crystals are compressed on the (110) facet, with hexagonal projection on the $(\bar{1}01)$ facet and square projection on the (100) facet. All the samples are desolvated, unless denoted with '@ guest molecules'. The behaviour column only lists the phenomenon that requires the highest energy if multiple phenomena are involved, *e.g.*, in the shock experiments of films, bond breakage usually also involves pore collapse, amorphisation may also involve crystal fragmentation. Hence, the compression pressure P is the pressure that corresponds to the listed behaviour, but in some cases, it is the peak pressure in the experiment and therefore denoted with '<'. Data with asterisks are values estimated from testing curves and images, and '—' means data unavailable.

Quasi-static experiment	Material	d/μm	E_{load}/GPa	E_{unload}/GPa	P/GPa	Behaviour	E_{ab}/kJ g^{-1}	Ref.
In situ nanocompression on single crystal under TEM	ZIF-8@MeOH_hexagonal	1.2	—	—	= 0.8*	Shatter	—	312
	ZIF-8_hexagonal	1.2	4.6 ± 0.2	41 ± 4	= ~3	Amorphisation	—	312
	ZIF-8_square	1.2	3.2 ± 0.3	45 ± 6	= ~2	Amorphisation	—	312
	ZIF-8_hexagonal	0.55	3.9 ± 0.5	75 ± 9	—	—	—	312
	UiO-66	0.5*	—	—	<~2	Elastic	~0.1	24
			7.8	82*	<10	Bond breakage	~4	24 and 313
	UiO-67	0.4*	5.1	84*	<11*	Bond breakage	4.5*	313
	UiO-abdc	0.5	3.9	70*	<7*	Bond breakage	3*	313
	MOF-801	0.3	2.1	53*	<3.5	Pore collapse	1.0*	313
Hydraulic pelletisation on powder	ZIF-8	1.2	—	—	= ~1.1	Amorphisation	—	312
	UiO-66	0.5*	—	—	= 0.4	Bond breakage	—	24
DAC compression on powder	ZIF-8 (Z1200)	0.5*	—	—	= 0.34	Amorphisation	—	314
	ZIF-8	0.049	—	—	<1.60	Elastic	—	315
			—	—	<39.15	Amorphisation	—	

Shock experiment	Materials	d/μm	t/μm	v/km s^{-1}	P/GPa	Phenomenon	E_{ab}/kJ g^{-1}	Ref.
Gas gun impact on powder	CuBTC@ H_2O	20–50	50	—	= 0.5	Pore collapse	—	316
	CuBTC-Fc@ H_2O	20–50	50	—	= 3.2	Pore collapse	—	

Shock experiment	Materials	d/μm	t/μm	v/km s^{-1}	P/GPa	Phenomenon	E_{ab}/kJ g^{-1}	Ref.
Laser flyer plate impact on film (PVA polymer bonded)	ZIF-8	1.2	45	0.75	<2.5	Fragmentation	—	317
				1.3	<5	Amorphisation	—	
				1.6	<8	Bond breakage	—	
	ZIF-8	2	20	0.6	= 2.1	Bond breakage	0.2*	21 and 318
				0.75	—	Bond breakage	0.51	
				1.0	—	Bond breakage	1.0*	
				1.3	—	Bond breakage	1.5*	
				1.6	—	Bond breakage	2.7	
				1.9	—	Bond breakage	5*	
	ZIF-8	2	75	0.6	—	Bond breakage	0.2*	318
				0.75	—	Bond breakage	0.33*	
				1.0	—	Bond breakage	0.6*	
				1.3	—	Bond breakage	1.1*	
				1.6	—	Bond breakage	1.4*	
				1.9	—	Bond breakage	2.9*	

Shock simulation	Materials	v/km s^{-1}	P/GPa	Phenomenon	E_{ab}/kJ g^{-1}	Ref.
MD simulation of shock	MOF-5	0.1	—	Pore collapse	—	319
		1	—	Bond breakage	—	
	ZIF-8	—	= 0.63	Pore collapse	—	320

(1.1 *vs.* 2–3 GPa), probably due to the local stress concentration between crystals.[312] A similar observation was made for UiO-66 in its different bond breakage pressures under single-crystal and powder compressions.[24] Since ZIF-8 is anisotropic, the loading modulus measured in the hexagonal projection is slightly higher than in the square projection (4.6 *vs.* 3.2 GPa), while the unloading moduli are similar due to the similar response of the dense amorphous phase in two projections.[312] The comparison between the moduli of ZIF-8_hexagonal samples of two different crystal sizes (d = 1.2 *vs.* 0.55 μm) suggests that smaller crystals are easier to compress, and they are also less elastic upon unloading, as evidenced by their higher unloading modulus.[312]

Among the four isostructural UiO-type MOFs with different linker lengths (MOF-801 < UiO-66 < UiO-67 < UiO-abdc), UiO-66 exhibits the highest modulus of 7.8 GPa, followed by UiO-67 at 5.1 GPa and UiO-abdc at 3.9 GPa. MOF-81 exhibits the lowest modulus at 2.1 GPa and fully collapses at 3.5 GPa despite it having the shortest linker, probably due to the high concentration of defects in its structure.[313] These results are in line with the DFT calculations by Banlusan *et al.*, which show that the average Young's modulus increases with shorter organic linkers (IRMOF-1 < IRMOF-10 < IRMOF-16) although the simulation is within the elastic regime.[321] The bond breakage and mechanochemical behaviour of UiO-66 were revealed by extended X-ray absorption fine structure (EXAFS) spectroscopy measurements after compression.[24] As shown in Table 5.7, such phenomenon results in an extremely high energy absorption density of up to 4 kJ g^{-1}, which is comparable to the energy released in a typical explosion of TNT, and orders of magnitude higher than other energy absorption mechanisms. However, bond breakage might not occur in some MOFs under similar compression conditions, for example, it is found that ZIF-8 loses porosity and long-range order but maintains its local structure around its bridging Zn(II) ions.[312] This is probably due to its relatively dense structure and compactness of the 2-methylimidazole (mIm) ligand, as well as the possible free rotation of the Zn–mIm–Zn linkages (Figure 1.8(d) in Chapter 1) subject to compression.[24]

The first dynamic experiment was a gas-gun planar impact on a large sample of CuBTC,[316] where the material was found to be crushed at 0.5 GPa but was enhanced up to 3.2 GPa if ferrocene was included as the guest molecule (*i.e.*, the CuBTC-Fc sample in Table 5.7). This work was followed by pulsed laser-driven flyer plate impact studies on polymer-bonded ZIF-8 film in its desolvated state,[317,318] as shown in Figure 5.15(c and d). For the ZIF-8 (t = 45 μm) sample shown in Table 5.7, massive bond scission, structural collapse, and loss of local symmetry were observed as the impact speed exceeds 1.6 km s^{-1}, with a shock pressure of around 8 GPa.[317] Zhou *et al.* quantified the shock wave energy absorption by measuring the input and transmitted energy using a high-speed optical interferometer (photon Doppler velocimeter, PDV), and found that ZIF-8 film is much more efficient in absorbing shock energy than a standard polymer material poly(methyl methacrylate) (PMMA), and that the amount of energy absorbed increases with impact speed and film thickness, as shown in Table 5.7, although a thinner film can trigger multiple absorption mechanisms and may exhibit a higher energy absorption density under the same conditions.[318] These experiments are complemented by the simulation work of Banlusan *et al.* on MOF-5, the deformation of which involves chemical reactions when the impact speed is over 1 km s^{-1},[319,322] and recently on ZIF-8, which shows a transition from elastic to pore collapse when the shock pressure is above 0.63 GPa.[320]

Three energy absorption mechanisms are typically involved for MOF materials under pressure or shock: (1) compaction, which is a process of squeezing out void space between crystals through the movement, deformation, and potential fracture of crystals; (2) pore collapse, which is a process of reducing internal porosity with the collapse of free pore volume inside the structure; and (3) endothermic chemical bond destruction, which is associated with the catastrophic collapse of the pore structure. Note that the void collapse of MOFs can be either a physical or chemical process, for example, a physical process can involve the rotation of a linker around a metal site, and a chemical process can involve the breakage of a metal–linker bond or the linker itself. The response is related to the flexibility or rigidity of the framework, meaning that it is easier for soft frameworks to maintain their local structure during compression, which also includes the reversible switching of flexible MOFs (such as MIL-53), while rigid frameworks can undergo more substantial structural changes, including bond breakage at rigid sites (such as of UiO-66).

Compared to the liquid intrusion and structural transition mechanisms introduced in the previous sections, the uniaxial plastic deformation of MOFs absorbs energy at several orders of magnitude higher, from J g^{-1} to kJ g^{-1}. This enhancement mainly stems from the different working pressure, which is also orders of magnitude higher, at GPa, as opposed to MPa for the other two mechanisms. Such extreme pressure is not routinely encountered in practical applications, therefore the plastic deformation of MOFs mainly finds its applications under extreme dynamic conditions, such as lightweight ballistic defence or space debris applications, where the collision of solids can be at km s^{-1}, while the liquid intrusion and structural transition mechanisms can work at substantially lower speeds, at m s^{-1}, commonly seen in applications such as vehicle crashworthiness, sports protection, vibration attenuation, *etc.* Due to the limitation of crystal size, the miniaturisation of dynamic mechanical testing techniques down to the micro/nanoscale may advance this field further in the future, by enabling the measurement of dynamic load–displacement relationships or offering insight into the time-dependent response of plastic deformations.

5.7 Conclusions and Outlook

This chapter discussed the energy absorption application of MOFs, which is highly relevant to MOF mechanics as well as other important features of MOFs such as surface area, chemical properties, and guest–host interactions.

It is an interesting topic that connects different areas in the MOF community. This chapter focussed on the liquid intrusion of MOFs but also included other energy absorption mechanisms, and other nanoporous materials such as zeolites and silica, due to their strong relevance to this topic. By putting MOF energy absorption in the broader context of porous liquid intrusion, one can also find some review articles that summarise the forced liquid intrusion of different groups of materials,[2,101,323] highlight the fundamental questions related to confined water and framework changes,[183,324] and look into other applications beyond energy absorption.[325–327] This is an open area with exciting opportunities for both fundamental and applied research, and therefore it has become an active topic in recent years.

There are still lots of unknowns on the subject of MOF confined liquids, which is a complex materials system, especially in the dynamic domain. The interaction between water and MOFs is complicated and beyond conventional nanofluidic studies due to the heterogeneous interior of MOF pores.[159] Moreover, possible framework deformations triggered by mechanical pressure or intruded liquid molecules need to be considered due to the flexibility of MOF materials. The good news is that understanding the behaviour and properties of water and more complex liquids in MOFs is also important to many other processes, such as water adsorption, water harvesting, and liquid separation, amongst others.[328–331] Therefore, it is of great interest to invest more research in this direction to understand the nature of liquids under framework confinement. This kind of research often requires a combination of experimental and simulation methods[1,332,333] and interdisciplinary collaborations, especially when moving toward dynamic non-equilibrium studies where more diverse expertise needs to be incorporated.

MOF energy absorption has demonstrated its advantages in potential practical applications, such as high energy density and reusability to build efficient systems, rapid response to work at high strain rates, and distinctive thermal behaviour to prevent excessive heating, amongst others. They may find applications in many areas related to impact and vibrations, as well as other areas, such as to store and recover unused mechanical energy. Due to the nature of the liquid intrusion process, it has a relatively short path from fundamentals to applications, however the need does exist to address the challenges associated with practical applications. For example, the rational design of materials is required to guide engineering processes, which calls for more research to shed light on the governing parameters and principles of the intrusion behaviour that may involve framework structures, chemical moieties, crystal morphologies, and others. Practical challenges might also include the mass production of MOFs and liquid suspensions, the long-term stability and endurance of MOFs in liquids and under pressure, engineering questions for integrating novel materials in existing systems,[13,334–336] *etc.* Together with the knowledge gaps on the fundamental side, these are also outstanding opportunities for future research in this field.

References

1. Y. Sun, S. M. J. Rogge, A. Lamaire, S. Vandenbrande, J. Wieme, C. R. Siviour, V. Van Speybroeck and J.-C. Tan, *Nat. Mater.*, 2021, **20**, 1015.
2. G. Fraux, F.-X. Coudert, A. Boutin and A. H. Fuchs, *Chem. Soc. Rev.*, 2017, **46**, 7421.
3. J. Hutchinson, M. J. Kaiser and H. M. Lankarani, *Appl. Math. Comput.*, 1998, **96**, 1.
4. G. Lu and T. Yu, *Energy Absorption of Structures and Materials*, Elsevier, 2003.
5. M. N. Fazita, H. A. Khalil, A. N. A. Izzati and S. Rizal, in *Failure Analysis in Biocomposites, Fibre-Reinforced Composites and Hybrid Composites*, Elsevier, 2019, pp. 51–78.
6. L. Sun, R. F. Gibson, F. Gordaninejad and J. Suhr, *Compos. Sci. Technol.*, 2009, **69**, 2392.
7. L. J. Gibson and M. F. Ashby, *Cellular Solids: Structure and Properties*, Cambridge University Press, Cambridge, 2nd edn, 1997.
8. Y. Sun and Q. Li, *Int. J. Impact Eng.*, 2018, **112**, 74.
9. W. Li, Y. Luo, M. Li, F. Sun and H. Fan, *Int. J. Mech. Sci.*, 2018, **140**, 241.
10. A. Baroutaji, M. Sajjia and A.-G. Olabi, *Thin Wall. Struct.*, 2017, **118**, 137.
11. V. Rai, H. Ghasemnejad, J. W. Watson, J. Gonzalez-Domingo and P. Webb, *Eng. Struct.*, 2019, **199**, 109620.
12. J. Song, Y. Chen and G. Lu, *Thin Wall. Struct.*, 2012, **54**, 65.
13. Y. Sun, Y. Li, C. Zhao, M. Wang, W. Lu and Y. Qiao, *Int. J. Damage Mech.*, 2018, **27**, 439.
14. M. Li, S. Barbat, R. Baccouche, J. Belwafa and W. Lu, *Composites, Part B*, 2020, **193**, 108047.
15. C. W. Isaac and C. Ezekwem, *Compos. Struct.*, 2021, **257**, 113081.
16. R. Brooks, in *Comprehensive Composite Materials*, ed. A. Kelly and C. Zweben, Pergamon, Oxford, 2000, vol. 6, pp. 341–363.
17. B. Liu, T. Xu, X. Xu, Y. Wang, Y. Sun and Y. Li, *Int. J. Impact Eng.*, 2016, **90**, 26–36.
18. X. Zhang, Y. Wang, B. Ding and X. Li, *Small*, 2020, **16**, 1902842.
19. J.-H. Lee, P. E. Loya, J. Lou and E. L. Thomas, *Science*, 2014, **346**, 1092.
20. J. Hyon, O. Lawal, O. Fried, R. Thevamaran, S. Yazdi, M. Zhou, D. Veysset, S. E. Kooi, Y. Jiao and M.-S. Hsiao, *Mater. Today*, 2018, **21**, 817.
21. X. Zhou, Y. Miao, K. S. Suslick and D. D. Dlott, *Acc. Chem. Res.*, 2020, **53**, 2806.
22. I. Beurroies, M. Boulhout, P. L. Llewellyn, B. Kuchta, G. Férey, C. Serre and R. Denoyel, *Angew. Chem., Int. Ed.*, 2010, **49**, 7526.
23. Y.-R. Miao and K. S. Suslick, in *Adv. Inorg. Chem.*, Elsevier, 2018, vol. 71, pp. 403.
24. Z. Su, Y.-R. Miao, G. Zhang, J. T. Miller and K. S. Suslick, *Chem. Sci.*, 2017, **8**, 8004.
25. J. C. Dixon, *The Shock Absorber Handbook*, John Wiley & Sons, 2008.

26. Y. Sun, Y. Li and J.-C. Tan, *Phys. Chem. Chem. Phys.*, 2018, **20**, 10108.
27. J. Rouquerol, G. Baron, R. Denoyel, H. Giesche, J. Groen, P. Klobes, P. Levitz, A. V. Neimark, S. Rigby and R. Skudas, *Pure Appl. Chem.*, 2011, **84**, 107.
28. S. Lowell, J. E. Shields, M. A. Thomas and M. Thommes, *Characterization of Porous Solids and Powders: Surface Area, Pore Size and Density*, Springer Science & Business Media, 2006.
29. J. Rouquerol, G. V. Baron, R. Denoyel, H. Giesche, J. Groen, P. Klobes, P. Levitz, A. V. Neimark, S. Rigby, R. Skudas, K. Sing, M. Thommes and K. Unger, *Microporous Mesoporous Mater.*, 2012, **154**, 2.
30. S. J. Gregg, K. S. W. Sing and H. W. Salzberg, *J. Electrochem. Soc.*, 1967, **114**, 279Ca.
31. E. W. Washburn, *Phys. Rev.*, 1921, **17**, 273.
32. A. Y. Fadeev and V. A. Eroshenko, *J. Colloid Interface Sci.*, 1997, **187**, 275.
33. V. Y. Gusev, *Langmuir*, 1994, **10**, 235.
34. F. Gomez, R. Denoyel and J. Rouquerol, *Langmuir*, 2000, **16**, 4374.
35. V. N. Bogomolov, *Phys. Rev. B*, 1995, **51**, 17040.
36. V. Eroshenko, R.-C. Regis, M. Soulard and J. Patarin, *J. Am. Chem. Soc.*, 2001, **123**, 8129.
37. V. Eroshenko, *WO patent*, WO1996018040, 1994.
38. V. D. Borman, A. A. Belogorlov, A. M. Grekhov, G. V. Lisichkin, V. N. Tronin and V. I. Troyan, *Tech. Phys. Lett.*, 2004, **30**, 973.
39. V. Borman, A. Belogorlov, A. Grekhov, G. Lisichkin, V. Tronin and V. Troyan, *J. Exp. Theor. Phys.*, 2005, **100**, 385.
40. V. D. Borman, A. M. Grekhov and V. I. Troyan, *J. Exp. Theor. Phys.*, 2000, **91**, 170.
41. V. D. Borman, A. A. Belogorlov, A. M. Grekhov, V. N. Tronin and V. I. Troyan, *J. Exp. Theor. Phys. Lett.*, 2001, **74**, 258.
42. T. Martin, B. Lefevre, D. Brunel, A. Galarneau, F. Di Renzo, F. Fajula, P. Gobin, J. Quinson and G. Vigier, *Chem. Commun.*, 2002, 24.
43. B. Lefevre, A. Saugey, J.-L. Barrat, L. Bocquet, E. Charlaix, P.-F. Gobin and G. Vigier, *J. Chem. Phys.*, 2004, **120**, 4927.
44. A. Tinti, A. Giacomello, Y. Grosu and C. M. Casciola, *Proc. Natl. Acad. Sci. U. S. A.*, 2017, **114**, E10266.
45. V. Eroshenko, *WO patent*, WO2001055616, 2001.
46. V. Eroshenko, I. Piatiletov, L. Coiffard and V. Stoudenets, *Proc. Inst. Mech. Eng., Part D*, 2007, **221**, 301.
47. V. Eroshenko, *US patent*, US8925697B2, 2015.
48. C. V. Suciu, T. Iwatsubo and S. Deki, *J. Colloid Interface Sci.*, 2003, **259**, 62.
49. C. V. Suciu, T. Iwatsubo and S. Deki, *JSME Int J., Ser. C*, 2004, **47**, 180.
50. C. V. Suciu, T. Iwatsubo, K. Yaguchi and M. Ikenaga, *J. Colloid Interface Sci.*, 2005, **283**, 196.
51. T. Iwatsubo, C. V. Suciu, M. Ikenagao and K. Yaguchio, *J. Sound Vib.*, 2007, **308**, 579.
52. C. Suciu and K. Yaguchi, *Exp. Mech.*, 2009, **49**, 383.

53. C. V. Suciu, S. Tani and K. Yaguchi, *Acta Mech.*, 2010, **214**, 195.
54. C. V. Suciu, Y. Kimura and S. Tani, *Adv. Mater. Res.*, 2010, **123**, 1003.
55. C. V. Suciu, S. Tani and K. Miyoshi, *J. Syst. Des. Dyn.*, 2010, **4**, 899.
56. C. V. Suciu and Y. Kimura, *J. Syst. Des. Dyn.*, 2012, **6**, 538.
57. C. V. Suciu, *Proc. Int. Conf. on Noise and Vibration Eng*, Leuven, 2010.
58. C. V. Suciu and T. Tobiishi, *J. Syst. Des. Dyn.*, 2012, **6**, 555.
59. H. Li, X. Xu and Y. Qiao, *J. Nanomech. Micromech.*, 2013, **3**, 43.
60. X. Kong and Y. Qiao, *Appl. Phys. Lett.*, 2005, **86**, 151919.
61. X. Kong, F. B. Surani and Y. Qiao, *J. Mater. Res.*, 2005, **20**, 1042.
62. F. B. Surani, X. Kong and Y. Qiao, *Appl. Phys. Lett.*, 2005, **87**, 251906.
63. X. Kong and Y. Qiao, *Philos. Mag. Lett.*, 2005, **85**, 331.
64. F. B. Surani, X. Kong, D. B. Panchal and Y. Qiao, *Appl. Phys. Lett.*, 2005, **87**, 163111.
65. A. Han, V. K. Punyamurtula, T. Kim and Y. Qiao, *J. Mater. Eng. Perform.*, 2008, **17**, 326.
66. V. D. Borman, A. A. Belogorlov, G. V. Lisichkin, V. N. Tronin and V. I. Troyan, *J. Exp. Theor. Phys.*, 2009, **108**, 389.
67. M. Soulard, J. Patarin, V. Eroshenko and R. Regis, in *Stud. Surf. Sci. Catal.*, Elsevier, 2004, vol. 154, pp. 1830.
68. G. Ortiz, H. Nouali, C. Marichal, G. Chaplais and J. Patarin, *Phys. Chem. Chem. Phys.*, 2013, **15**, 4888.
69. I. Khay, G. Chaplais, H. Nouali, G. Ortiz, C. Marichal and J. Patarin, *Dalton Trans.*, 2016, **45**, 4392.
70. V. D. Borman, A. A. Belogorlov and V. N. Tronin, *Colloids Surf., A*, 2018, **537**, 540.
71. A. Han, V. K. Punyamurtula and Y. Qiao, *Philos. Mag. Lett.*, 2008, **88**, 67.
72. V. K. Punyamurtula, A. Han and Y. Qiao, *Adv. Eng. Mater.*, 2007, **9**, 209.
73. W. Lu, V. K. Punyamurtula and Y. Qiao, *Int. J. Mater. Res.*, 2011, **102**, 587.
74. A. Ryzhikov, T. J. Daou, H. Nouali, J. Patarin, J. Ouwehand, S. Clerick, E. De Canck, P. Van Der Voort and J. A. Martens, *Microporous Mesoporous Mater.*, 2018, **260**, 166.
75. W. Lu, T. Kim, C. Zhao and Y. Qiao, *Int. J. Mater. Res.*, 2013, **104**, 594.
76. M. Trzpit, M. Soulard and J. Patarin, *Chem. Lett.*, 2007, **36**, 980.
77. M. A. Saada, S. Rigolet, J.-L. Paillaud, N. Bats, M. Soulard and J. Patarin, *J. Phys. Chem. C*, 2010, **114**, 11650.
78. M. Trzpit, M. Soulard, J. Patarin, N. Desbiens, F. Cailliez, A. Boutin, I. Demachy and A. Fuchs, *Langmuir*, 2007, **23**, 10131.
79. A. Ryzhikov, I. Khay, H. Nouali, T. J. Daou and J. Patarin, *Phys. Chem. Chem. Phys.*, 2014, **16**, 17893.
80. Y. Sun, W. Lu and Y. Li, *Appl. Phys. Lett.*, 2014, **105**, 121609.
81. Y. Sun, P. Li, Y. Qiao and Y. Li, *Sci. Rep.*, 2014, **4**, 6547.
82. Y. Sun, J. Xu, C. Zhao, P. Li and Y. Li, presented in part at the 2014 IRCOBI Conference, 2014.
83. L. Tzanis, M. Trzpit, M. Soulard and J. Patarin, *Microporous Mesoporous Mater.*, 2011, **146**, 119.

84. L. Tzanis, M. l. Trzpit, M. Soulard and J. l. Patarin, *J. Phys. Chem. C*, 2012, **116**, 20389.
85. L. Ronchi, A. Ryzhikov, H. Nouali, T. J. Daou, S. Albrecht and J. Patarin, *Microporous Mesoporous Mater.*, 2018, **255**, 211.
86. A. Ryzhikov, I. Khay, H. Nouali, T. Daou and J. Patarin, *RSC Adv.*, 2014, **4**, 37655.
87. L. Ronchi, *PhD thesis*, Mulhouse, 2017.
88. F. Cailliez, M. Trzpit, M. Soulard, I. Demachy, A. Boutin, J. Patarin and A. H. Fuchs, *Phys. Chem. Chem. Phys.*, 2008, **10**, 4817.
89. M. A. Saada, M. Soulard, B. Marler, H. Gies and J. l. Patarin, *J. Phys. Chem. C*, 2011, **115**, 425.
90. L. Tzanis, B. Marler, H. Gies and J. l. Patarin, *J. Phys. Chem. C*, 2013, **117**, 4098.
91. L. Ronchi, A. Ryzhikov, H. Nouali, T. J. Daou, S. Albrecht and J. Patarin, *Microporous Mesoporous Mater.*, 2017, **254**, 153.
92. L. Ronchi, H. Nouali, T. Daou, J. Patarin and A. Ryzhikov, *New J. Chem.*, 2017, **41**, 15087.
93. I. Khay, L. Tzanis, T. J. Daou, H. Nouali, A. Ryzhikov and J. Patarin, *Phys. Chem. Chem. Phys.*, 2013, **15**, 20320.
94. L. Tzanis, M. l. Trzpit, M. Soulard and J. l. Patarin, *J. Phys. Chem. C*, 2012, **116**, 4802.
95. L. Ronchi, A. Ryzhikov, H. Nouali, T. Daou and J. Patarin, *New J. Chem.*, 2017, **41**, 2586.
96. A. Ryzhikov, L. Ronchi, H. Nouali, T. J. Daou, J.-L. Paillaud and J. l. Patarin, *J. Phys. Chem. C*, 2015, **119**, 28319.
97. C. Baerlocher, L. B. McCusker and D. H. Olson, *Atlas of Zeolite Framework Types*, Elsevier, 2007.
98. E. L. First, C. E. Gounaris, J. Wei and C. A. Floudas, *Phys. Chem. Chem. Phys.*, 2011, **13**, 17339.
99. N. Desbiens, I. Demachy, A. H. Fuchs, H. Kirsch-Rodeschini, M. Soulard and J. Patarin, *Angew. Chem., Int. Ed.*, 2005, **44**, 5310.
100. Y. Grosu, M. Li, Y. L. Peng, D. Luo, D. Li, A. Faik, J. M. Nedelec and J. P. Grolier, *ChemPhysChem*, 2016, **17**, 3359.
101. G. Confalonieri, T. J. Daou, H. Nouali, R. Arletti and A. Ryzhikov, *Molecules*, 2020, **25**, 2145.
102. C. E. Kirschhock, M. De Prins, E. Verheijen, A. Ryzhikov, T. J. Daou, H. Nouali, F. Taulelle, J. A. Martens and J. Patarin, *Phys. Chem. Chem. Phys.*, 2016, **18**, 18795.
103. Y. Sun, J. Xu, Y. B. Li, B. Liu, Y. Wang, C. Liu and X. Chen, *Adv. Eng. Mater.*, 2013, **15**, 740.
104. Y. Sun, Z. Guo, J. Xu, X. Xu, C. Liu and Y. Li, *Mater. Des.*, 2015, **66**, 545.
105. F.-X. Coudert, F. Cailliez, R. Vuilleumier, A. H. Fuchs and A. Boutin, *Faraday Discuss.*, 2009, **141**, 377.
106. Y. G. Bushuev, G. Sastre, J. V. de Julián-Ortiz and J. Gálvez, *J. Phys. Chem. C*, 2012, **116**, 24916.
107. Y. G. Bushuev, Y. Grosu, M. A. ChorąŻewski and S. Meloni, *Nano Lett.*, 2022, **22**, 2164.

108. Y. G. Bushuev, Y. Grosu, M. A. Chorążewski and S. Meloni, *ACS Appl. Mater. Interfaces*, 2022, **14**, 30067.
109. N. Gokulakrishnan, J. Parmentier, M. Trzpit, L. Vonna, J. L. Paillaud and M. Soulard, *J. Nanosci. Nanotechnol.*, 2013, **13**, 2847.
110. Y. Gao, M. Li, Y. Zhang, W. Lu and B. Xu, *Proc. Natl. Acad. Sci. U. S. A.*, 2020, **117**, 25246.
111. B. Lefevre, A. Saugey, J.-L. Barrat, L. Bocquet, E. Charlaix, P.-F. Gobin and G. Vigier, *Colloids Surf., A*, 2004, **241**, 265.
112. M. Li, C. Zhan and W. Lu, *Appl. Phys. Lett.*, 2021, **119**, 181603.
113. Y. Gao, M. Li, H. Zhang, Y. Zhang, W. Lu and B. Xu, *Matter*, 2022, **5**, 266.
114. L. Xu, M. Li and W. Lu, *Langmuir*, 2019, **35**, 14505.
115. M. Li, L. Xu and W. Lu, *Langmuir*, 2020, **36**, 4682.
116. M. Li, L. Xu and W. Lu, *J. Appl. Phys.*, 2019, **125**, 044303.
117. L. Xu, M. Li and W. Lu, *J. Phys. Chem. C*, 2021, **125**, 5596.
118. M. Amabili, Y. Grosu, A. Giacomello, S. Meloni, A. Zaki, F. Bonilla, A. Faik and C. M. Casciola, *ACS Nano*, 2019, **13**, 1728.
119. N. Gokulakrishnan, T. Karbowiak, J. P. Bellat, L. Vonna, M.-A. Saada, J. L. Paillaud, M. Soulard, J. Patarin and J. Parmentier, *Colloids Surf., A*, 2013, **421**, 34.
120. M. Pillot, B. Lebeau, H. Nouali, T. J. Daou, J. Patarin and A. Ryzhikov, *Microporous Mesoporous Mater.*, 2019, **280**, 248.
121. V. K. Punyamurtula, A. Han and Y. Qiao, *J. Fluids Eng.*, 2009, **131**, 044501.
122. K. S. Park, Z. Ni, A. P. Côté, J. Y. Choi, R. Huang, F. J. Uribe-Romo, H. K. Chae, M. O'Keeffe and O. M. Yaghi, *Proc. Natl. Acad. Sci. U. S. A.*, 2006, **103**, 10186.
123. A. U. Ortiz, A. P. Freitas, A. Boutin, A. H. Fuchs and F.-X. Coudert, *Phys. Chem. Chem. Phys.*, 2014, **16**, 9940.
124. P. Z. Moghadam, A. Li, S. B. Wiggin, A. Tao, A. G. Maloney, P. A. Wood, S. C. Ward and D. Fairen-Jimenez, *Chem. Mater.*, 2017, **29**, 2618.
125. Y. Sun, Y. Li and J.-C. Tan, *ACS Appl. Mater. Interfaces*, 2018, **10**, 41831.
126. G. Ortiz, H. Nouali, C. Marichal, G. Chaplais and J. l. Patarin, *J. Phys. Chem. C*, 2014, **118**, 21316.
127. B. Mortada, G. Chaplais, H. Nouali, C. Marichal and J. Patarin, *J. Phys. Chem. C*, 2019, **123**, 4319.
128. Y. Grosu, S. Gomes, G. Renaudin, J.-P. E. Grolier, V. Eroshenko and J.-M. Nedelec, *RSC Adv.*, 2015, **5**, 89498.
129. A. F. Möslein and J.-C. Tan, *J. Phys. Chem. Lett.*, 2022, **13**, 2838.
130. C.-T. He, L. Jiang, Z.-M. Ye, R. Krishna, Z.-S. Zhong, P.-Q. Liao, J. Xu, G. Ouyang, J.-P. Zhang and X.-M. Chen, *J. Am. Chem. Soc.*, 2015, **137**, 7217.
131. K. Zhang, R. P. Lively, M. E. Dose, A. J. Brown, C. Zhang, J. Chung, S. Nair, W. J. Koros and R. R. Chance, *Chem. Commun.*, 2013, **49**, 3245.
132. B. Mortada, G. Chaplais, V. Veremeienko, H. Nouali, C. Marichal and J. Patarin, *J. Phys. Chem. C*, 2018, **122**, 3846.
133. M. Michelin-Jamois, C. Picard, G. Vigier and E. Charlaix, *Phys. Rev. Lett.*, 2015, **115**, 036101.

134. G. Ortiz, H. Nouali, C. Marichal, G. Chaplais and J. l. Patarin, *J. Phys. Chem. C*, 2014, **118**, 7321.
135. L. Tzanis, H. Nouali, T. J. Daou, M. Soulard and J. Patarin, *Mater. Lett.*, 2014, **115**, 229.
136. Y. Luo and B. Roux, *J. Phys. Chem. Lett.*, 2010, **1**, 183.
137. I. Khay, T. Daou, H. Nouali, A. Ryzhikov, S. Rigolet and J. Patarin, *J. Phys. Chem. C*, 2014, **118**, 3935.
138. A. Ryzhikov, H. Nouali, T. J. Daou and J. Patarin, *Phys. Chem. Chem. Phys.*, 2018, **20**, 6462.
139. J. Huve, T. J. Daou, H. Nouali, J. Patarin and A. Ryzhikov, *New J. Chem.*, 2020, **44**, 273.
140. L. Ronchi, A. Ryzhikov, H. Nouali, T. J. Daou and J. Patarin, *J. Phys. Chem. C*, 2018, **122**, 2726.
141. L. Ronchi, A. Ryzhikov, H. Nouali, T. J. Daou and J. Patarin, *Energy*, 2017, **130**, 29.
142. R. Arletti, L. Ronchi, S. Quartieri, G. Vezzalini, A. Ryzhikov, H. Nouali, T. J. Daou and J. Patarin, *Microporous Mesoporous Mater.*, 2016, **235**, 253.
143. A. Ryzhikov, I. Khay, H. Nouali, T. Daou and J. Patarin, *Microporous Mesoporous Mater.*, 2016, **221**, 1.
144. G. Confalonieri, A. Ryzhikov, R. Arletti, H. Nouali, S. Quartieri, T. J. Daou and J. Patarin, *J. Phys. Chem. C*, 2018, **122**, 28001.
145. Y. Sun, J. Xu, Y. Li, X. Xu, C. Liu and X. Chen, *J. Nanomater.*, 2013, **2013**, 249369.
146. J. Zhao, P. J. Culligan, J. T. Germaine and X. Chen, *Langmuir*, 2009, **25**, 12687.
147. L. Liu, X. Chen, W. Lu, A. Han and Y. Qiao, *Phys. Rev. Lett.*, 2009, **102**, 184501.
148. A. Han, W. Lu, T. Kim, V. K. Punyamurtula and Y. Qiao, *Smart Mater. Struct.*, 2009, **18**, 024005.
149. F. B. Surani and Y. Qiao, *J. Appl. Phys.*, 2006, **100**, 034311.
150. X. Kong, F. B. Surani and Y. Qiao, *Phys. Scr.*, 2006, **74**, 531.
151. A. Han, V. K. Punyamurtula and Y. Qiao, *Appl. Phys. Lett.*, 2008, **92**, 153117.
152. A. Han, W. Lu, T. Kim, X. Chen and Y. Qiao, *Phys. Rev. E: Stat., Nonlinear, Soft Matter Phys.*, 2008, **78**, 031408.
153. E. W. Washburn, *Proc. Natl. Acad. Sci. U. S. A.*, 1921, **7**, 115.
154. Z. Hu, Y. Chen and J. Jiang, *J. Chem. Phys.*, 2011, **134**, 134705.
155. K. M. Gupta, Z. Qiao, K. Zhang and J. Jiang, *ACS Appl. Mater. Interfaces*, 2016, **8**, 13392.
156. G. Confalonieri, A. Ryzhikov, R. Arletti, S. Quartieri, G. Vezzalini, C. Isaac, J.-L. Paillaud, H. Nouali and T. J. Daou, *Phys. Chem. Chem. Phys.*, 2020, **22**, 5178.
157. Y. Marcus, *J. Chem. Soc., Faraday Trans.*, 1991, **87**, 2995–2999.
158. G. Fraux, A. Boutin, A. H. Fuchs and F.-X. Coudert, *J. Phys. Chem. C*, 2019, **123**, 15589.
159. F.-X. Coudert, *Acc. Chem. Res.*, 2020, **53**, 1342.

160. G. Fraux, *PhD thesis*, PSL Research University; Chimie ParisTech, 2019.
161. T. Kim, W. Lu, A. Han, V. K. Punyamurtula, X. Chen and Y. Qiao, *Appl. Phys. Lett.*, 2009, **94**, 013105.
162. L. Liu, X. Chen, T. Kim, A. Han and Y. Qiao, *New J. Phys.*, 2010, **12**, 033021.
163. S. S. Han, S.-H. Choi and W. A. Goddard III, *J. Phys. Chem. C*, 2011, **115**, 3507.
164. L. Y. Chen, *J. Struct. Biol.*, 2013, **181**, 71.
165. Y. Zhang, N. Li, R. Luo, Y. Zhang, Q. Zhou and X. Chen, *J. Phys. D: Appl. Phys.*, 2015, **49**, 025303.
166. Y. Zhang, R. Luo, Q. Zhou, X. Chen and Y. Dou, *Appl. Sci.*, 2018, **8**, 1065.
167. Y. Sun, C. Xu, W. Lu and Y. Li, *RSC Adv.*, 2017, 7, 971.
168. W. Lu, T. Kim, V. K. Punyamurtula, A. Han and Y. Qiao, *J. Mater. Sci.*, 2011, **46**, 4053.
169. A. Han, W. Lu, V. K. Punyamurtula, X. Chen, F. B. Surani, T. Kim and Y. Qiao, *J. Appl. Phys.*, 2008, **104**, 124908.
170. X. Chen, G. Cao, A. Han, V. K. Punyamurtula, L. Liu, P. J. Culligan, T. Kim and Y. Qiao, *Nano Lett.*, 2008, **8**, 2988.
171. A. Han and Y. Qiao, *J. Am. Chem. Soc.*, 2006, **128**, 10348.
172. V. K. Punyamurtula and Y. Qiao, *Microporous Mesoporous Mater.*, 2007, **103**, 35.
173. V. K. Punyamurtula and Y. Qiao, *Mater. Lett.*, 2008, **62**, 2928–2930.
174. Y. Sun, C. Xu and Y. Li, *RSC Adv.*, 2018, **8**, 9070.
175. Y. Nakayama, *Introduction to Fluid Mechanics*, Butterworth-Heinemann, 2018.
176. R. H. Tunuguntla, R. Y. Henley, Y.-C. Yao, T. A. Pham, M. Wanunu and A. Noy, *Science*, 2017, **357**, 792.
177. J. K. Holt, H. G. Park, Y. Wang, M. Stadermann, A. B. Artyukhin, C. P. Grigoropoulos, A. Noy and O. Bakajin, *Science*, 2006, **312**, 1034.
178. R. Arletti, G. Vezzalini, S. Quartieri, F. Di Renzo and V. Dmitriev, *Microporous Mesoporous Mater.*, 2014, **191**, 27.
179. R. Arletti, E. Fois, L. Gigli, G. Vezzalini, S. Quartieri and G. Tabacchi, *Angew. Chem., Int. Ed.*, 2017, **56**, 2105.
180. P. Lotti, R. Arletti, G. D. Gatta, S. Quartieri, G. Vezzalini, M. Merlini, V. Dmitriev and M. Hanfland, *Microporous Mesoporous Mater.*, 2015, **218**, 42.
181. R. Arletti, G. Vezzalini, A. Morsli, F. Di Renzo, V. Dmitriev and S. Quartieri, *Microporous Mesoporous Mater.*, 2011, **142**, 696.
182. S. Quartieri, G. Montagna, R. Arletti and G. Vezzalini, *J. Solid State Chem.*, 2011, **184**, 1505.
183. G. Vezzalini, R. Arletti and S. Quartieri, *Acta Crystallogr., Sect. B: Struct. Sci., Cryst. Eng. Mater.*, 2014, **70**, 444.
184. R. Arletti, L. Leardini, G. Vezzalini, S. Quartieri, L. Gigli, M. Santoro, J. Haines, J. Rouquette and L. Konczewicz, *Phys. Chem. Chem. Phys.*, 2015, **17**, 24262.
185. K. W. Chapman, G. J. Halder and P. J. Chupas, *J. Am. Chem. Soc.*, 2008, **130**, 10524.

186. S. A. Moggach, T. D. Bennett and A. K. Cheetham, *Angew. Chem.*, 2009, **121**, 7221.
187. J. Im, N. Yim, J. Kim, T. Vogt and Y. Lee, *J. Am. Chem. Soc.*, 2016, **138**, 11477.
188. A. J. Graham, D. R. Allan, A. Muszkiewicz, C. A. Morrison and S. A. Moggach, *Angew. Chem., Int. Ed.*, 2011, **50**, 11138.
189. A. J. Graham, A.-M. Banu, T. Düren, A. Greenaway, S. C. McKellar, J. P. S. Mowat, K. Ward, P. A. Wright and S. A. Moggach, *J. Am. Chem. Soc.*, 2014, **136**, 8606.
190. C. L. Hobday, R. J. Marshall, C. F. Murphie, J. Sotelo, T. Richards, D. R. Allan, T. Düren, F.-X. Coudert, R. S. Forgan, C. A. Morrison, S. A. Moggach and T. D. Bennett, *Angew. Chem., Int. Ed.*, 2016, **55**, 2401.
191. R. P. Lively, M. E. Dose, J. A. Thompson, B. A. McCool, R. R. Chance and W. J. Koros, *Chem. Commun.*, 2011, **47**, 8667.
192. A. Nalaparaju, X. Zhao and J. Jiang, *J. Phys. Chem. C*, 2010, **114**, 11542.
193. P. Bai, M. Y. Jeon, L. Ren, C. Knight, M. W. Deem, M. Tsapatsis and J. I. Siepmann, *Nat. Commun.*, 2015, **6**, 5912.
194. R. Krishna and J. M. Van Baten, *ACS Omega*, 2020, **5**, 28393.
195. S. Gravelle, H. Yoshida, L. Joly, C. Ybert and L. Bocquet, *J. Chem. Phys.*, 2016, **145**, 124708.
196. J. C. Díaz, I. D. Gil-Chávez, L. Giraldo and J. C. Moreno-Piraján, *J. Chem.*, 2010, **7**, 483.
197. S. Li, *J. Membr. Sci.*, 2001, **191**, 53.
198. I. Khay, G. Chaplais, H. Nouali, C. Marichal and J. Patarin, *RSC Adv.*, 2015, **5**, 31514.
199. P. Zajdel, D. G. Madden, R. Babu, M. Tortora, D. Mirani, N. N. Tsyrin, L. Bartolomé, E. Amayuelas, D. Fairen-Jimenez and A. R. Lowe, *ACS Appl. Mater. Interfaces*, 2022, **14**, 26699.
200. W. Cai, T. Lee, M. Lee, W. Cho, D.-Y. Han, N. Choi, A. C. Yip and J. Choi, *J. Am. Chem. Soc.*, 2014, **136**, 7961.
201. A. A. Tiba, A. V. Tivanski and L. R. MacGillivray, *Nano Lett.*, 2019, **19**, 6140.
202. A. F. Möslein, L. Donà, B. Civalleri and J.-C. Tan, *ACS Appl. Nano Mater.*, 2022, **5**, 6398.
203. M. Trzpit, M. Soulard and J. Patarin, *Microporous Mesoporous Mater.*, 2009, **117**, 627.
204. M. Trzpit, M. Soulard and J. Patarin, *J. Mater. Sci.*, 2009, **44**, 6525.
205. A. A. Darhuber and S. M. Troian, *Annu. Rev. Fluid. Mech.*, 2005, **37**, 425.
206. F. Mugele, M. Duits and D. Van Den Ende, *Adv. Colloid Interface Sci.*, 2010, **161**, 115.
207. O. V. Ievtushenko, V. A. Eroshenko, Y. G. Grosu, J.-M. Nedelec and J.-P. E. Grolier, *Phys. Chem. Chem. Phys.*, 2013, **15**, 4451.
208. A. Han and Y. Qiao, *J. Mater. Res.*, 2011, **22**, 644.
209. Y. Qiao and A. Han, *Philos. Mag. Lett.*, 2007, **87**, 25.
210. Y. Qiao, V. K. Punyamurtula, A. Han, X. Kong and F. B. Surani, *Appl. Phys. Lett.*, 2006, **89**, 251905.

211. B. Xu, Y. Qiao, T. Park, M. Tak, Q. Zhou and X. Chen, *Energy Environ. Sci.*, 2011, **4**, 3632.
212. L. Coiffard and V. Eroshenko, *J. Colloid Interface Sci.*, 2006, **300**, 304–309.
213. Y. Grosu, O. Ievtushenko, V. Eroshenko, J. M. Nedelec and J. P. E. Grolier, *Colloids Surf., A*, 2014, **441**, 549.
214. W. Lu, T. Kim, A. Han, X. Chen and Y. Qiao, *Langmuir*, 2009, **25**, 9463.
215. W. Lu, T. Kim, A. Han, X. Chen and Y. Qiao, *J. Chem. Phys.*, 2011, **134**, 204706.
216. W. Lu, T. Kim, A. Han and Y. Qiao, *Mater. Chem. Phys.*, 2012, **133**, 259.
217. B. Xu, Y. Qiao, Q. Zhou and X. Chen, *Langmuir*, 2011, **27**, 6349.
218. W. Lu, A. Han, T. Kim, V. K. Punyamurtula, X. Chen and Y. Qiao, *Appl. Phys. Lett.*, 2009, **94**, 023106.
219. T. Kim, W. Lu, H. Lim, A. Han and Y. Qiao, *Appl. Phys. Lett.*, 2011, **98**, 053106.
220. L. Tao, C.-Y. Lin, S. Dou, S. Feng, D. Chen, D. Liu, J. Huo, Z. Xia and S. Wang, *Nano Energy*, 2017, **41**, 417.
221. L. Chen, C. A. Morrison and T. Düren, *J. Phys. Chem. C*, 2012, **116**, 18899.
222. K. Tan, S. Zuluaga, Q. Gong, P. Canepa, H. Wang, J. Li, Y. J. Chabal and T. Thonhauser, *Chem. Mater.*, 2014, **26**, 6886.
223. R. E. Osta, M. Frigoli, J. Marrot, N. Guillou, H. Chevreau, R. I. Walton and F. Millange, *Chem. Commun.*, 2012, **48**, 10639.
224. P. Küsgens, M. Rose, I. Senkovska, H. Fröde, A. Henschel, S. Siegle and S. Kaskel, *Microporous Mesoporous Mater.*, 2009, **120**, 325.
225. H. Furukawa, F. Gandara, Y. B. Zhang, J. Jiang, W. L. Queen, M. R. Hudson and O. M. Yaghi, *J. Am. Chem. Soc.*, 2014, **136**, 4369.
226. H. Zhang, D. Liu, Y. Yao, B. Zhang and Y. S. Lin, *J. Membr. Sci.*, 2015, **485**, 103.
227. H. Zhang, M. Zhao and Y. S. Lin, *Microporous Mesoporous Mater.*, 2019, **279**, 201.
228. H. Zhang, M. Zhao, Y. Yang and Y. S. Lin, *Microporous Mesoporous Mater.*, 2019, **288**, 109568.
229. K. A. Cychosz and A. J. Matzger, *Langmuir*, 2010, **26**, 17198.
230. Y. Grosu, G. Renaudin, V. Eroshenko, J.-M. Nedelec and J.-P. Grolier, *Nanoscale*, 2015, **7**, 8803.
231. N. C. Burtch, H. Jasuja and K. S. Walton, *Chem. Rev.*, 2014, **114**, 10575.
232. X. Liu, Y. Li, Y. Ban, Y. Peng, H. Jin, H. Bux, L. Xu, J. Caro and W. Yang, *Chem. Commun.*, 2013, **49**, 9140.
233. H. Zhang, J. James, M. Zhao, Y. Yao, Y. Zhang, B. Zhang and Y. S. Lin, *J. Membr. Sci.*, 2017, **532**, 1.
234. S. Tanaka and Y. Tanaka, *ACS Omega*, 2019, **4**, 19905.
235. M. De Toni, R. Jonchiere, P. Pullumbi, F.-X. Coudert and A. H. Fuchs, *ChemPhysChem*, 2012, **13**, 3497.
236. C. Zhang, C. Han, D. S. Sholl and J. Schmidt, *J. Phys. Chem. Lett.*, 2016, **7**, 459.
237. A. Han, W. Lu, V. K. Punyamurtula, T. Kim and Y. Qiao, *J. Appl. Phys.*, 2009, **105**, 024309.

238. Y. Grosu, M. Mierzwa, V. A. Eroshenko, S. Pawlus, M. Chorażewski, J.-M. Nedelec and J.-P. E. Grolier, *ACS Appl. Mater. Interfaces*, 2017, **9**, 7044.
239. M. Chorążewski, P. Zajdel, T. Feng, D. Luo, A. R. Lowe, C. M. Brown, J. B. Leão, M. Li, M. Bleuel, G. Jensen, D. Li, A. Faik and Y. Grosu, *ACS Nano*, 2021, **15**, 9048.
240. O. V. Ievtushenko, V. A. Eroshenko, Y. G. Grosu, J.-M. Nedelec and J.-P. E. Grolier, *Phys. Chem. Chem. Phys.*, 2013, **15**, 4451.
241. T. Karbowiak, C. Paulin, A. Ballandras, G. Weber and J.-P. Bellat, *J. Am. Chem. Soc.*, 2009, **131**, 9898.
242. T. Karbowiak, M.-A. Saada, S. Rigolet, A. Ballandras, G. Weber, I. Bezverkhyy, M. Soulard, J. Patarin and J.-P. Bellat, *Phys. Chem. Chem. Phys.*, 2010, **12**, 11454.
243. T. Karbowiak, C. Paulin and J.-P. Bellat, *Microporous Mesoporous Mater.*, 2010, **134**, 8.
244. T. Karbowiak, G. Weber and J.-P. Bellat, *Langmuir*, 2014, **30**, 213–219.
245. L. Coiffard, V. A. Eroshenko and J.-P. E. Grolier, *AlChE J.*, 2005, **51**, 1246.
246. Y. Grosu, A. Giacomello, S. Meloni, L. González-Fernández, M. Chorazewski, M. Geppert-Rybczynska, A. Faik, J.-M. Nedelec and J.-P. Grolier, *J. Phys. Chem. C*, 2018, **122**, 14248.
247. A. R. Lowe, W. S. Y. Wong, N. Tsyrin, M. A. Chorążewski, A. Zaki, M. Geppert-Rybczyńska, V. Stoudenets, A. Tricoli, A. Faik and Y. Grosu, *Langmuir*, 2021, **37**, 4827.
248. A. Alexiadis and S. Kassinos, *Chem. Rev.*, 2008, **108**, 5014.
249. M. l. Trzpit, S. Rigolet, J.-L. Paillaud, C. Marichal, M. Soulard and J. Patarin, *J. Phys. Chem. B*, 2008, **112**, 7257.
250. A. Laouir, L. Luo, D. Tondeur, T. Cachot and P. Le Goff, *AlChE J.*, 2003, **49**, 764.
251. Y. Qiao, V. K. Punyamurtula and A. Han, *J. Power Sources*, 2007, **164**, 931.
252. L. Liu, H. Lim, W. Lu, Y. Qiao and X. Chen, *Appl. Phys. Express*, 2013, **6**, 015202.
253. W. Olthuis, B. Schippers, J. Eijkel and A. van den Berg, *Sens. Actuators, B*, 2005, **111**, 385.
254. G. T. Gray III, *ASM Handbook*, 2000, vol. 8, pp. 462.
255. A. Phan, C. J. Doonan, F. J. Uribe-Romo, C. B. Knobler, M. O'Keeffe and O. M. Yaghi, *Acc. Chem. Res.*, 2010, **43**, 58.
256. A. Lowe, N. Tsyrin, M. Chorążewski, P. Zajdel, M. Mierzwa, J. B. Leão, M. Bleuel, T. Feng, D. Luo, M. Li, D. Li, V. Stoudenets, S. Pawlus, A. Faik and Y. Grosu, *ACS Appl. Mater. Interfaces*, 2019, **11**, 40842.
257. B. Xu, Y. Qiao and X. Chen, *J. Mech. Phys. Solids*, 2014, **62**, 194.
258. M. Li and W. Lu, *Phys. Chem. Chem. Phys.*, 2017, **19**, 17167.
259. L. Guillemot, A. Galarneau, G. Vigier, T. Abensur and É. Charlaix, *Rev. Sci. Instrum.*, 2012, **83**, 105105.
260. A. Schneemann, V. Bon, I. Schwedler, I. Senkovska, S. Kaskel and R. A. Fischer, *Chem. Soc. Rev.*, 2014, **43**, 6062.
261. S. Kitagawa and M. Kondo, *Bull. Chem. Soc. Jpn.*, 1998, **71**, 1739.
262. S. Horike, S. Shimomura and S. Kitagawa, *Nat. Chem.*, 2009, **1**, 695.

263. S. Kitagawa, R. Kitaura and S.-I. Noro, *Angew. Chem., Int. Ed.*, 2004, **43**, 2334.
264. S. B. Peh, A. Karmakar and D. Zhao, *Trends Chem.*, 2020, **2**, 199.
265. J. Troyano, A. Legrand and S. Furukawa, *Trends Chem.*, 2021, **3**, 254.
266. D. Bousquet, F.-X. Coudert, A. G. J. Fossati, A. V. Neimark, A. H. Fuchs and A. Boutin, *J. Chem. Phys.*, 2013, **138**, 174706.
267. F.-X. Coudert, A. Boutin, A. H. Fuchs and A. V. Neimark, *J. Phys. Chem. Lett.*, 2013, **4**, 3198.
268. M. Alhamami, H. Doan and C.-H. Cheng, *Materials*, 2014, **7**, 3198–3250.
269. S. Bourrelly, P. L. Llewellyn, C. Serre, F. Millange, T. Loiseau and G. Férey, *J. Am. Chem. Soc.*, 2005, **127**, 13519.
270. L. Alaerts, C. E. A. Kirschhock, M. Maes, M. A. Van Der Veen, V. Finsy, A. Depla, J. A. Martens, G. V. Baron, P. A. Jacobs, J. F. M. Denayer and D. E. De Vos, *Angew. Chem., Int. Ed.*, 2007, **46**, 4293.
271. P. L. Llewellyn, P. Horcajada, G. Maurin, T. Devic, N. Rosenbach, S. Bourrelly, C. Serre, D. Vincent, S. Loera-Serna, Y. Filinchuk and G. Férey, *J. Am. Chem. Soc.*, 2009, **131**, 13002.
272. P. G. Yot, Z. Boudene, J. Macia, D. Granier, L. Vanduyfhuys, T. Verstraelen, V. Van Speybroeck, T. Devic, C. Serre, G. Férey, N. Stock and G. Maurin, *Chem. Commun.*, 2014, **50**, 9462.
273. P. G. Yot, Q. Ma, J. Haines, Q. Yang, A. Ghoufi, T. Devic, C. Serre, V. Dmitriev, G. Férey, C. Zhong and G. Maurin, *Chem. Sci.*, 2012, **3**, 1100.
274. I. Beurroies, R. Denoyel, B. Kuchta, P. Llewellyn and C. Serre, *Poromechanics V*, Vienna, 2013.
275. J. Rodriguez, I. Beurroies, T. Loiseau, R. Denoyel and P. L. Llewellyn, *Angew. Chem., Int. Ed.*, 2015, **54**, 4626.
276. J. Rodriguez, I. Beurroies, M. V. Coulet, P. Fabry, T. Devic, C. Serre, R. Denoyel and P. L. Llewellyn, *Dalton Trans.*, 2016, **45**, 4274.
277. I. Beurroies, D. Presle, J. Rodriguez and R. Denoyel, *Poromechanics VI*, Paris, 2017.
278. B. Seoane, S. Sorribas, Á. Mayoral, C. Téllez and J. Coronas, *Microporous Mesoporous Mater.*, 2015, **203**, 17.
279. M.-A. Springuel-Huet, A. Nossov, Z. Adem, F. Guenneau, C. Volkringer, T. Loiseau, G. Férey and A. Gédéon, *J. Am. Chem. Soc.*, 2010, **132**, 11599.
280. F. Salles, G. Maurin, C. Serre, P. L. Llewellyn, C. Knöfel, H. J. Choi, Y. Filinchuk, L. Oliviero, A. Vimont, J. R. Long and G. Férey, *J. Am. Chem. Soc.*, 2010, **132**, 13782.
281. G. Kumari, N. R. Patil, V. S. Bhadram, R. Haldar, S. Bonakala, T. K. Maji and C. Narayana, *J. Raman Spectrosc.*, 2016, **47**, 149.
282. R. N. Widmer, G. I. Lampronti, S. Chibani, C. W. Wilson, S. Anzellini, S. Farsang, A. K. Kleppe, N. P. M. Casati, S. G. Macleod, S. A. T. Redfern, F.-X. Coudert and T. D. Bennett, *J. Am. Chem. Soc.*, 2019, **141**, 9330.
283. M. Fisch, A. Lanza, E. Boldyreva, P. Macchi and N. Casati, *J. Phys. Chem. C*, 2015, **119**, 18611.
284. J. D. Evans, V. Bon, I. Senkovska, H.-C. Lee and S. Kaskel, *Nat. Commun.*, 2020, **11**, 2690.

285. D. J. Cerasale, D. C. Ward and T. L. Easun, *Nat. Rev. Chem.*, 2022, **6**, 9.
286. P. Ramaswamy, J. Wieme, E. Alvarez, L. Vanduyfhuys, J.-P. Itié, P. Fabry, V. Van Speybroeck, C. Serre, P. G. Yot and G. Maurin, *J. Mater. Chem. A*, 2017, **5**, 11047.
287. P. G. Yot, K. Yang, V. Guillerm, F. Ragon, V. Dmitriev, P. Parisiades, E. Elkaïm, T. Devic, P. Horcajada, C. Serre, N. Stock, J. P. S. Mowat, P. A. Wright, G. Férey and G. Maurin, *Eur. J. Inorg. Chem.*, 2016, **2016**, 4424.
288. P. G. Yot, L. Vanduyfhuys, E. Alvarez, J. Rodriguez, J.-P. Itié, P. Fabry, N. Guillou, T. Devic, I. Beurroies and P. L. Llewellyn, *Chem. Sci.*, 2016, **7**, 446.
289. S. Leubner, R. Stäglich, J. Franke, J. Jacobsen, J. Gosch, R. Siegel, H. Reinsch, G. Maurin, J. Senker, P. G. Yot and N. Stock, *Chem. – Eur. J.*, 2020, **26**, 3877.
290. M. Wahiduzzaman, N. Reimer, J.-P. Itié, N. Stock, G. Maurin and P. G. Yot, *Polyhedron*, 2018, **155**, 144.
291. P. G. Yot, M. Wahiduzzaman, E. Elkaim, P. Fertey, P. Fabry, C. Serre and G. Maurin, *Dalton Trans.*, 2019, **48**, 1656.
292. S. Krause, J. D. Evans, V. Bon, I. Senkovska, P. Iacomi, F. Kolbe, S. Ehrling, E. Troschke, J. Getzschmann, D. M. Többens, A. Franz, D. Wallacher, P. G. Yot, G. Maurin, E. Brunner, P. L. Llewellyn, F.-X. Coudert and S. Kaskel, *Nat. Commun.*, 2019, **10**, 3632.
293. S. Krause, J. D. Evans, V. Bon, I. Senkovska, S. Ehrling, U. Stoeck, P. G. Yot, P. Iacomi, P. Llewellyn, G. Maurin, F.-X. Coudert and S. Kaskel, *J. Phys. Chem. C*, 2018, **122**, 19171.
294. S. Krause, J. D. Evans, V. Bon, I. Senkovska, S. Ehrling, P. Iacomi, D. M. Többens, D. Wallacher, M. S. Weiss, B. Zheng, P. G. Yot, G. Maurin, P. L. Llewellyn, F.-X. Coudert and S. Kaskel, *Chem. Sci.*, 2020, **11**, 9468.
295. S. Henke, M. T. Wharmby, G. Kieslich, I. Hante, A. Schneemann, Y. Wu, D. Daisenberger and A. K. Cheetham, *Chem. Sci.*, 2018, **9**, 1654.
296. P. Vervoorts, C. L. Hobday, M. G. Ehrenreich, D. Daisenberger and G. Kieslich, *Z. Anorg. Allg. Chem.*, 2019, **645**, 970.
297. M. Pera-Titus, *ChemPhysChem*, 2014, **15**, 1581.
298. J. Wieme, S. M. J. Rogge, P. G. Yot, L. Vanduyfhuys, S.-K. Lee, J.-S. Chang, M. Waroquier, G. Maurin and V. Van Speybroeck, *J. Mater. Chem. A*, 2019, **7**, 22663.
299. P. Iacomi, J. S. Lee, L. Vanduyfhuys, K. H. Cho, P. Fertey, J. Wieme, D. Granier, G. Maurin, V. Van Speybroeck, J.-S. Chang and P. G. Yot, *Chem. Sci.*, 2021, **12**, 5682.
300. J.-C. Tan, P. Jain and A. K. Cheetham, *Dalton Trans.*, 2012, **41**, 3949.
301. R. D. Shannon, *Acta Crystallogr., Sect. A: Cryst. Phys., Diffr., Theor. Gen. Crystallogr.*, 1976, **32**, 751.
302. B. Cordero, V. Gómez, A. E. Platero-Prats, M. Revés, J. Echeverría, E. Cremades, F. Barragán and S. Alvarez, *Dalton Trans.*, 2008, 2832.
303. A. L. Allred, *J. Inorg. Nucl. Chem.*, 1961, **17**, 215.

304. A. V. Neimark, F.-X. Coudert, C. Triguero, A. Boutin, A. H. Fuchs, I. Beurroies and R. Denoyel, *Langmuir*, 2011, **27**, 4734.
305. A. López-Olvera, J. A. Zárate, E. Martínez-Ahumada, D. Fan, M. L. Díaz-Ramírez, P. A. Sáenz-Cavazos, V. Martis, D. R. Williams, E. Sánchez-González, G. Maurin and I. A. Ibarra, *ACS Appl. Mater. Interfaces*, 2021, **13**, 39363.
306. A. Ghoufi, K. Benhamed, L. Boukli-Hacene and G. Maurin, *ACS Cent. Sci.*, 2017, **3**, 394.
307. M. T. Wharmby, F. Niekiel, J. Benecke, S. Waitschat, H. Reinsch, D. Daisenberger, N. Stock and P. G. Yot, *Nanomaterials*, 2020, **10**, 1698.
308. D. I. Kolokolov, H. Jobic, A. G. Stepanov, M. Plazanet, M. Zbiri, J. Ollivier, V. Guillerm, T. Devic, C. Serre and G. Férey, *Eur. Phys. J.: Spec. Top.*, 2010, **189**, 263.
309. N. Rosenbach, H. Jobic, A. Ghoufi, F. Salles, G. Maurin, S. Bourrelly, P. L. Llewellyn, T. Devic, C. Serre and G. Férey, *Angew. Chem., Int. Ed.*, 2008, **47**, 6611.
310. F. Salles, D. I. Kolokolov, H. Jobic, G. Maurin, P. L. Llewellyn, T. Devic, C. Serre and G. Ferey, *J. Phys. Chem. C*, 2009, **113**, 7802.
311. X. Wang, J. Eckert, L. Liu and A. J. Jacobson, *Inorg. Chem.*, 2011, **50**, 2028.
312. Z. Su, Y.-R. Miao, S.-M. Mao, G.-H. Zhang, S. Dillon, J. T. Miller and K. S. Suslick, *J. Am. Chem. Soc.*, 2015, **137**, 1750.
313. Y.-R. Miao, Z. Su and K. S. Suslick, *J. Am. Chem. Soc.*, 2017, **139**, 4667.
314. K. W. Chapman, G. J. Halder and P. J. Chupas, *J. Am. Chem. Soc.*, 2009, **131**, 17546.
315. Y. Hu, H. Kazemian, S. Rohani, Y. Huang and Y. Song, *Chem. Commun.*, 2011, **47**, 12694.
316. Q. Wei, H. Xu, X. Yu, T. Shimada, M. Rearick, D. Hickmott, Y. Zhao and S. Luo, *J. Appl. Phys.*, 2011, **110**, 056102.
317. Z. Su, W. L. Shaw, Y.-R. Miao, S. You, D. D. Dlott and K. S. Suslick, *J. Am. Chem. Soc.*, 2017, **139**, 4619.
318. X. Zhou, Y.-R. Miao, W. L. Shaw, K. S. Suslick and D. D. Dlott, *J. Am. Chem. Soc.*, 2019, **141**, 2220.
319. K. Banlusan and A. Strachan, *J. Phys. Chem. C*, 2016, **120**, 12463.
320. K. Banlusan and V. Amornkitbamrung, *J. Phys. Chem. C*, 2020, **124**, 17027.
321. K. Banlusan and A. Strachan, *J. Chem. Phys.*, 2017, **146**, 184705.
322. K. Banlusan, E. Antillon and A. Strachan, *J. Phys. Chem. C*, 2015, **119**, 25845.
323. G. Cao, *J. Micromech. Mol. Phys.*, 2022, 1.
324. F.-X. Coudert, A. Boutin and A. H. Fuchs, *Commun. Chem.*, 2021, **4**, 106.
325. X. Chen, B. Xu and L. Liu, *Appl. Mech. Rev.*, 2014, **66**, 050803.
326. A. Giacomello, C. M. Casciola, Y. Grosu and S. Meloni, *Eur. Phys. J. B*, 2021, **94**, 163.
327. A. Le Donne, A. Tinti, E. Amayuelas, H. K. Kashyap, G. Camisasca, R. C. Remsing, R. Roth, Y. Grosu and S. Meloni, *Adv. Phys.: X*, 2022, **7**, 2052353.

328. J. Canivet, A. Fateeva, Y. Guo, B. Coasne and D. Farrusseng, *Chem. Soc. Rev.*, 2014, **43**, 5594.
329. M. J. Kalmutzki, C. S. Diercks and O. M. Yaghi, *Adv. Mater.*, 2018, **30**, 1704304.
330. D. Lenzen, J. Zhao, S.-J. Ernst, M. Wahiduzzaman, A. Ken Inge, D. Fröhlich, H. Xu, H.-J. Bart, C. Janiak, S. Henninger, G. Maurin, X. Zou and N. Stock, *Nat. Commun.*, 2019, **10**, 3025.
331. J. Lu, H. Zhang, J. Hou, X. Li, X. Hu, Y. Hu, C. D. Easton, Q. Li, C. Sun, A. W. Thornton, M. R. Hill, X. Zhang, G. Jiang, J. Z. Liu, A. J. Hill, B. D. Freeman, L. Jiang and H. Wang, *Nat. Mater.*, 2020, **19**, 767.
332. A. J. Rieth, K. M. Hunter, M. Dincă and F. Paesani, *Nat. Commun.*, 2019, **10**, 4771.
333. J. Wolanin, L. Michel, D. Tabacchioni, J. M. Zanotti, J. Peters, I. Imaz, B. Coasne, M. Plazanet and C. Picard, *J. Phys. Chem. B*, 2021, **125**, 10392.
334. Y. Sun, X. Xu, C. Xu, Y. Qiao and Y. Li, *Sci. Rep.*, 2016, **6**, 26694.
335. C. Zhan, M. Li and W. Lu, *Compos. Commun.*, 2020, **20**, 100357.
336. M. Li, J. Li, S. Barbat, R. Baccouche and W. Lu, *Compos. Struct.*, 2018, **200**, 120.

Subject Index

Figures and Tables in **BOLD**